Advances in Soil Science

Advances in Soil Science

B.A. Stewart, Editor

Advances in Soil Science

Volume 17
Soil Restoration

Edited by R. Lal and B.A. Stewart

With Contributions by
J.K. Cronk, J.P. Curry, D. Dent, W.T. Frankenberger, J.A. Good, N.N. Goswami, F.M. Hons, L.R. Hossner, R. Lal, T.J. Logan, W.J. Mitsch, T.J. Nimlos, R. Prasad, W.E. Sopper, B.A. Stewart, and E.T. Thompson-Eagle

With 97 Illustrations

Springer-Verlag
New York Berlin Heidelberg London Paris
Tokyo Hong Kong Barcelona Budapest

Rattan Lal
Department of Agronomy
Ohio State University
Room 202
2021 Coffey Road
Columbus, OH 43210-1086
USA

B.A. Stewart
USDA Conservation and Production Research Lab
Bushland, TX 79012
USA

ISSN: 0176-9340

Printed on acid-free paper.

Production managed by Karen Phillips
Typeset by Asco Trade Typesetting Ltd., Hong Kong.

9 8 7 6 5 4 3 2 1

ISBN-13:978-1-4612-7684-5 e-ISBN-13:978-1-4612-2820-2
DOI:10.1007/978-1-4612-2820-2

Preface

Soil degradation is clearly one of the most pressing problems facing mankind. A continuation of soil degradation will eventually lead to a loss in crop productivity even though fertilizers and other inputs often result in increased yields in the short term. Soil degradation also leads to environmental pollution. A decrease in soil quality invariably leads to a decrease in water quality, and often in air quality.

While there is a clear consensus that soil degradation is a major problem, the literature on this subject leaves numerous baffling questions. If statistics on land degradation are correct, there is a definite cause for concern, and present a mammoth challenge for agricultural scientists. There are those that say the scientific community has over dramatized this issue, and created a credibility problem. Consequently, Volume 11 of *Advances in Soil Science* was organized by Dr. Rattan Lal who is recognized as a leading authority on the subject. The objective of Volume 11 was to assess the types and processes of soil degradation and establish some of the major cause-effect relationships. Volume II documented the seriousness of soil degradation in many parts of the world. Therefore, it seemed immediately important to devote a volume to the principles and technologies for restoring degraded soils to a productive status. While the land resources are limited, world population is rapidly increasing, particularly in developing countries. Dr. Rattan Lal has again assumed the leadership in selecting leading authorities to address these critical issues. The scientific principles for restoring many degraded soils are known and practical technologies are available in many cases. This Volume will analyze and summarize the scientific information on this important topic, assessing its importance and identifying additional research needs.

I want to thank Dr. Lal for his leadership in developing this Volume, the authors for their excellent contributions, and the Springer-Verlag staff for their kind assistance and counsel. Finally, and most importantly, I thank the readers for their acceptance and use of *Advances in Soil Science*. Future volumes will continue to include a mix of single topic volumes with guest editors and volumes covering a wide array of soil science topics.

B.A. Stewart

Contents

Contributors

J.K. Cronk, School of Natural Resources and Environmental Science Program, Ohio State University, 2021 Coffey Road, Columbus, Ohio 43210, U.S.A.

J.P. Curry, Department of Environmental Resources Management, University College Dublin, Belfield, Dublin 4, Ireland.

D. Dent, School of Environmental Sciences, University of East Anglia, Norwich N4R 7TJ, U.K.

W.T. Frankenberger, Jr., Department of Soil and Environmental Sciences, University of California, Riverside, California 92521, U.S.A.

J.A. Good, Department of Environmental Resources Management, University College Dublin, Belfield, Dublin 4, Ireland.

N.N. Goswami, Department of Agronomy, Indian Council of Agricultural Research, New Delhi, 110012, India.

F.M. Hons, Department of Soil and Crop Sciences, Texas A&M University, College Station, Texas 77843, U.S.A.

L.R. Hossner, Department of Soil and Crop Sciences, Texas A&M University, College Station, Texas 77843, U.S.A.

R. Lal, Department of Agronomy, Ohio State University, 2021 Coffey Road, Columbus, Ohio 43210, U.S.A.

T.J. Logan, Department of Agronomy, Ohio State University, 2021 Coffey Road, Columbus, Ohio 43210, U.S.A.

W.J. Mitsch, School of Natural Resources and Environmental Science Program, Ohio State University, Columbus, Ohio 43210, U.S.A.

T.J. Nimlos, School of Forestry, University of Montana, Missoula, Montana 59812, U.S.A.

R. Prasad, C-41, IARI Campus, Indian Agricultural Research Institute, New Delhi 110012, India

W.E. Sopper, Institute for Research on Land and Water, Pennsylvania State University, University Park, PA 16802, USA.

B.A. Stewart, U.S. Department of Agriculture, Agricultural Research Service, Bushland, Texas 79106, U.S.A.

E.T. Thompson-Eagle, Department of Soil and Environmental Sciences, University of California, Riverside, California 92521, U.S.A.

Need for Land Restoration

R. Lal and B.A. Stewart

I. Introduction

Sustainable management of natural resources involves the concept of "using, improving, and restoring" the productive capacity and life-support processes of soil—the most basic of all natural resources. The objective is not only to minimize soil degradation but to reverse the trend through restorative measures of soil and crop management. The soil quality and its productive capacity must be enhanced beyond preservation (status quo) through soil-building measures, e.g., preventing soil erosion and enhancing development of the rooting depth, replenishing nutrients harvested in crops and animals through judicious use of mineral fertilizer and organic amendments and effective nutrient recycling practices, encouraging biological activity of soil fauna, and improving soil organic matter content. The land use or farming system to be adopted must be "soil-restorative" rather than "soil-depletive," "fertility-mining," or "soil-degrading." In addition, soil should not be misused as a dumping ground for toxic wastes. Although soil has a built-in resilience, there is a limit to the abuse that it can withstand.

1992 by Springer-Verlag New York Inc.
Advances in Soil Science, Volume 17

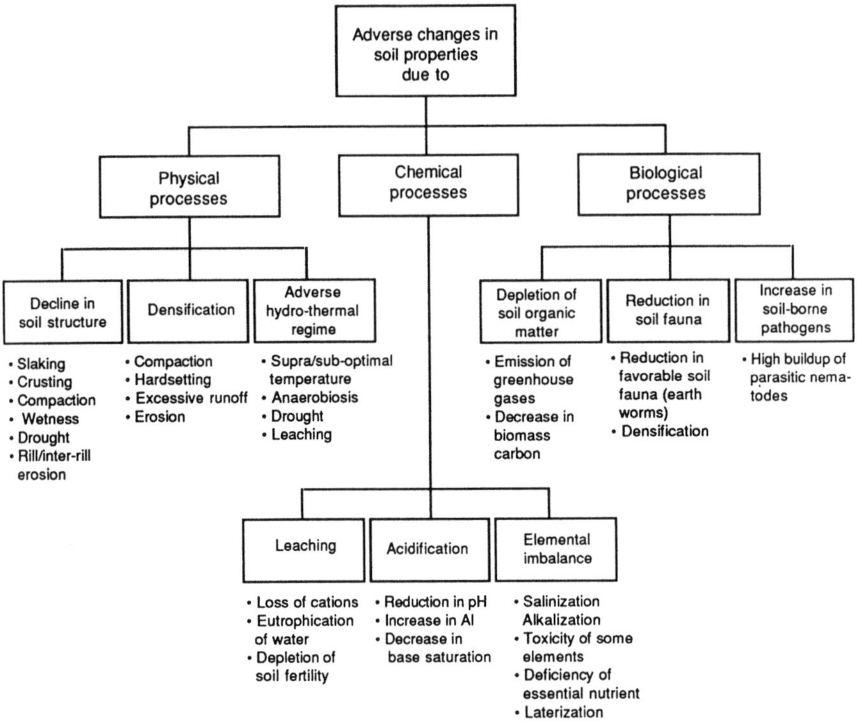

Figure 1. Processes of soil degradation

II. Basic Concepts of Soil Degradation

Soil degradation implies diminution of its productive capacity through intensive use leading to adverse changes in soil properties. Processes leading to soil degradation may be physical, chemical, or biological (Fig. 1). Important among these factors are decline in soil structure, compaction, reduction in infiltration capacity, depletion of soil organic matter and reduction in biomass carbon, salt imbalance, and build-up of soil-borne pathogens. The rate of soil degradation by different processes is greatly accentuated by using land for whatever it is not capable of and by unsuitable methods of soil and crop management. There are several factors that set in motion various soil-degradative processes. These factors may be natural or anthropogenic (Fig. 2). Natural factors include climate, vegetation, parent material, terrain, and hydrology. Among important anthropogenic factors are population density, land use, and the development of roads, waterways, and the industrial complex.

Land scarcity and demographic pressure are the driving forces responsible for bringing marginal lands under cultivation, with attendant problems

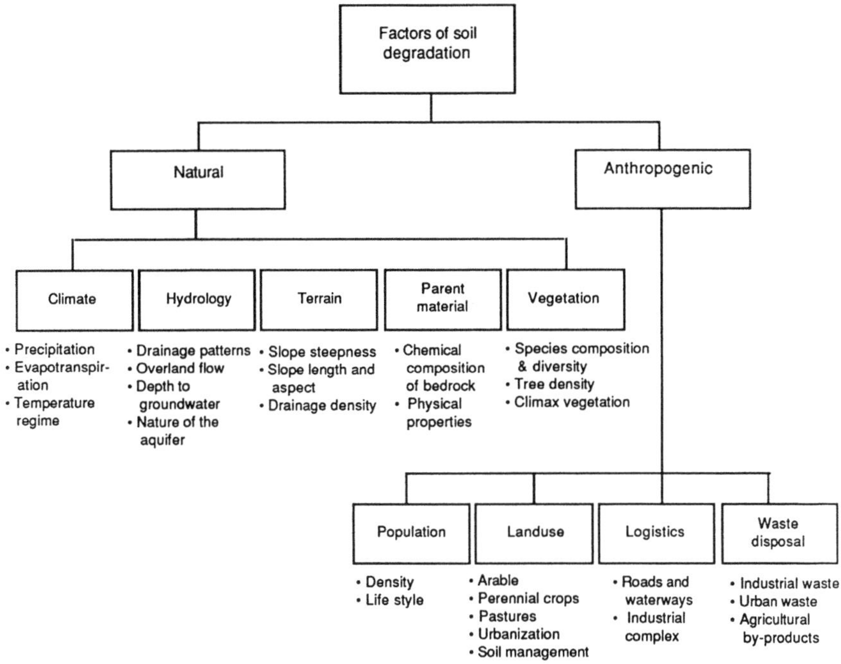

Figure 2. Factors responsible for soil degradation

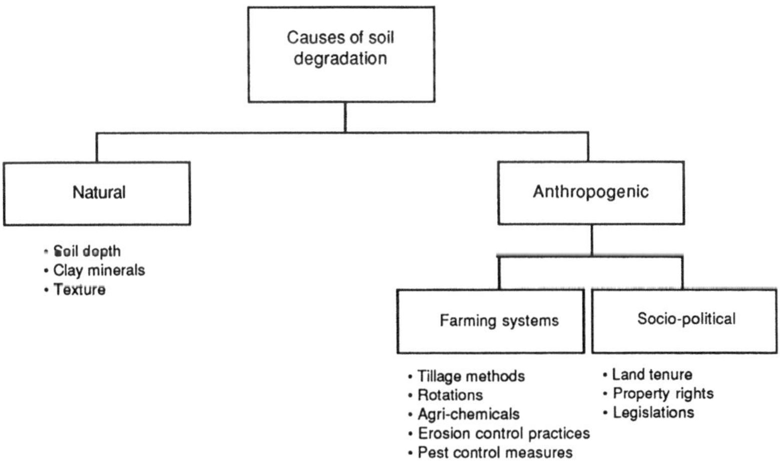

Figure 3. Causes of soil degradation

of severe soil and environmental degradation. Resource-poor farmers and landless laborers are forced to cultivate lands that are too steep, too shallow, or too dry for cultivation, and by methods that are ecologically nonviable. Some highly weathered and impoverished soils are being intensively cultivated without the fallow period required for restoration of soil fertility and enhancement of soil structure (Okigbo, 1987). Consequently, soil degradation sets in resulting in widespread occurrence of sheet and gully erosion (Lal, 1984), and encroachment by *Imperata cylindrica.* There are vast areas of *Imperata*-infested land in Asia, west Africa, and tropical America.

The effects of these factors can be accentuated by several natural or anthropogenic causes (Fig. 3). Principal causes of soil degradation are deforestation, tillage methods, farming systems, use of agrichemicals, etc. Social and political factors also play an important role.

III. Global Extent of Soil Degradation

The world's arable land resources are finite. Seventy-eight percent of the total earth's surface area is unsuitable for agricultural purposes. Out of the 22% of the land that is agriculturally suitable, 13% has low productive capacity, 6% a medium, and only 3% is characterized with a high capacity for an intensive crop production (Buringh, 1989).

At present 5 to 7 million hectares of arable land (0.3% to 0.5%) are lost every year through soil degradation. The projected loss by the year 2000 is 10 million hectares annually (0.7% of the area presently cultivated). The world is now losing some 23 billion t of top soil per year from uplands in excess of new soil formation (Brown, 1984). By the year 2000, productivity of about one-third of the world's arable land may be severely impaired due to accelerated erosion (UNEP, 1982). In addition, a total of 3770 million hectares are prone to desertification (Mabbutt, 1978; UNEP, 1984). There are about 323 million hectares of salt-affected soils in the world (Beek et al., 1980).

Accelerated erosion is a serious problem in several ecologically sensitive regions, e.g., the Himalayan-Tibetan ecosystem, the Andean region, the Caribbean, eastern Africa, and other densely populated regions with severe land shortage. Steeplands, comprising a large percentage of the total land area in these regions, are over-exploited and grossly misused. Rivers draining the Himalayan region (e.g., Ganges, Mekong, Irrawdy, Brahmaputra) have a high sediment load. In India, 150 million hectares are subject to accelerated soil erosion (FAO/UNEP, 1983). Siltation of reservoirs in northern India is about 200% more than anticipated in their design (Dent, 1984). In Nepal, 63% of the Shivalik zone, 26% of the Middle Mountain zone, 48% of the Transition zone, and 22% of the High Himalayas are subject to severe erosion. The Upper Indus basin in Pakistan is severely eroded. About 46 million hectares of the loess plateau of China are subject to severe erosion, raising the bed of Yellow River by as much as 10 cm annually. Severe erosion is also observed in the watersheds of Yangtze, Huaihe, Pearl, Liaolie, and Songhua Rivers (Dent, 1984). About 39 million hectares or 8% of the Amazon Basin is characterized by soils of high erodibility (Sanchez et al., 1982). Brown (1981) observed that as much as 1 billion t of topsoil is lost from Ethiopian highlands each year. Finn (1983) reported that an average rate of soil erosion from Madagascar is 25 to 40 t ha^{-1} yr^{-1}. FAO/UNEP (1983) estimated that a total of 87% of the Near East and Africa north of the equator are subjected to accelerated erosion.

Wind erosion is equally severe in arid and semi-arid regions, e.g., the West African Sahel, western India, Pakistan. In southern Tunisia, Floret and Le Floch (1973) and Le Houerou (1977a) observed that wind erosion rates of 10 mm of topsoil removed per year are common. Wind-blown dust from the Sahara causes air-pollution and "sand rains" in the Caribbean (Rapp, 1974), and in northern Europe (Le Houerou, 1977b). It is estimated that between 25 and 37 million t of African soil are annually blown across the Atlantic Ocean (Prospero and Carlson, 1972). A total of 16.6 million km^2 of the world's arid regions, 17.1 million km^2 of the semi-arid regions, and 4.0 million km^2 of the subhumid regions are subject to desertification (Mabbutt, 1978). The global area subject to desertification is estimated to be 37.7 million km^2. The global loss to desertification is estimated at 6 million ha yr^{-1}, and the rural population severely affected by desertification is about 135 million (UNEP, 1984).

Similar to human population, many regions have experienced a dramatic increase in the animal population, e.g., the Sahel, Indian subcontinent. Gallais (1979) and NRC (1984) reported that the cattle population in western Sahel increased five fold during the 25 years preceding the 1968 drought. Uncontrolled and excessive grazing is responsible for depleting vegetation and denuding the landscape, causing shift in climax vegetation, soil compaction and hard-setting, and accelerated runoff and erosion. Nowhere else are the adverse effects of uncontrolled and excessive grazing more severe and obvious than in the Sahel. Grazing can cause degradation even if the stocking rate is moderate or low. Perrens (1986) reported that in Australia 55% of the total grazing area of 3.4 million km^2 in arid regions needs restorative measures for land degradation. In all, about 51% of the total land area of 5.2 million km^2 in Australia is in need of restorative measures against degradation.

Soil degradation and environmental pollution go hand-in-hand. Atmospheric concentrations of CO_2 and other gases have been steadily increasing over the past century or more (Batch, 1986). Tang et al. (1990) observed that terrestrial ecosystems play an important role in the global carbon budget. In addition to the effects of deforestation (Houghton et al., 1987), it is now widely believed that the world's soils play an important role in the global carbon budget. Soil misuse and overexploitation, causing rapid depletion of soil organic matter, can lead to emission of greenhouse gases into the atmosphere. Lal (1990) estimated that reduction of about 1% of the antecedent level of organic carbon content of the top 15-cm layer of soils of the tropics can lead to an annual emission of about 128 billion t of C into the atmosphere. It is a serious environmental hazard of soil misuse.

Both wind and water erosion are among major pollutants of the environment. The dust ejected in the atmosphere by wind erosion is a major health hazard and a risk to civil aviation. The quantity of dust added to the atmosphere each year includes: volcanic dust, 4 million t; anthropogenic dust, 296 million t; and smoke, 40 to 60 million t (Bryson, 1974a, b; Kovda, 1980).

Transport of agricultural chemicals into the world's rivers is increasing (Stewart and Rohlich, 1977). The recovery of nitrogenous fertilizers by crops is usually less than 50%, and a maximum of 10% is recovered by a succeeding crop. The unrecovered fertilizer is easily transported into natural waters.

IV. Land Hunger

There are several global issues that highlight the need for restoring degraded soils. First among these is the unprecedented growth of population, especially in countries that have limited resources and are of marginal utility. The world population was 1 billion in 1800, 2.5 billion in 1950, 4.7 billion in 1983, will be 6.2 billion in 2000, and 9.3 billion in 2050, and is expected to level off at about 11 billion by the end of the twenty-first century. Population growth rates for 1980–85 were 3.01%, 2.30%, 2.20% and 1.20% per year for Africa, Latin America, South Asia, and East Asia, respectively (McNamara, 1985). The population of sub-Saharan Africa, a region with a perpetual food crisis, is expected to increase from 450 million now to 680 in 2000, 1.2 billion in 2025, 1.7 billion in 2050, and 2.0 billion by the year 2100. Population in developing countries, increasing on average by about 2.0% per year between 1986 and 2000, will contribute to about 90% of the increase in global population (World Bank, 1988; Population Reference Bureau, 1986). This growth rate translates into about 1.1 billion additional mouths to be fed in developing countries in the 15-year period ending in the year 2000. The agricultural production by the year 2000 will have to be 50% to 60% more than in 1980 to maintain the same level of food intake.

As recently as the 1970s, a considerable proportion of increase in food production was achieved by bringing new land under agricultural production. Presently, however, the reserves of potentially arable prime agricultural land are hard to find. Furthermore, land resources are unevenly distributed. Whatever potentially arable land exists is located in regions with robust economies, e.g., North America. Densely populated Asia, home to 75% of the world's population, has little additional land to convert to arable use. Similarly, comparatively little is available in Europe. There is a possibility of an additional 200 million hectares in North America, 290 million hectares in South America, and 340 million hectares in Africa (Revelle, 1976; Buringh, 1981; Dudal, 1982). Most of the available land in Africa and South America is located within fragile and ecologically sensitive regions, e.g., tropical rainforest (TRF), acid savannas, drought-prone Sahel. Large proportions of the presently cultivated land are not compatible with sustainable agriculture (USAID, 1983). If the arable land area is maintained at 1.45 billion hectares, the per capita arable land will progressively decline from about 0.3 ha now to 0.23 ha in 2000, 0.15 ha in 2050, and 0.14 ha by the year 2100. These calculations are based on the assumption that neither new land is brought under cultivation nor existing land is taken out of production due to soil degradation.

Potentially productive agricultural land is either inaccessible, too steep, too shallow, or is in regions with too little or too much water, and essential inputs for crop production are not available. Bringing new land under production through deforestation of TRF has severe ecological, environmental, and sociopolitical implications. The actual extent of deforestation in the tropics is still the subject of debate (Myers, 1981; Moore et al., 1988). In addition to loss of biodiversity and potentially valuable genetic resources, conversion of TRFs presumably contributes a large proportion of total global emissions of CO_2 (Houghton et al., 1987; Tirpak, 1988; Lashoff, 1988). However, the exact values are not known. The type, amount, and rate of gaseous emission also depend on the method of deforestation. The effects differ among techniques, e.g., slash and burn, chain-saw clearing, and deforestation by bulldozers and chemical poisoning of trees. The effects may also differ among different land uses following deforestation (Lal, 1987a, b).

Over the century ending in 1984, a considerable amount of new land has been brought under production. Land area has been increased by 172% in arable land use and 210% in pasture. In comparison, there has been a decrease of 11.4% in forest land and 21.9% in other land uses. There is some land that can still be brought under production. Most of the potential land available in the Amazon basin, Congo basin, and Sumatra is relatively infertile, covered by TRFs, and should be preferably left alone.

The production of the minimum dietary requirements from 0.14 ha of per capita arable land can be met by technological innovations that may bring about a quantum jump in food production. Population in large areas of Asia, Africa, and South America already exceeds the carrying capacity of the land. FAO (1984), in its Agro-Ecological Zones project, evaluated that in 1975, 38% of the total land area was carrying more people than could be fed with low inputs. In comparison, 22% of the total land area was carrying more people than could be fed even with high input.

The land population scenario should be assessed in terms of three challenges. First, the available land resources are unevenly distributed. Carrying capacity of the land is not a problem in regions of North America, Europe, and Australia. Second, regions with high demographic pressures are also characterized by low available land reserves and poor resources to use high-input technologies. Several countries of South Asia and Central America under this category are also characterized by a severe rate of soil degradation. Third, socio-economic, anthropological, and political considerations are often overwhelming and do not readily permit the adoption of improved land-restorative technologies.

Irrigation played a major role in increasing food production during the 50-year period ending in 1990. The world irrigated land was 8 million hectares in 1800, 48 million hectares in 1900, 92 million hectares in 1949, 198 million hectares in 1970, 235 million hectares in 1980 (Szabolcs, 1986). Presently, irrigated land accounts for 18% of the cultivated land, but it produces 33% of the world's food. However, the current rate of expansion has slowed to less than 1% per year (CAST, 1988). Both, availability of irrigable land and good quality water are severe constraints to further expansion.

Rapid urbanization is another major cause of land scarcity. About 233000 people are added to the human population daily; amounting to a total increase of 80 million people per year. An average of 0.1 ha is needed per capita for accommodation and living. This means an average of 8 million hectares of arable land are taken out annually for nonagricultural purposes.

V. Need for Soil Restoration

Soil is a finite and nonrenewable resource. Potentially arable land resources are limited and cannot meet the needs of projected increase in human and animal population. Therefore, not only should the desired high net output in production be achieved with a minimum of soil degradation, but degradative trends must also be reversed. The soil quality and its productive capacity must be restored and improved by preventing soil erosion, promoting high biological activity of soil fauna, improving soil organic matter content, and replacing the nutrients harvested in crops and animals through chemical fertilizers and organic amendments supported by effective nutrient recycling mechanisms. The productive efficiency of a system must be evaluated in terms of its effect on the natural resources, e.g., change in soil organic matter reserves, pH, nutrient reserves, exchangeable cations, plant-available water capacity, or effective rooting depth. Suitable farming systems are those that enhance soil quality. Fertility-mining and soil-degrading, low-input systems must be stopped.

VI. Conclusions

Soil degradation is caused by adverse changes in the soil's physical, chemical, and biological processes. Degradative trends and rates of these processes are regulated by a range of interacting anthropogenic and natural factors and causes. Land scarcity and demographic pressure are the driving forces responsible for bringing marginal lands under cultivation. Principal processes responsible for soil degradation are accelerated erosion by wind and water, desertification, sodication, and urbanization.

Scarcity of prime agricultural land and improving the environment are two principal motives for restoring degraded lands. In many parts of Asia and Africa, regions with high demographic pressure, large tracts of land are being (ab)used beyond their carrying capacity.

Technology for soil restoration exists, as is documented by technical papers presented in this volume. However, adoption of these restorative measures is subject to sociopolitical and economic pressures. Legislation and coercive measures are rarely successful. An effective strategy should be to involve the public in these efforts. Public participation is important to the success of this much needed global priority of restoring lands degraded because of our mistakes.

References

Batch, W. 1986. The gases and their influence on climate. *Natural Resources and Development* 24:90–124.

Beek, K.J., W.A. Blokhuis, P.M. Driessen, N. VanBreemen, R. Brinkman, and L.J. Pons. 1980. Problem Soils: Their reclamation and management. ILRI Publ. No. 27, pp. 47–72. ILRI, Wageningen, Netherlands.

Brown, L.R. 1981. World population growth, soil erosion and food security. *Science* 214:995–1002.

Brown. 1984. State of the World, 1984. World Watch Institute, Norton, New York

Bryson, R.A. 1974a. Climate change and agricultural responses. A Statement on Research and Technological Priorities Between Now and the Year 2000. Inst. Environ. Stud., Univ. of Wisconsin, Madison.

Bryson, R.A. 1974b. A perspective on climatic change. *Science* 184:753–760.

Buringh. 1981. An assessment of losses and degradation of productive agricultural land in the World. FAO, Rome, Italy.

Buringh, P. 1989. Availability of agricultural land for crop and livestock production. *In:* D. Pimentel and C.W. Hall (eds.) *Food and Natural Resources*, pp. 70–85. Academic Press, San Diego, CA.

Council for Agric. Science-Technology (CAST) 1988. Effective use of water in irrigated agriculture. Report No. 113, Amer. Iowa.

Dent, F. J. 1984. Land degradation: present status, training and education needs in Asia and the Pacific. UNEP Investigations on Environmental Education and Training in Asia and the Pacific. FAO Reg. Off., Bangkok.

Dudal, R. 1982. Land degradation in a world perspective. *J. Soil Water Conserv.* 37:245–247.

FAO. 1984. *Land, Food and People.* FAO, Rome, Italy.

FAO/UNEP. 1983. Guidelines for the control of soil degradation. FAO, Rome, Italy.

Finn, D. 1983. Land use and abuse in the East African region. *Ambio* 12:296–301.

Gallais, J. 1979. La situation de l'elevage bovin et le problem des eleveurs en Afrique occidentale et centrale. *Cah. Outre-Mer* 32:113–138.

Griggs, D. 1985. *The World Food Problem, 1950–1980.* Blackwell, Oxford.

Houghton, R.A. 1987. Terrestrial metabolism and atmospheric CO_2 concentrations: independent geophysical and ecological estimates of seasonal carbon flux address global change. *BioScience* 37:672–678.

Houghton, R.A., R.D. Boone, J.R. Hobbie, J.E. Melillo, C.A. Palm, B.J. Peterson, G.R. Shaver, G.M. Woodwell, B. Moore, D.L. Skole, and N. Myers. 1987. The flux of carbon from terrestrial ecosystem to the atmosphere in 1980 due to change in land use: Geographical distribution of the global flux. *Tellus* 398:122–139.

Kovda, V.A. 1980. *Land Aridization and Drought Control.* Westview Press, Boulder, Colorado, p. 277.

Lal, R. 1984. Soil erosion from tropical arable lands and its control. *Adv. Agron.* 37:183–248.

Lal, R. 1987a. Managing soils of sub-Saharan Africa. *Science* 236:1069–1076.

Lal, R. 1987b. Conversion of tropical rainforest: agronomic potential and ecological consequences. *Adv. Agron.* 39:173–264.

Lal, R. 1990. Managing soil carbon in tropical agro-ecosystems. EPA Workshop on Sequestering Carbon in Soils, Corvallis, Oregon, 26–28 Feb.

Lashoff, D. 1988. Global climate scenarios related to agriculture. U.S. EPA Workshop on climate change. Washington, D.C.

Le Floret, C., Le Floch, E. 1973. Production sensibilite et evolution de la vegetation et du milieu en Tunisie Pre-saharienne. CPEP Doc. No. 71, Montpellier, France.

Le Houerou, H.N. 1977a The Scapegoat. Ceres 10:14–18

Le Houerou, H.N. 1977b. The nature and causes of desertification. In "Desertification" (M.H. Glantz, ed) pp. 17–38, Westview Press, Boulder, Co.

Mabbutt, J.A. 1978. The impact of desertification as revealed by mapping. *Environ. Conserv.* 5:45–56.

McNamara, R.S. 1985. The challenges for sub-Saharan Africa. Sir John Crawford Mem. Lect., Washington, D.C.

Moore, B., M. Cildes, C. Vorosmarty, D. Skole, J. Medillo, B. Peterson, E. Rastetter and P. Steudder 1988. Biological cycles ecology: Towards a science of biosphere. M.B. Ramble, L. Margulis and R. Fester (eds), New York Academic Press Inc: pp. 113–141.

Myers, N. 1981. The hamburger connection: How Central America's forests become North America's hamburgers. *Ambio* 10:3–8.

NRC. 1984. Environmental changes in West African Sahel. Nat. Acad. Press, Washington, DC 9611.

Okigbo, B.N. 1987. Shifting cultivation. *FAO Soils Bull.* 57, Rome, Italy

Perrens, S.J. 1986. Conversion of forest land to annual crops: Australian experience. Landuse, Watersheds, and Planning in the Asia-Pacific Region. RAPA Rep. 1986/3, pp. 112–137, FAO Off. Asia Pac., Bangkok.

Population Reference Bureau. 1986. World Population data sheets. Washington, D.C.

Prospero, J. and T. Carlson. 1972. Vertical and areal distribution of Saharan dust over the western equatorial North Atlantic Ocean. *J. Geophys. Res.* 77:5255–5265.

Rapp, A. 1974. A Review of Desertization in Africa (Water, Vegetation, Man). Secretariat Int. Ecol., Stockholm.

Revelle, R. 1976. The resources available for agriculture. *Scientific American* 235:64–168.

Sanchez. P.A., D.A. Bandy, H. Villachica, and J.J. Nicholaides 1982. Amazon Basin Soils: Management for continuous crop production. Science 216:821–827.

Stewart, K.M. and G.A. Rohlich. 1977. Eutrophication: A review. Rep. State Water Qual. Control Board, Sacramento, California.

Szabolcs, I. 1986. Agronomic and ecological impact of irrigation on soil and water quality. *Adv. Soil Sci.* 4:189–218.

Tang. P.O., I.Y. Fung and T. Takahashi. 1990. Observational constraints on the global atmospheric CO_2 budget. *Science* 247:1431–1438.

Tirpak, D. 1988. Links between agriculture and climate change. U.S.EPA workshop on agriculture and climate change. March 1988, Washington, D.C.

UNEP. 1982. Worlds Soil Policy. U.N. Environ. Program, Nairobi, Kenya.

UNEP. 1984. General Assessment of Progress in the Implementation of the Plan of Action to Combat Desertification. 1978–84. Mimeo. Governing Council. 12th Session, Nairobi, Kenya.

U.S. Agency for International Development. 1983. Country Environmental Profiles, Washington, D.C. World Bank. 1988. World Development Report. 1988. Washington, D.C.

Reclamation of Chemically Degraded Soils

T.J. Logan

I. Introduction

Land degradation has become a major global concern in recent years as a result of increasing demands on the land for food production and waste disposal. Man is learning that the resiliency of soil is finite and that soil degradation is not easily reversed, if ever. The focus of land degradation in this century has been on soil erosion as increasing areas of forest, grassland, and wetland have been cleared for crop production. Soil erosion represents the most complete form of land degradation—the removal of the soil resource itself—and eroded sediment deposited on adjacent lands and drainageways can lead to further degradation. In recent years, the contribution of soil erosion to global carbon emissions has been recognized as equally important to that of deforestation and fossil fuel burning (Tans et al., 1990). Soil erosion remains the focus of conservation efforts in the developed world and many of the resources of agencies like the U.S. Soil Conservation Service are devoted to reducing soil loss to "tolerable" limits (Follett and Stewart, 1985). Other developed countries have similar programs. Soil erosion from land clearing and improper management in the

Advances in Soil Science, Volume 17

Table 1. Types of chemically degraded land

Source of degradation	Major processes	Major contaminants
Soil erosion	Soil removal	Subsurface acidity Subsurface $CaCO_3$
Mine spoil	Soil removal Oxidation	Excess acidity Toxic metals Excess salts
Industrial waste	Contaminant accumulation	Excess acidity Excess alkalinity Salt, toxic metals Toxic organics
Acid rain	Accelerated weathering Nutrient depletion	Excess acidity Toxic matals
Dredge spoil	Oxidation Mineralization	Excess acidity Toxic metals
Ore smelters	Metal and acid deposition	Excess acidity Toxic metals
Sewage sludge[a]	Contaminant acumulation	Toxic metals Excess nutrients
Manure[a]	Oxidation; mineralization Nutrient accumulation	Salts; toxic metals Excess nutients
Petroleum spills	Contaminant accumulation Mineralization	Toxic organics
Fly ash disposal	Contaminant accumulation	Toxic metals, salts
Coastal land	Oxidation; mineralization	Excess acidity Toxic metals
Salinity and sodicity	Contaminant accumulation	Excess alkalinity Salts, sodium
Oil shale waste disposal	Contaminant accumulation	Salts, toxic metals Excess alkalinity
Nuclear waste	Contaminant accumulation	Radionuclides
Landfills	Contaminant accumulation	Toxic metals Toxic organics Salts Excess acidity and alkalinity

[a] At rates greatly in excess of nutrient utilization by crops

developing world—and in particular those of the subtropics and tropics—has reached massive proportions and is the subject of worldwide attention.

Of no less importance, but often overlooked, is the impact of chemical degradation on soil. Chemical degradation is defined here as the accumulated negative impact of chemicals and chemical processes on those properties that regulate the life processes in the soil (Logan, 1989). The soil here is viewed as a living organism, and as a living organism has a complex system of self-regulation. A "healthy" soil has important chemical and biological attributes including nutrient supply, acid and base buffer capacity, organic matter decomposition, pathogen destruction, toxic metal inactivation, and toxic organic inactivation and degradation. These attributes are well expressed in a "healthy" soil, but their capacities are finite and can be overwhelmed by mismanagement.

I have previously reviewed the major causes of chemical soil degradation and the impacts on the short-term and long-term "health" of the soil (Logan, 1989). I identified important chemical processes in soil, including chemical weathering, buffering of soil acidity, redox regulation, cation and anion exchange capacity, adsorption-desorption, precipitation-dissolution, and complexation, and emphasized the central role of soil organic matter (SOM) in many of these processes. I reviewed major causes of chemical soil degradation, including nutrient depletion, acidity and toxic aluminum, pyrite oxidation in soils and mine spoils, land disposal of wastes, subsidence of organic soils, acid and trace metal deposition on soil from metal ore smelters, dredge spoil disposal, and radionuclide soil contamination. These and other causes of chemically degraded land are summarized in Table 1.

In this paper, I discuss general principles of soil reclamation with respect to chemical degradation, and I review engineering and ecological approaches to reclamation of chemically degraded soils.

II. Principles of Soil Reclamation

If soil degradation is the accumulated negative impact on the life processes of the soil, then soil reclamation is the reduction or elimination of those impacts so as to restore the soil to "health." Using the analogy of human health, it is necessary to do two things to restore soil to a healthy state: remove the source of chemical degradation and treat the symptoms of degradation. At this juncture, several issues must be faced. One is the edaphic versus pedologic definition of soil, with edaphology expanded to consider uses of soil other than for plant growth. From a practical standpoint, it is feasible to restore soil conditions to a state that is adequate for a given land use, but not for a more intensive one. In either case, the soil will not be restored to its original, pedalogic condition. The approach to steady-state in soil is a slow process, and it is reasonable to assume that restoration would occur at similar rates. The thermodynamics of soil systems suggest that it is easier to degrade soil than to restore it, and that degradation occurs at a far more rapid rate than reclamation. Ultimately, given the high capital, monetary, and societal costs of soil reclamation, some judgement will be made as to the required extent of reclamation. This will depend on the short- and long-term uses for the land. Reclamation of chemically degraded land must consider not only the effects of contamination on the soil itself, but also the off-site impacts of mobilized soil (erosion) or soil contaminants (runoff and leaching). Trace levels of metals and xenobiotic organics may have little effect on soil processes and soil ecology and yet will require remediation because of the low threshold concentrations established for drinking water quality, or the ability for pollutants immobilized in surface soil to be mobilized from eroded sediment in aquatic environments. In this paper, I will argue that, at the very least, the soil must be restored to biological health so as to provide the self-regulation of biological systems required for sustaining the restorative process.

There are two general approaches that have been used to reclaim chemically degraded soils: (1) engineering approaches and (2) ecological approaches. Engineering approaches rely exclusively on external measures for soil restoration while ecological approaches attempt to stimulate inherent soil processes to restore the soil to some acceptable steady-state condition.

A. Engineering Approaches to Remediation of Chemically Degraded Soil

The engineering approaches to reclamation of chemically degraded soil are used in cases of extreme degradation, where other approaches are unfeasible, or unacceptably slow, or where the resources available for reclamation are great. An example of this situation are the U.S. Superfund sites which have been identified by the U.S. Environmental Protection Agency as posing the greatest environmental risk (U.S. EPA, 1986). Reclamation costs for Superfund sites are often in the tens of millions of U.S. dollars, and engineering reclamation approaches have been almost exclusively used.

Engineering approaches are those which entail removal, immobilization, or chemical transformation of chemical contaminants from a site, or physical reconstruction of a chemically degraded site (Table 2). Removal techniques include whole soil excavation for reburial or treatment, in situ mobilization, and soil washing. Immobilization technologies include in situ fixation of chemical contaminants by adsorption, precipitation, or complexation (Sims and Sims, 1986); in situ vitrification, in which high temperatures are used to fuse the soil matrix into a glass-like mass and thus reduce the potential for contaminant leaching (Timmerman et al., 1989); and chemical immobilization of excavated soil or mobilized contaminants. Finally, biotic and abiotic degradation can be used to break down organic contaminants in situ and from soil and water removed from the site.

Table 2. Engineering approaches to reclamation of chemically degraded land

Approach	Contaminants	Processes
In situ immobilization	Metals	P, A, C, R
In situ mobilization	Metals, organics, salts	S, D, R, E
In situ degradation	Organics	M, PD
In situ burial	All	None
Soil washing	All	S, D, E
Soil removal and reburial	All	None
Vitrification	Metals, salts	P
Vacuum extraction	Organics	V
Steam flooding	Metals, organics, salts	S, E
Pumping and leaching	Metals, organics, salts	S, D
Electroosmosis	Metals	D, S, E, R
Electroacoustic extraction	Metals	D, S, E, R

P, precipitation; A, adsorption; C, complexation; R, oxidation/reduction; S, solubilization; D, desorption; E, extraction; M, microbial degradtion; PD, photolysis

Excavation and reburial are designed to remove the contaminated soil from an area of high environmental exposure, such as over a water supply aquifer, to a more protected or less exposed depository. Reburial usually involves placement of the contaminant in a landfill equipped with a protective liner and controlled drainage. This is one of the disposal options being used for Rhine River dredged sediments in Rotterdam (Nijssen, 1988).

In situ mobilization is used in conjunction with controlled drainage to extract part of the immobilized pollutant, so-called "pump-and-treat" technology. Possible mobilization techniques include vacuum extraction of volatile compounds (Malmanis et al., 1989), steam flooding (Hunt et al., 1988; Mori, 1990), leaching with water, acid, chelating agent (e.g., the synthetic chelator EDTA), organic solvent (Griffin and Chou, 1980), or surfactant (Chawla et al., 1990); oxidation of reduced species such as Cr^{3+} and metal sulfides; electro-osmosis in which an applied direct current electrical field produces fluid flow in soils with charged solids (Shapiro et al., 1990); and electro-acoustic extraction in which a combination of direct current electric field and acoustic waves are used to enhance mobility of immobilized organic and inorganic soil contaminants (Hinchee et al., 1990).

Soil washing is an energy-intensive process that uses water and inorganic or organic solvents to remove part of the immobilized pollutants in an excavated contaminated soil (Nash and Traver, 1989). The extracted pollutants are then concentrated for further treatment in the case of inorganic contaminants, or incinerated or biodegraded in the case of organics. Excavated soil can also be incinerated directly to destroy tightly bound organic contaminants.

In situ immobilization techniques are designed to reduce the solubility of soil contaminants by enhancing their sorption, precipitation, or complexation. Trace metals can be immobilized by precipitation with lime (Logan and Cassler, 1989), cement, or pozzolan (Cullinane et al., 1986), and metal sorption is also enhanced by increasing soil pH. Other precipitating ligands, such as sulfate or phosphate might be used in conjunction with pH adjustment. Oxyanion metals, such as CrO_4^{2-} or SeO_4^{2-}, are not immobilized at high pHs, and alternative strategies would have to be used for these metals. Redox manipulation offers some possibilities for enhancing immobilization (or, conversely, mobilization) of metals. For example, oxidation of Cr^{6+} to Cr^{3+} results in significant reduction of Cr solubility (Bartlett and Kimble, 1976; Eary and Rai, 1988), and metal sulfides are generally much less soluble than more oxidized forms (Lindsay, 1979). The problem with redox manipulation in reclamation of chemically degraded soil is the difficulty in maintaining a specific redox couple in face of varying environmental conditions. Trace metals can be complexed by the addition of organic materials containing metal-binding humic and fulvic substances or even solid resins. Complexation is enhanced by soil pHs > 6 (Stevenson, 1982).

Immobilization of toxic organics in soil is achieved primarily by partitioning of the organic solute into hydrophobic soil organic matter (Hassett and Banwart, 1989). The extent of partitioning (K_d) is a function of the intrinsic partition coefficient of the compound (K_{oc}), and the organic carbon content of the soil (f_{oc}):

$$K_d = K_{oc} \times f_{oc}$$

This relationship holds well for the high f_{oc} found in surface soils, but partitioning becomes much more solute and solvent specific at the low f_{oc} values found in subsurface environments (Murphy et al., 1990). Organics can also be immobilized and chemically transformed by metals on soil clays (Mortland, 1986). Dioxin was shown to be catalytically polymerized by Cu on smectite; the Cu acted to form organic free radicals which then polymerized on the clay surface (Boyd and Mortland, 1985).

The best recent example of an engineered biodegradation technology is the use of genetically altered bacteria for the rapid decomposition of petroleum contaminating the beaches of Alaska and Texas from off-shore oil spills (Miller, 1990). Extensive research is under way to isolate or "bioengineer" micro-organisms for degradation of specific organics, but their application to in situ remediation has been limited (Sims and Sims, 1986; Thibault and Elliott, 1980). Microbial degradation of extracted contaminants is more feasible at the present because of the greater environmental controls these systems offer (Surprenant et al., 1988). In the case of contaminants such as petroleum, there is an extensive body of knowledge in the petroleum industry in land treatment systems in which petroleum is incorporated shallowly into surface soil and indigenous heterotrophic micro-organisms utilize the petroleum as substrate (Lynch and Genes, 1989).

Photodegradation has also been used for the reclamation of soils contaminated by PCBs (Occhiucci and Patacchiola, 1982), dioxins (Crosby et al., 1971), and kepone (Dawson et al., 1980). Photodegradable compounds are those with strong to moderate absorption in the 290 + nm range (Sims and Sims, 1986).

Advances in the development of engineering approaches to reclamation of chemically degraded land are expected in the next decade as a result of extensive research by U.S. federal agencies such as EPA and Department of Energy (DOE). The U.S. Defense Department is also developing technologies for clean-up of waste-contaminated sites at their installations. While these developments are expected to greatly improve our ability to manage the most intractable problems of chemical land degradation, their cost will make them unacceptable for most reclamation projects. In most cases, manipulation of ecological processes will have to be used.

B. Ecological Approaches to Remediation of Chemically Degraded Soil

Ecological remediation involves the manipulation of inherent soil processes to immobilize, mobilize, transform, or degrade contaminants, and approaches can include any of the following: (1) landscape stabilization; (2) liming strongly acid soils; (3) acid neutralization of alkaline soils; (4) organic matter addition; (5) fertilization; (6) establishment of vegetative cover.

1. Landscape Stabilization

Before long-term restoration of degraded soils can be achieved, it is essential that the landscape be stabilized against physical degradation, primarily erosion or slope failure. An essential component of surface mine reclamation is the grading of the slope prior to surface treatment and revegetation. This is particularly true in the eastern coal belt of the U.S. where highwalling, the practice of removing the underlying coal seam from an exposed face, results in cliff faces of up to 20 m in height. Other stabilization practices include cut-off ditches and a variety of terraces. Rapid establishment of vegetative or mulch surface cover is also important in protecting the landscape against erosion, but adequate landscape modification will reduce the dependence of erosion control on rapid establishment of surface cover.

Land shaping is practised on capped landfills to encourage runoff and reduce percolation through the landfill contents. This requires that the slope is adequately vegetated to reduce erosion. Vigorously growing vegetation will also increase evapotranspiration and decrease percolation.

Badly degraded land may require the placement of a cap of soil, organic material (see below), or other material (e.g., fly ash) on the surface to provide an adequate rooting medium for revegetation. Increasing thickness of topsoil increased yields of maize (*Zea mays*) and wheat (*Triticum aestivum*) on a sodic coal mine spoil in North Dakota (Halvorson et al., 1987). The response to topsoil amendments may be both chemical and physical, but it is often easier to modify the chemical properties of a degraded soil by amendments than to change critical soil physical properties such as water-holding capacity.

Declining availability of high quality topsoil in many areas is forcing use of alternative materials for seed zone development; see section below for use of organic amendments in reclamation.

2. Liming in Reclamation

Extreme acidity is common in degraded soils as a result of erosion, base leaching, organic matter oxidation, pyrite oxidation, or acid deposition. The positive effects of liming include adsorption and precipitation of toxic metals, enhanced complexation of metals by SOM, increased nutrient bioavailability, and enhanced biological activity.

Soil minerals and SOM contain variable quantities of pH-dependent charge as a result of dissociation of surface hydroxyls and water (Stevenson, 1982). Liming results in an increase in metal adsorption as described by a characteristic S-shaped curve. This is a consequence of dissociation characteristics of the sorbent (e.g., oxide, aluminosilicate, carbonate, or SOM) and hydrolysis of the metal which gives rise to species of lower positive charge with increasing pH:

$$SOH = SO^- + H^+ \quad \text{(where S = unit sorbent)}$$

$$M^{2+} + OH^- = MOH^+ \quad \text{(where M = metal)}$$

$$MOH^+ + OH^- = M(OH)_2^0$$

In the case of the oxyanions (e.g., AsO_4, MoO_4, CrO_4, SeO_4), adsorption decreases with increasing pH. This complicates attempts to immobilize metals in soils contaminated by both cationic and ionic metals. Such mixed metal systems are common in landfills, dredge spoils, and high-metal, sewage sludge-contaminated soils.

Liming can precipitate toxic macrometals such as Fe, Al, and Mn, and trace metals if they occur at high enough concentrations. It is also likely that trace metals are coprecipitated in contaminated soils as solid phases with the macrometals. The major ligands precipitating with the metals are OH, CO_3, SiO_3, PO_4, and SO_4. Of these, the oxides, silicates, and phosphates are likely to be the most stable under normal environmental conditions. This suggests that metal immobilization with liming might be enhanced by the addition of stabilizing ligands. This is essentially the approach used in the various stabilization/fixation processes with cement, cement kiln dust, pozzolon, silicates, or other reactants (Surprenant et al., 1988), and the N-Viro sewage sludge stabilization process (Burnham et al.,

1990) in which metals are immobilized by a combination of precipitation and complexation when alkaline cement kiln dust is added to the sludge. For in situ immobilization and reclamation of the site, however, these processes are too extreme as they result in a physically rigid mass. What is needed is selective addition of metal-precipitating ligands in conjunction with liming to produce a thermodynamically stable solid without greatly changing the properties of the soil. Preliminary research at a number of institutions (Marcus Pavan, Instituto Agronomico do Parana, Parana State, Brazil; Malcolm Sumner, University of Georgia) have shown positive effects on crop growth of phospho-gypsum, a by-product of phosphorus fertilizer manufacture. This material may offer some potential for trace metal immobilization in metal-contaminated acid soils.

Liming also enhances SOM complexation of metals by increasing dissociation of surface functional groups. In the normal soil pH ranges of 4 to 8, these are primarily -COOH and more acidic -OH groups (Perdue, 1985). Complexation is a more significant mechanism than ion exchange for immobilization of trace metals because of the much higher complexation constants for trace metals relative to macrometals, as opposed to ion exchange in which selectivity for the trace metals does not compensate for the higher concentrations of macrometals.

Nutrient bioavailability is often enhanced by liming very acid soils. Lime adds Ca and increases the availability of Mg and K by lowering the competition for cation exchange sites by Al^{3+} at pHs below 5. Phosphorus bioavailability increases when soil pH increases from acid values to about 6.5 (Lindsay, 1979) as a result of decreased Fe and Al solubility with increasing pH. Liming also increases mineralization of soil organic N by providing a more optimum environment for soil micro-organisms. This presumes, however, that the C:N ratio in the soil is low enough to produce net N mineralization (Stevenson, 1982).

Liming very acid soil is important in stimulating soil microfauna. Soil bacteria and actinomycetes, in particular have higher activities at pHs near neutral, while fungi can function at more acid pHs (Atlas and Bartha, 1981). In addition to the direct effect of high H^+ activities, microorganisms are inhibited by high concentrations of macro and trace metals.

Liming is very important in reclamation in providing a chemically acceptable plant-rooting environment. Plant roots will not penetrate subsoils that contain high levels of exchangeable Al, and failure to provide a deeper rooting environment places the seeded species under greater nutrient and moisture stress. Deep incorporation or placement of lime is often difficult in rocky soils or mine spoils. For this reason, more soluble materials such as Na_2CO_3, Na and Ca silicates, and gypsum ($CaSO_4.2H_2O$) may offer some promise in moving lime to the subsurface. Several studies (Oates and Caldwell, 1985; Pavan et al., 1984; Reeve and Sumner, 1972) have shown that gypsum can penetrate and neutralize acid subsoils. Hammel et al. (1985) found that surface application of gypsum increased maize yields 35% 3 years after application; this compared with an increase of 56% with lime incorporated to a depth of 1 m. The increase was attributed to deeper rooting and greater water utilization. In addition, Pavan et al. (1984) have shown that gypsum can inactivate toxic Al^{3+} by forming $AlSO_4^+$ complexes. Adams and Hathcock (1984) have shown that liming acid subsoils may stimulate crop growth by either reducing Al toxicity or by increasing Ca levels. Radcliffe et al. (1986) also found that gypsum improved subsurface structure by increasing root activity.

3. Reclamation of Alkaline Soils

Although acidification is a more common occurrence in chemical land degradation, there are instances in which soil is degraded by disposal or build-up of strong alkali. Soils are naturally buffered in the alkaline pH range by precipitation of $CaCO_3$ which gives equilibrium soil pHs of 8 to 8.5 (Lindsay, 1979). Soil pHs in excess of these values are usually associated with strong alkali, such as NH_3 (a short-term condition because of rapid nitrification), and Na and K salts. An example of alkaline wastes placed on land are the red muds produced by the extraction of Al from bauxite with concentrated NaOH. The residue is a Na-saturated mixture of clays, primarily kaolinite and Fe oxides, with a pH in excess of 12. The muds are typically placed in shallow lagoons and are difficult to reclaim because of the high alkalinity and salt content and the dispersed nature of the solids. A more recent example of alkaline wastes is flue gas desulfurization by-product (FGD), a by-product of alkaline scrubbing of SO_2 from coal-powered electrical generating plants in the U.S. This material is a mixture of gypsum, fly ash, and unreacted CaO, and often has pHs >9 to 10 (Mattigod et al., 1990; Eary et al., 1990).

Neutralization of excess alkalinity in these materials can be achieved by direct addition of acid, application of elemental S which is oxidized to H_2SO_4, or by application of gypsum. The neutralized salts must be leached to reduce salt content for establishment of an active biological system. A more natural, but slower, approach to alkali neutralization is by carbonation in which the final product is $CaCO_3$. This is best achieved by application of a degradable organic substrate such as manure, sewage sludge, sludge compost, papermill sludge, etc. Establishment of vegetative cover will also aid in neutralization of alkalinity by root respiration.

4. Organic Amendments for Soil Reclamation

Organic matter in one form or another has long been used to rejuvenate degraded land (Sutton and Vimmerstedt, 1973; Hornick 1982; Sopper and Kerr, 1979; Berry, 1985; Franks et al., 1982). Organic amendments contain varying contents of undegraded organic materials (e.g., plant residue; wood chips, and tree bark in compost; paper fibers in papermill sludge) and humus. The undegraded material plays an important role in reclama-

Table 3. Potential organic amendments for reclamation of chemically degraded land

Material	Types	Characteristics	Role in reclamation
Plant residues	Straw Leaves Wood chips Shredded bark Sawdust	High in degradable OM	Mulch Stimulate biota Improve physical properties
Manure	Beef Dairy Poultry Swine Horse	High in degrabable OM Nutrient source	Stimulate biota Add nutrients
Sewage sludge	Anaerobic Aerobic Lime-stabilized Waste-activated Papermill	Nutrient source Lime source (lime sludges)	Stimulate biota Add nutrients Increase pH (lime sludges)
Composts	Manure Sewage sludge Leaf MSW Mushroom	High in stable OM Nutrient source	Improve physical properties Stimulate biota Add nutrients
MSW	Garbage	High in degradable OM Variable nutrient content	Stimulate biota Erratic nutrient supply Overall effects may be netative
Peat	Sphagnum Muck	High in stable OM	Improve physical properties Stimulate biota

MSW, municipal solid waste

tion by stimulation of biological activity, production of CO_2, and release of nutrients. The humus aids in neutralization of alkalinity, binding of toxic metals, and partitioning of toxic organics. Besides these direct chemical and biological effects, organic amendments improve the physical condition of the degraded soil by increasing water-holding capacity and promoting the formation of stable structure. This increases the potential for successful revegetation.

Types of organic amendments that might be used for reclamation are numerous (Table 3). Selection of a particular material will depend on local availability, transportation and application costs, and local regulations governing land application of wastes. In general, materials that contain considerable contents of stable organic matter (compost, peat) are more desirable for reclamation of severely degraded land where conditions may be severely limiting to establishment of vegetative cover. Physical properties of the material may also be important. In a field study of reclamation of pyritic mine spoil in Ohio, composted sewage sludge was found to segregate on the steep slopes as a result of runoff, with the nutrient-rich fine humus migrating to the toe of the slope, and the carbon-rich coarser wood chips remaining on the top of the slope (Hale, 1982). The result was greater vegetative growth on the lower slope. Likewise, in studies with digested papermill sludge, Hoitink et al. (1982) found that the fibrous nature of the paper waste gave it desirable slope stabilizing properties when applied in the field to abandoned coal mine land. Watson and Hoitink (1985) showed that the high free $CaCO_3$ content of papermill sludge enhanced its ability to reclaim acidic mine spoil by maintaining pH at 7.6 three years after application of 150 to 300 t/ha of the material to spoil with a pH of 3.4. Haghiri and Sutton (1982) found that both composted sewage sludge (179 to 716 t/ha) and papermill sludge (67 to 112 t/ha) were as effective as a 20-cm layer of limed topsoil in revegetation of acidic coal-mine spoil. The papermill sludge was more effective than sludge at equivalent rates. A relatively recent product, cement kiln dust-stabilized sludge, has been shown to have "soil-like" physical properties that make it an ideal material for reclamation (Logan, 1990). Its high lime content is also desirable for reclamation of highly acidic soils, and preliminary studies (Prezzotto and Logan, 1990) have shown the material to be superior to equivalent amounts of lime and fertilizer in promoting grass establishment on pyritic mine spoil.

5. Nutrient Additions

Chemically degraded soils are not always low in nutrients, but nutrient availability may be reduced by suppressed biological activity or by severe physical conditions such as drought stress. Nutrient additions need to be timed with revegetation to avoid runoff or leaching losses. For this reason, slow-release organic nutrient sources may be more effective than inorganic chemical fertilizer. To enhance nutrient availability, soil pH should be adjusted at the same time. A problem common to many reclamation projects is the difficulty in placing or incorporating fertilizer, lime, organic matter, or other amendments at sufficient depth to promote deep rooting. Nutrients such as lime and phosphate that react strongly with the soil are particularly difficult to move to lower depths.

6. Revegetation

Reclamation cannot be considered complete until the site has been revegetated. Vegetative cover is needed to protect the soil from erosion, but vegetative cover can also be viewed as an ecological test of the success of reclamation.

Major considerations in selection of species for revegetation are: (1) native species, (2) proposed land use, and (3) ecological and environmental constraints. Restoration ecology has as a primary goal reestablishment of the plant and animal communities native to the area (Jordan et al., 1987). Few long-term studies are available to indicate if this is a realistic goal in most cases. Leopold's attempts to reconstruct natural communities at the University of Wisconsin Arboretum have yet to produce a finished replica of the natural model 50 years after restoration was started (Jordan et al., 1987). While restoration of native-like communities may be a desirable goal in some cases, in others the goal of revegetation is stabilization of the land surface and reestablishment of a biologically active surface soil. The initial species may be selected for tolerance to high levels of chemical contaminants in the soil or to harsh physical conditions such as drought, wetness, or temperature stress. Nutrient demands must also be considered. Revegetation may involve multiple species (mixtures of annual and perennial grasses and legumes together with trees) in the initial planting, or sequential planting in which the initial revegetation is with a fast-growing annual species, followed by establishment of perennial species and/or trees. I have observed *Casuarina* (Table 6) planted into a grass sod for reclamation of lateritic nickel mine spoil in the Dominican Republic. The grass species was not identified, but was planted from stolens. Selection of reclamation species should consider differential responses to soil amendments and their effects on competition. Kerr and Sopper (1982) found that hardwoods were easier than conifers to establish on sewage sludge-amended coal-mined land in eastern Pennsylvania because the understory vegetation crowded out the slower growing conifers. However, the use of contact herbicides may permit planting into existing vegetation without redisturbance of the site while still reducing competition from the established vegetation.

Selection of species for revegetation must, by necessity, be a local consideration. Tables 4 to 6 provide a listing of grass, legume, and tree species used in reclamation. The listings are based on actual reclamation studies or on recommendations for reclamation. These should be considered as general guides, and it is wise to follow the admonition of Thoreau to "consult with Nature in the outset for she is the most extensive and experienced planter of us all." The grass species (Table 4) are often used to produce rapid ground cover, particularly where high levels of nitrogen are available, as with the application of sewage sludge or N fertilizer. Reclamation mixes often include both annual and perennial grasses, the annual grasses providing most of the initial ground cover and the perennial grasses taking over in subsequent growing seasons. Legumes are effectively excluded from grass-legume mixtures where N levels are high, but are effective first

Table 4. Grass species used in reclamation

Scientific name	Common name	Characteristics	Attributes for reclamation
Agrostis gigantea	Red top	Perennial	Not adapted to extremes
Agrostis stolonifera	Creeping bentgrass	Perennial	Low drought tolerance
Agrostis tenuis	Bentgrass	Perennial	Low nutrient requirements Acid tolerant
Agropyron cristatum	Crested wheat	Perennial	Acid tolerant Low fertility tolerant
Alopercurus pratensis	Meadow foxtail	Perennial	Not adapted to extremes
Andropogon spp.	Bluestem	Perennial	Drought tolerant
Avena sativa	Oats	Annual	Drought tolerant Acid tolerant Low fertility tolerant
Bromus inermis	Bromegrass	Perennial	Acid tolerant Low fertility tolerant
Dactylis glomerata	Orchard grass	Perennial	Requires high fertility
Festuca rubra	Red fescue	Perennial	Low nutrient requirements
Festuca or uninacea	Tall fescue	Perennial	Requires high fertility Low drought tolerance
Festuca pratensis	Meadow fescue	Perennial	Not adapted to extremes
Lolium perenne	Perennial rye	Perennial	Not adapted to extremes
Panicum virgatum	Switchgrass	Perennial	Acid tolerant
Phleum pratense	Timothy	Perennial	Not adapted to extremes
Poa pratensis	Kentucky bluegrass	Perennial	Acid tolerant Drought tolerant
Poa annua	Annual bluegrass	Annual	Low fertility tolerant
Secale cereale	Annual rye	Annual	Acid tolerant Low fertility tolerant
Tripsacum dactylodes	Eastern Gama	Perennial	Acid tolerant

Source: Bradshaw and Chadwick (1980); Peters, (1988); Franks et al. (1982)

Table 5. Legume species used in reclamation

Scientific name	Common name	Characteristics	Attributes for reclamation
Calopogonium mucunioides	Calopo	Perennial vine Tropical	Acid tolerant
Centrosema pubescens	Centro	Perennial vine Tropical	Acid tolerant
Coronilla varia	Crown vetch	Perennial	Acid tolerant Prolific seed producer
Desmodium uncinatum	Silver leaf desmodium	Perennial Tropical	Acid tolerant
Lablab purpureus	Lablab	Tropical	Acid tolerant
Lespedeza bicolor	Bicolor	Perennial	Acid tolerant
Lespedeza cuneata	Sericea	Perennial	Acid tolerant
Lespedeza japonica	Japanese	Perennial	Acid tolerant
Lotus corniculatus	Birdsfoot trefoil	Perennial	Widely adapted
Macroptilium atropurpureum		Tropical	Acid tolerant
Medicago sativa	Alfalfa	Perennial	Not widely adapted
Melilotus alba	Sweet clover	Annual/perennial	Widely adapted
Phaseolus atropurpureus	Siratro	Perennial vine Tropical	Acid tolerant
Pueraria phaseoloides	Kudzu	Tropical vine	Acid tolerant
Stylosanthes humilis	Stylo	Perennial vine Tropical	Acid tolerant
Trifolium hybridum	Alsike clover	Perennial	Acid tolerant
Trifolium pratense	Red clover	Annual/perennial	Not widely adapted
Trifolium repens	White clover	Annual/perennial	Widely adapted

Sources: Bradshaw and Chadwick (1980); Peters, 1988; Archer et al., (1988)

Table 6. Tree species used in reclamation

Scientific name	Common name	Characteristics	Attributes for reclamation
Acacia spp.	Acacia	Legume; tropical	Drought tolerant
Acer pseudoplatanus	Sycamore	Deciduous	Drought tolerant
Albizzia spp.	Albizzia	Legume; tropical	Acid tolerant Drought tolerant
Alnus glutinosa	Black alder	Legume	Acid tolerant
Alnus incana	Grey alder	Legume	
Betula papyrifera	Birch	Deciduous	Low nutrient requirements Acid tolerant
Betula pubescens	Birch	Deciduous	Low nutrient requirements
Cassia obtusifolia	Cassia	Legume; tropical	Acid tolerant
Casuarina equisetifolia	Casuarina	Legume; tropical	Acid tolerant
Coriaria arborea	Coriaria	Legume	Acid tolerant Drought tolerant
Crotalaria anagyroides	Crotalaria	Legume; tropical	Acid tolerant
Eleagnus sylvatica	Russian olive	Evergreen	Drought tolerant
Eleagnus umbellulata	Autumn olive	Evergreen	Drought tolerant
Fagus sylvatica	Beech	Deciduous	Acid tolerant Drought tolerant
Fraxinus americana	White ash	Deciduous	Wet and drought tolerant
Fraxinus excelsior	Ash	Deciduous	Drought tolerant
Fraxinus pennsylvanica	Green ash	Deciduous	Alkali sensitive
Juniperus virginiana	Eastern red cedar	Evergreen	Acid and drought tolerant Low nutrient requirements
Larix leptolepis	Japanese larch	Conifer	Acid and drought tolerant Low nutrient requirements
Leucaena leucocephala	Leucaena	Legume; tropical	Drought tolerant
Picea mariana	Black spruce	Conifer	

Table 6. (cont.)

Scientific name	Common name	Characteristics	Attributes for reclamation
Pinus banksiana	Jack pine	Conifer	Acid and drought tolerant Low nutrient requirements
Pinus caribbea	Caribbean pine	Conifer	Acid and drought tolerant Low nutrient requirements
Pinus echinata	Shortleaf pine	Conifer	Acid and drought tolerant Low nutrient requirements
Pinus negra	Austrian pine	Conifer	Acid and drought tolerant Low nutrient requirements
Pinus rigida	Pitch pine	Conifer	Acid and drought tolerant Low nutrient requirements
Pinus strobus	White pine	Conifer	Acid and drought tolerant
Pinus sylvestris	Scots pine	Conifer	Acid and drought tolerant Low nutrient requirements
Pinus taeda	Loblolly pine	Conifer	Acid and drought tolerant
Pinus virginiana	Virginia pine	Conifer	Acid and drought tolerant Low nutrient requirements
Platanus occidentalis	Western plane	Deciduous	Wetness tolerant
Populus spp.	Hybrid poplar	Deciduous	Acid tolerant
Prosopis spp.	Mesquite	Legume	Drought tolerant
Prunus pumila	Sand cherry	Deciduous	
Robinia fertilis	Bristly locust	Legume	Acid and drought tolerant

Table 6. (cont.)

Scientific name	Common name	Characteristics	Attributes for reclamation
Robinia pseudoacacia	Black locust	Legume	Acid and drought tolerant
Salix spp.	Willow	Deciduous	Acid, drought, and wetness tolerant
Sesbania grandiflora	Sesbania	Legume; tropical	Acid tolerant
Thuja occidentalis	Northern white cedar	Conifer	Drought and wetness tolerant

Source: Bradshaw and Chadwick (1980); Archer et al. (1988); Kerr and Sopper, (1982); Borovsky and Brooks (1982)

cover where there is no opportunity to amend the soil with N. In most cases, the soil or seed must be inoculated with *Rhizobium* for effective establishment. Tree species must be selected for their tolerance to subsoil conditions, rate of growth and canopy formation, and their ability to compete with lower story species (McLeod et al., 1986). Chemical weed control may have to be used to establish trees in a grass or legume stand. A further consideration with tree establishment in degraded soils is subsurface tillage. Berry (1986) reported significant growth response with subsoiling by loblolly pine established in a borrow pit and amended with either sewage sludge or lime and fertilizer. The response is primarily one of drought stress rather than nutrient availability or toxicity. In another study involving multiple species establishment, Smith et al. (1986) found that, although most species responded to increasing inputs of wood residue and N fertilizer, species diversity peaked at input rates lower than those that gave greatest total biomass production. This is an important tradeoff that should be considered in any reclamation strategy. An interesting finding of this study, which involved revegetation of swelling bentonite clay spoils, was the superior response of sod-forming species (rhizomatous, stoloniferous, or shallow-rooted tufted). This was attributed to the general resistance of these species to root breakage associated with clay swelling.

The ability to initiate and maintain vegetative cover will be the final, most important, test of any reclamation project. In the U.S., federal and state funded mine reclamation projects usually require demonstration of vegetative cover for periods up to 5 years. Closed landfills, particularly if they contain hazardous or radioactive wastes, will have to be maintained with appropriate vegetative cover for indefinite periods. Given the rapid rate at which public environmental concerns change, particularly with respect to expenditure of tax revenues, it is essential that reclaimed sites have conditions that will provide them with inherent resiliency. I believe that this will be achieved best by the establishment of an active biological system in the form of adequate vegetative cover. Stresses on the reclaimed site should be minimized to the extent possible—these lands will probably never have the capacity for sustained productivity that an intact soil would. From the standpoint of chemical degradation, the importance of inherent chemical buffering—acidity, nutrient availability, redox, exchange, etc.—cannot be overemphasized as a means of maintaining the reclaimed state.

III. Conclusions

1. Chemical degradation of land is likely to become a more widespread problem as mining, industrial, and agricultural pressures on the land increase.
2. Chemical degradation can be reversed by appropriate treatment, but the extent and rate of remediation will depend on the extent and type of contamination, importance of the soil resource, and the associated water and biological systems, as well as the resources committed to reclamation.
3. Engineering approaches to chemical degradation are expensive, and highly disruptive of the soil system, but may be required for more recalcitrant contaminants or for more rapid reclamation of a valuable resource.
4. Ecological approaches to chemical degradation involve manipulation of inherent soil processes to immobilize, mobilize, transform, or degrade the contaminant. Ecological processes are likely to take longer than engineering approaches to achieve remediation.
5. Ultimate reclamation will only occur when an active biological system is established in the form of vegetative cover and soil micro and macro fauna.
6. The established biological system must be maintained by reducing land use stresses on the reclaimed site.

References

Adams, F. and P.J. Hathcock. 1984. Aluminum toxicity and calcium deficiency in acid subsoil horizons of two coastal plains soil series. *Soil Sci. Soc. Am. J.* 48:1305–1309.

Archer, I.M., N.A. Marshman, and W. Salomons. 1988. Development of a revegetation programme for copper and sulphide-bearing mine wastes in the humid tropics. In: W. Salomons and U. Forstner (eds.) *Environmental Management of Solid Wastes*, pp. 166–184. Springer-Verlag, Berlin.

Atlas, R.M. and R. Bartha. 1981. *Microbial Ecology: Fundamentals and Applications.* Addison-Wesley, Reading, PA.

Bartlett, R.J. and J.M. Kimble. 1976. Behavior of chromium in soils: II. Hexavalent forms. *J. Environ. Qual.* 5:383–386.

Berry, C.R. 1985. Subsoiling and sewage sludge aid loblolly pine establishment on adverse sites. *Reclamation Reveg. Res.* 3:301–311.

Berry, C.R. 1986. Reclamation of severely devastated sites with dried sewage sludge in the Southeast. In: D.W. Cole, C.L. Henry and W.L. Nutter (eds.). *The Forest Alternative*, pp. 497–507. University of Washington Press. Seattle, WA.

Borovsky, J.P. and K.N. Brooks. 1982. Performance of woody plant species on iron-ore overburden material irrigated with sewage effluent in Minnesota. In: W.E. Sopper, E.M. Seaker, and R.K. Bastian (eds.) *Land Reclamation and Biomass Production with Municipal Wastewater and Sewage Sludge*, pp. 252–263. Pennsylvania State University Press, University Park, PA.

Boyd, S.A. and M.M. Mortland. 1985. Dioxin radical formation and polymerization on Cu(II)-smectite. *Nature* (London) 316:532–535.

Bradshaw, A.D. and M.J. Chadwick. 1980. *The Restoration of Land.* Blackwell Scientific Publications, Oxford.

Burnham, J.C., T.J. Logan, and G.F. Bennett. 1990. Cement kiln dust stabilization of municipal wastewater sludge: microbiology, chemistry and product development. Haztech Int. 90 Symp. Am. Chem. Soc. Houston, TX.

Chawla, R.C., M.S. Diallo, J.N. Cannon and J.H. Johnson. 1990. In-situ treatment of soils contaminated with hazardous organic wastes using surfactants: a critical analysis. In: H.S. Muralidhara (ed.) *Solid/Liquid Separation: Waste Management and Productivity Enhancement*, pp. 356–367. Battelle Press, Battelle Memorial Institute, Columbus, OH.

Crosby, D.G., A.S. Wong, J.R. Plimmer, and E.A. Woolson. 1971. Photodecomposition of chlorinated dibenzo-p-dioxins. *Science* 73:748.

Cullinane, M.J., L.W. Jones, and P.G. Malone. 1986. Handbook for stabilization/solidification of hazardous waste. U.S. EPA Hazardous Waste Engineering Research Laboratory, Cincinatti, OH. EPA-540/2-86/001.

Dawson, G.W., B.W. Mercer, and C.H. Thompson. 1980. Strategy for the treatment of spills on land. *Control of Hazardous Materials Spills.* Vanderbilt University, Nashville, TN.

Eary, L.E. and D. Rai. 1988. Chromate removal from aqueous wastes by reduction with ferrous iron. *Environ. Sci. Tech.* 22:972–977.

Eary, L.E., D. Rai, S.V. Mattigod, and C.C. Ainsworth. 1990. Geochemical factors controlling the mobilization of inorganic constituents from fossil fuel combustion residues: II. Review of the minor elements. *J. Environ. Qual.* 19: 202–214.

Follett, R.F. and B.A. Stewart (eds.). 1985. Soil Erosion and Crop Productivity. Am. Soc. Agron., Madison, WI.

Franks, W.A., M. Persinger, A. Iob, and P. Inyangetor. 1982. Utilization of sewage effluent and sludge to reclaim soil contaminated by toxic fumes from a zinc smelter. In: W.E. Sopper, E.M. Seaker, and R.K. Bastian (eds.) *Land Reclamation and Biomass Production with Municipal Wastewater and Sewage Sludge*, pp. 219–251. Pennsylvania State University Press, University Park, PA.

Griffin, R.A. and S.J. Chou. 1980. Proc. Sixth Annual Res. Symp. U.S. Environmental Protection Agency. Office of Research and Development, Cincinnati, OH. EPA/600/9-80-010. pg. 549.

Haghiri, F. and P. Sutton. 1982. Vegetation establishment on acidic mine spoils as influenced by sludge application. pp. 433–446. In: W.E. Sopper, E.M. Seaker and R.K. Bastian (eds.) *Land Reclamation and Biomass Production with Municipal Wastewater and Sewage Sludge*, pp. 433–446. The Pennsylvania State University Press, University Park, PA.

Hale, A.M. 1982. Little Kyger Creek demonstration plot program. In: N.E. Smeck and P. Sutton (eds.) *Proc. Abandoned Mine Reclamation Symposium*, pp. 6-1–6-24. All-Ohio Chapter Soil Conserv. Soc. Am., Ohio State Univ., Columbus, OH.

Halvorson, G.A., A. Bauer, S.A. Schroeder and S.W. Melsted. 1987. Corn and wheat response to topsoil thickness and phosphorus on reclaimed land. *J. Environ. Qual.* 16:73–76.

Hammel, J.E., M.E. Summner, and H. Shahandeh. 1985. Effect of physical and chemical profile modification on soybean and corn production. *Soil Sci. Soc. Am. J.* 49:1508–1511.

Hassett, J.J. and W.L. Banwart. 1989. The sorption of nonpolar organics by soils and sediments. pp. 31–44. In: B.L. Sawhney and K. Brown (eds.) *Reactions and Movement of Organic Chemicals in* Soils. Soil Sci. Soc. Am., Madison, WI.

Hinchee, R.E., H.S. Muralidhara, F.B. Stulen, G.B. Wickramanayake, and B.F. Jirjis. 1990. Electroacoustic soil decontamination process for in-situ treatment of contaminated soils. In: H.S. Muralidhara (ed.) *Solid/Liquid Separation: Waste Management and Productivity Enhancement*, pp. 370–383. Battelle Press, Battelle Memorial Institute, Columbus, OH.

Hoitink, H.A.J., M.E. Watson, and P. Sutton. 1982. Reclamation of abandoned mine land with papermill sludge. In: N.E. Smeck and P. Sutton (eds.) *Proc. Abandoned Mine Reclamation Symposium*, pp. 5-1–5-6. All-Ohio Chapter Soil Conserv. Soc. Am., The Ohio State Univ., Columbus, OH.

Hornick, S.B. 1982. Crop production on waste amended gravel spoils. In: W.E. Sopper, E.M. Seaker, and R.K. Bastian (eds.) *Land Reclamation and Biomass Production with Municipal Wastewater and Sewage Sludge*, pp. 207–218. Pennsylvania State University Press, University Park, PA.

Hunt J.R., N. Sitar, and K.S. Udell. 1988. *Water Resources Res.* 24(8):1259–1269.

Jordan, W.R., M.E. Gilpin, and J.D. Aber. 1987. Restoration ecology: ecological restoration as a technique for basic research. In: W.R. Jordan, M.E. Gilpin, and J.D. Aber (eds.) *Restoration Ecology: A Synthetic Approach to Ecological Research*, pp. 3–21. Cambridge University Press, New York.

Kerr, S.N. and W.E. Sopper. 1982. Utilization of municipal wastewater and sludge for forest biomass production on marginal and disturbed land. In: W.E. Sopper, E.M. Seaker, and R.K. Bastian (eds.) *Land Reclamation and Biomass Production with Municipal Wastewater and Sewage Sludge*, pp. 75–87. Pennsylvania State University Press, University Park, PA.

Lindsay, W.L. 1979. *Chemical Equilibria in Soils.* Wiley, New York, pp. 295–297.

Logan, T.J. 1989. Chemical degradation of soil. *Advances Soil Sci.* 1:187–222.

Logan, T.J. 1990. Chemistry and bioavailability of metals and nutrients in cement kiln dust-stabilized sewage sludge. Proc. Conf. on The Status of Municipal Sludge Management for the 1990s. Water Pollution Control Fed., Alexandria, VA.

Logan, T.J. and D.E. Cassler. 1989. Correcting widespread cadmium contamination. *Water Environ. Technol.* 1:312–315.

Lynch, J. and B.R. Genes. 1989. Land treatment of hydrocarbon contaminated soils. pp. 163–174. In: P.T. Kostecki and E.J. Calabrese (eds.) *Petroleum Contaminated Soils*, vol. 1: Remedation techniques, environmental fate, risk assessment. Lewis Pubs., Chelsea, MI.

Malmanis, E., D.W. Fuerst, and R.J. Pineiwski. 1989. Proc. Sixth National RCRA/Superfund Conf. and Exhibition, pp. 538–541. New Orleans, LA.

Mattigod, S.V., D. Rai, L.E. Eary, and C.C. Ainsworth. 1990. Geochemical factors controlling the mobilization of inorganic constituents from fossil fuel combustion residues: I. Review of the major elements. *J. Environ. Qual.* 19: 88–201.

McLeod, K.W., C.E. Davis, K.C. Sherrod, and C.G. Wells. 1986. Understory response to sewage sludge fertilization of loblolly pine plantations. In: D.W. Cole, C.L. Henry, and W.L. Nutter (eds.) *The Forest Alternative*, pp. 308–323. University of Washington Press, Seattle, WA.

Miller, S.S. 1990. In a faraway state. *Environ. Sci. Technol.* 24:1286–1289.

Mori, P.N. 1990. In-situ hot air/steam extraction of volatile organic compounds. In: H.S. Muralidhara (ed.) *Solid/Liquid Separation: Waste Management and Productivity Enhancement*, pp. 386–396. Battelle Press, Battelle Memorial Institute, Columbus, OH.

Mortland, M.M. 1986. Mechanisms of adsorption of nonhumic organic species by clays. In: P.M. Huang and M. Schnitzer (eds.) *Interactions of Soil Minerals with Natural Organics and Microbes*, pp. 59–76. Soil Sci. Soc. Am., Madison, WI.

Murphy, E.M., J.M. Zachara, and S.C. Smith. 1990. Influence of mineral-bound humic substances on the sorption of hydrophobic organic compounds. *Environ. Sci. Technol.* 24:1507–1516.

Nash, J.H. and R.P. Traver. 1989. Field studies of in situ soil washing. In: P.T. Kostecki and E.J. Calabrese (eds.) *Petroleum Contaminated Soils, vol. 1: Remediation Techniques, Environmental Fate, Risk Assessment*, pp. 157–161. Lewis Pubs., Chelsea, MI.

Nijssen, J.P.J. 1988. Rotterdam dredged material: approach to handling. In: W. Salomons and U. Forstner (eds.) *Environmental Management of Solid Wastes*, pp. 243–281. Springer-Verlag, Berlin.

Oates, K.M. and A.G. Caldwell. 1985. Use of by-product gypsum to alleviate soil acidity. *Soil Sci. Soc. Am. J.* 49:915–918.

Occhiucci, G. and A. Patacchiola. 1982. Sensitized photodegradation of adsorbed polychlorobiphenyls (PCBs). *Chemosphere* 11:255–262.

Pavan, M.A., F.T. Bingham, and P.F. Pratt. 1984. Redistribution of exchangeable calcium, magnesium and aluminum following lime or gypsum applications to a Brazilian Oxisol. *Soil Sci. Soc. Am. J.* 48:33–38.

Perdue, E.M. 1985. Acidic functional groups in humic substances. In: G.R. Aiken, D.M. McKnight, R.L. Wershaw, and P. MacCarthy (eds.) *Humic Substances in Soil, Sediment, and Water*, pp. 493–526. Wiley, New York.

Peters, T.H. 1988. Mine tailings reclamation. Inco Limited's experience with the reclaiming of sulfide tailings in the Sudbury Area, Ontario, Canada. In: W. Salomons and U. Forstner (eds.) *Environmental Management of Solid Wastes*, pp. 152–165. Springer-Verlag, Berlin.

Prezzotto, M. and T.J. Logan. 1990. Reclamation of pyritic mine spoil with cement kiln dust-stabilized sewage sludge. Project Report. Agronomy Department, Ohio State University, Columbus, OH.

Radcliffe, D.E., R.L. Clark, and M.E. Sumner. 1986. Effect of gypsum and deep-rooting perennials on subsoil mechanical impedence. *Soil Sci. Soc. Am. J.* 50:1566–1570.

Reeve, N.G. and M.E. Sumner. 1972. Amelioration of subsoil acidity in Natal Oxisols by leaching of surface applied amendments. *Agrochemophysica* 4:1–6.

Shapiro, A.P., P.C. Renaud, and R.F. Probstein. 1990. In-situ extraction of contaminants from hazardous waste sites by electroosmosis. In: H.S. Muralidhara (ed.) *Solid/Liquid Separation: Waste Management and Productivity Enhancement*, pp. 346–353. Battelle Press, Battelle Memorial Institute, Columbus, OH.

Sims, R.C. and J.L. Sims. 1986. Cleanup of contaminated soils. Utilization, Treatment and Disposal of Waste on Land, pp. 257–277. Soil Sci. Soc. Am., Madison, WI.

Smith, J.A., E.J. DePuit, and G.E. Schuman. 1986. Wood residue and fertilizer amendment on bentonite mine spoils: II. Plant species responses. *J. Environ. Qual.* 15:427–435.

Sopper, W.E. and S.N. Kerr (eds.). 1979. *Utilization of Municipal Sewage Effluent and Sludge on forest and Disturbed Land.* Pennsylvania State University Press, University Park, PA.

Stevenson, F.J. 1982. *Humus Chemistry.* Wiley, New York, pp. 355–373.

Surprenant, N., T. Nunno, M. Kravett, and M. Breton. 1988. Halogenated-Organic Containing Wastes. Treatment Technologies. Noyes Data Corporation, Park Ridge, NJ.

Sutton, P. and J.P. Vimmerstedt. 1973. Treat stripmine spoils with sewage sludge. *Ohio Report* 58:121–123.

Tans, P.O., I.Y. Fung, and T. Takahashi. 1990. Observational constraints on the global atmospheric CO_2 budget. *Science* 247:1431–1438.

Thibault, G.T. and N.W. Elliott. 1980. Biological detoxification of hazardous organic chemical spills. Control of Hazardous Materials Spills, pp. 398–402. Vanderbilt Univ., Nashville, TN.

Timmerman, C.T., J.L. Buelt, and V.F. Fitzpatrick. 1989. In situ vitrification processing of soils contaminated with hazardous wastes. In: P.T. Kostecki and E.J. Calabrese (eds.) *Petroleum Contaminated Soils, vol. 1: Remediation Techniques, Environmental fate, Risk Assessment*, pp. 137–156. Lewis Pubs., Chelsea, MI.

U.S. Environmental Protection Agency. 1986. Superfund innovative technology evaluation (SITE) strategy and program plan. EPA/540/G-86-001. Washington, DC.

Watson, M.E. and H.A.J. Hoitink. 1985. Long term effects of papermill sludge in stripmine reclamation. *Ohio Report* 70:19–21.

Soil Fertility Restoration and Management for Sustainable Agriculture in South Asia

Rajendra Prasad and N.N. Goswami

I. Introduction

Cropland, range, and fisheries constitute the three major sources of food. The 6 million Mg each of marine and beef protein consumed annually compounds to 250 million Mg of grain equivalents (Mg ge) in primary terms (Gilland, 1979). The 1984 gross primary world food production was about 3000 million Mg ge, of which cropland provided 80% and range and fisheries 10% each. Even with modern aquaculture techniques, it seems unlikely

Advances in Soil Science, Volume 17

that the 80:10:10 proportions will change significantly during the remaining years of this century (Swaminathan, 1986). Most of the world's food need, therefore, has to be met from arable soils. The statement, "Long after our mines cease to give up their treasures of iron, coal and precious metals, the soil must continue to produce the food necessary for feeding the ever increasing population of the world," was made by U.S. President Calvin Coolidge in inaugurating the First International Congress on the Soil Science in Washington, D.C., and it still holds true (Kanwar, 1982).

On the global scale the Food and Agriculture Organisation of the United Nations (FAO, 1981) estimated a deficit in food need of the order of 52 to 100 million Mg yr^{-1}. The estimate of the global food deficit by the International Food Policy Research Institute (Paulino, 1986) for the year 2000 A.D. is of the order of 70 million Mg yr^{-1}. Thus, even with all the advances made in agriculture sciences, millions of people may remain hungry by the end of the twentieth century. Hunger and malnutrition have been called the causes of perhaps the most widespread human suffering in the world today (WFP, 1990).

South Asia, comprising Bangladesh, Bhutan, India, Nepal, Pakistan, and Sri Lanka produced 173 million Mg of food grain (cereals and pulses) in 1978 and 220 million Mg of food grain in 1988 (Table 1); thus, over a period of 10 years this represents an increase of of 4.7 million Mg. According to Paulino (1986) South Asia is likely to produce 323 million Mg of fond grain per year by the turn of the century; its consumption is estimated at 282 to 310 million Mg yr^{-1}. Thus, in the next 10 years, South Asia has to produce 103 million Mg yr^{-1} more food grain. Since not much additional arable land is available, this increase has to come from improved production. There is considerable scope to improve production technology since per hectare average yields of most cereals are only about one-third to one-fourth of those achieved in most developing countries. During 1961–80 in South Asia, of the increase in food grain production 23% was due to increased cropped area while 77% of the increase was due to improved production technology (Paulino 1986). The soils of South Asia have thus to bear the burden of feeding the increasing population and maintenance of the productivity of these soils has to be given due attention. This chapter reviews the results of studies made in this direction. Most of the data are from India since it is published and readily available; available data from other countries of South Asia are included.

II. Soils, Climate, and Crops of South Asia

A. Soils

Major soils of South Asia are Inceptisols, Entisols, Aridisols, Alfisols, and Vertisols. Some Mollisols and Ultisols are also found (NBSS 1985; FAO, 1971, 1973; Moorman and Panabokke, 1961).

Table 1. Cereal production and fertilizer consumption in South Asia in 1978 and 1988.

Country	Year	Cereal production (million Mg)	(Mg ha^{-1})	Fertilizer (million Mg)	Consumption (kg ha^{-1})
Bangladesh	1978	20.0	1.9	0.35	36.6
	1988	23.0	2.1	0.80	81.0
Bhutan	1978	0.15	1.4	—	0.3
	1988	0.19	1.6	—	0.3
India	1978	119.7[a]	1.4	5.13	28.3
	1988	156.5[b]	1.7	11.05	60.9
Nepal	1978	3.6	1.6	0.02	4.4
	1988	4.6	1.6	0.06	13.0
Pakistan	1978	14.7	1.4	0.88	35.2
	1988	18.8	1.7	1.74	67.2
Sri Lanka	1978	1.9	2.1	0.14	58.4
	1988	2.5	2.9	0.21	90.4
Total	1978	160.05	—	6.52	—
	1988	265.59	—	13.86	—

Source: Adopted from FAO (1989) and RAPA (1989)
Note: Pulses (legume grains) are the main source of protein and an important component of the vegetarian diet in India. India produced 12.2 million Mg pulses in 1978 and 13.7 million Mg pulses in 1988; figures for Pakistan were 0.8 and 0.55 million Mg pulses in 1978 and 1988, respectively.
[a] 1978–79
[b] 1988–89
Source for a and b: FAI (1990)

In the northern river plains of Bangladesh, India, and Pakistan Inceptisols (Ochrepts, Tropepts, and Aquepts) occur with Entisols (Orthrenths, Fluvents, Aquents). Due to annual floods and their deposits these alluvial soils are very productive and are the granary of South Asia. The occurrence and severity of floods increases as one moves eastwards, with Bangladesh suffering most due to floods. Predominant soils in the northern-most hilly regions of India, Pakistan, and Nepal are Udolls.

In the western part of India and southern part of Pakistan there are large areas under Aridisols (Orthids, Argids), which occur along with Entisols (Psamments, Fluvents, Aquents). These areas receive less than 400 mm rainfall and are deserts. Arable farming is difficult and rearing of sheep, camels, and other farm animals is practised.

In central India, the most dominant soils are Vertisols (mostly Usterts), which occur with Entisols (Orthents and Fluvents) and Alfisols (Tropepts). Vertisols, due to their high swelling clay content are difficult to manage during the rainy season, and dry rapidly after the rains stop. Vertisols form the major share of the cotton belt of India. Cotton, being a deep-rooted crop, grows well on these soils, because it can extract moisture from the lower layers of the soil profile.

In the southern peninsula, eastern, and hill regions of India, and in Pakistan and Sri Lanka, Alfisols (Ustalfs, Udalfs, and Aqualfs) dominate and these occur with Inceptisols (Ochrepts and Tropepts), Entisols (Fluvents), and some Ultisols (Udults). In the Southern most part of India (Kuttanad region of Kerala) acid sulfate soils with pH less than 3.0 are found. This area lies below sea level and is kept submerged in water for the major part of the year. In certain tracts, high toxicity of Fe, Al, and Mn is encountered. The soils are rich in organic matter (organic carbon 6% to 8%), which on decomposition produces acidity (Aiyer and Nair, 1985). Mostly paddy rice is grown on these soils.

B. Climate

The climate in South Asia varies from subtemperate in the north to tropical in the south. Most of the rain in Bangladesh, Bhutan, India, Nepal, and Pakistan is received by southwest monsoon, which reaches the Kerala coast by the end of May, advances along the Konkan coast in early June, and extends over the entire Indian subcontinent by the end of July. The rains continue up to the end of September, when the southwest monsoon recedes. Thus, there is a clear wet season (June–October) and a dry season (November–May) in a large part of South Asia. The northeast monsoon brings rains in November and December and is the main contributor of rainfall over the southeastern part of the Indian Peninsula and drybelt of Sri Lanka.

The areas of very heavy rainfall in the Indian sub-continent exist on the windward side of Western Ghats, the Khasi Hills, and the Himalayas. These are the source regions for many of the major river systems of the country, particularly the Himalayan region. Western parts of India and southern parts of Pakistan are the driest and have less than 400 mm annual rainfall, while the northeastern part and the Western Ghats of India are the wettest; Arunachal Pradesh in the northeast has an annual rainfall of 4142 mm, and Kerala in the Western Ghat region has an annual rainfall of 2996 mm (ICAR, 1980).

In Sri Lanka the rainfall in the dry zone is bimodal. The main rainy reason is from late September–December with a rainfall of about 750 mm received mostly from the northeast monsoon. The smaller season is around April with about 350 mm rainfall. In the wet zone of Sri Lanka the rainfall is well distributed throughout the year and is 2000 mm or more.

The maximum temperatures in South Asia are highest during April–May (above 40° C in western and central India). In June the highest maximum temperature values shift north-westwards. In January, the cold dry winds blow from the North-west and night temperatures in northern Pakistan and north-west India may be 5° C or below. There is snowfall in Jammu and Kashmir and other hilly regions during December–January.

C. Crops and Cropping Systems

Since most of the rain is received during a period of 3 to 4 months, there are distinctly two crop seasons, namely, the wet season, which is popularly known as *kharif* season in India and Pakistan and the dry season, which is known as *rabi* season. Major crops grown during wet *kharif* season are: cereals, including rice (*Oryza sativa* L.), corn (*Zea mays* L.), sorghum (*Sorghum bicolor* (L.) Moench), pearl millet (*Pennisetum typhoides* (Burm.f.) Stapf and C.E. Hubb), finger millet (*Eleusine coracana* Gaertn); pulses (grain legumes), including mung bean (*Vigna radiata* (L.) Wilczek), black gram (*Vigna mungo* (L.) Hepper), cowpea (*Vigna unguiculata* (L.) Walp), and pigeonpea (*Cajanus cajan* Milp); oilseeds, including peanut (*Arachis hypogaea* L.); fiber crops, including cotton (*Gossypium* sp.) and jute (*Corchorus* sp.). Major crops during the dry *rabi* season in northern South Asia, where there is a distinct winter season, are: cereals, including wheat (*Triticum aestivum* (L.) emend.Thell), barley (*Hordeum vulgare* L.), and oats (*Avena sativa* L.); pulses, including Chickpea (*Cicer arietinum* L.), field peas (*Pisum sativum*), and lentils (*Lens esculenta* Moench); oilseeds, including rapeseed mustard (*Brassica* sp.), safflower (*Carthamus tinctorius* L.), and sunflower (*Helianthus annus* L.). Sugarcane (*Saccharum officinarum* L.) is grown for 10 to 12 months in northern India (March–February) or for 18 months in peninsular India (July–December in the next year). A number of vegetable and fodder crops are also grown. In areas where irrigation facilities are available a short summer season (May–June) crop may also be grown. A number of wet season crops of northern India are grown in the south during December–April especially in the region receiving rains from the southeast monsoon.

The largest acreage in South Asia is under rice, which is the major staple food crop and is grown in wet as well as dry seasons (with irrigation). In northern India and Pakistan, wheat is the staple food crop and the "green revolution" in South Asia set in due to the introduction of high yielding dwarf varieties of wheat (Borlaug, 1971; Tandon and Narayan, 1990), which made rice-wheat double cropping feasible. Introduction of dwarf wheats has also led to the development of a large number of intensive multiple cropping systems in regions of South Asia, where wheat is the major cereal during the *rabi* season.

In the southern part of Western Ghats of India and in the wet region of Sri Lanka there are large plantations of cardamom (*Elettaria cardamomum* (L.) Maton), pepper (*Piper nigrum* L.), arecanut (*Areca catechu* L.), coffee (*Coffea arabica* L.), and rubber (*Hevea brasiliensis* Mull—Arg.). In the eastern hilly region of India and in Sri Lanka there are large tea (*Camellia thea*) plantations.

III. Soils Under Shifting Cultivation

A. Shifting Cultivation

Shifting cultivation leads to the most rapid deterioration of soil fertility. It is practised in the eastern and island regions of South Asia and known by various local names, such as, Chena in Sri Lanka and Jhum/Bewar/Dhya/Dippa/Erka/Kumri/Penda/Podu in India (FAO, 1984). The rapid exhaustion of the fertility of rough and ready fields, lying on slopes often too steep to hold either soil or moisture compels the tribal people living in the mountain tracts in Assam in India and similar parts in Sri Lanka to look for fresh lands to raise their food crops. In their trail, they leave behind abondoned patches of cultivation with badly hecked, charred, and lopped trees here and there. A kindly providence tries to cure the scar left by the improvidence of man (Chaturvedi and Uppal, 1960).

In a recent study (Handawela, 1989) in Sri Lanka the following emerged as critical reasons for shifting cultivators to shift:

1. Deteriorating surface soil tilth resulting in declining capacity of the soil to take in rainwater, store it in the soil, and furnish it to growing crops.
2. Increasing weed hazards.
3. Decimation of nutrient reserves released on burning of felled forest biomass.

However, in today's context of population pressure in South Asia and increasing demand for food and land for nonagricultural activities and for nonfood agriculture, shifting cultivation in its traditional form is no longer feasible. Therefore, shifting cultivation can be found in different regions at different phases, identified on the basis of land use factor (L) calculated by the expression: $L = (C + F)/C$, where C is cropping period and F is the fallow period. The different phases of shifting cultivation and kinds of agriculture practised for different land use factors are given below (Okigbo and Greenland, 1976):

Phase I, >10 land use factors: nomadic, herd rearing
Phase II, 5–10 land use factors: bush fallowing
Phase III, 2–4 land use factors: rudimentary sedentary agriculture
Phase IV, <2 land use factors: compound farming and extensive subsistence agriculture

B. Effects on Soil Fertility

The effects of slashing and burning on soil fertility are influenced by two factors: (1) the effect of burning per se, and (2) the addition of plant nutrients through ash.

In a study on a Rhodic Oxic Paleustalf at Vanathavillu Agricultural

Table 2. Temperature and intensity of burn in shifting cultivation at Venathavillu (Sri Lanka)

			Temperature (° C)			
Plot	Biomass surface coverage	Intensity of burn	Soil surface	1 cm below surface	2 cm below surface	5 cm below surface
First burn						
A	Thin	Medium	400	150	<100	<100
A	Moderate	Low	300	<150	<100	<100
B	Moderate	Medium	300	<175	<100	<100
C	Thin	Low	300	<100	<100	<100
After piling						
B	Thick	High Centre	350	300	300	250
		Edge	450	350	300	250
C	Thick	High Centre	450	350	300	250
		Edge	350	150	150	150

Source: Adopted from Andriesse (1989)

Experiment Station, Sri Lanka (Andriesse, 1989), it was observed that temperatures over 150° C at 1 cm or lower depths were found only when biomass was piled (Table 2). Since, from an earlier simulated burning study Andriesse and Koopmans (1984/85) had concluded that significant changes in soil properties occurred only when temperatures surpassed values of 150° C, the Sri Lankan study suggested no significant changes in soil chemical properties, although heat sterilization of the soil may have affected its microbiological population.

A comparison of nutrient content in the above-ground biomass and its ash showed that total N content in ash was reduced to one-tenth of that in biomass. However, P, K, Ca, and Mg concentrations in ash were 7, 4, 15, and 10 times that in biomass. Thus, the soils were enriched by the ash, which also increased soil pH. Changes in soil (ash) affected pH and available P, S, K, and Mg for up to 4 years after burning (Fig. 1); none of these nutrients showed a decline up to 4 years after burning. Thus, no evidence was found for decline in plant nutrients other than nitrogen. Poor nitrogen supply is therefore the main reason driving populations living in these areas to new lands; this was recently confirmed in another study in Sri Lanka (Handawela, 1989), where after 3 years of experimentation with the response of corn to N, P, K the study was restricted to N alone, since no response to P and K was obtained.

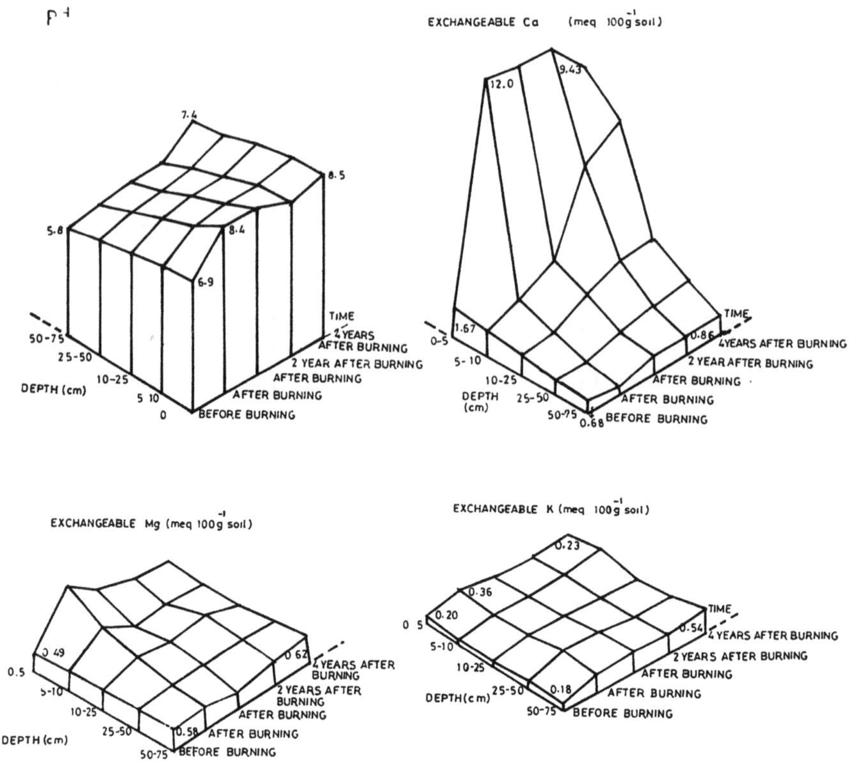

Figure 1. pH, Exchangeable Ca, Mg, and K in ash affected soil after burning of trees. (Adapted from Andriesse, 1989)

C. Alternatives

With the increasing population pressure several alternatives to shifting cultivation have been suggested (FAO, 1984). These include: (1) tree crop plantations, (2) agroforestry, (3) planted fallow system (tree and shrub fallows + arable crop sequences), (4) livestock production, and (5) special commercial horticulture. Results for studies on these alternative cropping systems are yet not available.

Experience with alley cropping in Sri Lanka (Sangakkara, 1989; Gunasena, 1989) is quite encouraging. Sangakkara (1989) reported an increase in soil organic carbon and total N due to surface application (as mulch) or incorporation of prunings of *Leucaena leucocephala* or *Gliricidia sepium* planted in alleys along contours spaced 8 to 10 m apart; the plant-to-plant spacing was 1 to 2 m. Crops grown in between tree alleys were corn, mung bean, cowpea, and sesame.

IV. Soils Under Intensive Cultivation

The ever increasing population pressure in South Asia has brought intensive cultivation of land to the forefront since there is little possibility of increasing land area under cultivation. Intensive farming received an impetus from the introduction of high yielding dwarf wheats from CIMMYT (Mexico) in India and Pakistan in the mid-1960s (Prasad and De, 1983). It was further accentuated by the availability of semidwarf high yielding rice varieties from the International Rice Research Institute (Philippines) (Pillai and Chatterjee, 1983). Under intensive cultivation two to three crops are being produced per year and the grain production of 10 to 14 Mg ha^{-1} yr^{-1} is being achieved year after year (AICARP, 1985; Prasad, 1983; Sinha and Swaminathan, 1979). Some of these intensive multiple cropping systems are: rice-wheat-mung bean; rice-wheat-corn; pearl millet-wheat-green gram; corn-wheat-mung bean; the first crop being produced during July–October, the second crop during November–April, and the third during May–June. These intensive cropping systems remove about 554 to 932 kg of nutrients (N, P, K) per hectare per year (Sharma and Prasad, 1980; George and Prasad, 1989a, b; Grewal and Singh, 1989). Increased application of chemical fertilizer has been the major input in achieving high production levels in intensive cropping systems; in 1978 South Asia consumed 6.52 million Mg of fertilizer (primary nutrients), while in 1988 this was doubled to 13.86 million Mg (Table 1). Sekhon and Puri (1986) estimated that the deficiency of N, P, and K was 40, 15, and 205.8 kg ha^{-1} yr^{-1} in Punjab; 28, 11.4, and 103.7 kg/ha^{-1} yr^{-1} in Uttar Pradesh; and 42, 11.9, and 82.5 kg ha^{-1} yr^{-1} in Bihar states of India. However, a recent study in Punjab (Brar and Singh, 1986) showed that over an 8-year-period (1973–74 to 1981–82) most soil samples from the fields of farmers practising intensive cropping and applying adequate amounts of fertilizers showed an improvement in organic C, available P, and available K. The effects of fertilizers and manures and other measures adopted to restore and maintain the fertility of South Asian soils under intensive cultivation are briefly discussed below.

A. Fertilizers and Manures

The establishment of long-term manurial experiments in India dates back to 1885 when the first experiment was undertaken at Kanpur. A little later two more long-term experiments were started, one at Pusa (Bihar) in 1908 and the other at Coimbatore in 1909. These and other long-term experiments initiated during the 1930s and 1940s in India revealed that for sustained crop production adequate P and K fertilization was necessary along with nitrogen (Nambiar and Abrol, 1989). Many of these experiments were not followed up by a detailed monitoring of soil fertility changes and the effects of most of them are now lost.

A comprehensive experiment was started in 1956 at Kanke, Ranchi, India on a Paleustalf (acid red loam) of pH 5.5 with different combinations of N, NP, NPK, farmyard manure (FYM), and lime (Lal and Mathur, 1988, 1989b; and Sarkar et al., 1989). Maize in the rainy season (July–October) and wheat in the winter (November–March) were grown each year. Lime was applied at the rate of 2.5 Mg ha^{-1}, once in 4 years. FYM was applied equivalent to N dose 15 days before the sowing of each crop. Levels of N, P_2O_5 and K_2O were 44 kg ha^{-1} each up to 1969–70. In 1970–71, the high yielding varieties of corn (Ganga Safed 2) and wheat (S227) were introduced and the level of N, P_2O_5 and K_2O was increased to 100, 60, and 40 kg ha^{-1}, respectively. N was supplied as ammonium sulfate, P as ordinary superphosphate, and K as muriate of potash.

The effects of some treatments are shown in Table 3. Continuous application of N as ammonium sulfate decreased the soil pH from 5.5 in 1956 to 3.9 in 1984; even application of P and K along with N did not prevent this fall in pH. However, continuous application of FYM improved soil pH, which was 5.9 in 1984. The application of lime along with NPK resulted in the highest pH of 6.3. Continuous application of N as ammonium sulfate also resulted in the highest exchange acidity (6.6 meg/100g soil), which was kept low at 0.3 and 0.4 with lime + NPK and FYM application, respectively. Base saturation was the lowest with continuous ammonium sulfate application and the highest with lime + NPK or FYM. Thus, application of lime and FYM helped in overcoming the soil acidity problem in Alfisols. Continuous application of FYM increased soil organic C content and water-holding capacity (Table 3). To some extent organic C content and water-holding capacity also increased with N, NP, NPK, and lime + NPK treatments. Available P in soils increased in plots receiving phosphate fertilizer. All forms of K (water soluble, exchangeable, nonexchangeable, and total) decreased in all the treatments; the decrease being least in plots receiving FYM, due to the addition of K in FYM (Lal et al., 1990). DTPA-extractable zinc and copper were highest in plots receiving FYM (Lal and Mathur, 1989a).

The effects of increased soil acidity due to N alone were seen in wheat yields (Fig. 2) which were reduced to zero by 1975 and onwards. With NP fertilization, wheat yields were reduced to zero by 1988, that is 13 years after plots receiving N alone. This could be due to continuous application of gypsum present in ordinary superphosphate. The application of lime along with NPK kept wheat yields at 5 Mg ha^{-1}. Corn yields were reduced to zero from 1982 onwards both in N and NP plots, while the yield was kept at 3.7 Mg ha^{-1} with NPK + lime (Lal and Mathur, 1988).

With the introduction of high yelding dwarf varieties of wheat in India in 1967, which had very high fertilizer requirements, the need for monitoring depletion of soil fertility was increasingly felt and the All India Coordinated Research Project on Long-Term Fertilizer Experiments was launched by the Indian Council of Agricultural Research, New Delhi at 11 centers

Table 3. Effect of fertilizer treatments on soil properties of an alfisol (Kanke, India)

Treatments	pH		Organic C (%)		Available P (ppm)		Available K (ppm)		Bulk density (g ml^{-1})		Water-holding capacity (%)	
	1964	1984	1964	1984	1964	1984	1970	1984	1970	1984	1970	1984
Check	5.7	5.8	0.540	0.536	5	3	155	118	1.44	1.54	31.8	33.2
N	4.7	3.8	0.514	0.558	5	11	110	66	1.46	1.54	38.3	39.3
NP	5.2	3.9	0.580	0.614	41	201	100	66	1.40	1.51	37.4	36.2
NPK	5.2	4.0	0.630	0.640	28	220	175	147	1.36	1.46	33.9	40.5
NPK + lime	6.5	6.3	0.520	0.608	27	85	150	98	1.48	1.50	31.6	42.4
FYM	5.8	5.9	0.735	0.732	36	20	175	169	1.43	1.32	33.5	47.2
Initial (1956)	5.5		0.529		7.5		360		1.45		31.5	

Source: Adapted from Sarkar et al. (1989)

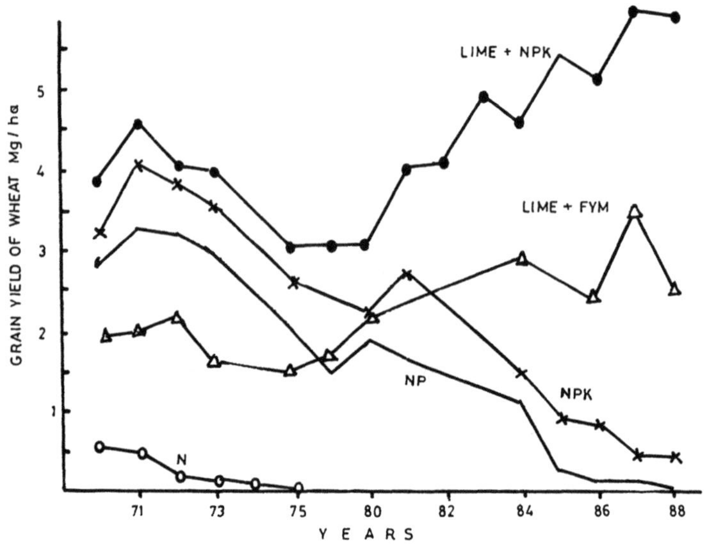

Figure 2. Grain yield of wheat (1970–88) as affected by chemical fertilizers, lime, and FYM. (Adapted from Sarkar et al. 1989)

located in 11 major agroclimatic regions of the country. Information on soils, important soil properties, crops grown, and fertilizer doses used is given in Table 4. Major soil fertility changes over the period 1971–87 as reported by Nambiar et al. (1989) along with the data from some other studies are discussed below.

1. Soil Reaction

The soil reaction was acid at three centers, namely, Ranchi (Haplustalfs), Palampur (Hapludalfs), and Bhubaneswar (Tropaquepts). Continuous application of N alone as urea depleted soil pH from 5.3 to 4.4 at Ranchi and from 5.8 to 4.8 at Palampur (Table 5). At Bhubaneswar the lowest pH (4.9) was recorded in plots receiving N and P. Application of lime improved soil pH at Ranchi and Palampur, while application of FYM did so at Bhubaneswar. At Pantnagar (Hapludolls) there was an increase in soil pH, mostly with NP treatment, where it increased from the initial value of 7.3 to 8.0. This was ascribed to a decrease in soil organic carbon due to cultivation, which was reduced from 1.48% to 0.76% with NP treatment after 15 years of cultivation.

Restoring soil fertility by liming on acid soils has been reported by several workers (Mandal et al., 1975; Prasad et al., 1976; Pradhan and Khera, 1976; Datta and Gupta, 1983). At Tadong (Sikkim) pH increased from 4.2 to 5.9, effective CEC from 5.1 to 6.8 C mol (p+) kg^{-1}, lime potential from

3.29 to 5.61, available P from 8.2 to 14.7 ppm, and available Ca from 187 to 461 ppm due to application of 2.5 Mg lime ha^{-1} (Gupta, et al. 1989).

2. Soil Organic Carbon

At most centers cultivation reduced organic carbon content in soil, the reduction being most in nonmanured soil. The application of fertilizer prevented this decline in soil organic carbon content due to the addition of increased root biomass. Continuous application of FYM increased soil organic carbon at all centers. Similar results were reported by several workers (Formoli and Prasad, 1979; Hesse and Misra, 1981; Sharma et al. 1984). However, at Ludhiana (Ustochrepts), Hyderabad and Bhubaneswar (Tropaquept), and Palampur (Hapludalfs) cultivation for 15 years increased soil organic carbon. The increase was most distinct on laterite soils of Bhubaneswar. This was due to provision of land cover and addition of organic matter mostly as root residues.

3. Available Plant Nutrients

In general, available NPK declined in nonmanured plots, while application of fertilizer increased available NPK in soil. Similar results were reported by Swarup and Singh (1989) from a study on an Aquic Natrustalf at Karnal, India. An increase in available P in soils due to P fertilization and application of FYM was also reported by Sharma et al. (1984) and Yaduvanshi (1988). In a P fixation study on an Inceptisol at Coimbatore, Subramanian and Kumarswamy (1989) found that soils receiving N alone had the highest P fixation capacity while those receiving NPK had much less P fixation capacity. Least P fixation capacity was found in soils receiving FYM along with NPK. This was explained as due to the complexing of Ca, Mg, Fe, and Al, and blocking of the fixation sites by the organic molecules. About 80% to 85% of the fixed phosphorus was transformed as Ca-P, 8%–14% as Al-P, 1% to 2% as Fe-P, and 1% to 5% as saloid P. A decline in available S was recorded at several centers in plots where ordinary superphosphate or FYM was not applied and significant reduction in rice yield due to S deficiency was noted at Barrackpore and Bhubaneswar. Similarly, yields of corn and wheat at Palampur, and, those of soybean and wheat at Jabalpur showed a decline due to deficiency of the micronutrients. A decline in available Zn due to continuous cropping on unfertilized as well as fertilized plots was noted at several centers and this decreased crop yields. Application of Zn and FYM reduced the fall in available Zn to some extent. Data from Pantnagar and Ludhiana are shown in Fig. 3.

A field experiment was started in 1971 with sugarcane on a Udic Haplustalf (red sandy loam) at Mandya, India, and soil samples were collected in 1982. Application of N, P, or K alone gave low cane yields and resulted in low organic C and poorer physical properties (Table 6). Application of N as ammonium-sulfate resulted in reduction in soil pH. However, applica-

Table 4. Some basic characteristics of the soils, crops grown, and fertilizer doses at the centers under All India Coordinated Research Projects on Long-Term Fertilizer Experiments (ICAR)

								(Fertilizer rate)		
Center no.	Location	Soil group/ association (taxonomic class)	Texture	Organic C (%)	pH	CEC (mEq 100 g^{-1} soil)	Crops	N	P_2O_5	K_2O
1.	Barrackpore	Recent alluvium (Eutrochrepts)	Sandy loam	0.71	7.1	19.0	Rice	120	60	60
							Wheat	120	60	60
							Jute	60	30	60
2.	Ludhiana	Alluvial (Ustochrepts)	Loamy sand	0.21	8.2	5.1	Corn	150	75	75
							Wheat	150	75	37
							Cowpea	20	40	20
3.	New Delhi	Old alluvium (Ustochrepts)	Sandy loam	0.44	8.1	10.6	Corn	120	60	40
							Wheat	120	60	40
							Cowpea	20	40	20
4.	Coimbatore	Medium black (Verticusto-chrepts)	Clay loam	0.30	8.2	25.8	Fignermillet	90	45	17
							Cowpea	25	50	0
							Corn	135	67	35
5.	Jabalpur	Medium black (Chromusterts)	Clayey	0.57	7.8	49.0	Soybean	20	80	20
							Wheat	120	80	40
							Corn (fodder)	80	60	20
6.	Bangalore	Red loam (Haplustalfs)	Sandy loam	0.55	5.5	5.8	Fingermillet	100	60	25
							Corn	120	60	25
							Corn (fodder)	25	50	25

7.	Hyderabad	Red loam (Tropaquepts)	Sandy clay loam	0.51	8.2	14.0	Rice	115	20	30
							Rice	115	20	30
8.	Ranchi	Red loam (Haplustalfs)	Silty clay loam	0.45	5.3	7.8	Soybean	25	60	40
							Wheat	80	60	40
							Potato	100	80	120
							Toria	25	30	20
9.	Bhubaneswar	Lateritic (Tropaquepts)	Sandy	0.27	5.5	4.0	Rice	100	60	60
							Rice	100	60	60
10.	Palampur	Submontane (Hapludalfs)	Silt loam	0.79	5.8	12.1	Corn	120	60	90
							Potato	120	100	100
							Wheat	90	90	45
11.	Pantnagar	Foothill (Hapludolls)	Silty clay loam	1.48	7.3	20.0	Rice	120	60	45
							Wheat	120	60	40
							Cowpea	0	0	0

Source: Adapted for Nambiar et al. (1989)

Table 5. Changes in soil pH as affected by continuous manuring and cropping at different research centers in India (1971–87)

Treatments	Ranchi	Bhuba-neswar	Palam-pur	Pant-nagar	Barrack-pore	Ludhiana	Coimba-tore	Jabal-pur	Hydera-bad
Period of study (no. of years)	15	14	15	14	16	16	12	15	15
Initial	5.3	5.5	5.8	7.3	7.1	8.2	8.2	7.8	8.2
Check	5.4	5.4	5.6	8.0	7.0	8.0	8.0	7.2	8.2
N	4.4	5.5	4.8	7.8	7.0	8.0	8.0	7.0	8.2
NP	5.1	4.9	5.0	8.0	7.0	8.0	8.0	7.1	8.3
NPK	4.5	5.6	5.0	7.8	6.9	8.0	8.0	7.2	8.3
NPK + lime	6.1	5.6	6.2	—	—	—	—	—	—
NPK + FYM	4.7	6.0	5.2	7.8	6.9	7.9	8.0	7.1	8.0

Source: Adapted from Nambiar et al. (1989)

Figure 3. Effect of NPK, FYM, and Zn fertilization on available zinc status of soils. (Adapted from Nambiar et al. 1989)

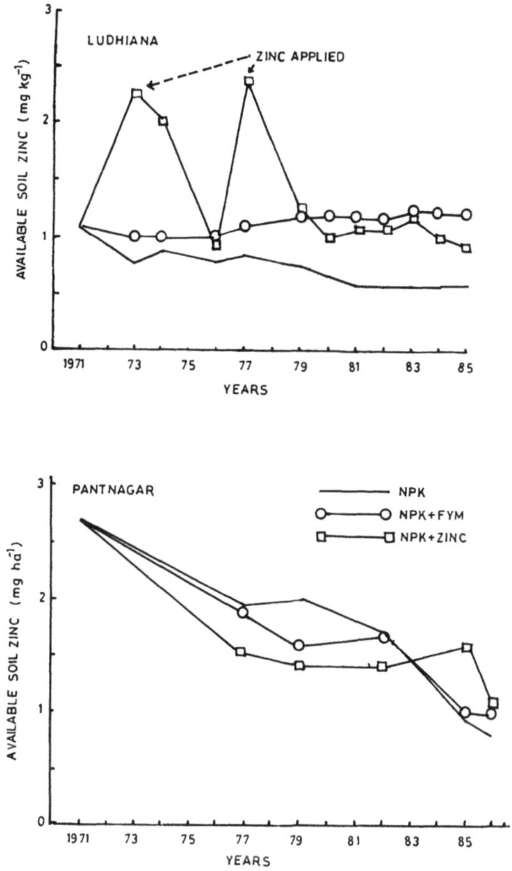

tion of the recommended dose of NPK resulted in high sugarcane yield and better soil physical and chemical properties. Combined application of NPK and FYM was the best treatment.

In a study conducted for 4 years (1975–76 through 1979–80) at Solan in Himachal Pradesh (Minhas and Mehta, 1984) on a gravelly loam soil, continuous application of 120 kg N as urea, 60 kg P_2O_5/ha as ordinary superphosphate and 60 kg K_2 O/ha as muriate of potash resulted in the highest wheat and corn yields. As regards soil properties at the end of the 4-year period of study there was a depression in soil pH, available N, K, and Zn, while organic carbon and available P were slightly increased (Table 7). Depression in available soil K and Zn was more distinct. Thus, there was some depletion in soil fertility despite recommended doses of fertilizer application due to harvesting of about 6.5 Mg ha^{-1} yr^{-1} of grain (corn + wheat). In this study inorganic P fractions in the soil at the start and at the end of the 4-year period were also estimated using a modified Chang and Jackson procedure according to Peterson and Corey (1966).

Table 6. Effect of manures and fertilizer on the yield of sugarcane and soil properties

Treatment	Av. cane yield (Mg ha^{-1}	Bulk density g cm^{-3}	Structural index	Maximum water-holding capacity (%)	Porosity (%)	pH	CEC meq 100 g^{-1} soil	Organic C (%)	Total N (%)	Available P_2O_5 (kg ha^{-1})	Available K_2O (kg ha^{-1})
Initial soil properties	—	1.84	30.2	30.4	34.2	6.8	14.3	0.46	0.037	23.6	276
Check	28	1.98	24.3	28.7	30.1	7.0	12.3	0.28	0.010	10.2	68
N[a] as Am. sulfate	38	1.93	27.1	29.2	33.1	6.4	12.6	0.41	0.020	13.2	137
N as urea	42	1.92	28.2	30.1	33.8	6.7	12.8	0.43	0.020	13.8	140
P	45	1.78	29.2	33.5	35.6	7.0	13.7	0.40	0.020	21.2	221
K	31	1.89	28.6	31.6	34.8	7.1	13.4	0.49	0.014	20.3	241
NPK	114	1.75	36.1	35.1	37.2	6.8	13.9	0.68	0.034	21.6	280
FYM at 25 Mg/ha	56	1.72	38.7	37.8	40.2	6.9	15.7	0.79	0.032	22.6	187
NPK + FYM at 25 Mg/ha	120	1.86	47.1	38.6	39.2	7.0	16.9	0.81	0.048	24.2	312

[a]NPK at 250 kg N ha^{-1}, 100 kg P_2O_5ha, and 125 kg K_2O ha^{-1} as urea, ordinary super phosphate and muriate of potash
Source: Adapted from Rabindra et al. (1985) and Rabindra and Gowda (1986)

Table 7. Grain yield of wheat and corn and soil chemical properties as affected by application of different levels of fertilizers

$N-P_2O_5-K_2O$ ($kg\ ha^{-1}\ yr^{-1}$)	Corn[a] ($Mg\ ha^{-1}$)	Wheat[b] ($Mg\ ha^{-1}$)	pH	Organic C (%)	Alkaline $KMnO_4$ extractable N ($kg\ ha^{-1}$)	P ($kg\ ha^{-1}$)	Exch. K ($kg\ ha^{-1}$)	Zn ($kg\ ha^{-1}$)
Control	1.2	1.0	7.5	1.45	323	59	220	9.4
30-15-15	1.7	1.6	7.2	1.45	320	70	243	8.8
60-30-30	2.3	2.2	7.0	1.50	379	79	254	8.0
90-45-45	3.0	2.7	6.9	1.38	367	76	224	8.6
120-60-60	3.3	3.2	6.9	1.53	357	85	224	6.8
Initial value			7.5	1.45	432	701	274	9.0

[a] Averaged over 5 year
[b] Averaged over 4 years
Source: Adapted from Minhas and Mehta (1984)

Table 8. Inorganic P fractions (ppm) in the soil as affected by fertilization

$N–P_2O_5–K_2O$ ($kg\ ha^{-1}\ yr^{-1}$)	Saloid P	Al-P	Fe-P	Ca-P
Control	1	35	105	356
30-15-15	4	45	135	378
60-30-30	4	48	185	318
90-45-45	6	48	175	318
120-60-60	7	52	210	356
Initial value	5	75	125	324

Source: Adapted from Minhas and Mehta (1984)

The results showed that Al-P was depleted in all the plots (Table 8). There was some depletion in Fe-P when no fertilizer was applied. However, when P was applied as chemical fertilizer the Fe-P increased with the level of P application. There was no depletion of Ca-P even in control plots.

Some studies have been made only with FYM. Results from a 45-year study (1932 through 1978–79) on a Vertisol (medium black clay loam, pH 8.0) are available from Pune, India (Khiani and More, 1984). FYM was applied at the rate of 6.2 Mg ha^{-1}. The mean yield of seed cotton with FYM was twice and that of sorghum with FYM 1.25 times of that in nonmanured plots. Organic C in soil after 45 years of FYM application was nearly twice (1.14%) that in nonmanured plots (0.56%). There was a significant increase in total N, alkaline $KMnO_4$ hydrolysabale (available) N, 0.5 *M* $NaHCO_3$ extractable P, 1 *N* ammonium acetate exchangeable K, maximum water-holding capacity, and moisture retention at 0.33 bar due to FYM application (Table 9). A number of earlier studies in India have

Table 9. Physicochemical properties of a vertisol and mean yield of cotton and sorghum grown in rotation under rainfed conditions

Treatment	No manure	FYM	LSD (0.05)
Seed cotton (Mg/ha)	0.15	0.31	0.03
Sorghum grain (Mg/ha)	0.84	1.08	0.22
Organic C (%)	0.56	1.14	0.04
Total N (%)	0.05	0.06	0.01
Alkaline $KMnO_4$ hydrolysable N (ppm)	75.0	92.3	2.26
0.5 M $NaHCO_4$ extractable P (ppm)	11.1	14.5	0.56
1 N Am. acetate exchangeable K (ppm)	290	333	47
Maximum water-holding capacity (%)	57.3	64.3	1.3
Moisture retention at 0.33 bar (%)	42.8	44.8	2.6

Source; Adapted from Khiani and More (1984)

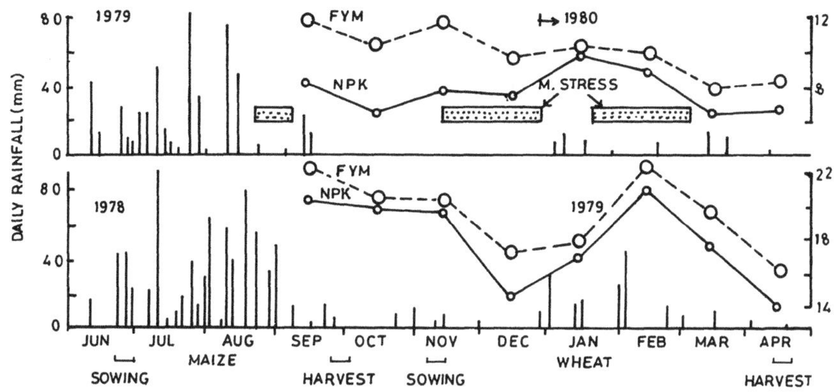

Figure 4. Soil-water content of 120-cm profile under NPK and FYM treatments for the period starting from maturity stage of maize till wheat harvest in a good and a poor rainfall year. (Adapted from Grewal et al. 1985)

shown improvement in soil physical properties due to application of organic manures (Biswas et al., 1970, 1971; Chaudhary et al. 1979).

FYM application at the rate of 50 Mg ha^{-1} for 4 years resulted in higher moisture content in soil profile (Fig. 4) and better use of profile water on a sandy loam alluvial soil (Udic ustochrept of pH 8.4, organic C 0.32%) of rainfed Siwalik in the foothill region near Chandigarh, India (Grewal et al. 1985). Biswas and Ali (1969) also reported an increase in moisture-retention capacity of soils by the addition of organic manure.

Results on the rates of FYM application are available from two Alfisols at Almorah (Bhriguvanshi, 1988), where 0, 10, 30, and 50 Mg ha^{-1} of FYM was applied for 5 years to wheat, which was followed by finger millet. No FYM was applied to finger millet. After 5 years of application of FYM, soil samples were collected from the surface soil and analysed for pH, EC, organic C, water-holding capacity, available P, and available K; data are given in Table 10. With the exception of soil pH and available P, all other soil properties studied showed an increase with an increase in the rate of FYM application on both soils. Similar findings were reported from a pearl millet–wheat cropping sequence study on coarse loamy Ustochrept at Hisar (Gupta et al., 1988). Soil samples were collected after 18 years of application of treatments as O-day samplings. After the O-day sampling the required quantity of FYM was applied to the wheat crop and periodic soil sampling was done after 20, 52, 90, and 119 days. Changes in organic carbon, alkaline $KMnO_4$ hydrolysable N and Olsen's 0.5 *M* $NaHCO_3$ extractable P in the surface 0 to 15 cm soil layer are shown in Fig. 5. Organic C, $KMnO_4$-N, and Olsen's P were more at all stages of sampling with 90 Mg ha^{-1} yr^{-1} than with 45 Mg ha^{-1} yr^{-1} or FYM application.

Benbi and Singh (1988) reported that under rainfed conditions at

Table 10. Effects of rates of FYM application on some soil properties of two alfisols at Almora

FYM ($Mg\ ha^{-1}\ yr^{-1}$)	pH	EC (dSm^{-1})	Maximum water-holding capacity (%)	Organic C (%)	Available P (ppm)[a]	Available K (ppm)[b]
			Sandy loam soil			
0	7.0	0.17	41.9	0.36	18	96
10	6.9	0.29	47.2	0.43	23	127
30	6.9	0.32	51.3	0.47	26	132
50	7.0	0.34	52.5	0.59	30	150
LSD (0.05)	—	NS	2.5	0.07	3.5	26
			Clay loam soil			
0	7.2	0.22	42.8	0.43	29	105
10	7.0	0.22	53.5	0.51	36	180
30	7.0	0.32	53.7	0.52	35	200
50	7.2	0.40	57.5	0.57	31	245
LSD (0.05)	—	0.07	3.2	0.07	5.7	31

[a] Olsen's 0.5 *M* $NaHCO_3$ extractable
[b] 1 N Am. acetate exchangeable
Source: Adapted from Bhriguvanshi (1988)

Ludhiana application of FYM retarded the movement of nitrate to lower depths and a higher amount of nitrate was retained in the surface layers of the manured plots. This property of FYM can play a vital role in rainfed agriculture, specially during the early stages of the crop, when the rains are heavy and the root systems are not developed sufficiently to penetrate deep enough to utilize the leached nitrogen.

B. Green Manuring

Green manuring in rice fields is an age-old practice in South Asia and has received special attention for the contribution that it can make toward the nitrogen needs of rice, which has become the key input since the introduction of high yielding and semidwarf heavy nitrogen demanding varieties and due to the fact that the efficiency of fertilizer nitrogen in rice is low (Prasad, 1990). A larger number of crops and plant species have been used as green manure. The crops used as green manures include: sunnhemp (*Crotolaria juncea* L.), dhaincha (*Sesbania aculeata* (Wild) Poir), phillipasera (*Phaseolus trilobus* Ait.) mung bean, cowpea, cluster beans

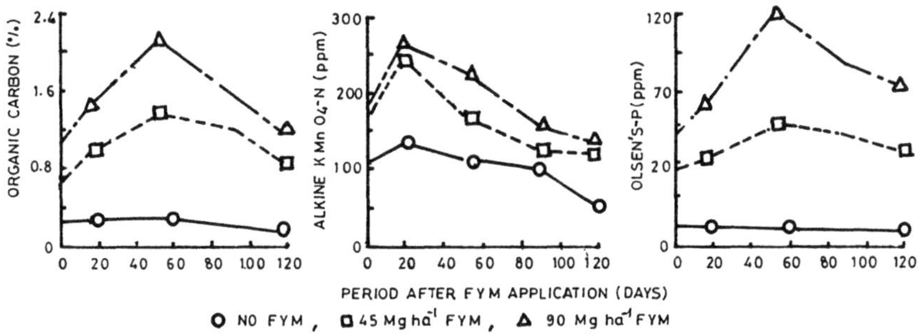

Figure 5. Periodic changes in organic carbon, alkaline $KMnO_4{}^{-N}$ and Olen's P of the surface soil after application of different levels of FYM. (Adapted from Gupta et al. 1988)

(*Cyamopsis tetragonoloba* L.), senji (*Melilotus alba* Medik.), and khesari (*Lathyrus sativus*), while the tree species and other plants used include *Thespesia populanea* L, *Cassia auriculata* L., *Pongamia glabra* Vent., *Azadirachta indica* A. Juss., *Calotropis gigantea, Jatropha gossypifolia* L., *Jatropha glandulifera* Roxb., *Glircidia meculata* (H.B. and K.) *Tephrosia purpurea* L., *Tephrosia candida* (Roxb) DC., *Cassia tora* L., and *Ipomea carnea* auct. (non-Jacq.) (Samad and Sahadevan, 1952; Sanyasi Raju, 1952).

Data from a large number of field experiments show a grain yield increase for rice equivalent to 25 to 60 kg N ha^{-1} or even more depending upon the green manure biomass incorporated and its total nitrogen content (Goswami et al., 1988; John et al., 1989). The subject has been recently reviewed by Abrol and Palaniappan (1988), Kulkarni and Pandey (1988), and Meelu and Morris (1988). Data on the effects on soil fertility are rather limited. Sahu and Nayak (1971) reported an organic C and total N content of 0.67% and 0.054%, respectively, in plots receiving green manure as compared with 0.58% and 0.021% in check plots. Bhardwaj and Dev (1985) also reported an increase in organic C and total N content in soil due to *Sesbania aculeata* green manuring. From a recent study Singh et al. (1990) working on a Fluvent at New Delhi reported an organic C content of 0.91% in the green-manured plots as compared with 0.69% in the fallow plots, just before transplanting rice (2 weeks after the incorporation of green manure).

In a study involving 19 experiments conducted on farmers' fields in Ropar and Patiala districts of Punjab for 2 years, summer green manuring was done with cowpea or *Sesbania aculeata* before planting corn in the succeeding rainy season (Bhandari et al., 1989). Cowpea green manure (4 Mg ha^{-1} dry matter) supplied 88, 8.4, and 54 kg ha^{-1} of N, P, K, respectively, while *Sesbania* (6 Mg ha^{-1} dry matter) supplied 114, 10, and 76 kg

ha^{-1} of N, P, K, respectively. Green manuring with cowpea or *Sesbania* nearly substituted 60 kg N ha^{-1} for the succeeding corn. The average values of organic C, available (Alkaline $KMnO_4$ hydrolysable) N, available Olsen's P, and available (exchangeable) K for green-manured plots after the harvest of corn were 0.57% and 167.7, 19.7, and 189 kg ha^{-1}, respectively, while the mean values of organic C, available N, P, and K, for plots receiving N (60, 90, 120 kg ha^{-1}) were 0.56% and 157, 17.3, and 177 kg ha^{-1}, respectively. The values of these characteristics in check plots were nearly the same as for N receiving plots. Thus, application of green manures built up soil fertility.

The incorporation of green manure after growing cowpea for 60 days during April–May in a sugarcane ratoon crop increased available N, P, and K in soil up to 60 cm depth as determined after the harvest of the second ratoon of sugarcane (Jafri and Pandit, 1986).

In a laboratory incubation experiment (Chahal and Khera, 1988), green manuring combined with submergence brought about a greater increase in water-soluble plus exchangeable iron (14 to 29 ppm) as compared with using the unsubmerged moisture regime (13 to 20 ppm). A decrease in iron held on organic sites and oxide surfaces followed by an increase in water-soluble plus exchangeable iron and iron held on inorganic sites was suggested as the possible reason for the increased iron availability with green manuring. Mandal and Mitra (1982) also reported a decrease in iron held on organic sites and an increase in iron held on inorganic sites with an increase in incubation period in the presence of manures.

Sharma and Bajpai (1989) reported that rainy season legumes, namely, black gram and mung bean could contribute about 30 kg N ha^{-1} to the succeeding wheat. Green manures had no significant influence on soil microbial population, however, fertilizer N up to 90 kg N ha^{-1} applied to wheat significantly increased the population of total bacteria and fungi.

C. Other Organic Residues

Hameed (1978) (quoted by J. Venkateswarlu, 1987) studied nutritional and physical parameters of the soil before and after 5 years of organic residue incorporation in both red sandy loam and red loamy sand soils. The organic residue incorporated either through cowpea or pearl millet was equivalent to 20 kg N ha^{-1}. Cowpea-pearl millet rotation was followed and only 20 kg P_2O_5 ha^{-1} or 20 kg N ha^{-1} was applied to the respective crops. The soil data are given in Table 11. All soil physical parameters and available N and P improved due to crop residue incorporation. Available K was not affected in sandy loam soil but was somewhat lowered in loamy sand.

Pandey et al. (1985) from Kanpur reported that incorporation of maize, rice, sorghum, and wheat straw in soil prior to growing rice showed improvement in bulk density, infiltration rate, stability index, and water re-

Table 11. Physical and chemical soil properties as affected by crop residue incorporation

	Red sandy loam		Red loamy sand	
Soil constituent	Initial	After 5 years of crop residue incorporation	Initial	After 5 years of crop residue incorporation
Mean weight diameter (mm)	2.63	3.04	3.21	4.19
Hydraulic conductivity (cm/ha)	4.19	10.50	2.34	8.00
Bulk density (g/cm^3)	1.81	1.69	1.73	1.65
Moisture content at field capacity (g/100 g)	15.7	19.0	12.4	12.9
Available N (kg/ha)	284	307	235	286
Available P_2O_5 (kg/ha)	5.5	14.0	6.6	14.0
Available K_2O (kg/ha)	256	254	269	234

Source: Adapted from Venkateswarlu (1987)

tention at 0.33 bar when soils were sampled after the harvest of succeeding wheat. In a study at Ranchi, incorporation of 2.5 Mg ha^{-1} yr^{-1} of a 1:1 mixture of wheat straw and water hyacinth (*Eichhornia* sp.) for 3 years increased soil organic C from 0.450% to 0.580%, total N from 0.005% to 0.065%, and available P from 35 to 59 ppm; the values with the incorporation of 5 Mg ha^{-1} yr^{-1} of mixture were 0.615%, 0.067%, and 72 ppm for organic C, total N, and available P, respectively. Joshi and Ghonsikar (1979) and Gupta (1981) also reported similar increases in organic C content of different soils treated with various crop residues.

In a study on an Aquic Hapludoll (Beni silty clay loam) at Pantnagar the effects of burning and incorporation of wheat straw (6 Mg ha^{-1}) on soil pH in a rice field after transplanting rice were studied (Sarkar et al. 1988). The addition of straw reduced soil pH, with the reduction being more when straw was incorporated than when it was burned at soil surface (Fig. 6).

An attempt to study the possibility of organic recycling of subabul (*Leucaena leucocephala* (Lamk) de Wrt) planted in association with winter sorghum grown on a Typic Chromustert at Solapur, India, was made by Narkhede and Ghugare (1987a, b). Subabul produced a dry matter of 1.97 Mg ha^{-1}, which contained 3.86% N. Thus, 76 kg N ha^{-1} was added by

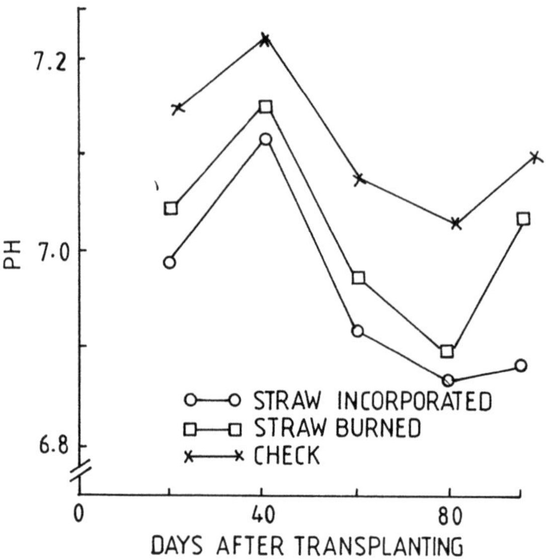

Figure 6. Effect of straw incorporation/burning on soil pH during rice growing season. (Adapted from Sarkar et al 1988)

subabul loppings; the lopping were applied on the soil surface. The addition of subabul loppings increased available alkaline $KMnO_4$ hydrolysable N in the surface 0 to 15 cm soil by 44.5% over the control. The increase in organic C content in soil was 11.9% over the control (Table 12). The addition of subabul loppings increased grain and straw yield, N and P uptake, and moisture use efficiency by sorghum and left more moisture in soil at sorghum harvest.

An incubation study on a red sandy loam Alfisol was conducted for a period of 90 days at Jhansi, India, by Murthy et al. (1990). Dry leaf meals (equivalent to 100 kg N ha^{-1}) of siris (*Albizia Lebek*), subabul, and neem (*Azadirachta indica*) containing 4.7%, 3.2%, and 2.6% total Kjaldahl N were mixed with 500 g soil (pH 7.3, organic C 0.22%, total N 0.05%, available N 0.011%) and incubated near field capacity moisture at 25 ± 5° C. Soil samples were collected after 30, 60, 75, and 90 days' incubation. Averaged over periods of sampling, soils receiving siris, subabul, and neem meal contained 0.60%, 0.58% and 0.56% organic C (LSD $P = 0.05$ was 0.02); total N values were 0.07%, 0.66%, and 0.069%, respectively (LSD $P = 0.05$ was 0.001). Thus, the addition of siris leaf meal, which contained the highest kjeldahl N, resulted in the highest content of organic C and total N in the soil. Neem leaf meal, which contained the lowest kjeldahl N, resulted in the lowest organic C content in soil, but the total N content in soil was significantly superior to that left after incorporation of subabul leaf meal. This could be partly due to the microbicidal properties of neem (Reddy and Prasad, 1975; Thomas and Prasad, 1982, 1983).

Table 12. Effect of organic recycling on sorghum and organic carbon and moisture content in soil

Characteristics	Control	Subabul loppings added
Sorghum		
Grain yield (Mg ha^{-1})	1.2	1.4
Stover yield (Mg ha^{-1})	3.3	3.6
N uptake (kg ha^{-1})	3.4	4.0
P uptake (kg ha^{-1})	1.4	1.7
Moisture use efficiency (kg grain mm–ha^{-1})	5.4	6.7
Soil		
Organic carbon (%)	0.42	0.47
Moisture content in 50 cm soil depth (mm)		
At sowing	232	234
At flowering	175	172
At harvest	132	136

Source: Adapted from Narkhede and Ghugare (1987b)

D. Growing of Trees and Grasses

Vegetation has a pronounced effect on many soil properties (Banerjee et al. 1985). When a population of one species is replaced by a plant of a different species, significant changes in soil properties are in the offing. Yadav and Sharma (1968) showed that soils under teak (*Tectona grandis*) had higher exchangeable Ca than those under sal (*Shorea robusta*). Singh et al. (1985) investigated the soil properties under teak, sal, and other species and observed maximum Ca in soil under teak. In a study by Nath et al. (1988) in a contiguous area of Kalimpong Division of West Bengal (India), a significant increase in pH and base saturation of the soil under teak was reported. After 28 years, pH and base saturation further increased and the soil was transformed from an Inceptisol to Mollisol. Sal even after 34 years did not change much of the properties of the original soil. Both Ca and Mg were generally recycled from lower to surface horizons by both sal and champa (*Michelia champaca*).

In a study on an alluvial soil at Kanpur, India, (Bhatia and Srivastava, 1984) soil samples were collected from bare fields as well as from fields growing corn or corn intercropped with black gram for the past 5 years or those under *Eucalyptus tereticornis* trees or *Cenchrus ciliaris* grass for the past 12 years. Data on organic C, water stable aggregates, suspension percentage, and other parameters related to soil erosion are given in Table 13.

Table 13. Organic carbon and erodibility indices in soil under field crop, grass, or tree cover

Land use	Soil depth (cm)	Organic carbon (%)	Water[a] stable aggregates 0.25 mm (%)	Suspension percentage[b]	Dispersion ratio	Erosion ratio
Bare soil	0–15	0.21	11.7	16.9	60.0	50.8
	15–30	0.17	6.6	14.7	63.1	75.2
Corn	0–15	0.30	19.3	17.3	44.8	44.8
	15–30	0.22	1.6	16.1	48.8	50.8
Corn +	0–15	0.34	28.2	15.8	38.0	38.8
black gram	15–30	0.29	15.7	15.1	42.8	43.8
Eucalyptus	0–15	0.58	47.0	4.9	14.6	9.5
	15–30	0.56	57.4	5.0	13.7	7.7
Grassland	0–15	0.61	50.7	4.1	9.9	7.7
	15–30	0.59	42.2	4.1	11.5	9.2

[a] Yoder (1936)
[b] Middleton (1930)
Source: Adapted from Bhatia and Srivastave (1984)

In the 0 to 15 cm surface soil layer organic C and water stable aggregates were the highest and suspension percentage, dispersion ratio, erosion ratio, and erosion index were the lowest in soil under *Cenchrus* grass, closely followed by soil under eucalyptus. Soils growing field crops were next in order and between the two fields continuously cropped, the soil under corn intercropped with black gram had more organic C and water stable aggregates and lesser values for parameters related to soil erosion than the soil under corn alone. Bare soil was the poorest. The same trend was observed in the 15 to 30-cm deep soil layer, with the exception that it had more water stable aggregates under eucalyptus than under *Cenchrus*. Thus, providing a crop cover improved the soil's organic C content and reduced the erodability of the soil. *Cenchrus* grass cover was most effective in this regard and it was closely followed by eucalyptus cover.

In a 2-year (1977–78 to 1978–79) field study (Upadhyay and Biswas, 1984) on an Ustochrept (sandy loam, pH 6.9, CEC 8.7 meq 100 g^{-1} soil, 0.22% organic C) at New Delhi, soybean, cowpea, cluster bean, moth bean (*Phaseolus aconitifolius* J.), and pearl millet were raised for fodder for 60 days during the rainy season (July–September) and were followed by barley during the winter season (October–April). These crops were grown under rainfed conditions. Soil aggregation expressed as mean weight diameter (Van Bavel, 1949) after barley (pooled over 2 years of study) was the highest under soybean (0.264 mm), significantly superior to that under all other crops. The mean weight diameter after cowpea, moth bean, cluster bean, and pearl millet was 0.234, 0.229, 0.210, and 0.209 mm (LSD $P = 0.05$ was 0.02 mm). In general, soil aggregation was better after legumes than after pearl millet.

Growing pea or chickpea as compared with maize, and green gram or black gram as compared with soybean increased organic C, total and valuable N, available P, total bacteria, actinomycetes, and fungi in an alluvial soil at Kanpur (Sharma et al., 1986). George and Prasad (1989a) working on a Fluvent at New Delhi found that cropping sequences which included one or two legumes removed lesser amounts of K from soil; K removal by cereals was two to three times that by legumes. Depletion in neutral N ammonium acetate extractable K was 100 kg ha^{-1} yr^{-1} in totally cereal based cropping, while it was 70 to 80 kg ha^{-1} yr^{-1} in the case of cereal-legume systems.

V. Soils Under Salinity or Sodicity

A. Spread

A discussion on South Asian soils would remain incomplete without a mention of salt-affected soils, although a detailed discussion on these is beyond the scope of this chapter and reader is referred to detailed treatises on the subject (Abrol et al., 1988; Kelley, 1951; USDA, 1954; Szabolcs, 1974). What follows is a very brief narration highlighting the main findings from the point of view of restoring soil fertility and productivity.

Productivity in large soil areas in South Asia is adversely affected by salinity or sodicity. The area under saline soils is 23.2 million ha in India, 10.4 million ha in Pakistan, 2.48 million ha in Bangladesh, 0.63 million ha in Burma, and 0.2 million ha in Sri Lanka. Sodic soils are found in India (0.57 million ha) and Bangladesh (0.54 million ha) (Massoud, 1977). Abrol and Bhumbla (1971), however, reported that in India about 2.5 million ha of land are affected by sodicity.

B. Distinguishing Characteristics

The distinguishing characteristic of saline soils from the agricultural standpoint is that they contain sufficient neutral soluble salts to adversely affect the growth of most crop plants. Barren spots and stunted plants may appear in cereal or forage crops growing on saline soils. For purposes of definition, saline soils are those which have an electrical conductivity of 4 dSm^{-1} at 25° C (USDA, 1954). This value is generally used the world over although the Terminology Committee of the Soil Science Society of America has lowered the boundary between saline and nonsaline soils to 2 dSm^{-1} in the saturation extract (Abrol et al., 1988). The pH value of the

Table 14. Soil salinity classes and crop growth

Soil salinity class	ECe (dSm^{-1})	Effect of crop plants
Nonsaline	0–2	Salinity effects negligible.
Slightly saline	2–4	Yields of sensitive crops may be restricted.
Moderately saline	4–8	Yields of many crops are restricted.
Strongly saline	8–16	Only tolerant crops yield satisfactorily.
Very strongly saline	>16	Only a few very tolerant crops yield satisfactorily.

Source: Adapted from Abrol et al. (1988)

saturation paste is always less than 8.2 and more often near neutrality. Soluble salts most commonly present are the chlorides and sulfates of sodium, calcium, and magnesium. Nitrates are rarely present in appreciable quantities. Leaves of plants growing in salt-infested soils may be smaller and darker blue-green in color than the normal leaves. Increased succulence often results from salinity, particularly if the concentration of the chloride ions in the soil solution is high. Symptoms of specific element toxicities such as marginal or top burn of leaves occur as a rule in woody plants. Chloride and sodium ions and boron are the elements most usually associated with toxic elements. Soil salinity classes in relation to crop growth are given in Table 14.

The chief characteristic of sodic soils from the agricultural standpoint is that they contain sufficient exchangeable sodium to adversely affect the growth of most crop plants. For the purpose of definition, sodic soils are those which have an exchangeable sodium percentage (ESP) of more than 15. The soils lack appreciable quantities of neutral salts but contain measurable to appreciable quantities of salts capable of alkaline hydrolysis, e.g., sodium carbonate. The electrical conductivity of saturation extracts is less than 4 dSm^{-1} at 25° C. The pH of the saturated soil paste is 8.2 or more and in extreme cases may be above 10.5. The relationship of ESP and sodicity hazard has been described as follows (Abrol et al., 1988): when the approximate ESP is < 15, the sodicity hazard is none to slight; at ESP 15–30, it is light to moderate; at 30–50, it is moderate to high; at 50–70, it is high to very high; and at > 70, it is extremely high.

C. Reclamation

Reclamation and management measures for saline soils aim at removing salts by scraping, flushing, or leaching, and growing salt-tolerant crops and

Figure 7. Effect of passage of different quantities of water on salt distribution. (Adapted from Khosla et al. 1979)

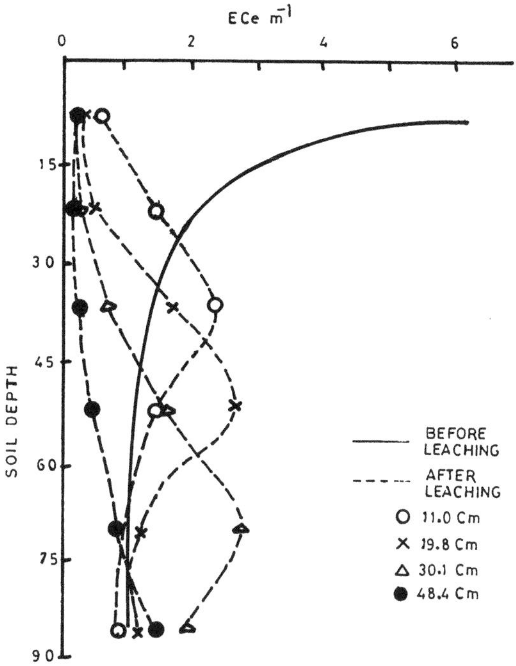

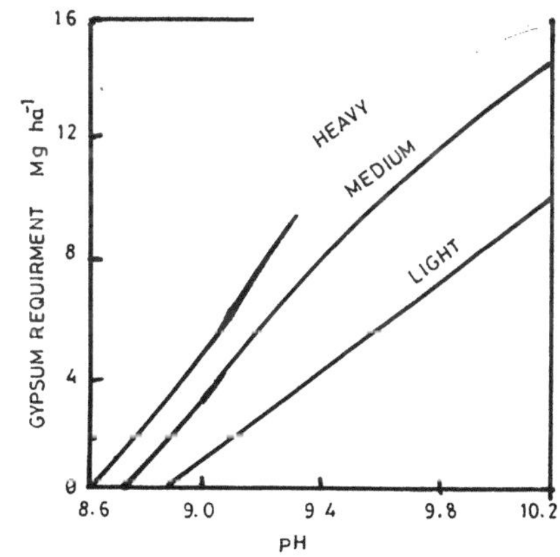

Figure 8. Relationship between pH of 1:2 soil-water suspension and the gypsum requirement of sodic soils of Indo-Gangetic plains. Light, medium, and heavy refer to soils with a clay content of approximately 10%, 15%, and 20%, respectively. (Adapted from Abrol et al. 1988)

their varieties (Bhattacharya, 1976). The effects of passage of different quantities of water on salt distribution are shown in Fig. 7.

Soil supplements such as gypsum, iron pyrites, and pressmud from sugarcane factories are used for reclaiming sodic soils. The relationship between pH and the gypsum requirement is shown in Fig. 8.

Paddy rice is an ideal crop. In a study conducted on an Aquic Natrustalf at Karnal, India (Singh and Abrol, 1988), for 11 consecutive years, rice

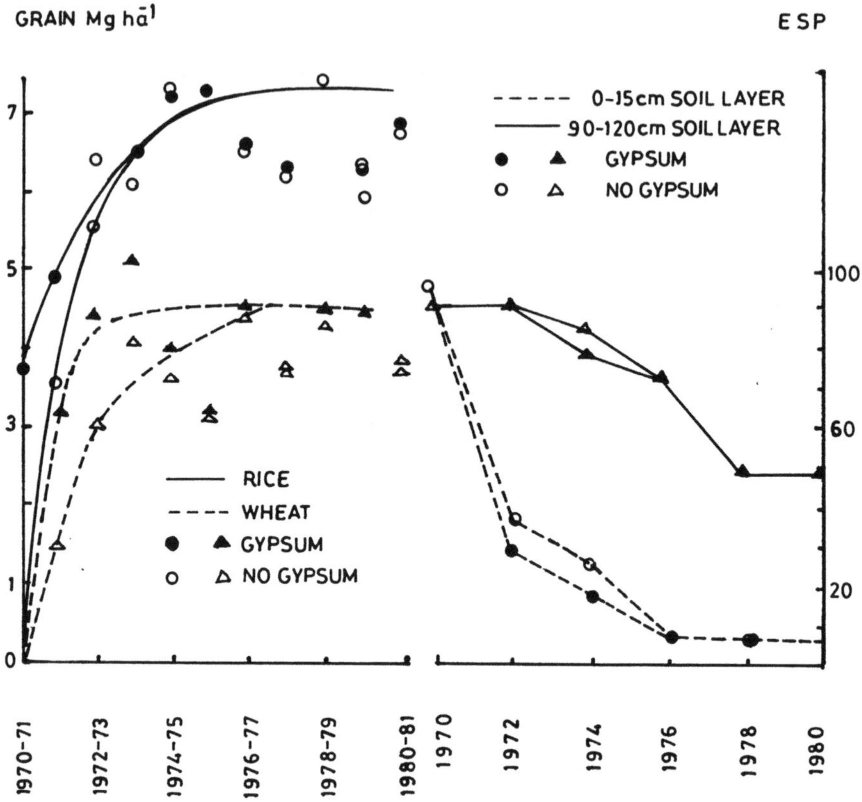

Figure 9. Effect of gypsum treatment on grain yield of rice and wheat and exchangeable sodium percentage (ESP) in a sodic soil over years. (drawn on the data by Singh and Abrol, 1988)

yields in plots receiving no gypsum increased from 0.2 to 6.7 Mg ha^{-1}; the increase in plots receiving 14.5 Mg ha^{-1} of gypsum (applied only once in 1970) was from 3.6 to 6.8 Mg ha^{-1} (Fig. 9). Rice yields were the same in gypsum-receiving and check plots from the fifth year onwards. Wheat yields in rice-wheat double cropping did not significantly differ in gypsum-applied and check plots from the sixth year onwards. In the surface 0- to 15-cm soil layer the ESP dropped from an initial 96% to 6% by 1980 (Fig. 9). The drop in ESP in the 15- to 30-cm layer was almost the same, while in the 90- to 120-cm layer ESP dropped from initial 90% to 49% by 1980. From the research center, Karnal (India), Swarup and Singh (1989) reported a reduction in soil pH from 9.2 to 8.5 and in ESP from 32 to 8 due to rice-wheat double cropping for 12 years.

Comparative efficiency of gypsum and pyrite in reducing sodicity and on rice and wheat yield has been reported by Verma and Abrol (1980a, b)

Table 15. Effect of gypsum and equivalent quantities of pyrite on ESP and crop yield

Treatment		ESP (0–15 cm)	ESP (15–30 cm)	Rice ($Mg\ ha^{-1}$)	Wheat ($Mg\ ha^{-1}$)
Check		76.5	92.4	3.8	0.2
Gypsum	7.1	33.4	75.1	6.7	1.5
($Mg\ ha^{-1}$)	14.2	32.4	79.2	6.8	3.1
	21.3	19.2	59.5	7.4	3.6
	28.4	13.6	56.5	7.2	4.2
Pyrite	3.6	64.1	90.2	5.7	0.1
($Mg\ ha^{-1}$)	7.2	52.3	86.4	6.0	0.5
	10.8	44.1	80.2	6.7	1.3
	14.4	38.8	80.3	6.9	1.3

Source: Adapted from Verma and Abrol (1980a, b)

(Table 15). Pyrite was only one-fourth as effective as gypsum. This was apparently due to lack of complete oxidation of pyrite in sodic soils. Thus, the growing of paddy rice and application of gypsum can improve the productivity of sodic soils.

VI. Summary and Conclusions

Research on restoring and maintaining soil fertility in South Asia has brought out the following:

1. In slash-and-burn shifting cultivation the major plant nutrient lost is nitrogen. With adequate nitrogen fertilization it could be possible to practise arable farming and to check further deforestation and subsequent erosion of land.
2. Alley cropping with fast growing trees and arable crops is a suitable alternative to shifting cultivation.
3. Growing of trees and perennial grasses such as *Cenchrus* can reduce soil erosion losses. Teak (*Tectona graudis*) builds up soil fertility, while sal (*Shores robusta*) does not.
4. On acid soils, continuous application of nitrogen fertilizer alone leads to a decline in soil pH; to some extent this decline can be checked by the simultaneous application of ordinary superphosphate. The application of lime or the continuous application of farmyard manure can prevent this decline in soil pH.
5. Cultivation reduces organic matter in soil. Fertilizer application prevents this decline in soil organic matter due to the addition of root biomass.

6. The continuous application of farmyard manure or other organic residue improves soil organic matter content.
7. Intensive agriculture involving growing of two to three crops a year and producing 10 to 14 Mg ha^{-1} of grain (or grain equivalents) results in heavy depletion of plant nutrients. Adequate balanced NPK fertilization needs to be practised to check the decline in soil fertility.
8. Intensive cropping can lead to deficiencies of secondary and micronutrients in soils. Deficiency of zinc is wide-spread in India and recommendations for adequate Zn and S fertilization are made for most soils and crops.
9. The application of farmyard manure and other organic residues helps considerably in maintaining soil fertility including micronutrients and improving the physical properties of the soil.
10. Part of the nitrogen needs (30–60 kg N ha^{-1}) of a crop can be met by green manuring with leguminous crops. Green manuring is practised in some areas and is recommended for others.
11. Sodic soils can be reclaimed by the application of gypsum and by growing paddy rice.
12. The continuous monitoring of soil fertility in South Asian soils and the judiciously combined mix of inorganic fertilizers and organic manures is suggested.

References

Abrol, I.P. and D.R. Bhumbla. 1971. Salt affected soils of India—distribution and formation. *FAO Rept. World Soil Resources.* 41:42.

Abrol, I.P. and S.P. Palaniappan. 1988. Green manure crops in irrigated and rainfed lowland rice-based cropping systems in South Asia. *Green Manure in Rice Farming.* International Rice Research Institute, Los Banos, Laguna, Philippines.

Abrol, I.P., J.S.P. Yadav, and F.I. Massoud. 1988. Salt Affected Spoils and their Management. *F.A.O. Soils Bull.* 39:131.

AICARP. 1985. Long range effect of continuous cropping and manuring with nitrogen, phosphorus and potassium in selected crop sequences. All India Co-ordinated Agronomic Research Project (ICAR) and University of Agricultural Sciences, Bangalore, India. Project Bull. 4:38.

Aiyer, R.S. and K.H. Nair. 1985. Soils of Kerala and their management. In: B.C. Biswas, D.S. Yadav and S. Maheshwari (eds.) *Soils of India and their Management*, pp. 208–224. Fertilizer Association of India, New Delhi, India.

Andriesse, J.P. 1989. Nutrient management through shifting cultivation. In: J. van der Heide (ed.) *Nutrient Management for Food Crop Production in Tropical Farming Systems*, Institute for Soil Fertility, pp. 29–62. Haren, The Netherlands and Universitas Brawijaya, Malang, Indonesia.

Andriesse, J.P. and T.T. Koopmans. 1984/85. A monitoring study on nutrient cycles in soils used for shifting cultivation under various climatic conditions in tropical Asia. I. The influence of simulated burning on form and availability of plant nutrients. Agriculture, Ecosystems and Environment, pp. 1–16 Elsevier, Amsterdam.

Banerjee, S.K., S.B. Singh, S. Nath, and D.K. Pal. 1985. chemical properties of soils under different old stands on upper forest hill of Kalimpong (Darjeeling), West Bengal. *J. Indian Soc. Soil Sci.* 33:788–794.

Benbi, D.K. and R. Singh, 1988. Effect of FYM and nitrogen application on the distribution of nitrate in soil profile under rainfed sub-humid condition. *J. Indian Soc. Soil Sci.* 36:563–566.

Bhandari, A.L., K.N. Sharma, M.L. Kapur, and D.S. Rana 1989. Supplementation of N through green manuring for maize growing. *J. Indian Soc. Soil Sci.* 37:483–486.

Bhardwaj, K.K.R. and S.P. Dev. 1985. Production and decomposition of *Sesbania cannabina* (Retz) in relation to its effect on the yield of wetland rice. *Trop. Agric.* (Trinidad) 62:233–236.

Bhatia, K.S. and K.K. Shukla. 1982. Effect of continuous application of fertilizers and manure on some physical properties of eroded alluvial soils. *J. Indian Soc. Soil Sci.* 30:33–35.

Bhatia, K.S. and A.K. Srivastava. 1984. Studies on soil characteristics related to erodibility under different types of land use. *J. Indian Soc. Soil Sci.* 32:201–204.

Bhattacharya, R.K. 1976. New salt tolerant rice varieties for coastal saline soils of Sunderbans, West Bengal. *Sci. Cult.* 42:122–123.

Bhriguvanshi, S.R. 1988. Long-term effect of high doses of farmyard manure on soil properties and crop yield. *J. Indian Soc. Soil Sci.* 36:784–786.

Biswas, T.D. and M.H. Ali. 1969. Retention and availability of soil water as influenced by soil organic carbon. *Indian J.agric. Sci.* 39:582–588.

Biswas, T.D., B.L. Jain, and A. Gupta. 1970. Role of phosphatic fertilizers in improving soil physical properties. *Bull. Indian Soc. Soil Sci.* 8:83–89.

Biswas, T.D., B.L. Jain, and S.C. Mandal. 1971. Cumulative effect of different levels of manures on the physical properties of soils. *J. Indian Soc. Soil Sci.* 19:31–38.

Borlaug, N.E. 1971. The green revolution: for bread and peace. *Bull. Atomic Scientists.* June, pp. 6–9, 42–48.

Brar, S.P.S. and B. Singh 1986. Changes in soil fertility under intensive agriculture. *Indian J. Ecol.* 13:152–157.

Chahal, D.S. and S.S. Khera. 1988. Effect of green manuring on iron transformation in soils under submerged and unsubmerged conditions. *J. Indian Soc. Soil Sci.* 36:433–438.

Chaturvedi, M.D. and B.N. Uppal 1960. A study in shifting cultivation of Assam. *Indian Council of Agricultural Research, New Delhi Res.*, Ser. 2:1–22.

Chaudhary, M.L., R.K. Singh, R. De, and T.D. Biswas. 1979. Effect of doses of nitrogen, alone and in combination with phosphorus in wheat-bajra rotation on some soil properties. *Bull. Indian Soc. Soil Sci.* 12:370–397.

Datta, M. and R.K. Gupta. 1983. Response of leguminous crops to pressmud and lime in Nagaland. *J. Indian Soc. Soil Sci.* 31:635–637.

FAI. 1990. *Fertilizer Statistics 1989–90.* The Fertilizer Association of India, New Delhi, p. 374.

FAO. 1971. Soil Resources of Bangladesh. The Food and Agriculture Organisation of the United Nations, Rome, Italy, FAO/UNDP Soil Survey Project, Pakistan. Tech. Rept. 3.

FAO. 1973. Calcareous Soils. Rep.FAO/UNDP Regional Seminar on Reclamation and Management of calcareous soils. Nov. 27–Dec., 1972. Cairo, Egypt. *FAO Soils Bull.* 21:261.

FAO. 1981. *Agriculture: Toward 2000.* Food and Agriculture Organization of the United Nations, Rome, Italy.

FAO. 1984. Improved Production Systems as An Alternative to Shifting Cultivation. *FAO Soils Bull.* 53:201.

FAO. 1989. *Fertilizer Year Book.* Food and Agriculture Organization of the United Nations, Rome, p. 127.

Formoli, G.N. and R. Prasad. 1979. Effect of farmyard manure and phosphorus and potassium fertilizer on soil properties in rice-wheat rotation. *J. Agric.Sci. (Camb.)* 92:359–362.

George, Mercy and R. Prasad. 1989a. Potassium removal by rice based cereal-cereal and cereal-legume intensive cropping systems and their effect on exchangeable potassium in soils. *J. Pot. Res.* 5:98–103.

George, Mercy and R. Prasad. 1989b. Studies on the effect of cereal-cereal and cereal-legume cropping systems on the productivity and fertility of soils. *Fert. News.* 34(5):21–25.

Gilland, B. 1979. *The Next Seventy Years: Population, Food and Resources.* Abacus, Tunbridge, Wells, U.K.

Gilland, B. 1985. Cereal yields in theory and practice. *Outlook on Agri.* 14(2):56–60.

Goswami, N.N., R. Prasad, M.C. Sarkar, and S. Singh. 1988. Studies on the effect of green manuring in nitrogen economy in a rice-wheat rotation using ^{15}N technique. *J. Agric. Sci. (Camb.)* 111:413–417.

Grewal, S.S. and B. Singh. 1989. Intensive farming in India. *Indian J. Agric. Sci.* 59:551–557.

Grewal, S.S., S.P. Mittal, K. Singh, and P. Singh. 1985. Effect of manure and fertilizer on soil-water retention and use by rainfed maize-wheat in Siwalik region. *J. Indian Soc. Soil Sci.* 33:728–731.

Gunasena, H.P.M. 1989. Interaction of free legumes in cropping systems in the intermediate zone of Sri Lanka. In: J. Vander Heide (ed.) *Nutrient Management for Food Crop Production in Tropical Farming Systems*, pp. 327–330. Institute for Soil Fertility (IB), Haren, the Netherlands and Universitas Brawijaya, Melang, Indonesia.

Gupta, R.P. 1981. Highlights of research—1980–81. I.C.A.R. All India Coordinated Research Project on Improvement of Soil Physical Conditions to Increase Agricultural Production of Problematic Areas. S.P.C. Bull. 1, p. 14. Indian Agricultural Research Institute, New Delhi-110012.

Gupta, A.P., R.S. Antil, and R.P. Narwal. 1988. Effect of farmyard manure on organic carbon, available N and P content of soil during different periods of wheat growth. *J. Indian Soc. Soil Sci.* 36:269–273.

Gupta, R.K., R.N. Rai, R.D. Singh, and R.N. Prasad. 1989. Effect of liming acid soil on yield of some crops and soil characteristics. *J. Indian Soc. Soil Sci.* 37:126–130.

Hameed, M.S. 1978. Physico-chemical changes with different systems of organic recycling in the semi-arid red soil region. M.Sc. Thesis. Andhra Pradesh Agricultural University, Hyderabad, India.
Handawela, J. 1989. Agriculture under simulated forest—an alternative to shifting cultivation. In: J. Vander Heide (ed.) *Nutrient Management for Fond Crop Production in Tropical Farming Systems*, pp. 375–380. Institute for Soil Fertility (IB), Haren, The Netherlands and Universitas Brawijaya, Melang, Indonesia.
Hegde, B.R., T.C. Chennappa, and B.K. Ananela Ram. 1987. Management alternatives for increased productivity of red soils: experience in Karnataka. Alfisols in the Semi-Arid Tropics: A constraints Workshop, pp.137–143. International Crops Research Institute for the Semi-Arid Tropics, Patancheru, AP.502 324, India.
Hesse, P.R. and R.V. Misra. 1981. Improving Soil Fertility Through Organic Recycling. FAO/UNDP Regional Project RAS/75/004. FAO. Field Document No.14.
ICAR. 1980. *Handbook of Agriculture*. Indian Council of Agricultural Research, New Delhi. p. 1303.
Jafri, S.M.H. and S.N. Pandit. 1986. Residual build up of N, P and K in soil by intercropping of cowpea as green manure with sequence. *J. Indian Soc. Soil Sci* 34:414–416.
John, P.S., R.K. Pandey, R.J. Buresh, and R. Prasad. 1989. Lowland rice response to urea following three cowpea cropping systems. *Agron. J.* 81:853–857.
Joshi, D.C. and C.P. Ghonsikar. 1979. Humification of added organic materials in some typical soils of Rajasthan. *J. Indian Soc. Soil Sci.* 27:261–267.
Kanwar, J.S. 1982. Managing soil resources to meet the challenges to mankind, Presidential Address, 12th Int. Congress Soil Sci., New Delhi, 8–16 Feb. (Plenary Session papers).
Kelley, W.P. 1951. *Alkali Soils*. Reinhold, New York.
Khiani, K.N. and D.A. More, 1984. Long-term effects of tillage operations and farmyard manure application on soil properties and crop yield in a vertisol. *J. Indian Soc. Soil Sci.* 32:392–393.
Khosla, B.K., R.K. Gupta, and I.P. Abrol. 1979. Salt leaching and the effect of gypsum application in a saline—sodic soil. *Agric. Water Management* 2:193–202.
Kulkarni, K.R. and R.K. Pandey. 1988. Annual legumes for food and as green manure in a rice-based cropping system *Green Manure in Rice Farming*. International Rice Research Institute, pp. 289–300. Los Banos, Lagume, Philippines.
Lal, S. and B.S. Mathur, 1988. Effect of long-term manuring and liming on crop yield and some physico-chemical properties of acid soil. *J. Indian Soc. Soil Sci.* 36:113–119.
Lal, S. and B.S. Mathur, 1989a. Effect of long-term application of manure and fertilizers on the DTPA-extractable micronutrients in acid soil. *J. Indian Soc. Soil Sci.* 37:588–590.
Lal, S. and B.S. Mathur. 1989b. Effect of long-term manuring and liming of an alifsol on maize-wheat and soil properties—I. Maize and wheat. *J. Indian Soc. Soil Sci.* 37:717–724.
Lal, S. and B.S. Mathur. 1989c. Effect of long term manuring and liming of an alfisol on maize, wheat and soil properties—II. Soil Physical properties. *J. Indian Soc. Soil Sci.* 37:815–817.

Lal, B.S. Mathur and K. Sinha. 1990. Effect of long-term manuring and timing of an alfisol on maize and timing of an alfisol on maize, wheat and soil properties—III. Forms of potassium. *J. Indian Soc. Soil Sci.* 38:21–26.

Mandal, L.N. and R.R. Mitra. 1982. Transformation of iron and manganese in rice soils under different moisture regimes and organic matter application. *Pl. Soil.* 69:45–56.

Mandal, S.C., M.K. Sinha, and H. Sinha. 1975. Acid soils of India and liming. *Tech.Bull.Indian Council Agric. Res., New Delhi* 51:29.

Massoud, F.I. 1977. Basic principles for prognosis and monitoring of salinity and sodicity. Proc. Int. Confer. on Managing Saline Water for Irrigation, pp. 432–454. Texas Tech. Univ., Texas, 16–20 August.

Meelu, O.P. and R.A. Morris. 1988. Green manure management in rice-based cropping systems. *Green Manure in Rice Farming*, pp. 209–222. International Rice Research Institute, Los Banos, Laguna, Philippines.

Middleton, H.E. 1930. Properties of soils which influence soil erosion. U.S.Dept. Agric., Washington, D.C. Tech. Bull. 178, pp. 16.

Minhas, R.S. and R.L. Mehta. 1984. Effect of continuous application of fertilizers on crop yield and some soil chemical properties under wheat-maize rotation. *J. Indian Soc. Soil Sci.* 32:749–751.

Moorman, V. and C.R. Panabokke. 1961. Soils of Ceylon—a new approach to the identification and classification of the most important soil groups of Ceylon. *Trop. Agricst.* 117:3–65.

Murthy, I.Y.L.N., C.R. Hazara, and Ashok Kumar. 1990. Effect of incorporation of tree leaves on soil fertility. *J. Indian Soc. Soil Sci.* 38:325–327.

Nambiar, K.K.M. and I.P. Abrol 1989. Long-term fertilizer experiments in Indian Agriculture. *Fert. News.* 34(4):11–20.

Nambiar, K.K.M., P.N. Soni, M.R. Vats, D.K. Sehgal, and D.K. Mehta. 1989. All India Coordinated Research Project on Long-Term Fertilizer Experiments (I.C.A.R.). Ann. Rept. 1985–86/1986–87, p. 152.

Narkhede, P.L. and R.V. Ghugare. 1987a. Organic recycling in drylands. I. Pruning technique for green matter production of subabul (*Leucaena leucocephela*). *J. Indian Soc. Soil Sci.* 35:412–416.

Narkhede, P.L. and R.V. Ghugare. 1987b. Orgranic recycling in drylands II. Effect of organic recycling of subabul on yield, nutrient uptake and moisture utilization by winter sorghum. *J. Indian Soc. Soil Sci.* 35:417–420.

Nath, S., M. Banerjee, G. Chattoraj, S.K. Ganguly, P.K. Das, and S.K. Banerjee. 1988. Changes in soil attributes consequent upon differences in forest cover in a plantation area. *J. Indian Soc. Soil Sci.* 36:515–521.

NBSS. 1985. *Soil Map of India.* National Bureau of Soil Survey and Land Use Planning (ICAR), Nagpur, India.

Okibo, B.N. and D.J. Greenland. 1976. Intercropping systems in Tropical Africa. Multiple cropping. American Society of Agronomy. Special Pub. 27.

Pandey, S.P., Hari Shankar, and V.K. Sharma. 1985. Efficacy of some organic and inorganic residues in relation to crop yield and soil characteristics. *J. Indian Soc. Soil Sci.* 33:179–181.

Patil, S.V. 1987. Cropping systems and agro-forestry in Alfisols and related soils in dryland areas of Karnataka. Alfisols in the Semi-Arid Tropics: International Crops Research Institute for the Semi-Arid Tropics, Patencheru, A.P. 502 324, India.

Paulino, L.A. 1986. Food in the Third World: Past Trends and Projections to 2000. International Food Policy Research Institute, Washington D.C., p. 76.

Peterson, G.W. and R.G. Corey. 1966. A modified Chang and Jackson procedure for routine fractionation of inorganic and soil phosphates. *Proc. Soil Sci. Soc. Am.* 30:563–565.

Pillai, K.G. and B.N. Chatterjee. 1983: Rice based cropping systems. Proc. Symp. Systems Approach to Fertilizer Industry New Delhi, 8–10 Dec. Fertilizer Association of India, New Delhi.

Pradhan, H.R. and M.S. Khera. 1976. Lime needs of soils of Sikkim based upon exchangeable aluminium test. *Bull. Indian Soc. Soil Sci.* 11:291–299.

Prasad, R. 1983. Increased crop production through intensive Cropping osystems. In: J.C. Holmes and W.M. Tahir (eds.) *More Food from Better Technology*, pp. 313–322. FAO, Rome.

Prasad, R. 1990. Effective nitrogen management. Proc. Sem. Fertilizer Scene in Nineties. Fertilizer Association of India, New Delhi, Dec. 6–9, p. SVI-2/1–8.

Prasad, R. and R. De 1983. Wheat based cropping systems. Symp. Systems Approach to Fertilizer Industry, New Delhi, 8–10 Dec. Fertilizer Association of India, New Delhi, p. AGS III-2/1–17.

Prasad, C.R., R. Prasad, K.K. Jha, and S.C. Mandal. 1976. Effect of continuous replication of manures, fertilizers and lime on some physico-chemical properties of acidic red loam soils of Bihar. *Bull Indian Soc. Soil Sci.* 11:253–259.

Rabindra, B. and H. Gowda. 1986. Long range effect of fertilizer, lime and manures on soil fertility and sugarcane yield on a red sandy loam soil (Udic Haplustalf). *J. Indian Soc. Soil Sci.* 34:200–202.

Rabindra, B., G.V. Narayanaswamy, N.A. Janardhan Gowda, and Shivanagappa. 1985. Long range effect of manures and fertilizers on soil physical properties and yield of sugarcane. *J. Indian Soc. Soil Sci.* 33:704–706.

RAPA. 1989. Selected Indicators of food and Agriculture Development in Asia—Pacific Region 1978–88. FAO. Regional Office for Asia and the Pacific, Bangkok, p. 197.

Reddy, R.N.S. and R. Prasad. 1975. Studies on the mineralization of urea and nitrification inhibitor treated urea in soil. *J. Soil Sci.* 26:304–312.

Sahu, B.N. and B.C. Nayak. 1971. Soil fertility investigation in Bhubaneswar long-term fertility trial. Proc. International Symposium in Soil Fertility Evaluation Indian Society of Soil Science, New Delhi.

Samad, A. and P. Sahadevan. 1952. Manuring for rice in Malabar *Madras Agri. J.* 39.160–172.

Sangakkara, U.R. 1989. The role of leguminous alley crops in the nitrogen nutrition of selected annual crops in small holder farming systems of Sri Lanka In: J. van der Heide (ed) *Nutrient Management for Food Crop Production in Tropical Farming Systems*, pp. 270–278. Institute for Soil Fertility (IB), Haren, The Netherlands and Universites Brawijaya, Malang, Idonesia.

Sanyasi Raju, N. 1952. Role of organic manures and inorganic fertilizers in soil fertility. *Madras Agric. J.* 39:130–147.

Sarkar, S., T.R. Rathore, R.S. Sachan, and B.P. Ghildyal. 1988. Effect of straw incorporation and burning in pH and organic carbon status of Tarai soils. *J. Indian Soc. Soil Sci.* 36:158–160.

Sarkar, A.K., B.S. Mathur, S. Lal, and K.P. Singh. 1989. Long-term effects of manure and fertilizers on important cropping systems in sub-humid, red and laterite soils. *Fert. News.* 34(4):71–80.

Sekhon, G.S. and D.N. Puri. 1986. The input-output balance of plant nutrients in some intensive cropping systems in India. *Potash Review.* 6:52–59.

Sharma, C.P. and P.D. Bajpai. 1989. Effect of kharif legumes on succeeding wheat crop and soil properties at different levels of nitrogen. *J. Indian Soc. Soil Sci.* 37:475–482.

Sharma, C.P., B.R. Gupta, and P.D. Bajpai. 1986. Residual effect of leguminous crops on some chemical and microbiological properties of soil. *J. Indian Soc. Soil Sci.* 34:206–208.

Sharma, K.N., B. Singh, D.S. Rana, M.L. Kapur, and J.S. Sodhi. 1984. Changes in soil fertility studies as influenced by continuous cropping and fertilizer application. *J. Agric. Sci. Camb.* 10:215–218.

Sharma, S.N. and R. Prasad. 1980. Nutrient (NPK) removal in rice-wheat rotation. *Fert. News.* 25(10):34–36, 44.

Singh, S., R. Prasad, B.V. Singh, S.K. Goel, and S.N. Sharma. 1990. Effect of green manuring, blue-green algae and neemcake coated urea on wetland rice (*Oryza sative* L.). *Biol. Fertil. Soils.* 9:235–238.

Singh, S.B. and I.P. Abrol, 1988. Long-term effect of gypsum application and rice-wheat cropping on changes in soil properties and crop yield. *J. Indian Soc. Soil Sci.* 36:316–319.

Singh, S.B., S. Nath, D.K. Pal, and S.K. Banerjee. 1985. Changes in soil properties under different plantations of the Darjeeling Forest Division. *Indian Forester* 111:90–98.

Sinha, S.K. and M.S. Swaminathan. 1979. The absolute maximum food production potential in India—an estimate. *Curr. Sci.* 48:425–429.

Srivastava, L.L., B. Mishra, and N.C. Srivastava. 1988. Recycling of organic waste in relation to yield of wheat and rice and soil *J. Indian Soc. Soil Sci. Sci.* 36:693–697.

Subramanian, K.S. and K. Kumaraswamy 1989. Effect of continuous cropping and fertilization on the phosphate fixation capacity of soil. *J. Indian Soc. Soil Sci.* 37:682–686.

Swaminathan, M.S. 1986. Building national and global nutrition security systems. In: M.S. Swaminathan and S.K. Sinha (eds.) *Global Aspects of Food Production*, pp. 417–449. Tycooly International, London, U.K. and International Rice Research Institute, Los Banos, Laguna, Philippines.

Swarup, A. and K.N. Singh. 1989. Effect of 12 years' rice/wheat cropping sequence and fertilizer use on soil properties and crop yields. *Field Crops Res.* 21:277–287.

Szabolcs, I. 1974. Salt Affected Soils in Europe. Martinus Nijhoff, The Hague, p. 63.

Tandon, H.L.S. and P. Narayan. 1990. Fertilizers in Indian Agriculture—Past, Present and Future (1950–2000). Fertilizer Development and Consultation Organisation, New Delhi, p. 160.

Thomas, J. and R. Prasad. 1982. Studies on mineralization of neem and sulphur coated and N-Serve treated urea. *Fert. News.* 27(10):38–43.

Thomas, J. and R. Prasad. 1983. Mineralization of urea, coated urea and nitrification inhibitor treated urea in different rice growing soils. *Z. Pflenzen. Boden.* 146:341–347.

TSI-FAI. 1988. Sulphur in Indian Agriculture. Proc.TSI-FAI Symp., 9–11 March,

New Delhi, p. 276. The Sulfur Institute, Washington, D.C. and the Fertilizer Association of India, New Delhi.
USDA 1954. Diagnosis and improvement of saline and alkali soils, In: L.A. Richards, (ed.). *Agric. Handbook 60*, p. 160.
Upadhayay, R. and T.D. Biswas. 1984. Influence of legumes on soil aggregation under dryland conditions. *J. Indian Soc. Soil Sci.* 32:358–360.
Van Bavel, C.H.M. 1949. Mean weight-diameter of soil aggregates as a statistical index of aggregation *Proc. Soil Sci. Soc. Am.* 13:20–23.
Venkateswarlu, J. 1987. Soil fertility management of red soils. Alfsols in the Semi-Arid Tropics—Constraints Workshop, pp. 115–121. International Crops Research Institute for the Semi-Arid Tropics, Patancheru, AP.502324, India.
Verma, K.S. and I.P. Abrol. 1980a. Effect of gypsum and pyrites on soil properties in a highly sodic soil. *Indian J. Agric. Sci.* 50:844–851.
Verma, K.S. and I.P. Abrol. 1980b. Effect of gypsum and pyrite on yield and chemical composition of rice and wheat grain in highly sodic soil. *Indian J. Agric. Sci.* 50:935–942.
Vijailakshmi, K. 1987. Soil management for increasing productivity in semi-arid red soils: Physical aspects. Alfisols in the Semi-Arid Tropics: A Constraints Workshop, pp. 109–113 International Crops Research Institute for the Semi-Arid Tropics, Patancheru, AP 502 324, India.
WFP 1990. *Food Aid Review*. World Food Programme, Rome, Italy pp. 85.
Yadav, J.S.P. and D.R. Sharma. 1968. A soil investigation with reference to distribution of sal and teak in Madhya Pradesh. *Indian Forester.* 94:897–902.
Yaduvanshi, H.S. 1988. Changes in available phosphorus in an alfisol under prolonged manuring and intensive cropping *J. Indian Soc. Soil Sci.* 36:719–724.
Yaduvanshi, H.S., Kanwar, E.S., and B.R. Tripathi. 1984. Nitrogen, phosphorus and potassium balance under continuous manuring in multiple cropping sequence in an acid hill soil. *J. Indian Soc. Soil Sci.* 32:97–101.
Yoder, R.E. 1936, A direct method of aggregate analysis of soils and a study of the physical nature of erosion losses. *J. Am. Soc. Agron.* 28:337–351.

Reclamation of Acid Sulphate Soils

D. Dent

I. Introduction

Through long usage, reclamation has come to mean winning of land from the sea or waste. In the case of acid sulphate soils, a more literal usage is apposite because many of these soils have been laid waste in the first place by ill-judged drainage.

Advances in Soil Science, Volume 17

Acid sulphate soils suffer extreme acidity as a result of the oxidation of pyrite (FeS_2). Pyrite accumulates in waterlogged soils that are both rich in organic matter and flushed by dissolved sulphate, usually from tidewater. Sulphidic sediments also accumulate in some brackish seas, for example, the bottom sediments of the Baltic some of which have been exposed by isostatic uplift.

Drainage of sulphidic soils or exposure of sulphidic sediments by excavation or as mine spoil allows oxygen into the system and pyrite is oxidized to sulphuric acid. Acid sulphate soils develop where the production of acid exceeds the neutralizing capacity of the parent material and the pH falls to less than 4. The variety of local names testifies to their distinctive and wretched nature—the Dutch name Katteklei has been used most often in earlier literature, while Chenery (1954) appears to have been the first to use the term "acid sulphate soil."

Van Breemen (1980) estimated that in recent coastal plains and tidal swamps worldwide, there are some 12–14 million ha, mostly in the tropics, where the topsoil is severely acid, or will become so if drained. In addition, there may be twice this area of potentially acid material overlain by shallow peat or alluvium. However, estimates of the extent and distribution of acid sulphate soils suffer more than most from scant field survey, even less reliable laboratory data, and also variable definition. In the Bangkok Plain, for example, van der Kevie and Yenmanas (1972) estimated an area of 760 000 ha of acid sulphate soils but Osborne (1985), defining extreme acidity as base saturation $<50\%$ and extractable aluminium >5 mEq/100g, estimates the area of extremely acid soils as only 226 400 ha.

Some acid sulphate soils have developed naturally as a result of changes in hydrology or relative sea level, for example those of the Bangkok Plain and the Baltic fringe. In Senegambia, a falling water table as a result of extended drought since 1971 has caused new acid sulphate soils on the low estuarine terraces and intertidal flats (Marius, 1985; Sadio, 1989). But extensive areas of acid sulphate soils have also developed as a result of deliberate land drainage.

On a world scale, acid sulphate soils are not extensive but significant areas occur in areas of critical population pressure, notably in West Africa and Southeast Asia where other land for subsistence food production or cash crops is not readily available, and they bring severe management problems wherever they are found. The hazards of a rising sea level, on the one hand, and drought-induced lowering of water tables, on the other, add a new urgency to devising wise strategies for their use.

II. Previous Reviews and Major Sources

Since these soils were first specifically recognised by van Kerckhoff (1856) and van Bemmellen (1863, 1886), a considerable literature has built up. There have been several comprehensive reviews, notably those of van

Beers (1962), Bloomfield and Coulter (1973), Metson et al. (1977), and Dent (1986). Most of the significant generalizations of the past 25 years have been published in the proceedings of international symposia on acid sulphate soils held in Wageningen (Dost 1973) Fort Collins, Colorado (Kittrick et al., 1982), Bangkok (Dost and van Breemen, 1982), and Dakar (Dost, 1988). New developments are published in an occasional newsletter by an active working group of the International Society of Soil Science[1].

This review concentrates on reclamation of these problem soils, which is still relatively neglected in the literature. It includes information won by major collaborative research and development projects in Thailand, Vietnam and Indonesia that is not yet fully reported in primary journals but which provides answers to several of the questions highlighted by previous reviews.

III. Formation and Characteristics of Acid Sulphate Soils

A. Soil-Forming Processes

1. Accumulation of Sulphides

Reduced compounds of sulphur accumulate in water-logged soils and sediments where there is easily decomposed organic matter and a supply of sulphates. Bacteria, breaking down this organic matter, reduce sulphate ions to sulphides and iron III oxides to iron II. Pyrite is the usual end-product of this process. It is not evenly distributed in the soil but concentrated in organic remains (Harmsen, 1954; Berner, 1970; Rickard, 1973; Wada and Bongkot Seisuwan, 1989).

The main source of sulphate ions is seawater and the most extensive potential acid sulphate environment is tidal swamp and marsh. The dense vegetation fuels sulphate reduction; the tides bring in sediment, renew the supply of sulphates and remove the equivalent alkalinity (Pons and van Breemen, 1982; Pons et al., 1982). The rate of accumulation of sulphides under mangrove swamp is of the order of 10 kg S m^{-3} of sediment per 100 years (Goldhaber and Kaplan, 1982; Dent, 1986).

A sulphidic material only becomes a potential acid sulphate soil when the potential acidity, represented by the pyrite, exceeds the soil's neutralizing capacity. Calcium carbonate is the main reserve of neutralizing capacity. One part by mass of pyrite sulphur is countered by three parts of calcium carbonate. Exchangeable bases also provide a buffer against acidity but not more than 0.5% by mass of sulphur can be neutralized from this soruce.

[1]Secretary, M.E.F. van Mensvoort, Department of Soil Science and Geology, Agricultural University, Wageningen, The Netherlands.

Sulphidic soils are usually dark grey and dark greenish grey clays with a soft, buttery consistency. They smell strongly of hydrogen sulphide. Peats that are flooded by brackish water also accumulate sulphides and, locally, sulphidic soils develop inland on waterlogged sites flushed by sulphate-rich water. Old sulphidic marine sediments and sulphide ore bodies also constitute potentially acid parent materials if they are brought to the surface.

2. Development of Acidity Following Drainage

Potential acid sulphate soils become acid as a result of drainage. Drainage brings oxygen into the soil and pyrite is oxidized, generating sulphuric acid. The reaction of pyrite with oxygen is slow, but pyrite is rapidly oxidized by Fe^{3+} ions. Fe^{3+} is thereby reduced but is regenerated by the bacterium *Thiobacillus ferrooxidans*. This catalytic oxidation of pyrite takes place only at pH values less than 4, because Fe^{3+} is soluble only under very acid conditions (van Breemen, 1976). The net result, with iron III hydroxide as the end product, may be expressed as:

$$FeS_2 + \frac{15}{4}O_2 + \frac{7}{2}H_2O \rightarrow Fe(OH)_3 + 2SO_4^{-2} + 4H^+$$

Red oxides and hydroxides of iron precipitate as mottles and ped coatings within the soil profile, and in drainage and floodwaters where they often provide the first visible indication of acid sulphate soils.

Gypsum, $CaSO_4.2H_2O$, is formed where the sulphuric acid is neutralized by calcium carbonate. Gypsum appears as dirty grey crystals in fissures, on ped faces and ditch sides but, being sparingly soluble, is readily leached.

Under severely acid, oxidizing conditions, pale yellow deposits of jarosite, $KFe_3(SO_4)(OH)_6$, develop around pores and on ped faces. Mottles and cutans of jarosite, contrasting with a grey or light brown (beurre marron) matrix, are one of the diagnostic features of acid sulphate soils but, sometimes, they are absent, even under the most acid conditions. Apparently jarosite does not form under conditions of poor drainage and high organic matter content (van Mensvoort and Tri, 1989).

Severe acidity promotes the weathering of aluminosilicate minerals, releasing Al^{3+} ions which are increasingly soluble at pH values less than 4. Al^{3+} activity increases tenfold for every unit fall of pH (van Breemen, 1973, 1976). It is the high concentration of Al^{3+} in solution, rather than acidity per se, that is chiefly responsible for the toxicity of acid sulphate soils and their drainage waters.

The process is reversed by flooding. In the presence of easily decomposed organic matter, the reduction of iron oxides and sulphate ions by bacteria removes acidity and, therefore, Al^{3+} is precipitated as a basic sulphate or hydroxide:

$$Fe(OH)_3 + 2H^+ + \frac{1}{4}CH_2O \rightarrow Fe^{2+} + \frac{11}{4}H_2O + \frac{1}{4}CO_2$$

$$SO_4^{2-} + 2H^+ + 2CH_2O \rightarrow H_2S + 2H_2O + 2CO_2$$

B. Consequences of Severe Acidity

1. Agronomic Problems

Acid sulphate soils present chemical, physical and biological problems which have been reviewed by Rorison (1973) and by Bloomfield and Coulter (1973). Chemical problems for crops in drained acid sulphate soils include: toxicity, mainly the increased solubility of aluminium; unavailability of phosphate, caused by iron and aluminium-phosphate interactions; low base status and nutrient deficiencies; and salinity.

Under flooded conditions, for example under wetland rice, acidity is reduced but new problems include: iron II toxicity; hydrogen sulphide toxicity; CO_2 and organic acid toxicity.

Aluminium toxicity, in particular, causes stunted root systems that cannot ramify the soil and withdraw water. Crops suffer water stress. Soil ripening is also arrested.

2. Soil Ripening

Physical ripening is the process whereby mud is transformed to firm soil by the irreversible loss of water (Pons and Zonneveld, 1965; Rijniersce, 1983). Freshly deposited (unripe) clayey sediments have a very open microstructure that is easily deformed and, therefore, mechanically weak. Pore space is high but hydraulic conductivity low except through root pores. Unripe material may lose water by consolidation under a great weight of sediment or through the extraction of water by plant roots. Drainage by gravity cannot exert enough force to draw water from the fine pores.

Withdrawal of water leads to the partial collapse of the microstructure, shrinkage, fissuring and an increase in mechanical strength of the material because of greater contact between individual particles.

The concept of soil ripening elaborated by Pons and Zonneveld encompasses these physical changes and also the chemical and biological changes needed to change fresh sediment into productive soil. In the case of acid sulphate soils, unfavourable conditions for most micro-organisms impede the release of nutrients from soil organic matter, while the stressed crops are especially susceptible to disease.

3. Engineering Problems

Engineering problems posed by acid sulphate soils include: blockage of drains by iron oxide(ochre) deposits; corrosion of metal and concrete by acid and sulphate ions; low bearing strength and uneven subsidence of physically unripe soils; and lack of colonization of earthworks by vegetation, rendering them susceptible to erosion.

Special, often costly, techniques are needed to deal with these problems (Dent, 1986).

4. Environmental Impact

Reclamation of potential acid sulphate soils by drainage obviously leads to a loss of that wetland habitat, but the ecological impact is by no means confined to the drained area. There are implications throughout the web of life that is dependent on the wetland—especially in the case of mangrove and salt marsh that are the base of food chains for shellfish and coastal fisheries.

Where acid sulphate soils develop, rapid and dramatic changes in water chemistry take place and these are exported from the reclaimed area with drainage and floodwaters. Pollution problems have been extensively studied in relation to acid drainage waters from mines (Temple and Koehler, 1954); acid floodwaters may adversely affect crops growing on adjacent, better soils (see section VD); flushes of severe acidity kill fish in both tidal and inland waters; and drinking water supplies may be contaminated by dissolved aluminum and suspended iron oxides.

C. Soil Variability

Acid sulphate soils are not all the same. In particular, there are differences between (1) *ripe acid sulphate soils* that have weathered and leached naturally to depths of more than 1 m; (2) *raw acid sulphate soils* that still have reserves of pyrite at shallow depth; and (3) *unripe sulphidic soils* that are still waterlogged but which will become acid if drained (potential acid sulphate soils). There are also different problems of reclamation and use in contrasting hydrological environments (section IIIC3).

Inadequate characterization of soil and site makes it difficult to interpret both farmers' experience and a lot of the published experimental data. Obviously, soil and hydrological surveys are needed before reclamation to establish the nature of the beast, its potential for reclamation, and to determine the best combination of treatments. Unfortunately, soil survey is particularly difficult in the case of acid sulphate soils because they occur in areas of low relief and the crucial characteristics may have no surface expression.

From the review of regional studies by Dent (1986) and more recent surveys (e.g., Vê et al., 1989 in Vietnam), it is clear that broad soil patterns are related to the extent of past and present brackish or saltwater flooding, and that the critical upper limit of pyrite accumulation is determined by the upper limit of continuous waterlogging. Depth to pyrite and, in drained soils, depth to extreme acidity is a function of differences in drainage status, especially between levees and backswamps. The severity of acidity or potential acidity varies with subtle variations in the accumulation of pyrite and other reduced sulphur compounds, in rates of sedimentation, and in the carbonate content of sediments.

Burrough et al. (1986), in a study of soil variability in the Mekong Delta, showed that the depth to extreme acidity is a reliable criterion for distinguishing mapping units at scales from 1:20 000 to 1:100 000. Other characteristics, for example salinity, have such short-range variability that they cannot be mapped reliably. Even acidity and depth to acidity may vary significantly over distances of a few metres. Where the sulphidic material is overlain by more-recent sediment or by peat, subsoil characteristics may be unrelated to the present surface. Interpretation of the soil pattern from its surface expression is frequently made more difficult by clearance of natural vegetation and modification of drainage.

1. Soil Classifications

From the point of view of reclamation, it is helpful to make some distinctions of both soil and environmental characteristics. Those of interest include: acidity and potential acidity; salinity; composition and texture; ripeness; permeability; depth and thickness of limiting horizons; depth and seasonal variations of the watertable; duration and depth of flooding; and quality of floodwaters.

Most of the critical soil characteristics are embodied in the classification of the International Institute for Land Reclamation and Improvement (Dent, 1986). Its main soil groups are presented in Table 1. Essentially, acid sulphate soils are identified by an oxidized pH <4. Raw acid sulphate soils have reserves of pyrite within 1 m that will generate more acidity by oxidation; ripe acid sulphate soils no longer have pyrite but have jarosite within 1 m which represents a substantial reserve of acidity; and acid aluminum soils are severely acid with high exchangeable aluminum but do not have reserves of pyrite or jarosite within 1 m.

These groups can be considered as stages in development from sulphidic (potential acid sulphate) soils, at present waterlogged but with pyrite in excess of neutralizing capacity so that they will become acid sulphate soils if they are drained; to raw unripe soils; to ripe acid sulphate soils; finally to ripe acid aluminum soils.

With the notable exception of the FAO/Unesco legend (1988), most recent morphological classifications distinguish between (actual) acid sulphate soils and potential acid sulphate soils. Soil Taxonomy (Soil Survey

Table 1. Major categories of potential acid sulphate soils and acid sulphate soils

	Organic soils		Sandy soils	Clayey soils			
Undrained not potentially acid	Unripe peat and muck		(Saline) sand	Unripe (saline) clay			
Potential acid sulphate soils	Unripe sulphidic peat and muck		Sulphidic sand	Unripe saline sulphidic clay			
Acid sulphatc soils	Raw acid sulphate peat and muck	Ripe acid sulphate peat and muck	Raw acid sulphate sand Acid sulphatc sand	Raw saline acid sulphate clay	Ripe acid sulphate clay with raw subsoil	Ripe acid sulphate clay	
Associated nonacid sulphate soils	Peat and muck with unripe subsoil	Ripe peat and muck	Sand		Ripe clay with unripe subsoil	Ripe clay	Ripe acid aluminum clay

Staff, 1987) distinguishes the sulfuric horizon (with a pH in water < 3.5 and jarosite mottles) from sulfidic material (reduced material containing 0.75% or more S and less than three times as much $CaCO_3$ equivalent) that will become a sulfuric horizon if oxidized. Mineral soils with the sulfuric horizon are placed in the Order Inceptisols—as Sulfaquepts, if the sulfuric horizon is within 0.5 m of the surface, or as Sulfic Tropaquepts or Sulfic Haplaquepts, if it occurs between 0.5 and 1.5 m from the surface. Potential acid sulphate soils are placed in the Order Entisols, as Sulfaquents if sulfidic material occurs within 0.5 m of the surface or as Sulfic Hydraquents where it occurs between 0.5 and 1 m. Acid sulphate peats are Sulfihemists and potential acid sulphate peats Sulfohemists.

Recent proposals to amplify Soil Taxonomy (Fanning and Witty, 1989):

1. Redefine the sulfuric horizon, taking account of acid sulphate soils that do not show jarosite mottles, as a layer > 15 cm thick with a pH in water < 3.8 and evidence that this acidity is caused by oxidation of pyrite (either jarosite mottles, underlying sulfidic material or 0.05% or more soluble sulphate);
2. Redefine sulfidic material simply as material that shows a drop of at least 0.5 pH units to < 3.8 during incubation for 8 weeks in moist, oxidized conditions;
3. Provide a new Great Soil Group of Sulfochrepts for acid sulphate soils that are not poorly drained (e.g., on mine spoil) and a new subgroup of Salorthidic Sulfaquepts for very saline acid sulphate soils.

Pons et al. (1989) have proposed, in addition, a range of subgroups to cater for soils with sulfidic material or sulfuric horizons at different depths (they suggest <0.5 m, 0.5–0.8 m, and 0.8–1.2 m) and for the occurrence of peaty, dark acid and dark neutral topsoils.

2. Laboratory Support

All morphological classifications are limited by their reliance on soil characteristics that are measurable in the field and, in particular, the reliance on pH as a measure of acidity. Osborne (1985) has pointed out that many existing soil surveys may be inadequate for land reclamation, noting wide differences in crop performance and response to liming even within mapped soil series in the Bangkok Plain, Thailand. For these extensive areas of ripe acid sulphate clays and acid aluminum clays, he established acidity classes based on KCl-extractable aluminum, base saturation, jarosite S, and pyrite S (jarosite S is estimated from acid-extractable S minus extractable Ca) Table 2. In Thailand, on ripe acid sulphate clays and acid aluminum clays, crop performance is correlated best with KCl-extractable Al and the reserve of acidity represented by jarosite S. However, these acidity classes are not valid for raw acid sulphate soils and potential acid sulphate soils because of the presence of unoxidized pyrite.

Table 2. Acidity classes for ripe acid sulphate soils, Bangkok Plain, Thailand (from Osborne 1985)

Parameter	Acidity class I	II	III	IV	V
Extr. aluminum me/100 g	<1	1.0–5.0	5.1–9.0	9.1–13.0	>13
% Al saturation (Al/Al + bases)	0–5.0	5.0–25.0	25.1–45.0	45.1–65	>65
% base saturation	>65	50.1–65.0	35.1–50.0	20.0–35.0	<20
Extr. calcium me/100 g					
0–20 cm	>12.4	7.5–12.4	2.5–7.4	1.0–2.4	<1.0
0–40 cm	>10	6.5–10.0	3.0–6.4	1.0–2.9	<1.0
Total sulphur %	<0.05	0.05–0.100	0.101–0.240	0.241–0.400	>0.400
Acid extr. S-extr. Ca me/100 g					
0–40 cm	<−5.0	−5.0–2.5	2.6–10.0	10.1–20.0	>20
40–100cm	<+5.0	5.0–15.0	15.1–25.0	25.1–35.0	>35

The lack of reliable, sophisticated laboratory facilities remains a problem for reclamation of extensive areas of acid sulphate soils in the tropics. However, a range of useful field tests is given by Dent (1986), and Konsten et al. (1988) detail a rapid titration method to determine lime requirement that is suitable for field laboratories.

The treatment of samples prior to analysis is critical, because of the very rapid chemical changes that follow oxidation of sulphidic materials and flooding of acid sulphate soils. If information is needed on the condition of the soil in the reduced state, then samples must be kept reduced—by exclusion of air and chilling or freezing as soon as possible. Osborne (1985) added a little toluene to each sample in the field to inhibit bacterial catalysis of oxidation or reduction. Where a dry sample is needed for analysis, freeze drying or rapid oven drying (Thomas and Varley, 1982) may be used. Dried samples should then be analysed without delay.

To simulate either oxidized or reduced conditions, enough time must be allowed for bacterial action and the response of soil minerals, so rapid oxidation of pyrite by hydrogen peroxide (van Beers, 1962; Konsten et al., 1988) overestimates potential acidity. To simulate oxidation, Dent (1986) recommended moist incubation of 500-g samples for 3 months. Ponnamperuma et al. (1973) using 2-kg samples, Sen (1988) using lysimeters, and

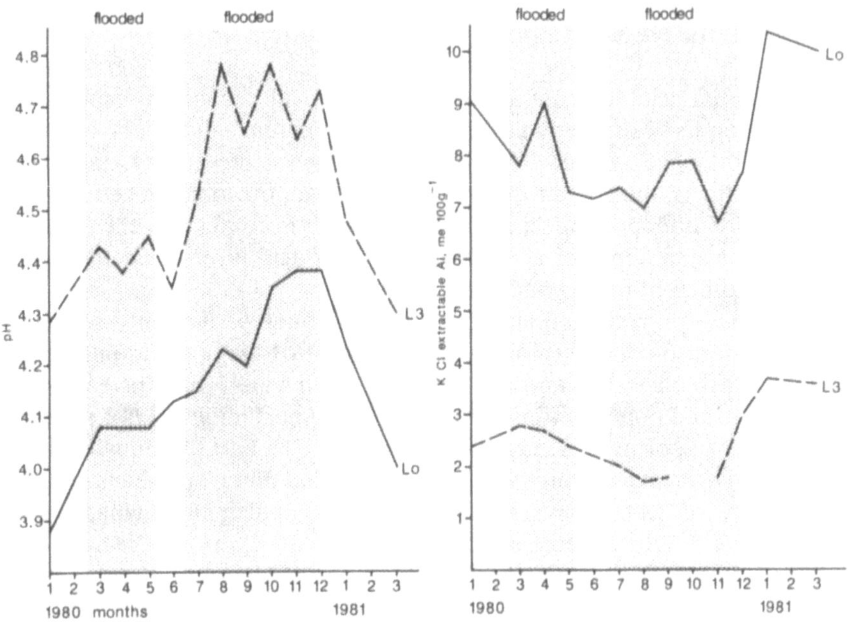

Figure 1. Effects of flooding on pH and extractable Al, Rangsit Series very acid phase, Ongkharak Station, Thailand. Lo, no lime; L_3, 15 t/ha applied 1973, mean of four samples per treatment. (From Osborne, 1985)

Osborne (1985) monitoring field conditions, demonstrate that several months may elapse before the full effect of flooding is felt in reducing pH and soluble aluminum levels (Fig. 1).

3. Hydrology

In acid sulphate soils, water management is the key to soil management. Reclamation must, therefore, take account of the interaction between soil, groundwater, and floodwater. It is useful to consider four broad environments: wet, big water surplus; wet, no water surplus; seasonally dry, more than 1 month dry season but with a seasonal water surplus; and dry, no water surplus.

Where there is a big water surplus, the main management problem is drainage; acidification can be limited by maintaining a high watertable. Where there is a dry season of more than a few weeks, acidity exerts increasingly severe constraints and the available water capacity, determined mainly by the thickness of any non-acid topsoil, becomes a critical soil characteristic. In dry areas, salinity rather than acidity may be the effective constraint on cropping; water for leaching is scarce and it may not be practicable to maintain a high watertable.

IV. Alternative Strategies for Reclamation

If a potential acid sulphate soil is drained, it will become severely acid. Large amounts of sulphuric acid, iron, and aluminum will then be released into the drainage system. The nearer the surface the acidity, the more acute its effects; the greater the reserves of pyrite, the longer its effects will last. Before attempting reclamation, developers need to know the detailed distribution and depth of acid and potentially acid layers, and the likely effects of different management options.

Strategies for reclamation should take account of not only soil constraints but also the length of the dry season, hydrology (including flooding, access to tidewater, and availability of fresh water), and the economic constraints and opportunities—not least the opportunities of not reclaiming. Better opportunities may exist elsewhere. The best option will be determined by the competing demands for the land and its products and the availability of technology, finance, material inputs, and management. Prospective developers face a series of choices:

They must decide whether to reclaim or not. If alternative land is available for development, the best policy is to leave acid sulphate soils well alone. This is the only course of action that does not preclude other options in the future. Especially in the case of tidal land, the value of mangroves (Fig. 2), salt marsh, and mud flats to commercial fisheries and wildlife should be recognized. Direct benefits can accrue from exploitation of natural products, including mangrove forestry (FAO, 1982) and from tourism. The mangrove and salt marsh fringe also provides the first line of coastal defence against waves and tidal surges. By trapping sediment, it can grow upwards with a rising sea level and thus protect low-lying coastlands. In recent years, the mangrove *Avicennia marina* has been actively planted in the Mekong Delta.

If reclamation is chosen, they must decide whether to adopt a minimum disturbance strategy (essentially to minimize oxidation of pyrite) or to attempt total reclamation (by drainage and neutralization). Several minimum-disturbance options are available, including:

- Flooded rice cultivation, which relies on reduction of acidity by flooding
- Shallow-water fish ponds (Fig. 3) and alternate rice-shrimp cropping, which can be very profitable options where there is unrestricted access to tidewater
- A controlled high watertable for dryland crops
- Forestry or grazing—low cost options where a high watertable cannot be maintained

Figure 3. Brackish-water fish ponds, Iloilo, The Philippines (photograph by Robert Brinkman)

Figure 2. *Rhizophora racemosa* forest, 30 to 40 m high, The Gambia. This is the usual pioneer species in West Africa, colonizing unripe mud within the limits of daily tidal flooding. It is characterized by a tangle of aerial roots and a dense, fibrous root mat below the surface

Where the water regime can be controlled, good responses to modest applications of lime and fertilizer are sometimes achieved. However, both the choice of crops and timing of operations are restricted and severe acidity will recur at any time that watertable control is not maintained.

Total reclamation is likely to be economic only where the initial acidity is moderate or has already been reduced by a long period of weathering and leaching.

V. Minimum-Disturbance Strategies

A. Tidal Rice

In many estuaries and deltas, fresh water is backed up by the tide during the rainy season. Where there is at least 100 days of fresh tidal flooding, rice (*Oryza sativa*) can be grown even where potential acidity occurs within 20 cm of the surface.

Driessen and Ismangun (1973) describe the system developed by Bandjarese farmers in Kalimantan which involves up to three transplantings of rice seedlings, to produce plants with many tillers and fair tolerance of toxicity and drought. Equally important, the farmers scrupulously avoid disturbing the raw acid subsoil and delay planting to allow leaching of salt and toxins from the topsoil. Yields with traditional varieties reach 1.5 to 2 t ha^{-1} if a high watertable and effective surface flushing are maintained.

In West Africa also, where tidal rice has been grown for more than 100 years, large seedlings are transplanted to combat the risks of salinity, toxicity, and uncontrolled water levels (Fig. 4). But the risks remain. Yields are usually well below 2 t ha^{-1} and early return of salinity can lead to complete crop failure.

Agyen-Sampong et al. (1988) report that yields of tidal rice can be

Table 3. Yields of rice, t/ha, West African Rice Research Station, Rokupr, with and without tidal flushing (from Bloomfield and Coulter 1973)

Block no.	Average under tidal regime 1935–43	Tidal flooding excluded				Tidal floolding restored				
		1944	1945	1946	1947	1948	1949	1950	1951	1952
22	1.9	0.9	0.0	0.0	0.0	0.8	2.0	1.7	2.6	3.3
23	2.3	1.2	0.2	0.0	0.0	1.3	2.7	2.4	2.6	3.0
24	2.6	2.0	0.7	0.2	0.2	1.1	3.0	3.0	2.7	—
25	2.0	1.7	0.1	0.1	0.2	2.0	3.3	2.7	2.9	3.5
26	2.6	0.8	0.2	0.0	0.0	1.0	2.1	3.2	1.9	3.9
27	3.2	1.1	0.6	0.0	0.0	0.9	2.3	2.6	2.5	3.4

Figure 4. Tidal rice cultivation, The Gambia. Large seedlings are transplanted into tidal land with an uncontrolled water level and risk of salinity and toxicity. The bund is for access only. *Avicennia* and *Rhizophora* mangroves in the background

almost doubled by selection of suitable long- or short-season varieties, mechanized weeding, and injection of urea as a nitrogen fertilizer.

The secret of the success of tidal rice is that the natural hydrology remains undisturbed. Tidal flooding prevents the oxidation of pyrite. Many attempts to extend rice cultivation into areas of permanent salinity by excluding tidewater have failed because this leads to drying of the soil and oxidation of pyrite (Dent, 1947; WARRS, 1959–62; and Hart, 1959 in Sierra Leone; Beye, 1973a; and Marius, 1982, 1985 in Senegal). Table 3 shows the effects of excluding tidewater at Rokupr Rice Research Station in Sierra Leone. From 1935 to 1943 the farm was tidal; from 1944 to 1947 tidal flooding was stopped by bunds and yields fell dramatically; then tidal flooding was restored and yields recovered equally dramatically.

B. Rice-Shrimp Cropping

Good use is made of brackish-water flooding in the dry season by the rice-and-shrimp cropping system developed in the Minh Hai Province of the Mekong Delta. Rainfed rice is sown in the wet season, then shrimps are raised in the flooded fields, feeding on phytoplankton, during the dry season.

The best production is achieved on low-lying land that can be flooded daily with brackish water rich in silt and shrimp fry (*Penaeus* and *Metapenaeus* spp.). Xuan et al. (1986) recommend building polders, not bigger than 10 ha, fitted with floodgates to facilitate the intake of tidewater and also drainage for harvesting, while keeping the soil wet at all times to prevent the oxidation of pyrite. Shrimps are harvested every 2 months in the dry season. Cumulative yields of up to 690 kg ha^{-1} yr^{-1} are claimed but a good farmers' yield is 150 kg ha^{-1} yr^{-1}.

As soon as the rains come, brackish water is let out of the field and, after flushing out the salt, rice is sown. Rice yields of up to 4 t ha^{-1} are reported and a good farmers' yield is 3 t ha^{-1}.

Perhaps 30 000 ha of the littoral fringe of the Mekong Delta is now under this system, which is not confined to acid sulphate soils.

C. Seasonally Flooded Rice

Acid sulphate soils that are seasonally flooded by deep water have been, traditionally, broadcast with floating rice. This grows as a dryland crop before flooding. Yields are low (usually well below 1.5 t ha^{-1}), partly because of aluminum toxicity before reducing conditions are established.

Alternatively, big seedlings can be transplanted as the floodwaters recede, thereby avoiding the period of toxicity. The introduction of high-yielding, short-duration varieties increases the opportunities for this method. Xuan et al. (1982) report yields of 4.6 to 6 t ha^{-1} on acid sulphate soils of the Mekong Delta from rice transplanted after deep flooding. To achieve such yields, nitrogen and phosphate fertilizer is applied and supplementary irrigation is needed if drought occurs before ripening.

D. Reclamation by Shallow Drainage

Farmers' experience has shown that acidity and toxicity can be held in check by control of the watertable—either by keeping the soil flooded for as long as possible, by preflooding before planting, or by intensive shallow drainage with blocks in the ditches to keep up the watertable in dry weather.

In the Mekong Delta, shallow drainage is used to assist the flushing of salts and acid that accumulate during the dry season. Fields are laid out in strips about 9 m wide, between which ditches are dug to 0.3 to 0.6 m, the slices of topsoil being spread over the intervening strips to build raised beds or "lips" (Figs. 5 and 6). Each ditch opens to a main drainage canal.

Severe acidification occurs during the dry season. Leaching begins with the first heavy rains. Drainage water is held in the ditches until it reaches the top of the raised beds, then it is drained to the river. This leaching cycle is repeated two or three times until the rising river floods the whole area.

Figure 5. Rice cultivation on raised beds (lips), Mekong Delta, Vietnam

Figure 6. Building a "lip." Topsoil from broad, shallow ditches is spread evenly between the ditches to build a raised bed about 9 m wide. Mekong Delta, Vietnam

Table 4. Examples of acid floodwaters from various places in the Mekong delta at the beginning of the rainy season (from Ni, 1984)

Place	Al ($mol\ m^{-3}$)	Fe ($mol\ m^{-3}$)	pH
Hoa An (in polder)	8.89	0.75	—
Tan Lap, Tien Giang	6.30	0.45	3.0
Extension farm TG	3.11	0.13	3.2
Lang Bien, Dong Thap	10.00	5.72	2.8
Min Hai	8.52	3.28	2.5
Cuu Long	5.29	1.90	3.6

Rice seedlings, 45 to 60 days old, are then transplanted into the flooded beds. Usually, nitrogen fertilizer is applied but not lime or phosphate.

Local yields on undrained acid sulphate soils are between 0.2 to 0.5 t ha^{-1}. Xuan et al. (1982) claim yields of about 4 t ha^{-1} after reclamation but Sen (1987) reports average farmers' yields, after 5 years of reclamation, of 2.3 t ha^{-1}.

Of course, the leached acid has to go somewhere. Where drains are isolated, most of the acid is immobilized by strong reduction in the ditch bottoms. Weeds are thrown into the ditches to speed up this process. Where the drains are connected to the rivers, most of the acid is removed and reclamation of large areas of acid sulphate soils has caused acidification of floodwaters affecting crops and soils in neighboring areas. This is an increasing problem as more acid areas are reclaimed, and some farmers have bunded their fields to protect them from acid floodwaters. Table 4 gives examples of acid floodwaters from the Mekong Delta.

The farmers' methods have been confirmed by pot experiments (Cate and Sukhai, 1964; Ponnamperuma et al., 1973) and lysimeters (Sen, 1988). Sen, working in the Philippines with a ripe clay and an unripe, saline, sulphidic clay, examined several drainage treatments over three rice crops. Between the first and second crop there was a 3-month dry period followed by 6 weeks leaching, then 1 week preflooding with the drains closed. During the growing season, 0.15 m of water was maintained over the surface. The full range of treatments was as follows:

- Deep drainage (to 0.8 m) during the leaching period
- Deep drainage with percolation of water through the soil during the growing season
- Shallow drainage (to 0.4 m) during the leaching period
- No drainage but surface flushing by removal of standing water every 3 days
- Surface mulching during the dry season
- Liming at zero, 1–1.25 t ha^{-1} and 2–2.25 t ha^{-1}

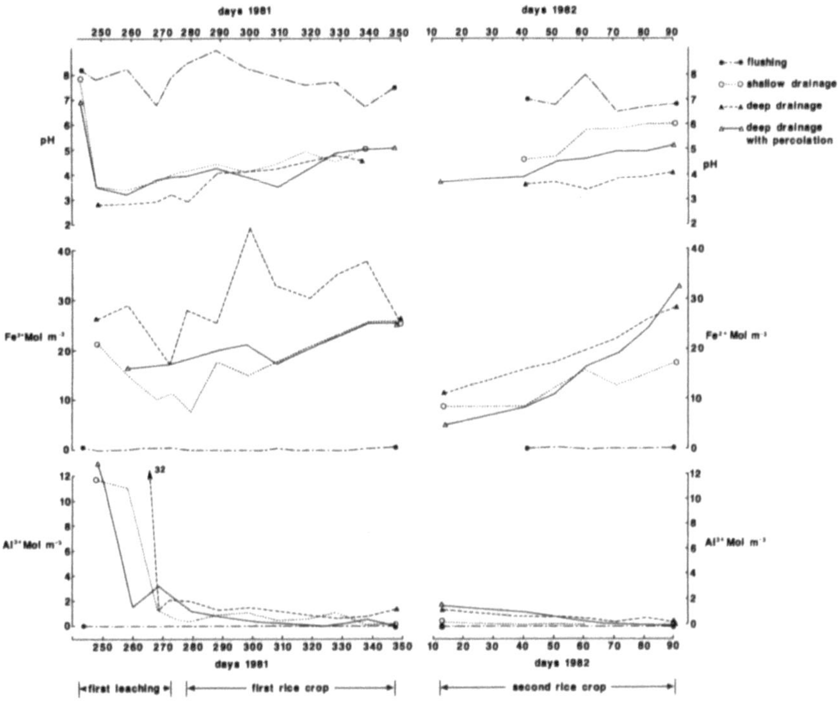

Figure 7. Lysimeter study of the effects of different water-table regimes on acidity and concentrations of soluble Al and Fe in a raw acid sulphate clay. (From Sen, 1987)

On the ripe clay, there were no differences in crop performance between treatments; but the sulphidic clay showed dramatic differences, with complete crop failure on drained lysimeters. Liming at the levels employed had no measurable effect in this very sulphidic soil.

Deep drainage of the sulphidic clay produced severe, deep-seated acidity and high Al^{3+} and Fe^{2+} concentrations. Once deep-seated acidity developed, pH recovered only slowly in the subsoil after flooding, taking 9 months to reach a new maximum level. Probably, reduction was retarded by the severe acidity and a lack of easily metabolized organic matter in the subsoil.

Figure 7 summarizes the results from different watertable regimes under the zero lime treatment. Optimum soil conditions and the best yields were achieved with the surface flushing treatment that avoided soil drainage. Surface mulch, by reducing evaporation in the dry season, lessened the oxidation of pyrite (see also Beye, 1973b).

Double cropping in the wet season also reduces pyrite oxidation and the second crop takes advantage of the reduced toxicity. The parallel with

Osborne's data (Fig. 1) from a ripe acid sulphate clay is clear. The dramatic fall in soluble aluminum during leaching while the pH of the soil remained below 3.5, or even below 3, is surprising but may be attributed to the efficacy of leaching under these experimental conditions.

E. Rainfed Rice

Where fresh floodwater is not available and it is not possible to maintain a constant high watertable because of a long dry season, severe acidity and salinity develop. Yet even under these constraints, rice is grown by a variety of local, indigenous systems.

Throughout West Africa where tidal creeks do not carry enough fresh water for tidal rice, small polders are made to exclude salt water. The topsoil is built into ridges which are leached entirely by rainwater, and the intervening furrows provide some water storage and drainage of surplus water. Rice is transplanted into the ridges. Oosterbaan (1982) describes this "bolanhas" system as used in Guinea Bissau.

Unreliable water supply, salinity, and acidity are severe constraints; yields are low (0.5 to 1.5 t ha^{-1}) and variable from place to place and season to season. However, Marius (1985) notes that in Casamance, Senegal, this traditional system of rice cultivation has been less affected by the severe drought of the past 20 years than have big polders that were drained and leached more deeply.

Ukkerman and van Gent (1989) report that yields in bolanha fields respond well to combined N and P fertilization. An increase up to 0.5 t ha^{-1} is attributed to fertilization with 30 kg N and 30 kg P_2O_5 ha^{-1}, and an increase of 1.1 t ha^{-1} to 90:60 fertilizer application. Liming at low rates, up to 0.5 t ha^{-1}, did not increase yield.

F. Controlled Watertable for Dryland Crops

Rice is the only major crop that can take advantage of the reduction of acidity by flooding. However, the production of acid can be limited by keeping the subsoil water-logged and, in this way, a limited range of dryland crops is grown on acid sulphate soils.

High watertable management encounters several difficulties, especially with perennial crops:

- The watertable must be strictly controlled. This may need sluices to maintain different levels in different parts of a polder, according to the depth at which acid sulphate conditions occur.
- During long dry periods the watertable will fall anyway and the operator may be faced with the alternatives of allowing acidity to develop or allowing brackish water into the drains to maintain the water level. This might be acceptable for coconut but most other crops would be lost.

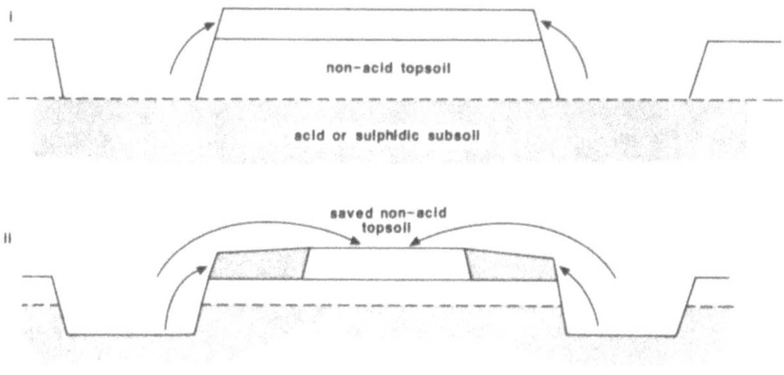

Figure 8. Alternative constructions of raised beds for dryland crops

- Where the groundwater is saline, there is an upward movement of salts into the rooting zone in dry weather.
- Where sulphidic material is close to the surface, drains cannot be deepened enough to permit leaching and ripening to produce a firm topsoil without exposing an acid layer. This is usually a problem in the lowest parts of the landscape. Some acidification of these areas may have to be accepted or combatted by spot liming.

1. Raised Beds

All these difficulties are exacerbated in tropical regions with high rates of evaporation and long droughts, where a greater thickness of non-acid topsoil is needed to maintain a crop. The farmers' response has usually been to build raised beds or mounds of topsoil, especially where watertable control is no more than rudimentary. Yields are never as good as on normal soils under comparable circumstances. For example, in the Mekong Delta, Sen et al. (1987) report average sugarcane yields on raised beds with shallow drainage of only 36.4 t ha^{-1}, still much better than an average of 13.6 t ha^{-1} on acid sulphate soils without effective watertable control.

Ideally, raised beds should be built entirely of non-acid topsoil excavated from shallow ditches. Where the raw acid or sulphidic layer is too close to the surface for this, then the centre of the raised bed (Fig. 8) should be built of saved topsoil, which may be insulated from rising acidity by a layer of sedge cut from the site prior to reclamation.

2. Oil Palm

Acid sulphate soils impair root development in oil palm, (*Eleis guineensis*) causing severe water stress and nutrient deficiencies, but spectacular responses to a controlled high watertable have been obtained (Bloomfield et al., 1968; Poon, 1974; Bloomfield and Powlson, 1977). Fig. 9 compares the

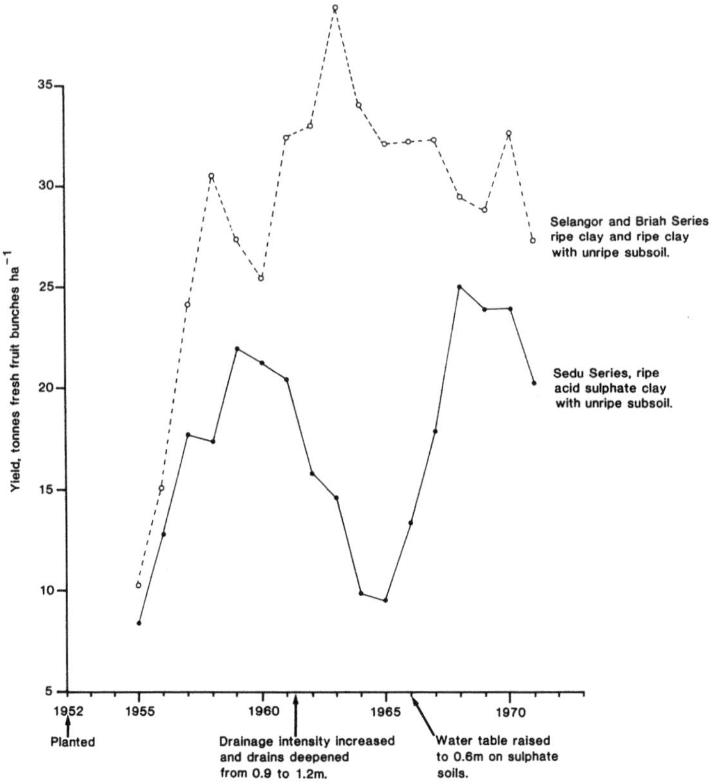

Figure 9. The Effects of watertable control on yields of oil palm, Sungei Sedu Estate, Selangor, Malaysia. The data are recalculated from Poon (1974) and represent estate management blocks in which significant areas of non-acid soils are included with the acid Sedu Series

yields obtained on acid and non-acid soils in Selangor, Malaysia. Deepening of the drains from 0.9 m to 1.2 m reduced yield. The recovery of yield following the raising of the watertable to 0.6 m is remarkable, although yield remained below that achieved on non-acid soils. However, Toh and Poon (1982) report that yields of oil palm established more recently on acid sulphate soils, with a high watertable maintained from first planting, have been similar to those on normal soils.

Estate practice is to make drains big enough to deal with normal periods of high rainfall but to maintain the water level at 0.6 m by weirs of used fertilizer bags packed with soil, or by wooden sluices. Even with the weirs in place, the watertable will fall during dry periods and the accumulated acid must be flushed out during the next wet period by opening the sluices before the watertable is allowed to build up again to the proper level (Fig. 9).

3. Coconut

Banjarese farmers in Southeast Kalimantan plant coconut (*Cocos nucifera*) on mounds in an 8 m × 8 m grid. Over several years, as the water demands of the crop increase, the mounds are built up into raised beds (Dreissen and Sudjadi, 1984). In the Baginese system, the watertable in the crop is gradually lowered by periodic deepening of the drainage ditches (van den Eelaart, 1982).

4. Citrus

Many citrus orchards have been developed on the most acid soils in the Southern Bangkok Plain of Thailand by building mounds gradually around the trees using material excavated from the ditches. This acid material is limed, however, and trees are irrigated by overhead spray from boats moving along the ditches. Periodically, the ditch water is drained to remove acid (J.F. Osborne personal communication).

5. Grassland

Grassland is grown successfully with a controlled high watertable on acid sulphate soils in temperate regions where a minimum of 0.2 m of topsoil of $pH > 4.5$ can be maintained. The first priorities are protection from flooding, and drainage of the topsoil. The watertable is maintained above the potentially acid layer by sluices in the drains, and outflow of surplus water to tidal rivers is through floodgates opening at low tide or by pumping (Pons, 1956; Dent, 1986).

Where extreme acidity can be avoided, grassland responds well to modest applications of phosphate and acid-tolerant strains of trefoil *Lotus pedunculatus* and clover *Trifolium repens* can be established. Useful grasses tolerant of moderately acid, slightly saline conditions include rye grass *Lolium perenne*, Harding grass *Phalaris tuberosa*, Bermuda grass *Cynodon dactylon*, and wheat grass *Agropyron elongatum*.

6. Forestry

Forestry remains an option for both potential acid sulphate soils under a tidal regime, and severely acid and saline acid soils where watertable control is not feasible or freshwater is inadequate for other crops. In central Thailand, with a 4- to 5-month dry season, *Casuarina junghuiana* is grown on ripe acid sulphate clays, planted on ridges of topsoil with the intervening furrows providing drainage and a degree of water storage. Production having exceeded local demand, other tree crops are being tried, including *Eucalyptus* spp.

Melaleuca spp. grow extensively on acid sulphate soils in Southeast Asia and North Australia. In the Mekong Delta, where 6 months flooding up to 2 m alternates with 6 months drought, about 10000 ha of *Melaleuca leucadendron* forest have been established over the past 10 years, both by broadcasting seeds onto flooded land and by transplanting seedlings (Brinkman and Xuan, 1991).

In Senegal, with a more severe and longer dry season of 9 months, Sadio (1989) reports trials with a range of forestry species on raw saline acid sulphate soils and saline soils. On severely acid soils, the most resilient species are *Melaleuca acacoides*, *M. viridiflora*, and *M. leucodendron*. *Eucalyptus camaldulensis* and *E. microtheca* also survived well and yielded significantly more than *Melaleuca*.

VI. Reclamation by Leaching and Liming

Liming was the first proven means of reclaiming acid sulphate soils (van Kerkhoff, 1856, Zuur, 1936; discussed by Pons, 1973). By neutralizing acidity, it immobilizes toxic Al^{3+} and Fe^{2+}, increases the availability of P and most other nutrients, and increases the rate of mineralization of N from organic matter.

There is no need to bring the pH to neutrality; raising the pH of the topsoil above 5 is enough to avoid most acid sulphate problems. But lime requirements vary enormously according to the reserves of acidity in the desired rooting depth (Table 5). Even after meeting the lime requirement of the topsoil, further applications of lime will be needed periodically to counter acidity that rises from the subsoil in dry seasons—depending on the rate of generation of acid in the subsoil and the upward flux into the rooting zone.

Early success in reclaiming inland polders in The Netherlands was achieved because their relatively thin acid layer was underlain by marl that could be brought up by deep ploughing. Similarly successful reclamation of ripe acid sulphate soils of the Bangkok Plain, Thailand, has been achieved because a long period of weathering and leaching has removed pyrite and sometimes also jarosite from the upper 1 m of the soil profile, and because marl can be quarried locally.

A. Rice

1. Ripe Acid Sulphate and Acid Aluminum Soils

Charoenchamratcheep et al. (1982), Maneewon et al. (1982), and Osborne (1985, 1986) report extensive experimental work on acid sulphate soils in Thailand where economic responses to application of lime have been obtained where the lime requirement to pH 5 is less than about 6.5 t ha^{-1}. Further applications are needed every 4 or 5 years to counter the rise of acidity from the subsoil. Table 6 shows that the residual effects on pH are slight after 5 years, even at a rate of 15 t ha^{-1}, but the depression of extractable Al levels is much more significant.

Table 5. Lime requirements for complete neutralization, in relation to total oxidizable sulphur content (from Dent, 1986)

Apparent density of the soil (g cm^{-3})	Lime requirement of a layer 10 cm thick (tonnes $CaCO_3$/ha)						Effective neutralizing capacity of a 10-cm layer containing no $CaCO_3$ (tonnes $CaCO_3$ equivalent/ha)[a]	
	Percentage oxidizable sulphur							
	0.5	1	1.5	2	3	4	Clayey soils	Sandy soils
0.6	9	19	28	37	56	74	11	2
0.7	11	22	33	44	65	87	13	2
0.8	12	25	37	50	74	99	14	2
0.9	14	28	42	56	84	112	16	3
1.0	16	31	47	62			19	3
1.1	17	34	51	68			20	3
1.2	19	37	56	74			22	4

[a] A neutralizing capacity of 18 mEq per 100 g is assumed for clayey soils, and 3 mEq per 100 g for sandy soils. These are estimates based on tidal soils in Northland, New Zealand. Between half and two thirds of this cation exchange capacity is available to neutralise acidity. Clay soils derived from strongly weathered rocks in tropical regions probably have lower neutralizing capacities.

Table 6. Residual effect of marl on Rangsit Series, very acid phase (ripe acid sulphate clay) at Ongkharak, Thailand (from Osborne, 1985)

t ha^{-1} $CaCO_3$ equivalent applied 1975	Nov. 1974	pH 0–20 cm mean of 12 monthly samples, 4 each plot 1980)	KCl-extractable Al (mEq 100 g^{-1}, 1980)
0	4.3	4.2	8.1
2.7	4.6	4.2	5.9
5.3	4.9	4.2	5.2
15.1	6.1	4.5	2.6

Table 7. Maximum response of rice yield to lime for 28 sites on the Bangkok Plain, 1985, by acidity class

Acidity class	Actual mean yield (t rice ha^{-1})	Response modelled by regression (t rice ha^{-1})
I	0.02	−0.04
II	0.48	0.56
III	0.92	0.88
IV	1.68	1.73

In 1984, the Department of Land Development, Thailand, began a series of liming experiments on farmers' fields over a wide area of Central Thailand. Results for 27 sites in the first year are reported by Osborne (1985). They are not easy to interpret. However, review of the 1985 data (Osborne, 1986) showed that exclusion of some non-acid soils with, so far, unexplained high responses to liming revealed a correlation between rice yield response and the acidity classes defined in Table 2.

Using the parameters defining acidity classes, multiple linear regression was used to model yield response to lime for 28 sites. The comparison between actual and predicted responses is given in Table 7.

At the end of 1988, the 5-year cycle of cropping following lime application had been completed for 24 sites, but studies are still in progress in more than 50 others. The provisional conclusion is that liming will not be profitable where the topsoil base saturation is already more than 50%. The main exception to this is the group of "terrace soils"—colluvial sandy clay on acid clay subsoil. Figures 10 and 11 show the pattern of rice yield response and current profitability. Mean maximum yields on ripe acid sulphate clays reach 3 t ha^{-1} for a lime application of 8.5 t ha^{-1}, and on ripe acid aluminum clays 3.5 t ha^{-1} for a lime application of 6.8 t ha^{-1}; but optimum lime applications are less, respectively 5.5 and 4.25 t ha^{-1} every 5 years.

Figure 12 depicts typical yield responses from soils of acidity class III and IV.

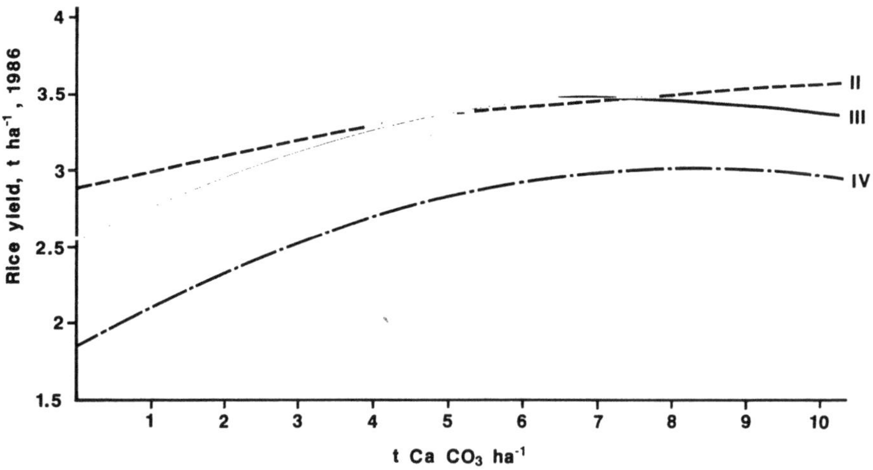

Figure 10. Mean rice yields at different rates of lime application, Bangkok Plain, Thailand 1986, following liming in 1984. (From J.F. Osborne, personal communication)

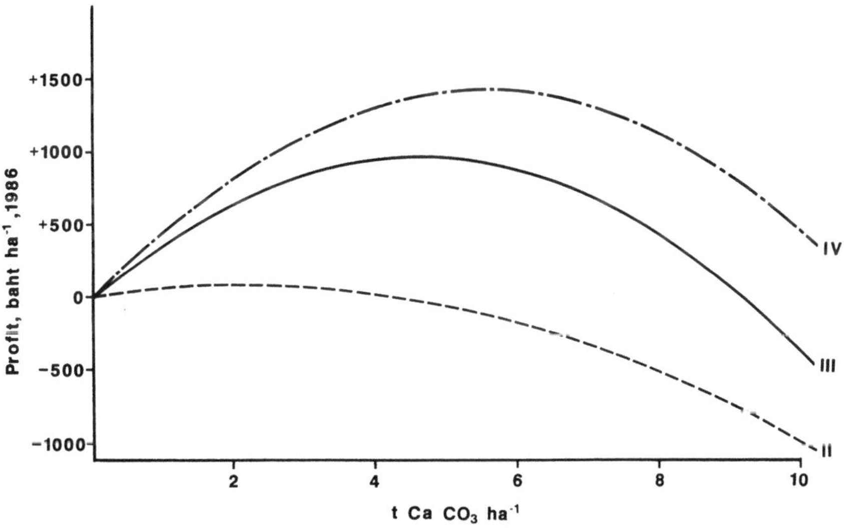

Figure 11. Profit patterns for acidity classes, Bangkok Plain, Thailand. (From J.F. Osborne, personal communication)

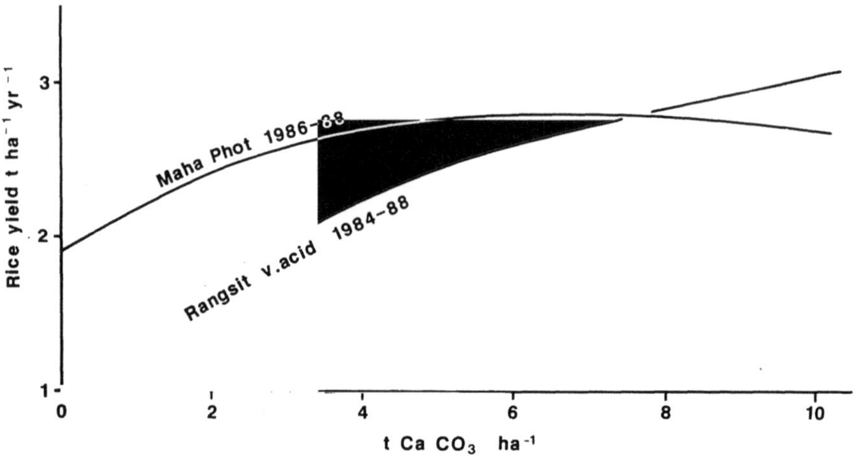

Figure 12. Quadratic regression of rice yield for different rates of lime application, Maha Phot Series ripe acid aluminum clay (acidity class III) 1986–1988 and Rangsit Series, very acid phase, ripe acid sulphate clay (acidity class IV) 1984–1988. (From J.F. Osborne, personal communication)

Most studies have demonstrated that, for rice, the combined effect on yields of liming and fertilization is much greater than lime or fertilizer alone and that applications much lower than the full lime requirement are effective (Maneewon et al., 1982; Williams, 1980). Because of the variability of soil, water regime, and management, it is difficult to draw general conclusions about fertilizer response, but P fixation is a major problem. Positive responses to P fertilizer are frequently reported, whatever the source of P. The fixation problem may be lessened by preflooding before application, to allow the peak values of soluble Al and Fe to subside and reduce the time available for fixation before transplanting the crop.

2. Raw Acid Sulphate Soils

In contrast to the proven effectiveness of liming on acid aluminum soils and ripe acid sulphate soils, the effectiveness of liming raw acid sulphate soils is doubtful. The lime requirement of the topsoil is very much higher than that of ripe soils—even hundreds of tons lime per hectare—and more acidity moves upwards continually from oxidation of sulphidic material in the subsoil. In these soils, careful management of the watertable to minimize further acidification must be the first priority.

Some field experiments have shown that liming in concert with NPK fertilization does improve yields, even at a low rate of application (0.5 t/ha, Hoa et al., 1986) but whether the treatment is economic is another matter. Figure 13 shows yield response to liming with 50:60:30 NPK application and dry season irrigation on a raw acid sulphate clay in Hau Giang, Mekong delta. Yields given are an average of a rainfed crop and the fol-

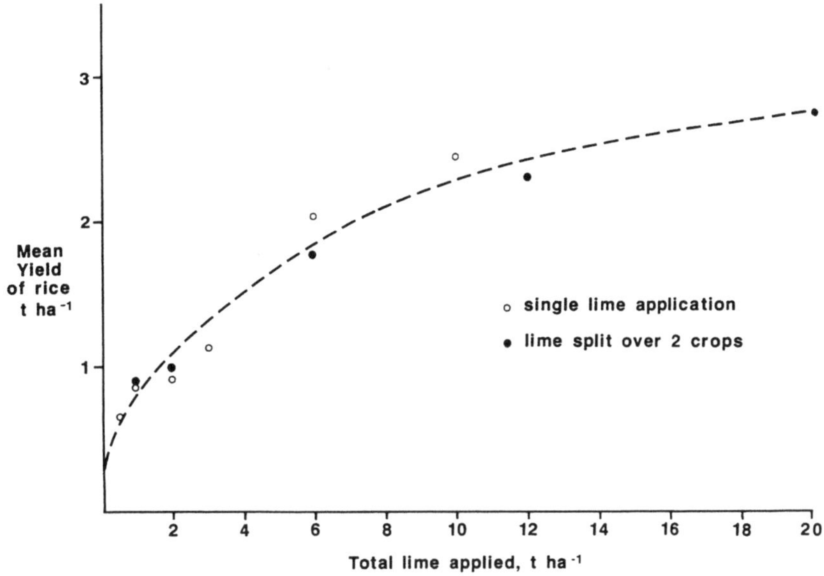

Figure 13. Response of rice yield to liming on a raw acid sulphate clay in Hoa Au, Vietnam. (Hoa and Brinkman, 1985)

lowing irrigated crop. Initial pH was 3.5 to 3.7; $EC_{1:5} < 1$ mS cm^{-1} throughout. Response was 1.1 t rice per t lime for the lowest rates of application. Subsequent investigation of Ca nutrition at this site has not revealed the reason for the good response to a low dressing of lime.

Brinkman (1982) has suggested that very small annual applications of lime (for example 100 or 200 kg ha^{-1}) applied at the preflooding stage could promote soil reduction by localized improvement of conditions for bacterial action. From such nuclei, reduction and consequent decrease in acidity, aluminum, and iron toxicity could proceed more quickly. However, this technique has not yet been tested rigorously.

3. Manganese Dioxide Application

Wen and Ponnamperuma (1966), Islam and Shah (1968), and Ponnamperuma and Solivas (1982) have advocated the application of manganese dioxide at a rate of about 100 kg ha^{-1} to depress iron toxicity but, again, this has not been evaluated adequately in field trials in iron-toxic soils.

B. Dryland Crops

Rice benefits from the reduction of acidity by flooding but for other crops there is no obvious way of combatting acidity other than by meeting the lime requirement of raising the pH of the rooting zone to at least 5.

Table 8. Yield response of dryland crops to lime application, Stauning, Denmark (from Larsen and Andersen, 1977)

Lime applied (t $CaCO_3$ ha^{-1})		pH		Yield (t ha^{-1})			
Topsoil	Subsoil	Topsoil	Subsoil	Barley	Oats	Grass	Sugar Beet
0	0	3.8	3.4	1.3	1.9	3.4	5.3
60	0	7.1	3.7	2.4	5.8	4.3	6.2
15	45	6.6	5.7	3.4	5.3	8.9	11.6

Field experiments in Denmark (Larsen and Andersen, 1977) have shown that much higher yields can be obtained by liming both the topsoil and the subsoil compared with liming the topsoil alone, especially for crops with a long growing season and, consequently, greater water requirements (Table 8).

But lime applied to the surface mostly stays there. Incorporating it deeper than plough depth is difficult and expensive. In the first place, suitable implements are not generally available; in the second place, the energy costs of deep cultivation are very high. Evans (1966) in Guyana, described "vertical mulching" for sugarcane—digging holes through the toxic soil layer and filling them with lime-rich filter mud. In the English Fenland, Smith et al. (1971) report successful operations with a single-blade subsoil raiser/mixer and a double-digging plough in which forward bodies remove the topsoil and following bodies work deeper. Andersen and Hendrick (1983) describe a chisel/slitter that can inject lime suspension along a slit in the subsoil. This needs much less energy than soil mixing.

C. Leaching

It is obvious that application of enough lime to neutralize all the potential acidity of most sulphidic and raw acid sulphate soils (Table 5) will usually be impracticable and hugely expensive. But it may be appropriate for local treatment of acidity, for example, on mine spoil (Kohnke, 1950; Barnhisel et al., 1982), to prevent pollution of water supplies; or for very profitable land uses such as fish ponds.

Many schemes have been undertaken to reduce the lime requirement of acid sulphate soils by deep drainage to oxidize the pyrite and to promote leaching of the ensuing acid. Most have proved disastrous (see, inter alia, Beye, 1973a; Marius, 1982; Loyer et al., 1988 in Senegal; Bloomfield et al., 1968; Bloomfield and Powlson, 1977 in Malaysia). Such proposals were flawed by extrapolation of results of laboratory oxidation and leaching experiments (e.g., Kivinen, 1950; Hart, 1963; Kanapathy, 1973) to quite different conditions in the field. Because of the overwhelming influences of hydrology on the behaviour of acid sulphate soils, and the relationships between the topsoil and the acid or sulphidic subsoil, pot experiments are

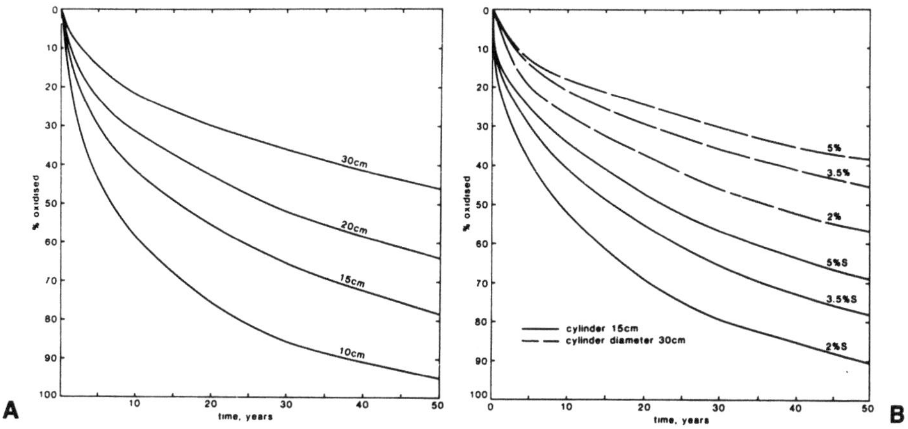

Figure 14. Rate of oxidation of pyrite in soils of different pyrite content and ped size. (From Dent and Raiswell, 1982)

of limited value in predicting field conditions or designing practical methods of reclamation.

A different approach was adopted by Dent and Raiswell (1982) who modelled pyrite oxidation on the assumption that the rate-limiting process is diffusion of oxygen through water-filled pores. Recognizing also that ripening soil fissures into coarse, prismatic peds, they modelled oxygen diffusion into cylinders of appropriate radius. Figure 14 shows the calculated rates of oxidation of pyrite in drained soils according to the initial content of pyrite. For example, an initial oxidizable sulphur content of 3.5% will be reduced by about half over 50 years. This is in line with a crude estimate of removal of oxidizable sulphur from polders in Northland, New Zealand, where weathering and leaching for 40 years has reduced the average lime requirement of the top 60 cm from 172 t ha^{-1} to 80 t ha^{-1} (Dent, 1986). These principles are now being extended and tested for land reclamation in Indonesia by Bronswijk and Ritsema (1987).

In the New Zealand polders, where drought is rarely prolonged, good grassland has been established by a moderate lime application (no more than 8 t ha^{-1}) after 25 years of weathering and leaching. Thereafter, it can be maintained by regular topdressing of superphosphate and maintenance liming.

In Denmark, Larsen and Andersen (1977) accelerated oxidation of pyrite by drainage and deep soil mixing to 0.7 m. Figures 15 and 16 show the results of their field experiments. With deep mixing, half of the pyrite was oxidized in the first year of the trial, although 7 years were needed to leach the resulting sulphate. Without mixing, pyrite oxidation and sulphate leaching proceeded at a steady rate throughout the trial.

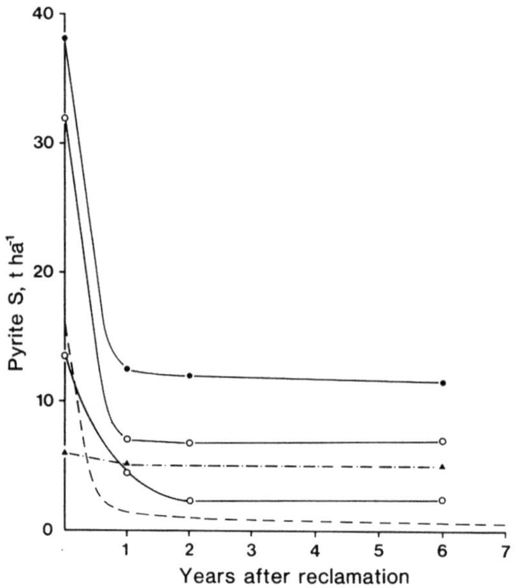

Figure 15. Rate of oxidation of pyrite in field experiments on deep soil mixing, Stauning, Denmark. (From Larsen and Anderson, 1977)

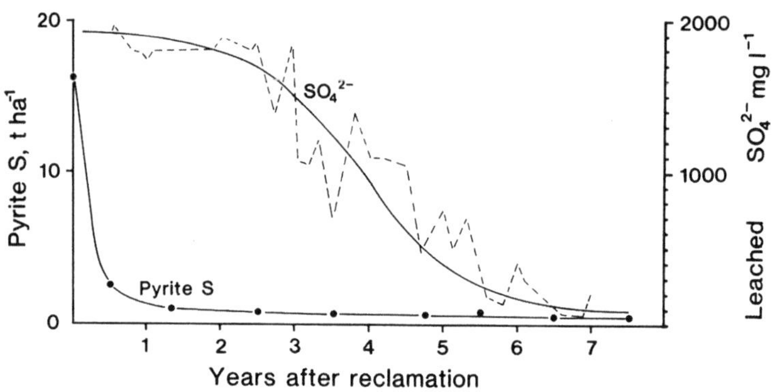

Figure 16. Rate of removal of pyrite and sulphates from the upper 0.7 m of a clay loam soil, Stauning, Denmark. Drainage 400 mm/year. (After Larsen and Andersen, 1977)

Table 9. Effects of a sequence of saltwater and freshwater leaching on acid sulphate soils from the Mekong Delta (from Ni et al 1988)

	Initial values		Final values, total		
	Total	Exchangeable	Saltwater	$\frac{1}{8}$ salt	Fresh
Bin Son, raw acid sulphate muck					
pH	3.0	—	3.7	3.3	3.2
EC_e	3.2	—	0.7	0.4	0.5
Al	13.4	11.4	5.8	8.0	12.1
Na	10.0	7.5	5.8	4.8	4.8
Ca	7.5	6.3	3.6	4.1	3.0
Mg	5.8	4.4	4.8	3.8	3.8
K	1.4	0.4	1.4	0.5	0.5
Doc Lieu, raw saline acid sulphate clay					
pH	3.2	—	3.8	3.6	3.6
EC_e	4.6	—	1.3	0.7	0.5
Al	11.5	9.1	3.8	6.3	7.1
Na	28.9	9.1	10.7	6.7	4.6
Ca	8.9	6.8	6.1	5.6	6.5
Mg	4.9	3.1	7.9	7.3	6.1
K	2.2	1.0	1.9	1.2	1.0

pH 1:1 water; EC_e mS cm^{-1}; cations mEq 100 g^{-1}, Al in *M* KCl, others in *M* $BaCl_2$

1. Seawater Leaching

Evans and Cate (1962) in Guyana and Hart et al. (1963, 1965) in Sierra Leone reported that leaching by seawater is much more effective than with freshwater. Seawater both neutralizes acidity and displaces part of the exchangeable aluminum from acid sulphate soils. Subsequent leaching of salt by freshwater does not seem to be hindered by exchangeable sodium, probably because of the accompanying exchangeable aluminum maintains a stable soil structure.

A sequence of leaching with saltwater followed by freshwater was monitored by Ni et al. (1988) in a pot experiment using topsoil samples from the Mekong Delta. Samples were leached with saltwater EC 37 mS cm^{-1} over 8 days so that the Na^+ added was equivalent to twice the extractable Al^{3+} content of the soil sample. Dilutions of $\frac{1}{2}$, $\frac{1}{4}$, and $\frac{1}{8}$ of saltwater, and freshwater alone were also used. Following saltwater leaching, freshwater was applied till the EC of the drainage water was <4 mS cm^{-1}. In all, about 75 mm of saltwater and 200 mm of freshwater were used for each sample.

Table 9 summarizes the results of the saline, $\frac{1}{8}$ dilution and freshwater treatments. It is clear that saltwater leaching is very effective in exchanging Al. The higher the salt concentration, the more Al is removed. However, prolonged flooding by seawater leads to precipitation of some of the exchanged Al, perhaps as a basic sulphate $Al(OH)SO_4$, rather than complete leaching.

In the interpretation of these results for practical land reclamation, the caveat about extrapolating the results of pot experiments to the field situation applies. In the field, the situation will be complicated by (1) continual generation of acidity by further oxidation of pyrite reserves, (2) by the spotty distribution of pyrite, and (3) by less effective drainage and less effective leaching. The latter point is emphasized by comparing the results of a similar pot experiment (Minh, 1986), equally effective in exchanging Al but where 500 to 600 mm of water was needed to reduce the EC of the drainage water to <4 mS mm^{-1}. Preliminary results from a field leaching experiment in Vietnam do indeed suggest that field leaching is less effective than pot leaching (Hanhart et al., 1989).

In the pot experiment of Ni et al. (1988), rice seedlings transplanted after completion of the leaching sequence showed symptoms of severe Al toxicity/P deficiency after 15 days of good growth. It is clear that more than one season of oxidation and leaching will be needed to reclaim soils of high pyrite content for dryland crops. Further field scale work would be instructive.

Not surprisingly, there is little indigenous farmers' experience of saltwater leaching. Most farmers in acid sulphate soils believe that salinity is the more severe problem. However, farmers in Min Hai and Tien Giang provinces of the Mekong Delta flood their land with saltwater just prior to the start of the rainy season, wait for the rain to dilute the salt, then grow their rice; and rice farmers in Pulau Petak, Indonesia, in Tien Giang, Vietnam and in Thailand sometimes apply 50 to 100 kg ha^{-1} of common salt to increase crop yields.

Saltwater leaching has been used successfully to reclaim acid sulphate fish ponds and a rice-shrimp cropping system is becoming increasingly popular among farmers in the Mekong Delta with direct access to tidewater. Rainfed rice is planted during the wet season, then shrimps are raised in the fields during the dry season by tidal flooding with brackish water. In addition to the valuable shrimp crop, the practice should reduce oxidation of pyrite during the dry season and replace exchangeable aluminum to the benefit of the wet season rice crop.

D. Reclamation for Fish Farming

Brackish water fish ponds are a very profitable land use in Southeast Asia, particularly for milkfish *Chanos chanos* and shrimps *Microbrachium* or *Penaeus monodon* (Fig. 3). But in acid sulphate soils, ponds develop acidity during excavation and during their periodic drainage and drying. This causes poor fish yield, poor response to fertilizers, especially to phosphate, and sometimes fish death when heavy rains follow a long dry period (Potter, 1976).

By thinking positively about the dry season, Brinkman and Singh (1982) developed a quick method of reclamation that has been widely adopted in the Philippines. The source of acidity is removed from the pond bottom by a repeated sequence (four or five times) of drying, tilling, flooding with brackish water, and draining. The larger bunds are also leached and a little lime (<0.5 t ha^{-1}) applied to the pond bottom to counter the rise of acidity from below.

In a representative case (Singh, 1985; Singh et al., 1988), reclaimed pond soil showed a rise in dry pH from 3.7 to 4.8, and to 5.7 after the first harvest. There was a tenfold decrease in exchangeable Al. Water quality improved with a rise of pH from 3.9 to 6.5, and decreases in soluble Al from 151 to 6.7 mmol m^{-3}, Fe from 167 to 24 mmol m^{-3}, and sulphate from 30 to 12 mol m^{-3} during reclamation. There were further significant improvements in water quality during the growing season. Most importantly, milkfish yield increased fivefold, from 112 to 530 kg ha^{-1}, and in subsequent years shrimps could be introduced successfully.

The high value of the crop outweighs the high cost of reclamation, yielding a benefit:cost ratio of 1.35 in the first year of operation.

E. Reclamation of Acid Mineral Workings

Mine tailings from the working of sulphidic ores are extremely acid and pollute surface waters with acid, aluminum, iron, and heavy metals. Coal mine spoil is also pyritic and pH values <3 commonly occur. Sometimes, the extraction of sand and gravel exposes pyritic material, leading to unacceptable levels of acidity in surface waters.

Conventionally, acid mine spoil has been reclaimed by liming. Determination of lime requirement must take account of reserves of acidity in the form of pyrite and jarosite (Barnhisel et al., 1982). However, the uneven distribution of pyrite within the spoil makes reliable control difficult without gross over-liming of the whole deposit. Singh et al. (1982) emphasize the advantage of covering pyritic materials with preweathered overburden. Pulford et al. (1988) have investigated a range of ways to inhibit the oxidation of pyrite in the spoil, either by preventing the conversion of Fe II to Fe III or preventing the oxidation of pyrite by Fe^{3+} ions. Addition of pulverized fuel ash from coal-burning power stations, chicken manure, and wood waste were all found to inhibit acid generation. Fuel ash probably combines with iron as iron III silicate; the organic amendments complex with iron, rendering it unavailable for oxidation of pyrite (researchers at the Macaulay Institute (Vaughan et al., 1984) used the same principle to keep field drainage systems from blocking with ochre deposits by emplacing sacks of pine bark).

Liming is also the conventional treatment for acidity in lakes, and in reservoirs built to contain acid drainage waters (Prokopovich, 1988). Continual inputs of lime are needed where there is a continuous input of acidity. An inexpensive alternative may be to reverse the process of acidification by promoting sulphate reduction on the lake bottom. Hunt (1987) describes the reclamation of an acid sulphate lagoon in pyritic sand workings by the addition of enough lime to raise the pH to a level where bacterial reduction can occur, and sewage sludge, as a source of organic matter. Sulphate reduction on the bed of the lagoon has maintained the pH of the water close to 7 despite continuous acid input, and the growth of algae in the lagoon may supply more decomposable organic matter to maintain the reduction continuously.

In addition to experiencing acidity and toxicity, vegetation on mine spoil commonly suffers a variety of nutrient deficiencies. As well as the usual acid sulphate problem of P fixation, Barnhisel et al. (1982) suggest that K, including that added as fertilizer, may be immobilized as jarosite. Shallow rooting, even on reclaimed soil, limits the volume of soil available to vegetation, so drought is a further hazard. Deep incorporation of amendments is difficult enough on ordinary acid sulphate soils. On coarse, stony mine spoil, very robust equipment is needed.

VII. Summary and Conclusions

Acid sulphate soils are not all the same. Before attempting reclamation, we need a detailed field survey and at least some critical laboratory analyses to know what we are dealing with. The first distinction to be made is obviously between acid sulphate soils and potential acid sulphate soils, but there are also big differences in the reserves of acidity within acid sulphate soils. Unfortunately, there is nearly always significant short-range variation that is difficult to map, especially if natural vegetation and drainage have been modified.

Probably, the worst raw and potential acid sulphate soils will be better not reclaimed; returns will be delayed, the cost of amelioration, of pollution, and degradation of the wetland environment will be high, as will be the losses of valuable existing or potential production in their wetland state. In their natural state under mangrove, swamp forest, or marsh, coastal wetlands are valuable as nurseries and feeding grounds of coastal fisheries, as a source of timber and local products, and as a coastal defence.

Concern over rising sea levels focusses attention on mangroves and coastal marshes which absorb the force of wind, wave and tides, protecting what are often densely settled coastal lowlands. Unlike manmade sea defences, they have the ability to grow upwards with the rising sea level, providing long-term protection of neighbouring areas. Strategies for responding to rising sea level might include not only conservation of the mangrove fringe but "dereclamation": deliberate sacrifice of areas of acid sulphate soils of low productivity and reestablishment of mangrove or marsh vegetation that can resume the land-building process.

Where land reclamation is judged to be appropriate, water management is the key to soil management. This is the over-riding principle of reclamation of acid sulphate soils. Therefore, development schemes should take account of the whole land and water system, not just locally but regionally—we are dealing with differences in elevation of less than 5 m maybe over distances of more than 100 km. Two examples of regional problems must suffice:

1. In the Mekong Delta there is increasing drainage and irrigation of raw acid sulphate soils inland, on the Plain of Reeds. This is depriving better, non-acid soils downstream of fresh water in the dry season, as well as damaging crops and soils in surrounding areas with acid floodwaters (van Mensvoort, 1987).
2. Forested peat domes overlie potential acid sulphate soils on many coastal plains in the humid tropics, for example, Guyana, southern Kalimantan, Sumatra, the southern Mekong Delta. They act as relatively high level reservoirs of fresh water, maintaining watertables in the surrounding areas, and providing a steady supply of water during dry spells. At present they are being destroyed by burning and encroachment for farming. There is a limit beyond which further destruction will render the surrounding reclamations untenable—through loss of dry season water supply and increased soil acidification through falling watertables (Pons, 1988).

By trial-and-error over several generations, farmers in many parts of the world have developed ways of raising crops on acid sulphate soils. Rice is the main crop because it can make use of the reduction of acid conditions by flooding, but dryland crops grown with some success include cassava, sweet potato, yams, pineapple, sugarcane, and coconut; grassland in temperate climates; and estate production of oil palm and rubber.

Farmers' methods depend entirely on a minimum-disturbance strategy,

1. Limiting the generation of acidity by:
   - Avoiding disturbance of the subsoil
   - Maintaining the highest possible watertable during dry periods
   - Preflooding, to allow reduction of acidity before planting the crop
   - Double cropping of rice (or even rice-shrimps as in Vietnam where freshwater is not available for a second rice crop), which shortens the period of soil drying
   - Mulching, which reduces evaporation in dry periods and encourages rapid reduction under flooded conditions
2. The use of large, transplanted seedlings rather than dry broadcasting
3. Leaching of acidity and salts without deep drainage by:
   - Frequent flushing of the surface with good quality water
   - Intensive shallow drainage, either by broad, shallow ditches between broad raised beds where flooding with good water can be practised,

as in the Mekong Delta, or by ridges and furrows where only rainwater is available, as in much of West Africa

4. Dryland crops are grown on mounds, ridges, or raised beds of topsoil, without deep drainage. Raised beds should be built only with ripe, brown mottled material, not grey sulphidic subsoil

Yields are never very wonderful under the traditional systems although worthwhile responses to N and P fertilizer, varietal selection, and improved weed and pest control have been reported. The risks of acidification, salinity, and deep flooding remain because small farmers cannot achieve more than local control of the water regime. Both floods and long dry periods commonly lead to the loss of a complete crop.

Many attempts to achieve better water control in large development schemes have failed disastrously because the soils were deeply drained in an attempt to eliminate salinity and to remove the source of acidity by oxidation and leaching. This must not be repeated. If total reclamation is attempted, it should be undertaken only if all the pyrite can be oxidized and the resulting acidity leached in a very few years. Simulation models are being developed to predict the effects of different water management strategies over long time periods.

A wider understanding of the variability of acid sulphate soils and of the processes operating in them reveals scope for better reclamation and management, by optimal combinations of techniques matched to the particular soil, hydrological and climatic conditions. Some of the farmers' constraints can be overcome by: regional control of flooding; better control of watertables, by systematic empoldering based on special-purpose soil, topographic, and hydrological survey—the target areas can be broken up into series of more homogeneous units in which the hydrology can be independently controlled; reclamation with brackish water followed by freshwater; careful monitoring of salinity, which may permit the use of brackish water to maintain water levels when fresh water is not available; provision of irrigation water; safe disposal of acid floodwaters, keeping irrigation and drainage waters separate; use of crop varieties that tolerate the prevailing constraints, and adaptation of the cropping calendar to seasonal availability of water; provision of physical infrastructure, inputs like agrochemicals and improved varieties, and technical support.

The benefits of liming to neutralize acidity have been demonstrated in acid aluminum soils and ripe acid sulphate soils where the lime requirement is only moderate (<8 t ha^{-1}). In raw acid sulphate soils, liming is impractical on grounds of cost and the huge amounts needed. Two amendments of potential value in reclaiming raw acid sulphate soils that have yet to be demonstrated under suitable field conditions include the annual application of homeopathic amounts of lime, not for neutralization but to initiate rapid, healthy reduction of acidity during preflooding of the rice crop; and application of manganese dioxide to avoid iron and H_2S toxicity during reduction.

Much more thought needs to be given to the integrated use of land resources, especially in coastal acid sulphate soils where a good method has been developed for reclamation of brackish water fish ponds, where tidewater is available to assist reclamation for other purposes, and where the potential value of the natural ecosystem is the greatest.

Acknowledgments

I am indebted to Mr J.F. Osborne and Mr M.E.F. van Mensvoort, who supplied unpublished material and project reports from Thailand and Vietnam, respectively. I should also like to thank Mrs F. Randell for patient and skillful work in processing the text, and Mr P. Judge and Ms S. Davies for the diagrams.

References

Agyen-Sampong, M., K. Prakah-Asante, and S.N. Fomba. 1988. Rice improvement in the mangrove swamps of West Africa. In Dost, pp. 163–175.

Andersen, D.L. and J.G. Hendrick. 1983. Subsoil lime injector. *Soil Sci. Soc. Am. J.* 47:337–339.

Balasuriya, I. 1987. Reclamation and development of rice cultivation on coastal low-lying lands of southern and western Sri Lanka. Tropical Agriculture Research Series 20. Tropical Agriculture Research Center, Min. Ag. For. Fish. Japan, pp. 213–223.

Barnhisel, R.J., J.L. Powell, G.W. Akin, and M.W. Ebelhar. 1982. Characteristics and reclamation of acid sulphate mine spoils. In Kittrick et al., pp. 225–234.

Beers, W.F.J. van 1962 Acid sulphate soils. Bull. 3. ILRI, Wageningen.

Bemmelen, J.M. van. 1863. *Bouwstoffen tot de kennis van de kleigronden der Provincie Groningen.* Rotterdam.

Bemmelen, J.M. van. 1886 Bijdrage tot de kennis van den alluvialboden in Nederland. *Verhanderlingen Academie Wentenschappen, Amsterdam* 25:33.

Berner, R.A. 1970. Sedimentary pyrite formation. *Am. J. Sci.* 288:1–23.

Beye, G. 1973a. Acidification of mangrove soils after empoldering in lower Casamance. Effects of type of reclamation system used. In Dost, vol. 2, pp. 359–372.

Beye, G. 1973b. A simple method of de-salting the 'tanne' soils of Casamance: straw mulching. *Agronomie Tropicale* 28(5):537–549.

Bloomfield, C. and J.K. Coulter. 1973. Genesis and management of acid sulphate soils. *Adv. Agron.* 25:278–326.

Bloomfield, C. and D.S. Powlson. 1977. The improvement of acid sulphate soils for crops other than padi. *Malaysian Agric. J.* 5(1):67–76.

Bloomfield, C., J.K. Coulter, and R. Kanaris-Sotiriou. 1968. Oil palms on acid sulphate soils in Malaya. *Trop. Agric. (Trinidad)* 45:289–300.

Rreemen, N. van. 1973. Soil forming processes in acid sulphate soils. In Dost, vol. 1, pp. 223–254.

Breemen, N. van. 1976. *Genesis and Solution Chemistry of Acid Sulphate Soils in Thailand*. Pudoc, Wageningen.
Breemen, N. van. 1980. Acid sulphate soils. Problem soils: their reclamation and management. Land reclamation and Water management. ILRI pub. 27, Wageningen, pp. 53–57.
Brinkman, R. 1982. Directions of futher research on acid sulphate soils. In Dost and van Breemen, pp. 12–20.
Brinkman, R. 1985. *V.H. 10 Project Mission Report 13*. Vakgroep Bodemkunde en Geologie, Landbouwuniversiteit, Wageningen.
Brinkman, R. and V.P. Singh. 1982. Rapid reclamation of brackish-water fishponds in acid sulphate soils. In Dost and van Breemen, pp. 318–330.
Brinkman, W.J. and Vo Tong Xuan. 1991. *Meleleuca leucadendron* s.l. a useful and versatile tree for acid sulphate soils and some other environments. *International Tree Crops Journal* (in press).
Bronswijk, J.J.B. and C.J. Ritsema. 1987. *Modelling physical and chemical processes in acid sulphate soils*, ICW note 1821. Instituut voor Cultuurtechniek en Waterhuishouding, Wageningen.
Burrough, P.A., M.E.F. van Mensvoort, and J. Bos. 1988. Spah'al analysis as a reconnaissance survey technique: an example from acid sulphate soil regions of the Mekong Delta, Viet Nam. In Dost, pp. 68–89.
Cate, R.B. and A.P. Sukhai. 1964. A study of aluminum in rice soils. *Soil Science* 98, 83–93.
Charoenchamratcheep, C.P., B. Tantisira, P. Chitnuson, and V. Sin-Aiem. 1982. Effects of liming and fertilizer application on acid sulphate soils improvement for rice production. In Dost and van Breemen, pp. 137–146.
Chenery, E.M. 1954. Acid sulphate soils in Central Africa. *Trans. 5th International Congress of Soil Science*, Leopoldville.
Dent, D.L. 1986. *Acid sulphate soils: a baseline for research and development*. ILRI pub. 39, Wageningen.
Dent, D.L. and R.W. Raiswell. 1982. Quantitative models to predict the rate and severity of acid sulphate development: a case study in The Gambia. In Dost and van Breemen, pp. 73–95.
Dent, J.M. 1947. Some problems of empoldered ricefields in Sierra Leone. *Empire J. Exp. Agric.* 15:206–212.
Dost, H. (ed.). 1973. *Acid sulphate soils: proceedings of the international symposium 13–20 August 1972 Wageningen*. Vols 1 and 2, ILRI pub. 18, Wageningen.
Dost, H. (ed.) 1988. *Selected papers of the Dakar symposium on acid sulphate soils*. ILRI pub. 44, Wageningen.
Dost, H. and N. van Breemen (ed.) 1982. *Proceedings of the Bangkok symposium on acid sulphate soils*. ILRI pub. 31, Wageningen.
Driessen, P.M. and Ismangun. 1973. Pyrite-containing sediments of Southern Kalimantan, Indonesia: their soils, management and reclamation. In Dost, vol. 2, pp. 343–348.
Driessen, P.M. and M. Sudjadi. 1984. Soils and specific problems of tidal swamps. In: *Workshop on research priorities in tidal swamp rice*. IRRI, Los Baños, Philippines.
Eelaart, L.J. van den. 1982. Problems in reclaiming and managing tidal lands of Sumatra and Kalimantan. In Dost and van Breemen, pp. 212–290.
Evans, H. 1966. Toxic sulphate soils and their improvement for sugarcane cultiva-

tion. Proc. Meet. Br. W. Indies Sug. Technol. Bookers Sugar Estates, Guyana, pp. 291–300.
Evans, H. and R.B. Cate. 1962. Studies in the improvement of problem soils in British Guiana. *World Crops* 14:370–373.
Fanning, D.S. and J.E. Witty. 1989. Proposed changes in the classification of acid sulphate soils in Soil Taxonomy. *Acid Sulphate Soils Newsletter* (ISSS, Wageningen) 2:10–13.
FAD (Food and Agriculture Organisation). 1982. *Management and utilization of mangroves in Asia and the Pacific*. Environment Paper 3, FAO, Rome.
FAO/UNESCO. 1988. *Soil map of the world, revised legend*. World Soil Resources Rept 60, Rome.
Goldhaber, M.B. and I.R. Kaplan. 1982. Controls and consequences of sulfate reduction rates in recent marine sediments. In Kittrick et al., pp. 19–36.
Hanhart, K., Duong Van Ni, and L.E. Quang Minh. 1989. The use of salt water for improvement of acid sulphate soils in Binh Son 3 state farm, Kien Giang Province, Vietnam. *V.H. 10 Project Mission Report 21*. Vakgroep Bodemkunde en Geologie, Lanbouwuniversiteit, Wageningen.
Harmsen, G.W. 1954. Observations on the formation and oxidation of pyrite in soil. *Plant and Soil* 5:324–348.
Hart, M.G.R. 1959. Sulphur oxidation in tidal mangrove soils in Sierra Leone. *Plant and Soil* 11(3):217–236.
Hart, M.G.R. 1963. Observations on the source of acid in empoldered mangrove soils II. Oxidation of soil polysulphides. *Plant and Soil* 19:106–114.
Hart, M.G.R., A.J. Carpenter, and J.W.O. Jeffrey. 1961. Problems in reclaiming saline mangrove soils in Sierra Leone. *Agron. Trop.* (Paris) 18:800–802.
Hart, M.G.R., A.J. Carpenter, and J.W.O. Jeffrey 1965. Problems in reclaiming saline mangrove soils in Sierra Leone. *Sols Africains* 10(1):71–75.
Hoa, Nguyen My and R. Brinkman. 1985. Effects of liming on the yield of rice on acid sulphate soil in Hoa An, Hau Giang Province. In Brinkman, pp. 44–47.
Hoa, Nguyen My, Diep Tan Tui, Le Thanh Tung, Bui Thi Ngo, and Vo-Tong Xuan. 1986. The effect of liming on the growth and yield of rice on an acid sulphate soil at Hoa An, Hau Giang, Vietnam. In van Mensvoort et al., p. 9.
Hunt, P. 1987. Sewage to the rescue of acid lakes. *Soil and Water* (Wellington) Autumn, pp. 20–21.
Islam, A. and U. Shan. 1968. Reclamation of kosh soil. *Pakistan J. Soil Sci.* 4(2):45–53.
Kanapathy, K. 1973. Reclamation and improvement of acid sulphate soils in West Malaysia. In Dost, vol. 2, pp. 383–390.
Keltjens, W.G. and M.E.F. van Mensvoort. 1984. *V.H. 10 Project Mission Report 12*. Vakgroep Bodemkunde en Geologie, Landbouwuniversiteit, Wageningen.
Kerkhoff, van. 1856. Quoted by Pons in Dost (1973), vol. 1, p. 6.
Kevie, W. van der and B. Yenmanas. 1972. *Detailed reconnaisance survey of southern central plain area*. Report SSR-89, Soil Survey Division, Land Development Dept., Bangkok.
Kittrick, J.A., D.S. Fanning, and L.R. Hossner (ed.). 1982. *Acid sulphate weathering*. SSSA Spec. Pub. 10, Madison, Wisconsin.
Kivinen, E. 1950. Sulphate soils and their management in Finland. Fourth Int. Congr. Soil Sci., Amsterdam II, pp. 259–262.
Kohnke, H. 1950. The reclamation of coal mine spoils. *Adv. Agron.* 2:318–349.

Konsten, C.J.M., W. Andriesse, and R. Brinkman. 1988. A field laboratory method to determine total potential and actual acidity in acid sulphate soils. In Dost, pp. 106–134.

Larsen, V. and S.A. Andersen. 1977. *Afvandings af pyritholdig jord og draenvandets kvalitet.* Det Danske Hedeselskab For Sagsvirksomheden Beretning nr 17. (Cultivation of soil containing pyrite and the quality of the drainage water. Danish Land Development Service.)

Loyer, J.Y., P. Bolviin, J.Y. le Brusq, and P. Zante. 1988. Les sols du domaine fluvio-marin de Casamance (Sénégal): Evolution récente et réevaluation de contraintes majeures pour leur mise en valeur. In Dost, pp.16–23.

Maneewon, M.N., N. Thawornwong, and B. Tantisira. 1982. Study on the rates of marl for rice production in acid sulphate soils in Thailand. In Dost and van Breemen, pp. 172–186.

Marius, C. 1982. Mangrove soils of Senegal and the Gambia. In Dost and van Breemen, pp. 103–136.

Marius, C. 1985. Mangroves du Sénégal et de la Gambie. Orstom Travaux et Documents 193, Paris.

Mensvoort, M.E.F. van. 1987. *V.H. 10 Project Mission Report 19.* Bodemkunde en Geologie, Landbouwuniversiteit, Wageningen.

Mensvoort, M.E.F. van, and L.E. Quang Tri. 1989. Morphology and genesis of actual acid sulphate soils without jarosite in the Hu Tien Plain, Mekong Delta, Vietnam. In Dost, pp. 11–15.

Mensvoort, M.E.F. van, H.W. Denecke, and N. van Breemen. 1986. *V.H. 10 Project Mission Report 14.* Vakgroep Bodemkunde en Geologie, Landbouwuniversiteit, Wageningen.

Mensvoort, M.E.F. van, K. Hanhart, and D. Meidertsma. 1988. *V.H. 10 Project Mission Report 20*, Vakgroep Bodemkunde en Geologie, Landbouwuniversiteit, Wageningen.

Metson, A.J., E.J. Gibson, J.E. Cox, and D.B. Gibbs. 1977. The problem of acid sulphate soils with examples from North Auckland, New Zealand. *New Zealand J. Science* 20:371–394.

Minh, Vo Quang. 1986. The use of salt or brackish water for the improvement of acid sulphate soils In van Mensvoort et al., pp. 15–16.

Ni, Duong Van. 1984. Effect of Al and Fe in high concentrations in the irrigation water on rice performance in a good alluvial soil In Keltjens and van Mensvoort, pp. 9–10.

Ni, Duong Van, M.E.F. Mensvoort, and Nam. 1988. Salt water leaching for improvement of acid sulphate soils. In van Mensvoort et al., pp. 3–10.

Oosterbaan, R.J. 1982. Natural and social constraints to polder development in Guinea Bissau. *Polders of the world*, vol. 1, pp. 146–160. ILRI, Wageningen.

Osborne, J.F. 1985. *End of assignment report to the acid sulphate soils improvement project.* Dept. Land Development Bangkok and Overseas Development Administration, London, vols. 1, 2.

Osborne, J.F. 1986. *Evaluation of test blocks 1985.* Overseas Development Administration project report–Acid sulphate soils improvement project, Ministry of Agriculture and Cooperatives, Bangkok.

Ponnamperuma, F.N., T. Attandana, and G. Beye. 1973. Amelioration of three acid sulphate soils for lowland rice. In Dost, vol. 2, pp. 391–406.

Ponnamperuma, F.N. and J.L. Solivas. 1982. Field amelioration of an acid sulphate soil for rice with manganese dioxide and lime. In Dost and van Breemen, pp. 213–222.

Pons, L.J. 1956, Some aspects of the improvement of pyrite–containing old marine soils in The Netherlands. *Trans 6ème Int. Cong. de la Science du Sol B*, pp. 727–734.

Pons, L.J. 1973. Outline of genesis, classification and improvement of acid sulphate soils. In Dost, vol. 1, pp. 3–27.

Pons, L.J. 1988. *Soil and landscape genesis in the Pulau Petak area of southern Kalimantan, Indonesia.* Research on acid sulphate soils in the humid tropics, mission report 6 ILRI, Wageningen.

Pons, L.J. and N. van Breemen 1982. Factors influencing the formation of potential acidity in tidal swamps. In Dost and van Breemen, pp. 37–51.

Pons, L.J. and I.S. Zonneveld. 1965. Soil ripening and soil classification. ILRI pub. 13, Wageningen.

Pons, L.J., N. van Breemen and P. Driessen. 1982. Coastal sedimentary environments influencing the development of potential soil acidity. In Kittrick et al., pp. 1–18.

Pons, L.J., M.E.F. van Mensvoort, and L.E. Quang Tri. 1989. A proposal for the classification of mineral Vietnamese acid sulphate soils according to Soil Taxonomy. *Acid Sulphate Soils Newsletter* (ISSS Wageningen) 2:6–10.

Poon, Y.C. 1974. The oxidation and reduction of acid sulphate soils and its effect on the growth of oil palm in peninsular Malaysia. M. Agric. Sci. thesis, University of Malaya.

Potter, T. 1976. *The problems of fish culture associated with acid sulfate soils and methods of their improvement.* 12th annual conf. Philippines federation of fish producers. Iloilo City Memo.

Prokopovich, N. 1988. Engineering impacts of acid sulphate deposits in California. In Dost, pp. 49–58.

Pulford, I.D., C.A. Backes and H.J. Duncan. 1988. Inhibition of pyrite oxidation in coal mine waste. In Dost, pp. 59–67.

Rickard, D.T. 1973. Sedimentary iron sulphate formation. In Dost, vol. 1, pp. 28–65.

Rijniersce, K. 1983 *A simulation model for physical ripening.* Rijksdienst voor de Ijsselmeerpolders flevobericht 203, Lelystad.

Rorison, J.W. 1973. The effect of acidity on the nutrient uptake and physiology of plants. In Dost, vol. 1, pp. 223–254.

Sadio, S. 1989. Pédogenese et potentialities forestieres des sols sulfates acides sales des tannes du Sine Saloum, Senegal. Thesis, Landbouwuniversiteit Wageningen.

Sen, Le Ngoc. 1987. *Water management aspects in acid sulphate soils in the Mekong Delta, Vietnam.* Final report TSD 302 NL project. Vakgroep Bodemkunde en Geologie Publicatie 988. Landbouwuniversiteit, Wageningen.

Sen, Le Ngoc. 1988. *Influence of water management and agronomic packages on the chemical changes and growth of rice in acid sulphate soils.* Thesis, Agricultural University, Wageningen.

Sen, Le Ngoc, Tran van Hoa, and L.E. Quang Minh. 1987. Farmers' experiences in reclaiming and using acid sulphate soils for agriculture in the Mekong Delta, Vietnam. In Sen, pp. 5–30.

Singh, R.N., W.E. Grube Jr., R.M. Smith and R.F. Keefer. 1982. Relation of pyritic sandstone weathering to soil and minesoil properties. In Kittrick et al., pp. 193–208.

Singh, V.P. 1985. Management and utilization of acid sulfate soils for aquaculture. FAO, Rome.

Singh, V.P., A.T. Poernomo, and R. Brinkman. 1988. Reclamation and management of brackish water fish ponds in acid sulphate soils: the Philippine experience In Dost, pp. 214–228.

Smith, J., R. Wickens, and S.J. Richardson. 1971. Soil mixing in the Fens. *6th Ann. Rept Arthur Rickwood Experimental Husbandry Farm* (Mepal, Ely, U.K.).

Soil Survey Staff. 1987. *Keys to soil taxonomy*. SMSS Tech. Man. no. 6. Cornell Univ., Ithaca N.Y.

Temple, K.L. and W.A. Koehler. 1956. *Drainage from bituminous coal mines*. West Virginia Univ. Engineering Expt. Sta. Bull. 25.

Thomas, P. and J.A. Varley. 1982. Soil survey of tidal sulphidic soils in the tropics: a case study. In Dost and van Breemen, pp. 52–72.

Toh, P.Y. and Y.C. Poon. 1982. Effects of water management on field performance of oil palms on acid sulphate soils in peninsular Malaysia. In Dost and van Breemen, pp. 260–270.

Ukkerman, H.R. and P.A.M. van Gent. 1989. Fertilizer trials in the bolanha rice cropping system used by the Balanta people of Guinea Bissau. *Acid Sulphate Newsletter* (ISSS/Agricultural University, Wageningen) 2:3–4.

Vaughan, D., R.E. Wheatley and B.G. Ord. 1984. Removal of ferrous iron from field drainage water by conifer bark. *J. Soil Sci.* 35:149–153.

Vê, Ngugeñ Bao et al. 1989. *Soil map of Trans Bassac area*, scale 1:250 000. Soil Dept, Univ. Can Thó, Vietnam Soil Survey and Winand Staring Centre, Wageningen, The Netherlands.

Wada, H. and B. Seisuwan. 1989. The process of pyrite formation in mangrove soils. In Dost, pp. 24–37.

Wen, L.Y. and F.N. Ponnamperuma. 1966. Chemical retardation of reduction of flooded soils and the growth of rice. *Plant and Soil* 25(3):347–360.

West African Rice Research Station. 1959–62. Investigations into tidal mangrove soils. *Annual Reports* 1959, pp. 31–36, 1960–62 pp. 28–32. Crown Agents, London.

Williams, C.N. 1980. Effects of lime, drainage, manganese dioxide and seedling condition on rice in acid sulphate soil in Brunei. *Exp. Agric.* 16(6):313–320.

Xuan, Vo-Tong, Nguyen Kim Quang, and LE, Quang Tri. 1982. Rice cultivation on acid sulphate soils in the Vietnamese Mekong delta In Dost and van Breemen, pp. 251–259.

Xuan, Vo-Tong, Nguyen van Sanh, Duong van Ni, Ngo Thiu Ut, and Huynh Min Hoang. 1986. The rice and shrimp cropping system on potential and actual acid sulphate soils in the Mekong delta. In van Mensvoort et al., pp. 11–12.

Zuur, A.J. 1936. Over de bodemkundige gesteldheid van de Wieringermeer. Dept of Communications, The Hague (discussed by L.J. Pons, in Dost (1973), vol. 1, pp. 3–26.

Restoring Land Degraded by Gully Erosion in the Tropics

R. Lal

I. Introduction

An eroded rill, on deepening and widening, becomes a gully. A gully is sufficiently deep that it would not be obliterated by normal tillage operations, whereas a rill is of lesser depth and would be smoothed by ordinary tillage. A gully is caused by a rapid expansion of the surface drainage system in an unstable landscape. SCSA (1982) defines a gully as "a channel or

Advances in Soil Science, Volume 17

Table 1. Gully names in different geographical regions

Name	Language/region
Gully	English
Ravine	French
Wadi	Arabic
Nulla or 'Cho	Indian
Donga	Southern Africa
Carcava, Arroyo	Spanish
Lavaka, Sakasaka	Madagaskcar
Bocorocas	Brasil

miniature valley cut by concentrated runoff but through which water commonly flows only during and immediately after heavy rains or during the melting of snow; it may be dendritic or branching or it may be linear, rather long, narrow, and of uniform width." Gully erosion is geographical ly a widespread problem and is, therefore, known by many names in different regions (Table 1). Gully erosion has been the subject of intensive studies conducted by geologists, geographers, hydrologists, agricultural engineers, and soil scientists. The literature abounds with discussions of the causes and mechanisms of gully formation and the methods of its control. Readers are referred to detailed reviews elsewhere (Peterson, 1950; Antevs, 1952; Ireland et al., 1939; Schumm and Hadley, 1957; Riquier, 1958; Thompson, 1964; Seginer, 1966; Headge, 1967; Piest and Spomer, 1968; USDA-SCS, 1973; Patton, 1973; Cooke and Reeves, 1976). The objective of this chapter is to highlight the problem of gully erosion in the tropics, outline the principal causes of gully formation, describe techniques for assessment of gully erosion, and evaluate methods for the control of gully erosion.

Gully erosion is usually common in arid and semi-arid regions characterized by denuded landscape and flash floods. Depending on the soil profile and rainfall characteristics, gullies are also common in humid tropics. Whereas sheet and rill erosion are surface processes (Figs. 1 and 2) and tunnelling or piping is a subsurface process (Fig. 3), the process of gullying is primarily a linear erosion process (Fig. 4). It is a linear process because the floor and interfluve crests of the gully have similar angles to that of the main slope. Because the gully sides have steeper slopes than the undissected landscape, surface erosion processes are also accelerated by gully initiation. Mass movement is common along gully sides. The quantity of sediments transported by interrill and erosion from uplands feeding into the gully may be more than that originating from the gully itself by several orders of magnitude. However, gullies are more a spectacular and damaging feature of the landscape. Gully erosion is more difficult and expensive to control than interrill and rill erosion.

Figure 1. Sheet or inter-rill erosion as evidenced by stones and gravels on soil surface

Figure 2. Rill erosion on freshly cultivated undulating terrain

Figure 3. Tunnelling or piping involve subsurface flow

Figure 4. Gully formation is a linear erosion process with a distinct longitudinal dimension

There are a few contrasts among interrill-rill erosion and gully erosion. The total area affected by gully erosion is comparatively small. For example, in Nigeria, Ofomata (1981) estimated that of the total 78 600 km^2 area of southeastern Nigeria, 45% is affected by sheet erosion but only 2% by gullies and ravines. The gully erosion is also more spectacular than interrill-rill erosion and is discontinuous within a landscape. Contrary to sheet and rill erosion, the damage done to land by gully erosion is permanent. The gully erosion also causes depreciation in land value by lowering the water table and depleting the available water reserves. Buildings and infrastructures are also undermined by rapidly advancing gullies.

II. Mechanisms of Gully Formation and Advance

Gullies are established by the deepening of rills and slumping of side slopes through the shearing effect of concentrated overland flow, increase in pore-water pressure, and decrease in soil strength along seepage lines close to the streams and rivers, and slumping due to excessive formation of tunnel or pipeflow. Once gullies are established, they form permanent locations for concentrating the overland flow. Consequently, progressive deepening and widening of the gully continues until the gully has adjusted to a new set of equilibrium conditions. Gully development in the vicinity of concentrated flow is facilitated in soils with predominantly coarse-textured A horizon abruptly overlying a compact, less permeable clayey B horizon. Gully initiation may be caused at the entrance or exit of a "tunnel" or a "pipe" flow. Burrowing activity of animals in the vicinity of a gully may extend the gully laterally. Tunneling is often an important mechanism for lateral gully expansion (Fig. 5).

After its initial incision a gully usually extends backwards and sideways through the development of secondary gullies. Gully extension may follow one or all of the stages: percoline–seepage line–tunnel–gully (Young, 1972). Gully erosion is caused by headward advance, upstream migration of secondary knick-points, widening of the gully channel by slumping and mass movement, and deepening by mobilizing or transporting sediments from the gully floor. The gully head extends backward through various mechanisms by (1) undercutting followed by slumping; (2) slumping caused by water moving through a vertical crack (Fig. 6a); (3) slumping and undercutting caused by tunneling or pipins in the vicinity of an active gully; (4) formation of new headcuts by rapidly expanding waterfall; (5) the point where a runoff drops over a vertical height is an important cause of gully head initiation, and the water fall can start the linear incision developing throughflow or interflow aided by burrowing activity and, in the vicinity of a stream, reduced soil strength causing slumping (Fig. 6b) (Bradford et al., 1978; Bradford and Piest, 1980, Roloff et al., 1981). The failure planes follow the natural cleavage planes within the soil mass.

Figure 5. Tunnelling leads to a lateral gully expansion

According to the threshold concept of gully formation proposed by Schumm (1973), there exists a critical slope for a given discharge that leads to gully initiation. This concept was further developed by Begin and Schumm (1979) who proposed an operational definition of the geomorphic threshold for gully initiation. This definition is used to delineate regions of the valley susceptible to gullying on the basis of drainage area, discharge, and flow depth. These factors are indicative of the shear stress inacted on the valley floor by the concentrated flow.

In geomorphological terms, the advance of a gully head can be considered a type of mass-transport process whereby slope-forming material is transferred from one elevation to another under the influence of gravity and flowing water. Once the gully head is formed, with a distinct water fall due to abrupt change in elevation, it advances backward (upslope) due to progressive failure and fissuring. Once fissures are developed, the mass of soil from gully bed to the head top is saturated and may develop a positive hydraulic pressure. Subsequent failure is caused by liquefaction whereby the saturated mass loses its shear strength and behaves like a fluid. Additional modes of failure of gully head may be:

Fall: free movement of soil at the gully bed
Slide: failure on discrete surfaces
Slip: failure on concave surface common with heavy-textured soils

Depending on the strength of the subsoil exposed, the gully may be "U" or "V" shaped. The U-shaped gullies are formed in a subsoil that has low

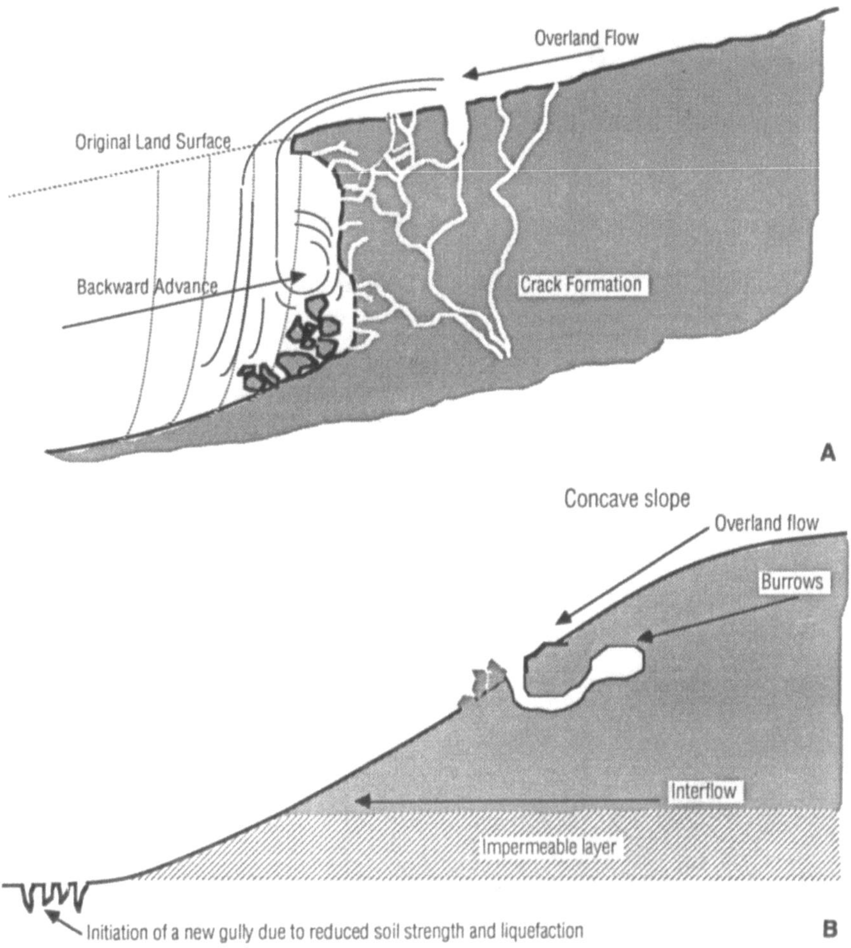

Figure 6a,b. Schematic of the processes of undercutting and slumping of a gully head by progressive failure

strength and is easily erodible. The V-shaped gullies are formed in a subsoil of high strength (Fig. 7). The morphology, shape, and types of gullies are discussed by Ireland et al. (1939) and Tejwani (1974).

III. Factors Affecting Gully Erosion

Gully erosion is determined by many factors. Some factors determine the potential hazard and others determine the intensity and rate of gully advance. In addition to anthropogenic factors, rainfall, vegetation cover, lithology, land form, and land use are also important physiographic factors affecting gully erosion. There are extrinsic and intrinsic thresholds re-

A

B

Figure 7a, b. Shape of the gully cross-section depends on strength of the subsoil material

sponsible for gully formation (Schumm, 1973, 1979). Extrinsic thresholds are those where an external variable changes progressively and eventually triggers an abrupt failure within the system, e.g., deforestation, road construction. Intrinsic thresholds are within the system and change independently of the external variables, e.g., piping. It is possible that both extrinsic and intrinsic thresholds may exist simultaneously (Patton and Schumm, 1975). These factors will not be discussed here but have been discussed by others (Patton and Schumm, 1975; Graf, 1979). Stocking (1981) also outlined extrinsic and intrinsic factors in relation to gully erosion.

Although gully erosion is accelerated due to anthropogenic factors, some geomorphologists question whether gullies are man-induced or natural (Schumm and Hadley, 1957; Denevan, 1967). The most obvious causes

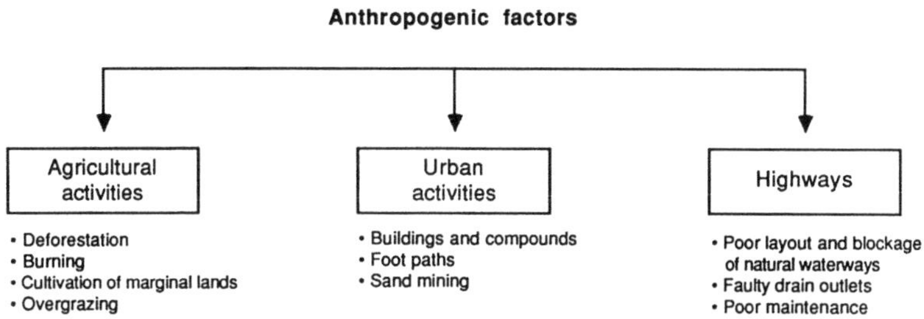

Figure 8. Anthropogenic factors responsible for gully erosion

of gully erosion include depletion of vegetation cover by deforestation and overgrazing and alteration in drainage patterns due to disturbances caused by road construction or any other natural or manmade causes.

IV. Anthropogenic Causes Responsible for Gully Erosion

In recent years, some parts of the tropics have been plagued by severe gully erosion, e.g., southeastern Nigeria. Although the initiation of gully erosion may be partly related to cyclic periods of erosion, deposition (Bryan, 1940; Stocking, 1980), and slope retreat (Beaty, 1959), the processes seem to be grossly accelerated by some anthropogenic factors; some of these are outlined in Fig. 8.

A. Overgrazing

In many regions of Africa and elsewhere in the tropics there are more cattle than people. Uncontrolled overgrazing has led to denudation of vegetation. High cattle population and overgrazing certainly constitute a major factor in semi-arid Africa (Fig. 9). Burning is another widely practised system of pasture renovation (Fig. 10). It leads to a rapid denudation and exposure of land to torrential rains. Cattle grazing in and around active gullies extend the gullies' knick point and dimensions. The destruction of protective vegetative cover and the trampling effect of animals on reduction in infiltration rate are important factors in increasing the rate and volume of runoff flow into the gully.

B. Demographic Pressure

While gully erosion may not be directly related to population pressure (Stocking, 1980), it affects the rate of gully advancement in many indirect ways:

Figure 9a, b. Overgrazing is a major factor of gully erosion

A B

Figure 10. Indiscriminate burning causes denudation of vegetation cover leading to flash floods at the onset of rains

Figure 11. Deforestation by heavy machinery causes excessive runoff

1. Intensive agriculture on marginal lands. Marginal steeplands prone to gully erosion which should otherwise be left alone are being cultivated because of the scarcity of land. Farmers are often forced to clear new lands, using inappropriate methods (Fig. 11), and grow crops even along the gully walls and in the vicinity of an active gully area. Food crops are grown right up to the gully edge (Fig. 12). The concentrated runoff, channeled into the gully by the ridge-furrow seedbed, or planting up-and-down the slope, creates secondary gully tributaries. The choice of crops that are labor-intensive or attract human traffic may also cause severe gully erosion. For example, cashew (*Anacardium occidental*) was introduced in southeastern Nigeria for the purpose of reafforestation and stabilization of gullies Cashew suppresses any undergrowth and encourages overland flow and sheet erosion. Manual harvesting of cashew fruits, and its transport, cause soil compaction and encourage gully erosion (Armon, 1984).
2. Foot paths. Uplanned land use can disturb the natural drainage ways. Nonavailability of water supplies in the rural communities necessitates villagers walking a distance of 1 to 2 km to the springhead. Sunken foot paths made up-and-down the slope become the focus of concentrated

Figure 12. Shortage of land causes an intensive landuse for food crop production even within an active gully

flow that eventually turns into gullies. This leads to the development of new foot paths that also are destined to a similar fate (Fig. 13).

3. Road construction. Road construction through steeplands, without adequate provision for drainage, is a major cause of gully erosion. Concentrated water flow in road ditches and wheel ruts leads to gully incision. Gregory and Park (1976) observed in Devon, England, how a gully was developed as a result of the faulty direction of a stormwater drain from the road into the stream channel. Whereas the development of this gully due to road drain took 30 years to develop in a mild English climate, it may hardly take a year under the harsh climatic conditions of the tropics. Although the road-caused gully erosion may occur anywhere in the world, the problem is particularly severe in developing countries due to neglect in maintenance and the lack of provisions for safe outlets for excess runoff (Fig. 14).
4. Drainage outlets from civil structures and public places. Unplanned urban development and lack of provisions for adequate and safe disposal of excessive runoff from public places is a serious cause of gully erosion. The land in schoolyards, market places, and playgrounds is severely compacted due to human traffic and it generates excessive runoff. The foot paths leading out of these compounds and natural depressions eventually turn into uncontrollable gullies, as has happened in eastern

Figure 13. Sunken foot paths are a major factor in gully formation

Nigeria (Grove, 1951; Floyd, 1965). Unplanned and unprotected drainage outlets from centers and factories have often turned into severe gullies.

5. Mining of sand. A gully bed is sometimes mined for sand and gravels to be used for construction. This practice is particularly common around the large urban centers in the gully-prone regions (Floyd, 1965). The demand for construction material increases with rapidly increasing population. The sand-mining of the gully bed increases gully erosion directly by increasing undercutting and indirectly through access roads up-and-down the slopes that eventually turn into gullies.

V. Watershed Factors in Gully Erosion

Watershed factors (Fig. 15) are intrinsic properties determined by the climate, landscape, parent material, profile characteristics, and soil characteristics. These properties interact with the anthropogenic factors in determining the susceptibility of a landscape to gully erosion.

Figure 14.a–c. Improperly designed drainage outlets from roads, water ways, and civil structures cause gully erosion

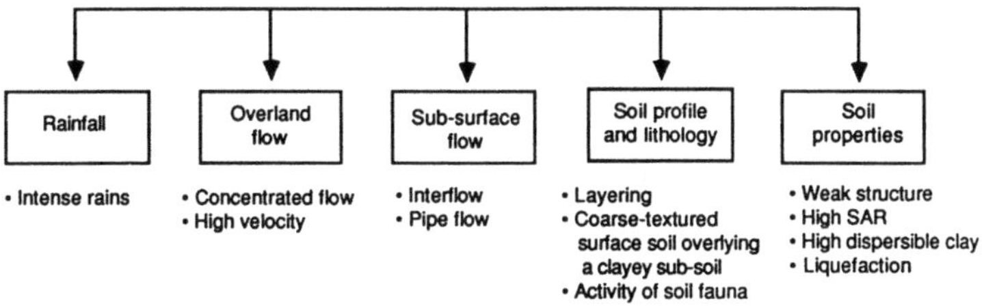

Figure 15. Soil and hydraulic factors in relation to gully formation

A. Rainfall

Rainfall is obviously an important factor, although it is not the total amount of rainfall but its distribution. Heavy rains concentrated in a short time, regardless of the total annual amount, can cause severe gullying. Thunderstorms of the tropics are characterized by sharp, high intensity peaks. Short-term intensities exceeding 200 mm/h are not uncommon (Fig. 16). Intense rains, coupled with soils prone to slaking and crusting, generate high runoff volume and concentrated flow. The shear stress generated by the concentrated flow causes gully erosion especially in semi-arid regions characterized by scanty vegetation cover.

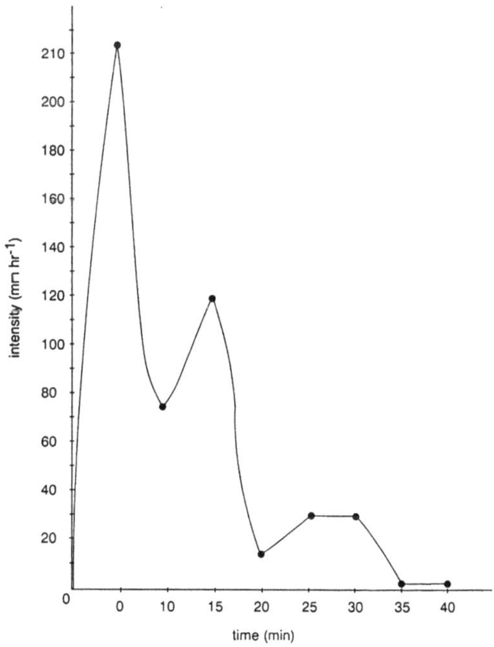

Figure 16. Rainfall intensities exceeding 200 mm h^{-1} for 5 to 10 min are common in several tropical environments

B. Vegetation Cover

The vegetation and land use within the watershed are also important in relation to the quantity and rate of runoff generated. The biomass at the valley floor is an important resistant factor (Graf, 1979). Relief plays an important role in gully development (Ofomata, 1966). A minimum critical slope is necessary for a gully to be initiated. The critical valley slope for gully initiation also depends on the area of the watershed feeding runoff into a channel way that may potentially become a gully. The larger the area the smaller the critical slope.

C. Soil Properties

Some soils are more prone to gully erosion than others. A soil with a coarse-textured highly permeable surface horizon with an abrupt transition to slowly permeable subsoil is normally prone to gully erosion. Soils derived from a well-graded, cohesionless loess parent material are also susceptible to severe gullying, e.g., the soils of the loess plateau in the Yellow River Valley, China. The presence or absence of groundwater in the region from which the gully is passing also affects the rate of gully advance. Faber and Imeson (1982) assessed soil factors that influence gully erosion. Subsoil containing high clay content, high moisture retention at low suction, and high shrinkage limit indicate that large volume changes could occur during wetting and drying cycles. However, gully erosion could occur in soils with an almost zero shrinkage limit, as is the case in soils of southeastern Nigeria (Armon, 1984).

D. Subsurface Flow

The matrix flow, seepage, or interflow are of diffused nature whereby water flows at a slow velocity with a high tortuosity factor. In contrast, the tunnel or pipeflow is concentrated, rapid, and follows a direct and preferential path. Furthermore, pipeflow is a widespread phenomenon in comparison with inteflow that primarily occurs in humid climates. Interflow is also more limited by the topographic factors than pipeflow. Soils with a high SAR and easily dispersible clay are prone to piping. In turn, soils that are conducive to piping and tunneling are also prone to gully erosion (Yair et al., 1980; Anderson and Burt, 1982). Roof collapse in pipeflow is the principal cause of gully initiation.

Interflow can also cause gully initiation through decreasing soil strength, increasing the probability of liquefaction and causing soil dispersion through alterations in soil structural attributes. It is the collapse of soil structure, either by physical or chemical processes, that renders soil particles vulnerable to entrainment by rapid flow.

VI. Measurement and Evaluation of Gully Erosion

The erosion caused by a gully can be assessed in two ways: by evaluating the sediments transported by the gully or by assessing the surface area affected by the gully erosion. The latter method evaluates the gully development in both sections by periodic surveying. The gully density can also be measured as km km^{-2} (Richter, 1980). Some of the commonly used techniques are briefly described below.

A. Aerial Photography and Reconnaissance Survey

Repeated aerial photographs are useful in evaluating the aerial extent and rate of development of gully erosion (Tuckfield, 1964; Stehlik, 1967). In Pakistan, McVean and Robertson (1969) used aerial photographs and a survey of the geology and soils to assess soil erosion in the mountainous regions. The survey showed that a vast proportion of the soils are severely eroded as indicated by truncated profile and skeletal texture. Kuo and Chien (1972) used an air-photo technique to study the occurrence of landslides in relation to slope, exposure, and land use in northeast Taiwan. Air-photos taken in 1962, 1965, and 1968 showed that more than 40% of the landslides were caused by typhoons especially in young plantations. In Brazil, Piedado and Carvalho (1981) used aerial photographs to assess the changes in the drainage network of two areas in Botucatu Municipality. This region is faced with a serious gully erosion problem. Aerial photographs taken in 1962, 1972, and 1977 showed increasing bifurcation ratios between first- and second-order channels in one area indicating continuing cycle of erosion. In another area, the decreasing bifurcation ratio indicated the state of equilibrium being attained due to the vegetation cover. In Kenya, Gachene and Barber (1983), on the basis of aerial surveys, prepared maps at 1:10 000 scale showing slope gradients and areas susceptible to gully erosion. Richter (1980) proposed a rating system for mapping gully

Table 2. Rating system for mapping gully erosion (Richter, 1980)

Degree of erosion	Density of gullies in km per km^2	Area affected in Sovakia	
		km^2	%
1. Slight	<0.1	25 977	53.1
2. Moderate	0.1–0.5	12 820	26.2
3. Increased	0.5–1.0	6 404	13.1
4. Severe	1.0–2.0	3 002	6.1
5. Highly severe	2.0–3.0	646	1.3
6. Highly severe	>3.0	108	0.2

Table 3. Rating system used for assessing gully erosion in Australia (Sargeant, 1984)

Category	Gully density
Low	<0.1 km km^{-2}
Medium	0.1–0.2 km km^{-2}
High	0.2–0.5 km km^{-2}
Very high	>0.5 km km^{-2}

erosion (Table 2). In Australia, Sargeant (1984) defined four categories of gully erosion on the basis of the gully density (Table 3).

B. Measuring the Rate of Gully Erosion

The rate of gully advance is physically measured by using fixed markers or reference points from which measurements can be made over different times. Commercially available profile meter or relief meters are also used to assess the volume of soil transported over a known time interval. A similar relief meter is also adopted for measuring sheet and rill erosion in row crops (Sallay and Prove, 1983).

C. Predicting Rate of Gully Advance

Mathematical formulae have been developed to predict the rate of gully development. Some of the widely used methods are those developed by USDA-SCS (1977). These formulae are based on the drainage area above the headcut, rainfall factor, soil properties through which the gully advances, land form and slope, runoff in relation to change in land use, and the change in land use. These formulae for estimating past and future rates of gully erosion are shown in Eqs. 1 and 2, respectively:

$$R_P = 1.5(W)^{0.46}(P_{0.5})^{0.20} \tag{1}$$
$$R_F = (A)^{0.46}(P)^{0.20} \tag{2}$$

where R_P and R_F represent the past and future average annual rate of gully head advance for a given reach in feet per year, W is average drainage area above headcut in acres, $P_{0.5}$ is the summation of 24-h rainfalls of 0.5 inch or greater occurring during the life of the gully, converted to an average annual basis, in inches, A is the ratio of average drainage area of a given upstream reach (W_f) to the average drainage area of the reach through which the gully has moved (W_p) (area in acres), and P is ratio of the expected long-term average annual inches of rain from 24-h rainfalls of 0.5 inch or greater (P_f) to the average annual inches of rain from 24-h rainfall of 0.5 inch or greater for the period, if less than 10 years, in which the gully head has moved (P_p). This model has been widely used in North America.

A few predictive models have been developed for the tropics on the basis of his observations in southern Africa. Socking (1981) developed the following regression equation to predict the rate of gully advancement (Eq. 3).

$$Y = 6.87 \times 10^{-3} P^{1.34} A_c^{1.0} H^{0.52} \quad (3)$$

where P is precipitation in mm, A_c is catchment area in km^2, and H is the height of headcut in meters. Stocking performed stepwise multiple regression and observed that other variables (e.g., antecedent precipitation, index of piping, vegetation cover, rainfall interception, population density, and slope of the approach channel) did not contribute significantly to the predictability of Eq. 3.

VII. Gully Erosion Control

Controlling gully erosion can be an elusive process. The rate of success in such schemes depends on the planning, design, and techniques employed. The ultimate success is governed by the proper diagnosis of the problem, steps taken to eliminate the causes, and on drastic changes in land use to stabilize the ecosystem.

The benefit/cost ratio of gully control must be carefully assessed. Some gully control measures are extremely expensive and resource-poor farmers cannot afford to invest. Furthermore, subsistence farmers, pre-occupied with food production for hand-to-mouth consumption, are not concerned with the stewardship appeal for preserving the resources for future generations. This means that gully preventive or control measures must produce short-term benefits in terms of increased yield, more land available for cultivation, and reliable crop yields through improved soil-water use. Above all, expensive measures of gully control and/or restoration have not been widely successful.

A sequence of steps recommended for restoring land degraded by severe gully erosion is shown in Fig. 17. Inventory of land, vegetation, hydrology and drainage pattern, climate, and land use is the first major step towards understanding the present status and potential risks of soil erosion. It is also important to assess capability and suitability of major landscape units and to evaluate various options and prioritize them with due considerations to socio-economic factors.

The first step in controlling gully erosion is fencing out the gully head to protect it from grazing cattle and/or wild animals. Second, diversion ditches or waterways should be installed to divert the surface runoff away from the gully head. The land use and soil management in the watershed

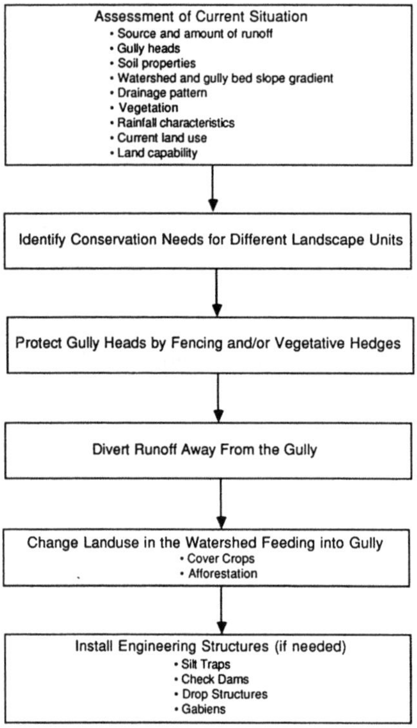

Figure 17. Sequence of steps in restoring land degraded by gully erosion

area feeding into the gully should be changed to soil-enhancing practices, e.g., planting cover crops and trees. A combination of biological and engineering measures is necessary.

A. Biological Measures

Biological measures are essential for stabilizing slope, improving soil structure, enhancing infiltration rate, and decreasing the runoff rate and amount. Prior to adopting vegetative measures, it may be necessary to undertake some land forming and shaping. Some of these practical steps are outlined by USDA-SCS (1973) and by Tejwani (1974; Tejwani et al., 1975). Stabilizing the eroding faces and bed of the gully is an important reclamative step. Establishing vegetation at the gully bed to provide more biomass is an important factor in decreasing the sediment-carrying capacity of the gully runoff. Shallow ravines on fertile lands can be reclaimed for agriculture by establishing grasses and trees (Khybri, 1974) and adding farmyard manure and as well as other needed soil amendments. Some useful legumes and grasses are listed in Table 4 and trees for controlling gully erosion in the tropics are listed in Table 5. The choice of appropriate species depends on soil, climate, and other environmental factors. Selec-

Table 4. Some grass and leguminous cover crops suitable for erosion control.

Grasses	Legumes
Andropogon gayanus	*Centrosema pubescens*
Axonopus micay	*Desmodium buergeri*
Brachiaria brizantha	*Desmodium ovalifolium*
Brachiaria decumbens	*Mucuna pruriens*
Brachiaria humidicola	*Mucuna utilis*
Brachiaria mutica	*Phaseolus aconitifolius*
Brachiaria ruziziensis	*Psophocarpus palustris*
Cenchrus ciliaris	*Pueraria phaseoloides*
Chrysophyllum macrophylla	*Stizolobium deeringianum*
Eragrostis curvua	*Stylosanthes guianensis*
Panicum antidotala	*Vigna catjang*
Panicum coloratum	
Panicum maximum	
Paspalum conjugatum	
Paspalum notatum	
Pennisetum clandenstinum	
Pennisetum clandestrium	
Pennisetum purpureum	
Thysanolaena maxima	

tion of appropriate species should, therefore, be done under local conditions.

B. Engineering Structures

Engineering structures for gully erosion control have been extensively described elsewhere (Hudson, 1971; Beasely, 1972; Heede, 1977; Starr, 1977; Braithwaite, 1980; Blaisdell, 1981; Sheng, 1981). There are a wide range of engineering structures, e.g., diversion channels, drop structure, gabiens (Fig. 18), check dams (Fig. 19), chutes. Engineering structures are expensive to install and maintain. Some of them have been used in Africa and elsewhere in the tropics. Drop structures require engineering skills and are often wrongly constructed and thus not always effective. Unplanned and poorly constructed structures are ineffective, fail shortly after installation, and cause untold damages to meager resources, land, and environments. Gabiens and check dams are also widely used, especially to protect gully development along highways. In a pine forest watershed in the Philippines, Florido (1985) evaluated the comparative effectiveness of different types of check dams. The data in Table 6 show that the depth of soil deposit was the highest behind the rock check dams and the least behind the hogwire check dams. The depth of soil deposited was partly related to the porosity. In

Table 5. Some trees suitable for different ecological regions of the tropics

Humid and Subhumid

Acacia auriculiformis
Acacia mangium
Acioa barteri
Afzelia bella
Albizia acle
Albizia chinensis
Albizia falcataria
Albizia falcataria Fosberg
Albizia ferruginea
Albizia gummifera
Albizia lebbek
Albizia minahassae
Albizia procera
Albizia spp.
Albizia zugia
Alchornea cordifolia
Andira inermis
Anthonotha macrophylla
Baphia nitida
Calliandra calothyrsus
Cassia siamea
Cassia spectabilis
Chlorophora excelsa
Cola nitida
Commiphora spp.
Cordia holstii
Dalbergia spp.
Daniellia oliveri
Dialium quineense
Entada abyssinica
Enterolobium cyclocarpum
Erythrina fusca
Erythrina glauca
Erythrina poeppigiana
Erythrina subumbrans
Eucalyptus citriodora
Eucalyptus grandis
Eucalyptus saligna
Eucalyptus tereticornis
Ficus spp.
Flemingia congesta
Gliricidia sepium
Gmelina arborea
Grevillea robusta
Inga edulis
Inga vera
Leucaena esculenta
Leucaena leucocephala
Maesopsis eminii
Pterocarpus spp.
Samanea saman
Schizolobium spp.
Sesbania bispinosa
Sesbania grandiflora
Syzigium malaccense
Terminalia superba
Trema orientalis

Semiarid

Acacia albida
Acacia saligna
Adansonia digitata
Albizzia lebbeck
Azadirachta indica
Bombax costatum
Casuarina equisetifolia
Cordia africana
Dalbergia sissoo
Grevilea robusta
Hyphaena thebaica
parkia irrea
Parkia spp.
Sclerocarya birrea
Sesbania aculeata
Tamarandus indica

Arid

Acacia albida
Acacia aneura
Acacia ehrenbergiana
Acacia laeta
Acacia nilotica
Acacia raddiana
Acacia senegal
Acacia seyal
Acacia sibriana
Ailanthus excelsa
Azadirachta indica
Balanites aegyptiaca
Boscia angustifolia
Calotropis procera
Combretum aculeatum

Commiphora africana
Euphorbia balsamifera
Maerua crassifolia
Parkinsonia aculeata
Prosopis cineraria
Prosopis juliflora
Salvadoria persica
Tamaris aphylla
Ziziphus mauritiana

Highlands

Acacia albida
Acacia auriculiformis
Acacia decurrens
Acacia mangium
Acacia melanoxylon
Acacia mollissima
Acacia tortolis
Acrocarpus framinifolius
Adhalhota species
Agathis loranthifolia
Albizia falcataria
Albizia lebbek
Albizia stipulata
Alnus acuminata
Alnus jorullensis
Aningeria adolfifreiderici
Calliandra calothyrsus
Casuarina sp.
Cedrela odorata
Colubrina arborescens
Cordia alliodora
Cupressus lusitanica
Cupressus macrocarpa
Cupressus toruloso
Dalbergia sissoo
Durico zibethinus
Ekebergia capensis
Erythrina abyssinica
Erythrina peaoppigian
Erythrina sp.
Eucalyptus camaldulensis
Eucalyptus darympleana
Eucalyptus delegatensis
Eucalyptus fastigata
Eucalyptus globulus
Eucalyptus maculata
Eucalyptus maidenii
Eucalyptus nitens
Eucalyptus viminalis
Gliricidia sp.
Gmelina arborea
Guilielma gaseqaes
Hygenea abyssinica
Inga edulis
Inga sp.
intsia sp.
Juniperus procera
Leucaena leucocephala
Maesopsis eminii
Mimosa scabrella
Moringa stenopetala
Pentaclethra macrophylla
Pinus caribaea
Pinus merkusii
Pinus michoacana
Pinus montesymae
Pinus patula
Pinus pseudastrobus
Pithecellobium sp.
Podocarpus gracillar
Populus deltoides
Prosopis cineraria
Pseudobambax ellipticum
Psidium guajava
Pterocarpus sp.
Schinus molle
Sesbania sp.
Trema orientalis
Triplochitan scleroxylon

Figure 18. Gabiens installed in an active gully

Figure 19. Rock check dams installed to stop gully expansion

Table 6. Mean annual soil deposit from gullies treated with different check dams at Upper Agno River Basin, Philippines (1976–80) (Florido, 1985)

Types of check dams	Mean annual depth of soil deposited (cm)	Mean annual sediment concentration (ppm)
Rock	2.00	12 690
Logs	1.16	14 700
Brush	1.92	4 900
Hogwire	0.52	768
Combination	0.78	2 336

addition to the depth of soil deposited, water samples collected on the downstream side of the check dams were analyzed for the sediment concentration. The highest sediment concentration was measured in the log check dams followed by rock, brush, hogwire, and a combination of check dams (Table 6). It is apparent from the sediment concentration that check dams per se were not effective. Furthermore, the soil had already been removed from the farmland or gully bed prior to its arrival at the point of measurement. Unless the runoff is diverted, gabiens and check dams may also fail and cause even lateral gully extension (Fig. 20). The material used in the construction of engineering structures should be carefully assessed. Soils containing dispersible clay fractions should not be used. In the eastern coast of New South Wales, Australia, Wickham (1976) suggested quidelines for the construction of an engineering structure using clay materials which display various modes of instability.

Biological measures are most effective when used in combination with engineering techniques. In the Philippines, Costales and Costales (1985) evaluated the effectiveness of several treatments for the control of gully erosion near the reservoir of the Binge Hydroelectric Plant. The data in Table 7 show that during the 6-month period of the first year (1981), the soil loss ranged from 14.2 t/ha in treatment with grass sod to 103 t/ha in control. By the second year, all vegetative treatments were equally effective and significantly different from the untreated control. The authors concluded that biological measures along with stone riprap were extremely effective in controlling stream bank and gully erosion.

There are several justifications for the use of grasses, legumes, shrubs, and trees. In addition to providing ecological diversity, there are important environmental reasons for the frequent and widespread use of vegetative measures within the landscape. Reversing the increasing trend in atmospheric carbon is one. Mitigating accelerating erosion and saving our soils is another.

Figure 20. Without vegetative measures to stabilize the soil, gabiens can also fail leading to a lateral expansion of the gully

Table 7. Effectiveness of different vegetative treatments along with stone riprap for gully control in the Philippines (Costales and Costales, 1985)

		Soil erosion (t ha^{-1})				
Treatment	Description	(1981)[a] Year 1	2	3	4	5
Sodding	Grass sod of Kikuyo bermuda and other creep plants	14.2[b]	27.8[b]	5.2[b]	0.8[b]	0.2[b]
Fascine	Trenches filled with the brushwood	75.1[a]	39.0[b]	6.6[b]	1.9[b]	0.4[b]
Wattling	Brushwoods interwoven between pegs	58.2[a,b]	48.6[b]	7.4[b]	1.6[b]	0.5[b]
Mulching plus cover crops	Residue mulch plus sowing to Centro, style, and other legumes, and application of Lurasol soil conditioner	74.9[a]	34.7[b]	6.2[b]	2.3[b]	0.3[b]
Control	No treatment	102.7[a]	160.6[a]	168.4[a]	84.2[a]	69.0[a]

[a]Data for 6 months from June to December, 1981.
Means bearing the same supper script are not significantly different.

VII. Conclusions

Gully erosion is caused by soil, hydrological, and landform factors related to watershed characteristics. Soil properties conducive to gullying are an abrupt textural break in surface and subsurface horizons, subsurface flow especially piping, high dispersible clay, and poor soil structure. The extent and rate of gully formation are exaggerated by anthropogenic factors, e.g., agricultural activities, deforestation, burning, overgrazing, road construction, foot paths, faulty drain outlets, and other engineering structures that lead to concentration of runoff. In most cases, gullies are formed by deepening of rills and slumping of sideslopes due to excessive formation of pipeflow. The burrowing activity of animals in the vicinity of a gully may extend the gully laterally. The rate of gully formation can be assessed either by evaluating the volume of sediment transported by the gully system or by assessing the surface area affected by it. Aerial photographs along with a standardized rating system are commonly used to evaluate severity of gullying. Mathematical formulae have also been developed by USDA-SCS to predict the rate of gully development. Gully formation can be controlled by engineering or biological measures. The former include diversion channels, drop structures, gabiens, and chutes. The latter are based on establishing the vegetation cover and include grasses and trees to stabilize the gully bed and walls.

References

Anderson, M.G. and T.P. Burt. 1980. The role of topography in controlling throughflow generation. *Earth Surface Processes* 3:331–344.

Antevs, E. 1952. Arroyo cutting and filling. *J. Geol.* 60:375–385.

Armon, M.N. 1984. Soil Erosion and Degradation in Southeastern Nigeria in Relation to Biophysical and Socio-economic Factors, Ph.D. dissertation, Univ. Ibadan, Nigeria, pp. 344.

Barker, T.C. 1990. Agroforestry in tropical highlands. In: K.G. MacDicken and N.T. Vergara (eds.) *Agroforestry: Classification and Management.* Wiley, New York, pp. 195–227.

Beasley, R.P. 1972. *Erosion and Sediment Pollution Control.* Iowa State University Press, Ames, Iowa, p. 320.

Beaty, C.B. 1959. Slope retreat by gullying. *Bull. Geol. Soc. Am.* 70:1479–1485.

Begin, Z.B. and S.A. Schumm. 1979. Instability of alluvial valley floors—a method for its assessment. *Trans. ASAE* 22:347–350.

Blaisdell, F.W. 1981. Engineering structures for erosion control. In: R. Lal and E.W. Russell (eds.) *Tropical Agricultural Hydrology* Wiley, Chichester, U.K., pp. 325–356.

Bradford, J.M. and R.F. Piest. 1980. Erosional development of valley gullies in the upper midwestern United States. In: D.R. Coates and J.D. Vitek, (eds.) *Threshold in Geomorphology* George Allen and Unwin, London, pp. 75–101.

Bradford, J.M., R.F. Piest, and R.G. Spomer. 1978. Failure sequence of gully headwalls in western Iowa. *Soil Sci. Soc. Am. J.* 42:323–327.
Braithwaite, P.G. 1980. Gully control in basalt soils. *Zimbabwe Agricultural J.* 77(6):253–258.
Bryan, K. 1940. Gully gravure—a method of slope retreat. *J. Geomorph.* 89–107.
Cooke, R.U. and R.W. Reeves. 1976. *Arroyos and Environmental Change in the American Southwest.* Oxford Univ. Press, Oxford.
Costales, E.F., Jr. and A.B. Costales. 1985. Stabilization of streambanks and riparian zones by riprap combined with selected vegetative engineering structures. *Sylvatrop Philipp. For. Res. J.* 10:17–33.
Denevan, W.M. 1967. Livestock numbers in nineteenth-century New Mexico and the problem of gullying in the Southwest. *Ann. Ass. Am. Geogr.* 57:691–703.
Ehlers, W., W.M. Edwards, and R.R. Van der Ploeg. 1980. Runoff controlling hydraulic properties of erosion susceptible grey-brown podzolic soils in Germany. In: M. De Boodt and D. Gabriels (eds.) *Assessment of Erosion* Wiley, Chichester, U.K., pp. 381–391.
Faber, Th. and A.C. Imeson. 1982. Gully hydrology and related soil properties in Lesotho. *ISHS Publ. No.* 137:135–144.
Florido, L.V. 1985. Check dams for the control of gully erosion in the pine forest watersheds. Silvatrop Philipp. *Res. J.* 10:9–16.
Floyd, B. 1965. Soil erosion and deterioration in eastern Nigeria: a geographical appraisal. *Nigerian Geogr. J.* 8:33–44.
Gachene, C.K.K., R.G. Barber. 1983. A preliminary report on an evaluation and detailed mapping of the erosion susceptibility at Kalolein, Kilifi District. In: D.B. Thomas, M.W. Senand, and W.M. Senga (eds.) *Soil and Water Conservation in Kenya.* Nairobi, Kenya.
Grabs, W.E. 1985. Potential and intensity of gully erosion in southeast Nigeria. Proc. Int. Symp. on Erosion, Debris Flow and Disaster Prevention. The Erosion Control Engineering Society, Tsukuba, Japan, pp. 57–62.
Graf, W.L. 1979. The development of montane arroyos and gullies. *Earth Surface Processes* 4:1–14.
Gregory, K.J. and C.C. Park. 1976. The development of a Devon gully and man. *Geography* 61:77–82.
Grove, A.T. 1951. Landuse and soil conservation in parts of Onitsha and Owerri Provinces. *Geol. Surv.* (Nigeria) Bulletin No. 21.
Headge, B.H. 1967. The fusion of discontinuous gullies. *Bull. I.A.H.S.* 12:42–50.
Heede, B.H. 1977. Gully control structures and systems. *FAO Conservation Guide* 1:181–222.
Hudson, N.W. 1971. *Soil Conservation.* B.T. Batsford, London.
Ireland, H.A., C.F.S. Sharpe, and D.H. Eargle. 1939. Principles of gully erosion in the Piedmont of South Carolina. *U.S. Dept. Agric. Tech. Bull.* 633.
Khybri, M.L. 1974. Ravine reclamation in Kota. *Soil Conservation Digest* 2(1):57–60.
Kuo, P.C., W.T. Chien. 1972. Air-photo study of landslides in Lo-tong watershed (NE Taiwan). *Technical Bulletin Experimental Forestry* (Taiwan University) No. 78.
MacDicken, K.G. 1990. Agroforestry management in the humid tropics. In: K.G. MacDicken and N.T. Vergara (eds.) *Agroforestry: Classification and Management.* Wiley, New York, pp. 98–149.

McVean, D.N., V.C. Robertson. 1969. An ecological survey of land use and soil erosion in the West Pakistan and Azad Kashmir catment of the River Jhelum. *J. Appl. Ecol.* 6:77–109.

National Academy of Sciences. 1979. *Tropical Legumes: Resources for the Future.* Washington, D.C.

Ofomata, G.E.K. 1966. The significance of relief to soil erosion in eastern Nigeria. *J. W. Afr. Sci. Ass.* 11:123–135.

Ofomata, G.E.K. 1981. The land resources of eastern Nigeria: a need for conservation. In: M. Uzo, (ed.) *Landuse and Conservation* University of Nigeria, Lagos, pp. 98–108.

Patton, P.C. 1973. Gully erosion in the semi-arid west. M.S. Thesis, Colorado State Univ., Fort Collins, Colorado.

Patton, P.C. and S.A. Schumm. 1975. Gully erosion, northern Colorado: a threshold phenomenon. *Geology* 3:88–90.

Peterson, H.V. 1950. The problem of gullying in western valleys. In: Trask, P.D. (ed.) Applied Sedimentation. Wiley, New York, pp. 407–434.

Piedado, G.C.R., W.A. Carvalho. 1981. Evolutive photo interpretation of the drainage network of gullied areas in Botucatu municipality, Brazil. *Cientifica* 9(2):191–196.

Piest, R.F. and R.G. Spomer. 1968. Sheet and gully erosion the Missouri Valley Loessial Region. *Trans. ASAE* 11:850–853.

Richter, G. 1980. Soil erosion mapping in Germany and Czechoslovakia. In: M. De Boodt and D. Gabriels (eds.) *Assessment of Erosion.* Wiley, Chichester, U.K., pp. 29–54.

Riquier, J. 1958. Les "Lavaka" de Madagascar. *Bull. Soc. Geogr. Marseilles* 69:181–191.

Roloff, G., J.M. Bradford, and C.L. Scrivner. 1981. Gully development in the deep loess hills region of Central Missouri. *Soil Sci. Soc. Am. J.* 45:119–123.

Sallay, H.M. and B.B. Prove. 1983. Erosion measurement techniques in row crops with application in canelands. Conf. Proc. Soil Erosion Research Techniques, 12–14 April, Toowoomba, Qld, Dept. of Primary Industries, Australia, pp. 43–50.

Sargeant, I.J. 1984. Gully erosion in Victoria. Proc. National Soils Conference, 13–18 May, Brisbanc, Australia.

Schumm, S.A. 1973. Geomorphic thresholds and complex response of drainage systems. In: M. Morisawa (eds.) Fluvial geomorphology. *Publications in Geomorphology* State University of New York, pp. 299–310.

Schumm, S.A. 1979. Geomorphic thresholds: the concept and its applications. *Trans. Inst. British Geographers* 4:485–515 (new series).

Schumm, S.A. and R.F. Hadley. 1957. Arroyos and the semi-arid cycle of erosion. *Am. J. Sci.* 225:161–174.

SCSA. 1982. Resource Conservation Glossary, Soil Conservation Society of America, Ankeny, Iowa.

Seginer, I. 1966. Gully development and sediment yield. *J. Hydrol.* 4:236–253.

Sheng, T.C. 1981. The need for soil conservation structures for steep cultivated slopes in the humid tropics. In: R. Lal, and E.W. Russell (eds.) *Tropical Agricultural Hydrology* Wiley, Chichester, U.K, pp. 357–374.

Starr, B.J. 1977. A practical guide to the gully control structure in the Queanbeyare district. *J. Soil Conserv. Serv. of New South Wales* 33(3):174–187.

Stehlik, O. 1967. On methods of measuring sheet wash and rill erosion. *Revue de Geomorph. Dynamique* 17:176.

Stocking, M.A. 1980. Examination of the factors controlling gully growth. In: M. De Boodt and D. Gabriels (eds.) *Assessment of Erosion* J. Wiley & Sens, Chichester, U.K. pp. 505–520.

Stocking, M.A. 1981. Causes and prediction of the advance of gullies. In: T. Tingsanchali and H. Eggers (eds.) *Proc. Southeast Asia Regional Symposium, Problems of Soil Erosion and Sedimentation.* AIT, Bangkok, Thailand, pp. 37–47.

Tejwani, K.G. 1974. Extent, classification and reclamation of gullied lands. *Soil Conservation Digest* 2(1):24–35.

Tejwani, K.G., R. Dayal, B. Verma, H.M. Gidwani, and S.C. Gupte. 1975. Classification, control and reclamation of ravine lands. In: Soil and Water Conservation Research 1956–1971. *Indian Council of Agricltural Research*, pp. 254–310.

Thompson, J.R. 1964. Quantitative effects of watershed variables on rate of gully head advancement. *Trans. ASAE* 7:54–55.

Tuckfield, C.G. 1964. Gully erosion in the New forest. Hampshire. *Am. J. Sci.* 263:616–627.

Verma, B., A.P. Patel. 1969. Erosion studies on soils of the Mahi ravines in Gujarat. *Indian J. Agric. Sci.* 39:515–522.

USDA-SCS. 1973. How to control a gully. *USDA Farmers Bulletin* 2171.

USDA-SCS (Engineering Div.) 1977. Procedure for determining rates of land damage, land depreciation and volume of sediment produced by gully erosion. *FAO Soil Conservation Guide* pp. 125–141.

Van Den Beldt, R.J. 1990. Agroforestry in semi-arid tropics. In: K.G. MacDicken and N.T. Vergara (eds.) *Agroforestry: Classification and Management.* Wiley, New York, pp. 150–194.

Wickham, H.G. 1976. Dispersible clay soils for soil conservation structures on the North Coast of New South Wales. *J. Soil Conserv. Serv. of New South Wales* 32(1):36–46.

Yair, A., R.B. Bryan, H. Lavee, and E. Adar. 1980. Runoff and erosion processes and rates in the Zin valley badlands, northern Negev, Israel. *Earth Surface Processes* 5:205–255.

Young, A. 1972. *Slopes*. Longman, London.

Reclamation of Indurated, Volcanic-Ash Materials in Latin America

T.J. Nimlos

I. Introduction

The 1974 energy crisis precipitated by the oil-producing nations focused world attention on dwindling petroleum reserves. The depletion of global soil reserves may be equally serious, but it has not attracted the same degree of international interest. Per-capita food production has been declining in Latin America since 1970, partly because land is being taken out of production; the per-capita number of hectares of arable land in developing nations will be reduced from 0.35 to 0.20 between 1975 and 2000 (Barney, 1980). Wind and water erosion are among the major reasons for the decline in food production and loss of productive land (Brown and Wolf, 1984).

A somewhat unique erosion problem that extends from Chile to Mexico

Advances in Soil Science, Volume 17

is water erosion in indurated, volcanic-ash soils. The chain of mountain ranges that includes the Andes of South America and the Sierras of Central America has been a zone of widespread volcanism. Some of the volcanic deposits are indurated and are relatively impermeable to roots, air, and water; porous soils on these deposits are very susceptible to erosion, especially when intensively cultivated. Once the porous soil is eroded, infiltration into the exposed indurated horizon is slow, and runoff and flooding become severe problems.

Land reclamation is essential to restore some semblance of the original agricultural productivity of these areas and to recreate proper hydrologic functioning, specifically to increase water infiltration and storage.

This chapter presents the nomenclature, origin, and properties of the indurated layers, the cause and impact of the associated erosion and procedures, and national programs for reclaiming indurated land. The literature reviewed, mostly in Spanish, for this paper was basically composed of obscure reports and graduate theses of local academic institutions. For instance, the best report on the distribution of the indurated materials and overlying soils in Nicaragua (Anonymous, 1971) is available only through interlibrary loan from Dartmouth College. Most research on the same topic in Mexico has been published as M.SC. theses at the Colegio de Postgraduados at Chapingo, Mexico. I have not attempted to cite all the reports, especially the theses, but the References section contains citations that list them.

II. Nomenclature

The nomenclature of indurated deposits is determined partly by cement type. The most common cements in indurated, volcanic ash are silica and carbonates. Indurations cemented with silica are softened by soaking in hot 0.5 *N* KOH and those cemented with carbonates by soaking in acid. Indurations cemented by both silica and carbonates require alternate treatments of KOH and acid. Occasionally weak indurations slake in water; they are often referred to as fragipans.

Induration nomenclature is very confusing. Goudie (1973) referred to silica- or carbonate-cemented materials as "duricrusts." The term suggests surficial indurations of pedogenic origin, although many indurated deposits in Latin America are more than 50 m thick and lack evidence of pedologic development. "Pan" is a general term applied to indurated earth materials but of such vague definition that it is not commonly used. Many local folk names, usually of Indian origin, have been applied to cemented layers (Nimlos, 1987). For instance, tepetate and talpetate are used in Mexico and Nicaragua, respectively, but have a common Nahuatl (a Central American Indian language) origin.

The nomenclature of silica-cemented horizons in Latin American countries is given in Table 1. Duripan may be the preferred name; it is defined

Table 1. Nomenclature of silica-cemented layers in various countries or by various tribes

Country or tribe	Term
United States	Duripan (fragipan)
Mexico	
Nahuatl	Tepetate
Otomi	Xido
Tarasco	Cheri
Nicaragua	Talpetate
Las Antillas	Pan
Columbia	Duripan (hardpan)
Ecuador	Cangahua (cangagua)
Peru	Harpan
Chile	Cancagua
	Moromoro
	Tosca

Table 2. Nomenclature for carbonate-cemented layers

Name	Used by
Petrocalcic	U.S. pedologists
Caliche	Geologists
Tertel	Chilean pedologists
Calcrete	Geologists and geographers

as "A mineral soil horizon that is cemented by silica, usually opal or microcrystalline forms of silica, to the point that air-dry fragments will not slake in water or HCl. A duripan may also have accessory cement such as iron oxide or calcium carbonate" (Soil Science Society of America, 1984). Duripan presence is a common differentiating characteristic in soil taxonomy (Soil Survey Staff, 1975). However, the term connotes the same pedologic emphasis as duricrusts. Geologists refer to silica-cemented deposits as silcrete (Goudie, 1973).

The confusion is not limited to the plethora of national terms. Cangahua (Ecuador) is sometimes spelled cangagua, and tepetate (Mexico) is also applied to hardened mine wastes. Cangahua (Clapperton and Vera, 1986) and tepetate (Nimlos, 1987) are names of geologic formations in Ecuador and Mexico, respectively. The Japanese terms for silica-cemented pans are "karo" (Martini, 1969) and "masa" (Leamy et al., 1984).

The nomenclature of carbonate-cemented layers is less complicated, partly because the process of induration is clearly pedogenic (Table 2). Petrocalcic may be the preferred name; it is defined as "A continuous, indurated calcic horizon that is cemented by calcium carbonate and, in some places, with magnesium carbonate. It cannot be penetrated with a

Table 3. Types of volcanic-ash indurations in Latin America

	Cements	
Induration genesis	Silica	Silica and carbonates
During deposition (geogenesis)	Ash-flow tuff or fluvial ash	Cannot exist
Pedogenesis and geogensis	Duripan in ash-flow tuff of fluvial ash	Petrocalcic/duripan in ash-flow tuff or fluvial ash
Pedocementation (pedogenesis)	Duripan	Not observed

spade or auger when dry, dry fragments do not slake in water, and it is impenetrable to roots" (Soil Science Society of America, 1984). Petrocalcic presence is also a common dierentiating characteristic of soil taxonomy (Soil Survey Staff, 1975).

III. Genesis of Indurated Volcanic-Ash Materials

Ash is indurated by two processes involving two major cements. The processes are termed "geogenic" and "pedogenic"; geogenic processes occur at the time the ash is deposited, while pedogenic processes include pedocementation caused by the eluviation and illuviation of cements after deposition. In some cases, both processes have acted on the same materials. The principal cements are silica and carbonates. The types of induration are tabulated in Table 3.

The first type of volcanic-ash induration in Latin America was formed strictly as a result of geologic processes and is usually an ash-flow tuff though it may include air-fall ash. The origin, nomenclature, and properties of ash-flow tuffs have been well covered (Chapin and Elston, 1979), as has the origin of tuffs in Ecuador (Vera and Lopez, 1986), Mexico (Nimlos, 1989), and Peru (Fenner, 1948). When a volcano erupts, a very hot mixture of gas, water, and ash is thrown upward, only to settle on the cone's flank. Because the mixture is hot, it has low viscosity and flows for long distances at high speeds. As the mixture cools, it consolidates into an ash-flow tuff, and the minerals fuse into an indurated mass. Further induration occurs if groundwater is present because silica dissolves and reprecipitates as a cement.

A typical tuff deposit consists of a series of flows stacked on top of each other. Individual flow thickness varies from 1 to 50 m. Since flow temperature and thickness vary, the degree of consolidation varies: some flows are

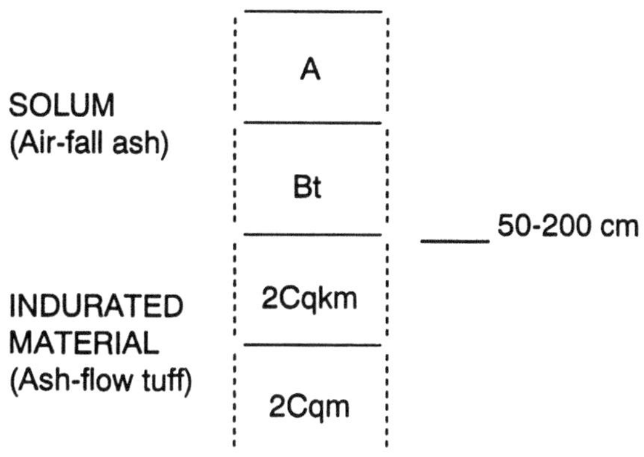

Figure 1. Typical morphology of ash soils with indurated horizons in Latin America

welded and have high strength and density, while others are so weakly consolidated that they slake in water and have densities as low as Andisols (ICOMAND, 1987). Ross and Smith (1961), for instance, have shown that density varies from 0.70 Mg/m^3 in unconsolidated tuffs to 2.35 Mg/m^3 in welded, highly consolidated tuffs. Ash-flow tuff deposits are quite thick; many in the Valley of Mexico, the basin where Mexico City is located, are more than 50 m thick. In Mexico and Nicaragua, the deposits usually have slope gradients of 2% to 5%, but in the Interandean Valley of Ecuador, the gradients are up to 70%. In areas of steep slopes, especially in Ecuador, deposits of ash reworked by mass movement are common (Crespo, 1987; Vera and Lopez, 1986); they are recognized by the presence of coarse fragments.

North of Guadalajara, Mexico (Jalisco state), another, less-extensive type of indurated ash, locally called fluvial ash, occurs that was deposited fluvially (Barrera, 1985), but it has not been well studied. These deposits are at least 30 m thick and can have relatively high strength and density; they appear to be dissected terraces. Apparently silica dissolved and reprecipitated as the terraces were being formed.

Carbonates occur in volcanic ash deposits only through pedogenic development and are not in the original flow. Therefore, ash-flow tuffs cemented by carbonates and silica do not exist.

The second type is an indurated geologic deposit that has been cemented additionally since deposition by pedogenic processes. Silica and carbonates eluviate from the overlying sola and illuviate in the tuff as cements.

Uncemented sola formed in air-fall ash overlie tuffs (or fluvial ash); they have been studied in Mexico (Nimlos, 1990), Nicaragua (Mickenberg, 1966), and Ecuador (Olmedo and Mejia, 1979) and are illustrated in Fig. 1. Since air-fall ash cooled in the atmosphere prior to deposition, it was not consolidated, and the sola formed in it have permeabilities, densities, and strength similar to other nonindurated soils. Sola thickness is usually 50 to 70 cm, but in moist sites it may be as much as 200 cm. They have a well-developed argillic horizon and have been classified as Durustalfs, Durxeralfs, and Durargids in Mexico (Nimlos, 1989), as Durandepts and various duric great groups and subgroups of Ustolls in Nicaragua (Anonymous, 1971), and as Andic Udic Haplustalfs in Ecuador (Olmedo and Mejia, 1979).

Illuvial silica cemented the upper part of the tuff into a 2Cqm horizon (and where carbonates accompanied the silica into a 2Cqkm). Silans occur in the pores, and illuvial silica probably occurs throughout the plasma.

Amorphous glass, a common constituent of volcanic ash, is relatively soluble compared with other silicates and is the source of silica for pedocementation. Silica translocates even in areas of 225 mm of precipitation, but a period of desiccation is necessary for dehydration (Chadwick et al., 1987a, b).

The thickness of silica pedocementation in ash-flow tuffs is difficult to determine in the field because pedogenic cementation frequently does not greatly augment geogenic induration and because illuvial silica is not easily recognized. Only when silica illuviation is advanced enough to form silans can it be clearly identified in the field. Silans in ash-flow tuffs occur to 1 m in some soils. Additional micromorphological studies are needed to determine the nature and thickness of silica illuviation.

Thus, the upper parts of these indurated horizons meet the morphological requirements of duripans because they are extremely firm, slake in hot KOH but not in water or acid and contain silans in the pores (Soil Survey Staff, 1975). However, the uncertainty about the lower boundary obfuscates the nomenclature. The advantage of calling these horizons "duripan" is that their genesis and morphology in other materials have been studied (Chadwick et al., 1987a, b), and local folk nomenclature can be avoided. Furthermore, the upper surface of the tuff deposit, not the lower limit of illuvial silica, is most important in management because it determines permeability, and it is this part of the tuff that requires reclamation where exposed by erosion.

The second cement type is a combination of silica and carbonates. It occurs in areas of low precipitation, less than 780 mm in Mexico. In areas of higher precipitation, the carbonates are leached out of the system. Unlike illuvial silica, illuvial carbonates (the 2Cqkm horizon) are readily identified in the field by their white color and reaction to acid. The carbonates are deposited as lamellae or are disseminated through the matrix. Thickness of the 2Cqkm increases with precipitation; in the Valley of Mexico, it is 40 cm thick where the precipitation is 372 mm and 200 cm where the precipitation is 750 mm. "Petrocalcic" is the correct term to apply to this horizon because it is a "hard, massive, continuous horizon" developed by pedogenic processes that includes carbonate pedocementation (Soil Survey

Staff, 1975). It is less preferable to call it a petrocalcic/duripan horizon because the horizon is easily identified by the illuvial carbonate not the illuvial silica, and the carbonates add greatly to its strength (Nimlos, 1989).

The third type of volcanic ash induration is formed through pedogenic processes only and is a true duripan; it is not extensive in Latin America. It occurs north of Guadalajara (Jalisco state), Mexico, and appears to be formed in fine-textured materials of lacustrine origin. The duripan is about 30 cm thick, has coarse prismatic structure, and is low strength (less than 3 kg/cm^2) compared with other indurated, volcanic-ash deposits. Erosion is not a problem in these soils because they are on flat slopes, and reclamation is not required. A duripan/petrocalcic horizon, one cemented by both pedogenic silica and carbonates, may occur in Latin America but local soil scientists have not reported it.

This chapter concentrates on the second induration type, that of both geologic consolidation and pedocementation, because erosion on this type is clearly the most serious. Other types of indurated, volcanic ash soils in Latin America include the Nadis soils of Chile (Valdes, 1969) which are apparently cemented by silica and iron.

IV. Classification of Indurated Materials

There are at least four classification schemes for indurated materials in Latin America. Each has some major deficiencies that stem in large measure from a lack of data. For a scheme to be successful, strength must be the major differentiating criterion as it is the most important physical property, especially with regard to reclamation.

An ancient Mexican taxonomy classified tepetate into red, yellow, and white classes. White tepetate (tepetate blanco) is cemented with carbonates and silica and is the most common type of tepetate in Mexico and the hardest (Nimlos, 1989). This scheme has been discarded because there is no clear relationship between red or yellow color and strength. A more recent scheme has been proposed in Mexico (C. Ortiz, Chapingo, Mexico, personal communication) that would classify tepetate into tepetate without carbonates, tepetate with carbonate lamellae, and tepetate with disseminated carbonates. As with most work on tepetate, this scheme has been tried only in the Valley of Mexico and is deficient for a number of reasons. Since there has never been a study of tepetate throughout Mexico, its carbonate morphology, content, and distribution in the profile is not known. Classifying the material without knowing the characteristics of the population is risky. Also, in some cases, both morphologies of carbonates, disseminated and laminar, occur in the same layer. Finally, while the system may be applicable to Mexico, where carbonates are the rule, it will not be of much value in other countries where carbonates are generally absent, as is the case in Nicaragua and Ecuador.

Ecuadorian taxonomies are based on genesis. An early system (Spillmann, 1931 as reported by Crespo, 1987) recognizes two classes: aeolian and lacustrine. Aeolian deposits are ash-flow tuffs interbedded with air-fall ash, that occur on well-drained slopes. Lacustrine deposits are of the same material but are found in small basins; they are of very limited extent. This system is largely ignored by Ecuadorian soil scientists. A more recent genetic taxonomy is that of Vera and Lopez (1986). They propose a complicated genesis scheme that also needs field testing.

IV. Properties of Indurated Materials

The important properties of exposed, indurated, volcanic-ash materials for assessing reclamation are strength, density, and hydraulic conductivity. Strength determines the energy required to break up the induration. The unconfined compressive strength of Mexican tepetate varies from 1.0 to 160 kg/cm^2 when air-dried (about 6% moisture) and is higher when disseminated carbonates are present (Nimlos, 1989). Strength decreases with increasing moisture content, especially in samples of high strength (Nimlos and Hillery, 1990), as shown in Fig. 2.

The Soil Conservation Service (U.S. Department of Agriculture) has developed new field procedures for estimating consistency (rupture resist-

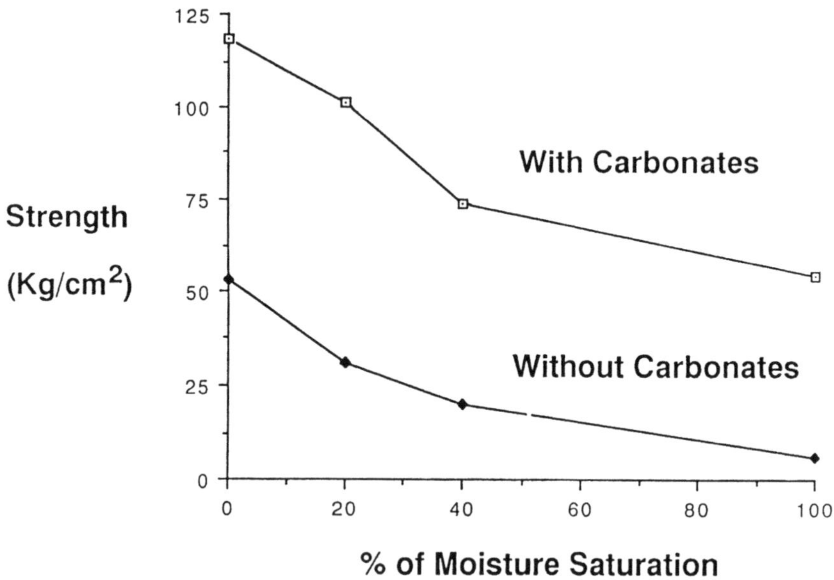

Figure 2. Strength of tepetate at various levels of moisture saturation

ance) (R. Grossman, SCS, Lincoln, Nebraska, personal communication). These procedures subject soil samples at various moisture contents to pressure tests by squeezing between the fingers, by standing on them, and by dropping a geologic hammer on them. When those procedures were applied to 18 samples of tepetate of known unconfined compressive strength the actual strength was estimated correctly in 15. Thus, these tests may prove useful for reclamation engineers as a rapid field technique for predicting strength.

The density of 18 samples of Mexican tepetate varied from 1.00 to 2.11 Mg/m^3 with a mean of 1.38 Mg/m^3. It is related to strength; the correlation coefficient between density and strength in samples without carbonates was 0.80 (Nimlos, 1989). Limited hydraulic conductivity data are available on indurated, volcanic materials. The saturated conductivity values of eight samples of Mexican tepetate varied from 1.5×10^{-7} to 36.0×10^{-7} m/s, which is characteristic of impermeable soils that are not indurated (Nimlos and Hillery, 1990).

These data show that indurated, volcanic-ash materials have higher strength and density and lower permeability than most soils.

VI. Distribution and Extent of Indurated Materials

Indurated volcanic materials occur throughout the Pacific Rim of Latin America, from Chile to Mexico, but they have been studied most in Mexico, Nicaragua, and Ecuador. Their extent in these three countries is shown in Table 4. Silica cemented layers have also been reported in the Lesser Antilles (McConaghy, 1969), Peru (Zavaleta, 1969), and Chile (Valdes, 1969).

A. Mexico

The bulk of tepetate research in Mexico has been in the Valley of Mexico, home of the Colegio de Postgraduados and the University of Mexico, and the adjacent state of Tlaxcala. This area represents only a small portion of the total area of tepetate. A national inventory of tepetate and its strength has never been made. However, reconnaissance soil surveys (order 3) of

Table 4. The extent of indurated volcanic ash materials in three countries

Country	Area ($km^2 \times 100$)	Reference
Mexico	80	Nimlos (1989)
Nicaragua	2.4	Marin et al. (1971)
Ecuador	20	Crespo (1987)

Figure 3. Tepetate distribution in Mexico

Mexico, prepared by the Secretaria de Programacion y Presupuesto at a scale of 1:1 000 000 show tepetate as duric phases. Figure 3 presents the area where tepetate occurs according to these maps. This area is also known as the neovolcanic zone because it is an area of intense; recent vulcanism.

Tepetate occurs in the central part of Mexico over a wide range of precipitation. The mean annual precipitation at Jalapa in the eastern limit of its distribution is 1454 mm, but in most areas it is less than 700 mm. If the precipitation map is laid over the soil map, it is apparent that 90% of the duric phases on the soil map are in areas of less than 780 mm of precipitation, the upper limit at which carbonates occur. Thus, most tepetate in Mexico is probably cemented with carbonates and silica, and most tepetate is of very high strength. Only in a few cases, at high elevations, is tepetate cemented by silica alone, and it is of such low strength that manual reclamation is feasible.

Figure 4. Talpetate distribution in Nicaragua

B. Nicaragua

A semi-detailed soil survey has been made of part of Nicaragua at a scale of 1:50,000 (Anonymous, 1971). It shows that talpetate occurs mostly south and east of Managua (Fig. 4). Whereas the parts of Mexico where tepetate occurs is generally dry, talpetate distribution in Nicaragua is in areas of higher precipitation. For instance, the mean annual precipitation in Managua and Grenada is 1161 and 1430 mm, respectively. Thus, carbonate-cemented talpetate is not common.

Nicaragua probably has more variable talpetate morphology than any other country in Latin America. In the lacustrine lowlands east of Managua, where slope gradients are less than 3%, a unique association of Alfisols/Vertisols is present. The terrain is a lacustrine basin of low relief. On the slight rises, less than 2 feet above the depressions, are soils with dark surfaces that have been called Mollisols but which become hard and massive when dry. I would classify them as Ustalfs. The boundary to talpetate is abrupt, and there are either no carbonates or very few.

In the slight depressions are Pellusterts that have a more gradual boundary to horizontally fractured talpetate that contains some carbonate lamellae in the upper 40 cm of talpetate. Apparently, the high clay content of the Pellustert prevents the leaching of carbonates and enhances the development of lamellae. I estimate the talpetate strength under both types of soils to be 20 to 40 kg/cm^2. Soil erosion in this association is very limited because of the very gentle slope gradient, and exposed talpetate is rare.

South of Managua on gentle slopes (gradients of 8% to 30%) of the Cordillera Los Marrabios foothills a second type of talpetate occurs. Soils in this area are duric subgroups of Ustalfs (Soil Survey Staff, 1975). Talpetate in this area is low strength (estimated at 3 to 6 kg/cm^2) and mostly discontinuous, and the solum/talpetate boundary is gradual. Current understanding is that the talpetate formed during a drier period and is now weathering. Fragments of this type disintegrate in 6 months if exposed to the atmosphere. Soils with this type of talpetate are being intensively cultivated and are severly eroded.

C. Ecuador

The soils of Ecuador have been mapped by a French research group (ORSTOM-Office de la Recherche Scientifique et Technique Outre-Mer) in conjunction with the Ecuadorian Ministry of Agriculture (Gonzales et al., 1986) at a scale of 1:200 000. Cangahua was mapped (Fig. 5) as Duriudolls and Duriustolls. Some of the Duriustolls are cemented with carbonates and silica, but they are thought to be of limited distribution (G. Del Posso, personal communication).

Tuffs have been studied more in Ecuador (Estrada, 1942; Clapperton and Vera, 1986) than in any other Latin American country.

VII. Soil Erosion on Indurated Materials

The erosion of soils on indurated materials in Latin America has been widespread; it has been so severe in some areas that few relict profiles remain. Many watersheds in the Valley of Mexico, for instance, have lost all uncemented soil over two-thirds of the land surface and have exposed the indurated horizons (Nimlos and Ortiz, 1987).

The history of soil erosion in Mexico (Nimlos and Ortiz, 1987) and Ecuador (de Noni, 1986) has been covered. Demographic and edaphic factors account for the erosion. The climate in Latin America is so favorable for growing crops that agriculture started early, between 2000 and 4000 years ago in the Valley of Mexico and about the same time in Ecuador. With the conversion from a nomadic to agricultural lifestyle came a number of population cycles of growth and decline. But the population peak of each cycle was greater than the peak of the previous cycle. As populations increased cultivation was extended on to steeper, more marginal land. Thus, the pressures of increasing population and cultivation, grazing,

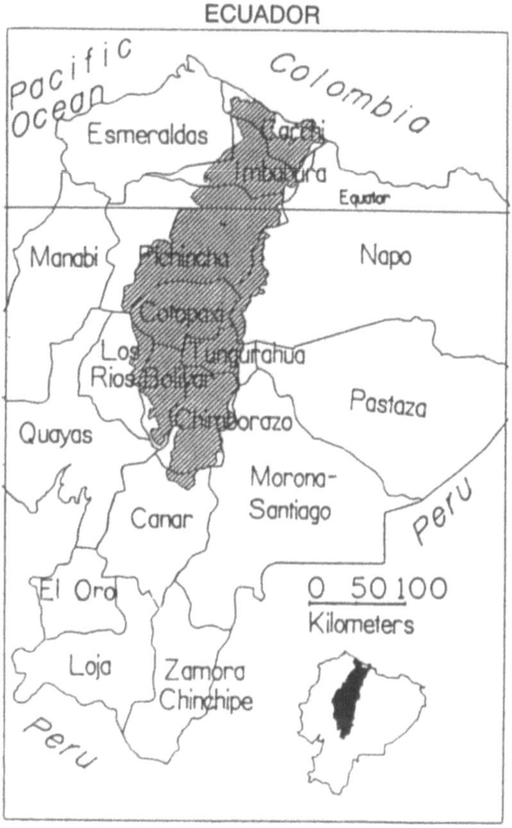

Figure 5. Cangahua distribution in Ecuador

forest cutting, and other activities associated with a sedentary culture greatly augmented soil erosion. These pressures are continuing today, and erosion is taking land out of production at the very time population growth is at an all time high; some suburbs of Mexico City are growing at the annual rate of 9% (Nimlos and Ortiz, 1987).

The porous soil on the indurated layers erodes differently than the indurated layers themselves. The Bt horizon and, more importantly, the indurated horizons restrict water percolation, and water accumulates in the profile; excess water flows over the surface. These morphological features are particularly significant in thin soils over the indurated layers; in Mexico, most sola are between 50 and 70 cm thick. The sola erode as a profile unit not as sheet erosion. Rather than eroding as a surface layer, the whole profile down to the indurated horizon washes away. The scar advances upslope in the shape of a horseshoe until entire slopes are denuded of uncemented sola, and the indurated horizon is exposed.

Once the sola is gone, erosion of the indurated horizon takes a very different form. If the induration is a petrocalcic horizon, erosion is extremely slow; the petrocalcic horizon strength is so high that particles are not detached by moving water. Duripans erode much faster than petrocalcic horizons because their strength is lower and because they occur at higher elevations where slope gradients are greater. Huge gulleys often form in tuffs with duripans.

VIII. Reclamation

Lands of exposed, indurated layers require reclamation to return them to some semblance of the agricultural productivity and hydrologic behavior of the porous soil prior to erosion. Farmers historically reclaimed tepetate in Mexico with wooden hoes by breaking chunks from the matrix and crushing them or allowing them to disintegrate by weathering (Nimlos and Ortiz, 1987). This practice is successful only where the strength is low, less than about 10 kg/cm^2 (air-dried), and the precipitation is sufficient for adequate plant growth and weathering. The practice is currently being applied in at least two instances in Latin America. In Nicaragua, precipitation is so high that carbonate pedocementation is negated and strength on upland sites, where erosion is a problem, is relatively low. Chunks of talpetate that work their way to the surface during cultivation disintegrate within 6 months. In Ecuador, a few kilometers southeast of Ambato, where slope gradients are 50% to 70%, farmers break chunks of cangahua with metal hoes and build terraces with the chunks as terrace lips. However, these instances of manual reclamation are the exception rather than the rule; in most cases induration strength is so great that manual reclamation is precluded and it can only be accomplished with large tractors.

Two countries have instituted reclamation programs, Mexico and Ecuador. Mexico is the most experienced Latin American country in the reclamation of indurated, volcanic ash materials. Their program was initiated in response to an environmental problem in the Valley of Mexico and adjacent watersheds caused by massive erosion (Valero, 1985). With the erosion came flooding which caused huge mud flats in the bottom of the Valley. Winter winds blew the sediment into Mexico City aggravating a serious air pollution problem (due mostly to industrial and automobile emissions) for the city's 20 million inhabitants.

Although some mechanical reclamation began in the mid-1960s the program began in earnest in 1972 when the federal government established the Comision del Lago Texcoco to improve soil and water management in the Valley. Between 1976 and 1985, the Comision built 1000 dams, 1000 km of terraces, 800 km of ditches (manually), and 314 600 pits for tree planting (manually) and ripped 3960 km. They also planted is million trees and 1 million nopales cactus in the areas treated (Valero, 1985.)

The result of this Herculean effort has been a dramatic reduction in erosion and flooding. Two-thirds of the Valley bottom that once was mud flats has been converted to pasture that supports 1000 head of cattle; complete restoration is expected soon. Much of the barren slopes of tepetate have been converted to plantations or crops. Although the program enjoyed broad public support, it was scaled back in 1985 when Mexico's foreign debt forced Draconian reductions in social programs.

There were two main types of mechanical reclamation in the Mexican program: ripping and terracing (Nimlos and Ortiz, 1987). Ripping-involved breaking the induration with crawler tractors of at least 180 horsepower. In densely populated areas, three bars spaced about 60 cm apart, ripped to about 50 cm and the reclaimed land was planted to annual crops including beans (*Phaseolus* spp.), corn (*Zea mays*), and squash (*Cucurbita* spp.). In more remote areas ripping was with a single bar to about 1 m, and it was planted to trees including pine (*Pinus* spp.), eucalyptus (*Eucalyptus* spp.), and cypress (*Cupressus* spp.).

Bench or storage terraces were built depending on the demand for cropland. Where demand was great ripped land was terraced into benches for cropping and where the demand was less severe contour terraces were constructed for water storage.

The other institutionalized reclamation program is currently in progress in Ecuador and is named the Community Land Use Management (CLUM) project. It is funded by the Agency for International Development and is jointly administered by CARE and the Ecuadorian Ministry of Agriculture. CLUM's objective is to improve the lot of poor Ecuadorian farmers by assisting them in the application of conservation practices and sustainable agriculture. The project employs a cadre of extensionists who work with small, subcommunal groups of farmers much as Soil Conservation Service technicians and agriculture extensionists in the U.S. The project is located in the Interandean Valley south of Quito: it started in three provinces in 1985 but has proven so popular that it has expanded to seven.

Conservation practices are low-input and labor intensive (manual) and include the agronomic practices of strip cropping, crop rotation, contour cultivation, green manuring, etc., and structural measures, hillside ditches (contour furrows), and bench terraces, for erosion control. The project area consists of very steep slopes with gradients commonly between 50% and 70%. Agriculture has been very hard on the land; erosion is severe and exposed cangahua widespread. The structural measures, therefore, are built directly in the indurated materials. Fortunately much of the cangahua in this area is low strength (less than 5 kg/cm^2-air dried) so that they can be installed manually.

Terrace construction normally follows a set pattern. Hillside ditches, about 30 cm high, are built on the contour and after a few growing seasons are combined into bench terraces whose dimensions vary with slope gradient but are commonly about 1 m high and 2 m wide. They are usually installed by the farmer and his family; although sometimes they are built by hired manpower; construction rates are about 20 m^2 per man-day. Land of exposed cangahua is worth about 50 000 sucres per hectare but when terraced its value is 700 000 sucres per hectare, a 13-fold increase.

There is some mechanical reclamation in Ecuador but it is on an ad hoc basis and is not associated with the CLUM project (de Noni, 1986). It involves the construction of contour furrows in cangahua of low strength with rubber-tired tractors. Institutionalized programs of land reclamation in other Latin American countries have not been reported.

IX. Summary

One common type of soil erosion in the Pacific Rim portion of Latin America is in volcanic ash soils with indurated horizons (duripan or petrocalcic). The horizons are formed by pedocementation superimposed on geologic consolidation and are thus doubly indurated. The geologic consolidation occurs when the ash-flow tuffs are deposited. Subsequently silica or silica and carbonate eluviate from the sola, formed in air-fall ash, into the upper portion of consolidated ash-flow tuffs, cementing them further. Another, less common type of ash deposit, fluvial ash, has been recognized only in Mexico. The indurated horizons impede water movement and enhance overland flow and erosion of the overlying sola because of their high strength and density and low hydraulic conductivity. The sola erode as a profile rather than sequentially by horizons. Erosion in the indurations is extremely slow when carbonate cements are present, but when they are absent gulleys form rapidly. Land of indurated material exposed by erosion has historically been reclaimed manually and still is today but only in a few cases. Mexico is the only country in Latin America that has had an organized reclamation program; it included ripping to create growing media and terracing to reduce runoff.

References

Anonymous. 1971. *Soil Survey of the Pacific Region of Nicaragua*, 3 vols. Published in Managua, Nicaragua by the Alliance for Progress and the Parsons Corp. in cooperation with Marshall and Stevens and the International Aero Services Corp.

Barney, O.G. 1980. *The Global Report to the President, Entering the Twenty-First Century*. Council on Environmental Quality and the Dept. of State, Washington, D.C.

Barrera, R.O.R. 1985. Tectonica y dinamica fluvial de Los Altos de Jalisco. Revista Cuatrimestral, Instituto de Geografia y *Estadistica* 1(3):67–120.

Brambila, M. 1940. The tepetate soils of Mexico. Proc. 6th Pacific Science Congress, Univ. of Cal. Press. Berkeley, CA 4:869–871.

Brown, L.R. and E.C. Wolf. 1984. *Soil Erosion: Quiet Crisis in the World Economy*. Paper 60. Worldwatch Institute, Washington.

Caujolle-Gazet A. and C. Luzuriaga. 1986. Estudio de un tipo de cangahua en el Ecuador: Posibilidades de mejoramiento mediante el cultivo. *Docum. de Invest.* 6:59–67. Centro Ecuatoriano de Investigacion Geografica.

Chadwick, O.A., D.M. Hendricks, and W.D. Nettleton. 1987a. Silica in duric soils: I. A depositional model. *Soil Sci. Soc. of Am. J.* 51(4):975–982.

Chadwick, O.A., D.M. Hendricks, and W.D. Nettleton. 1987b. Silica in duric soils: II. Mineralogy. *Soil Sci. Soc. of Am. J.* 51(4):982–985.

Chapin, C.E. and W.E. Elston. 1979. Ash-flow tuffs. *Geol. Soc. of Am.* Special Paper 180.

Clapperton C. and R. Vera. 1986. La secuencia glacial del cuaternario en el Ecuador: Una interpretacion al modelo de W. Sauer. *Paisajes Geograficos* 16:3–20. (Quito, Ecuador).

Crespo, E. 1987. Slope stability of the Cangahua Formation, a volcaniclastic deposit from the Interandean depression of Ecuador. M.Sc. thesis, Cornell Univ., Ithaca, New York.

de Noni, G. 1986. Breve vision historica de la erosion en el Ecuador. *Docum. de Invest.* No. 6:59–67. Centro Ecuatoriano de Investigacion Geografica. Quito, Ecuador.

Estrada, A. 1942. Contribucion geologico para el conocimento de la cangahua del la region interandina y del cuaternario en general en el Ecuador. *Anales de la Universidad Central, Quito*. 66:405–488.

Fenner, C.N. 1948. Incandescent tuff flows in southern Peru. Bull. *Geol. Soc. Am.* 59:879–893.

Gonzales A., A., F. Maldonado P., and L. Mejia V. 1986. *Memoria explicative del mapa general de suelos del Ecuador*. Sociedad Ecuatoriana de la Ciencia del Ecuador. Quito, Ecuador.

Goudie, A. 1973. *Duricrusts in Tropical and Subtropical Landscapes*. Clarendon Press, Oxford, England.

ICOMAND. 1987. *The ICOMAND proposal, 1 May 1987*. M.L. Leamy (ed.). New Zealand Soil Bureau, Private Bag, Lower Hutt, New Zealand.

Leamy, M.L., G.D. Smith, F. Colmet-Daage, and M. Otowa. 1984 . The morphological characteristics of Andisols. In *Andosols*, K.H. Tan, Ed., Van Nostrand Reinhold Co., New York.

Lowdermilk, W.C. 1953. *Conquest of the Land Through 7,000 Years*. U.S. Dept. Ag., Ag. Info. Bull. No. 99.

Machette, M.N. 1985. *Calcicsoils of the Southwestern United States*. Special Paper 203, Geological Society of America.

Marin, E.J., E. Ubeda, and J. Viramonte. 1971. *Contribucion al conocimiento de la genesis del talpetate*. Catastro e Inventario de Recursos Naturales. Managua, Nicaragua.

Martini, J.A. 1969. *Distribucion geografica y caracteristicas de los suelos derizados de cenizas volcanicas en Centroamerica*. Panel sobre Suelos Derivados de Cenizas Volcanicas de America Latina. CATIE, Turrialba, Costa Rica.

McConaghy, S. 1969. *Distribucion geografica y caracteristicas de suelosderivados de cenizas volcanicas en las Antillas*. In Panel sobre Suelos Derivados de Cenizas Volcanicas de America Latina. CATIE, Turrialba, Costa Rica.

Mickenberg, N. 1966. *Genesis, clasificacion y levantamiento cartograficode los suelos de Nicaragua*. Informe 2021, FAO, Rome.

Nimlos, T.J. 1987. The nomenclature of indurated horizons in volcanic ash soils. (in Spanish). Presented at the First SimposiaNacional sobre Uso y Manejo de Tepetates para el Desarrollo Rural. Tlaxcala, Mexico.

Nimlos, T.J. 1989. The density and strength of Mexican tepetate (duric materials). *Soil Sci.* 147(1):23–27.

Nimlos, T.J. 1990. Morphology, genesis and classification of soils formed over Mexican tepetate. Soil Survey Horizons 30(3):72–77.

Nimlos, T.J. and P.A. Hillery. 1990. Moisture/strength relations and hydraulic conductivity of Mexican tepetate. *Soil Sci.* (in press) 150(1):425–430.

Nimlos, T.J. and C. Ortiz S. 1987. Tepetate: the rock mat. *J. of Soil and Water Cons*. 42(2):83–86.

Olmedo, J.L. de and L. Mejia. 1979. Caracterizacion de un suelo desarrollado sobre cenizas volcanicas de la sierra ecuatoriana. *Anales de Edafologia y Agrobiologia* 38 (1/2):67–81.

Ross, C.S. and R.L. Smith. 1961. *Ash Flow Tuffs: Their Origin, Geologic Relations and Identification*. USGS Prof. Paper 366.

Soil Science Society of America. 1984. *Glossary of Soil Science Terms*. Published by Soil Science Society of America, 677 South Segoe Road, Madison, WI. 53711

Soil Survey Staff. 1975. *Soil Taxonomy: A Basic System of Soil Classification for Making and Interpretating Soil Surveys*. SCS-USDA Ag. Hbk. 436.

Spillmann, F. 1931. *Die saugetiere Ecuadors im Wandel der Zeit*, Part 1. Universidad Central del Ecuador, Quito.

Valdes, A. 1969. *Distribucion geografica y caracteristica de los suelos derivados de cenizas volcanicas de Chile*. In Panel sobre Suelos Derivados de Cenizas Volcanicas de America Latina. CATIE, Turrialba, Costa Rica.

Valero, J.M. 1985. *Rescate de una ciudad devestata: Plan Texcoco*. Informacion Cientifica y Technologica 7(107):17–19.

Vera, R. and R. Lopez. 1986. El origin de el cangahua. *Paisajes Geograficos* 16:21–28.

Williams, B.J. 1972. Tepetate in the Valley of Mexico. *Annals of the Assn. of Am. Geog*. 62(4):618–626.

Zavaleta, A. 1969. *Distribucion geografica y caracteristica de los suelos derivados de cenizas volcanicas del Peru*. In Panel sobre Suelos Derivados de Cenizas Volcanicas de America Latina. CATIE, Turrialba, Costa Rica.

Soil Faunal Degradation and Restoration

J.P. Curry and J.A. Good

I. Introduction

Accumulating evidence that animal activity is not merely desirable, but essential, for the functioning of restored ecosystems points to the need to explicitly take soil animals into account when planning restoration programs (Majer, 1989a). The need for soil faunal restoration will depend on the

Advances in Soil Science, Volume 17

extent to which the soil and its biological community has been degraded in the first place, and the degree and kind of soil reclamation which has been carried out. The assumption that invertebrates can eventually recolonize without assistance is probably valid in most cases, but there is considerable scope for intervention to accelerate the establishment of desirable species in many situations. The scope for intervention will depend on the overall objectives of the reclamation project. If the aim is to establish a self-sustaining relatively stable ecosystem on severely limited sites, then the emphasis would probably be on ameliorating site conditions to the extent that a biological community capable of maintaining a low level of biological activity under adverse conditions could be sustained. However, if the objective is to restore degraded land to a high level of productivity, there will be greater scope for the active encouragement of faunal groups known to be important in developing soil structure and maintaining soil fertility.

Since frequently the aim of rehabilitation is to establish communities resembling those of corresponding undisturbed habitats, the composition and role in soil processes of the fauna in such habitats will be reviewed briefly. The main factors causing degradation of the soil fauna, methods for faunal restoration, and the potential of invertebrates as indicators of soil rehabilitation and subsequent management will then be considered. While vertebrates such as rodents can have locally important effects on soil properties, this review will be confined to invertebrates for logistic reasons.

II. The Composition of the Fauna

The soil fauna is extremely variable in composition (see Wallwork, 1970, for review). Major factors influencing its composition include climate, soil, and habitat type, and degree of disturbance. Table 1 gives mean population densities and biomass in a selection of sites encompassing a range of climatic zones, habitats, and soil types. Total invertebrate biomass ranges from little more than 1 g m^{-2} in arid soils to over 100 g m^{-2} in earthworm-dominated temperate mull soils. Microfauna (mainly Protozoa and Nematoda) inhabiting the water film on soil particles predominate in arid soils; macrofauna which require capillary free moisture reach their greatest abundance in mull soils under mesic climatic conditions, while mesofauna (mainly Acari, Collembola, and Enchytraeidae) predominate in more acidic raw humus (mor) sites. Forest soils with well-developed surface organic layers support denser and richer microarthropod faunas than do grasslands, although the total faunal biomass is generally higher in grassland. Acidic soils have low faunal biomass, reflecting the scarcity of earthworms. Invertebrate biomass rarely exceeds 50 g m^{-2} in cultivated land except when soil management favors earthworms.

Earthworm biomass can exceed 100 g m^{-2} in temperate, mesic, mineral soils; they also may form the dominant component of the faunal biomass in humid tropical soils, while under drier tropical conditions termites, and to a lesser extent ants, are often the dominant invertebrate group. Enchytraeid worms, Collembola, and tipulid (Diptera) larvae are the main groups in wet, acidic tundra soils. Faunal biomass rarely exceeds 10 to 20 g m^{-2} in soils where earthworms are scarce.

Nematode population densities may range from 0.2 million m^{-2} in arid soils to over 30 million m^{-2} in mesic habitats; corresponding biomass estimates being 0.7 and 20 g m^{-2} (Wasilewska, 1979; Sohlenius, 1980). Population density and biomass in the order of 9 million individuals and 4 g m^{-2} appear to be typical of European grassland and cultivated soils. Protozoa are rarely included in faunal studies; biomass estimates range from 0.3 g m^{-2} in semi-arid prairie to more than 5 g m^{-2} in European grassland and cultivated soils (Stout and Heal, 1967; Elliott and Coleman, 1977; Clarholm, 1984).

The activities which determine the ecosystem roles of the main invertebrate groups are indicated in Table 2. Feeding habits vary considerably within and between groups, but 60% to 90% of the biomass is detritivorous, feeding on organic matter and associated microflora. Herbivores normally comprise less than 30% and predators/parasites less than 20% of the invertebrate biomass (Breymeyer, 1978; 1980). Normally, only 1% to 2% of the invertebrate biomass occurs above ground, but this can include considerable numbers of sap-feeding aphids and other Hemiptera, leaf-feeding caterpillars and beetles, shoot-mining Diptera, thrips, grasshoppers, and other herbivores in the vegetation layer. The soil surface fauna may include foraging termites and ants, litter- feeding isopods and millipedes, microbivorous Collembola and mites, predatory carabid and staphylinid beetles, predatory phytoseiid and bdellid mites, parasitic Hymenoptera, and polyphagous Dermaptera. Animal dung and carrion, leaf litter, rotting logs, tree trunks with moss or fern covering, tree bark, and organic accumulations in hollows and crevices of trees, etc., are important microhabitats for surface fauna. Large numbers of epigeic (surface-living) earthworms may be found in above-ground habitats associated with trees in tropical rainforests (Lee, 1985).

III. Influence of Fauna on Soil Fertility

A. Litter Decomposition and Nutrient Cycling

Invertebrates in general (with the exception of termites) are only able to utilize plant structural polysaccharides to a limited degree, and their direct role in primary litter decomposition is small. However, the total quantities of dead plant material ingested by the soil fauna can be quite significant: e.g., 33% to 55% of litter input in Swedish arable land (Paustian et al., 1990), and virtually the entire litterfall in temperate deciduous woodland

Table 1. Mean population densities and biomass of some invertebrates in selected soils (–, not estimated)

Habitat	Numbers						Biomass ($g\ m^{-2}$) (F, fresh mass) (D, dry mass)
	Nematoda (× 10^6)	Enchytraeidae (× 10^3)	Earthworms	Acari (× 10^3)	Collembola (× 10^3)	Other Arthropoda (× 10^3)	
Old grassland, Co. Kildare, Ireland (Curry, 1969)	—	—	364	106.1	105.4	10.6	c.88 F
Moorland alluvial grassland, England (Coulson and Whittaker, 1978)	3.3	120	390	36	40	0.51	35.5 D
Juncus moorland (peat), England (Coulson and Whittaker, 1978)	3.9	200	4	45	23	3.05	7.2 D
Old grassland on fen peat, Sweden (Persson and Lohm, 1977)	—	23.8	133	106.9	108.7	7.4	10.6 D
Grass ley, Sweden (Andrén et al., 1990)	8.5	6.3	39	12.3	17.8	c.4	5.4 D
Lucerne ley, Sweden (Andrén et al., 1990)	9.2	10.5	120	16.0	32.5	c.5	11.6 D
Lightly grazed sheep pasture, NSW, Australia (King and Hutchinson, 1976; Hutchinson and King, 1980b)	0.3	6	74	24.5	20.6	0.5	37.5 F
Tundra meadow, Alaska (Bunnell et al., 1975)	0.7	56.7		26.2	93.8	0.1	c.3 D

Herb grassland, USSR tundra (Chernova et al., 1975)	1.0	20.5	150	13	31.8	0.3	100.4 F
Humid savannah, Ivory Coast (Athias et al., 1974)	—	0.6	230	17.5	17.8	6.8	54.7 F
Miombo savannah woodland, Zimbabwe (Dangerfield, 1990)	—		6	—	—	191	12.5 F
Oak forest (mull) Denmark, (Bornebusch, 1930)	—	0.3	122	0.9	0.5	1.0	76.8 F
Coniferous forest Sweden (Persson et al., 1980)	4.4	16.2	—	684	60	2.4	1.7 D

Table 2. The main groups of soil invertebrates and their ecosystem roles

Group	Organic matter decomposition/ mineralization	Soil mixing/ affecting soil properties	Predation/ parasitism	Herbivory	Disease transmission
Microfauna					
Protozoa	+		+		
Nematoda	+		+	+	+
Mesofauna					
Enchytraeidae	+	+			
Acari	+		+	+	+
Collembola	+			+	
Protura	+				
Diplura	+		+		
Pauropoda	+				
Symphyla	+			+	
Macrofauna					
Lumbricidae	+	+			
Mollusca	+			+	
Isopoda	+				
Diplopoda	+			+	
Chilopoda			+		
Araneae			+		
Coleoptera	+		+	+	
Lepidoptera				+	
Diptera	+		+	+	
Thysanoptera				+	
Hymenoptera	+	+	+		
Hemiptera			+	+	+
Dermaptera	+		+		
Orthoptera	+			+	
Neuroptera			+		
Isoptera	+	+		+	

when earthworms are abundant (Satchell, 1967). Termites are often the main litter consumers in tropical soils; isopods and millipedes can ingest considerable amounts of litter in deciduous woodland and rough grassland, while enchytraeid worms are the main detritus consumers in acid moorland. Dung beetles, coprophagous dipterous larvae, and earthworms consume considerable amounts of vertebrate dung.

Most of the ingested organic matter passes through the invertebrate gut relatively unchanged chemically, but much fragmented and more amenable to microbial decomposition, especially when it is incorporated into the soil. Litter decomposition is generally found to be significantly retarded in experiments where invertebrates are excluded (Seastedt, 1984), and a surface layer of raw organic matter can develop rapidly when earthworms are suppressed by pesticides (Raw, 1962; Clements, 1982). Likewise, in the absence of an adequate coprophagous fauna, vertebrate dung can accumulate on pasture with consequent problems such as nutrient immobilization, pasture fouling, sward deterioration, and nuisance flies (Ferrar, 1973; Hughes et al., 1978).

Soil animals may influence rates of nutrient cycling directly through excretion and tissue turnover, and indirectly by altering microbially mediated nutrient transformation rates. Some estimates for annual return of N to the soil through earthworm tissue turnover and excretion are 100 kg ha^{-1} in English woodland (Satchell, 1963), and 109 to 147 kg ha^{-1} in New Zealand pasture (Keogh, 1979). In addition, worm casts generally have higher mineral N and assimilable P, K, Ca, and Mg levels than unworked soil. About 30% of net N mineralization in Swedish pine forest occurs via the fauna (Persson, 1983), and about 37% in short grass prairie, about 83% of this being attributable to amebae and nematodes (Hunt et al., 1987). Invertebrates appear to ingest relatively high proportions of microbial production in most ecosystems, e.g., 30% to 60% in Swedish pine forest (Persson et al., 1980) and 30% to 90% in arable land (Paustian et al., 1990), and there is evidence from laboratory and field microcosm studies that this may be an important mechanism for releasing nutrients immobilized in microbial tissue.

B. Invertebrates and Soil Properties

Earthworms can have a major influence on soil structure through burrowing, ingestion of soil and plant residues, and egestion of soil mixed with comminuted and partly digested plant residues in the form of casts which can comprise a high proportion of soil aggregates in the surface layers of mull soils (Kubiena, 1953; Lavelle, 1978). Most earthworm activity occurs in the top 10 to 20 cm soil layer which can be completely worked within a few years in humid tropics (Lavelle, 1974) and 25 to 60 years in mesic temperate soils (Barley, 1959; Boström, 1988; Curry and Bolger, 1985). Anécique (deep burrowing and surface-feeding) species such as *Lumbricus terrestris* L. and *Aporrectodea longa* (Ude.) have a particularly important role in soil mixing through incorporation of plant residues and transporting soil from deep in the profile to the surface.

Surface casting creates voids in the soil, thereby reducing bulk density and porosity, while subsurface casting redistributes soil components and alters pore size distribution. Large channels (2 to 11 mm diameter) influence soil aeration and root penetration; those opening to the surface strongly influence water infiltration rates, while medium sized pores created by casting in the soil enhance water-holding capacity (Hoogerkamp et al., 1983; Syers and Springett, 1983; Zachman et al., 1987). Incorporation of lime, fertilizers, and pesticides into the soil by earthworms can make an important contribution to the productivity of permanent pasture (Stockdill and Cossens, 1966; Springett, 1983). Enhanced microbial activity associated with earthworm casts results in increased mineralization rates, but also increased potential for gaseous N loss through denitrification (Svensson et al., 1986; Boström, 1988). Moderate to marked increases in plant growth have been attributed to earthworms in several experiments with plants grown in small containers or enclosures (Hopp and Slater, 1948; van Rhee, 1965; Curry and Boyle, 1987), and in areas where earthworms had recently become established (Stockdill and Cossens, 1966; Hoogerkamp et al., 1983).

Ants can modify their habitat through the construction of mounds, underground chambers, and galleries from mineral and organic materials, and they can make a significant contribution to soil mixing under dry conditions where earthworm activity is limited (Humphreys, 1981).

Termite mounds may contain quantities of soil ranging from 10 to 45 t ha^{-1} in arid lands and open savannah to 3000 t ha^{-1} in large *Macrotermes* mounds in tropical forest (Lee and Wood, 1971b; Lepage, 1972). Subsoil appears to be preferred for mound construction, leading to a slow process of profile inversion as surface soil horizons are formed from mound erosion (Lee and Wood, 1971a, b). Termite mounds tend to have more silt and clay than surrounding soil, and also higher concentrations of C and N and exchangeable bases including Ca, Mg, and K arising from admixture with feces (Lee and Wood, 1971a; Gupta et al., 1981). In addition to using soil for mound construction, humivorous species may consume considerable amounts of soil in tropical savannahs. Termite mound soil tends to be more compacted and to have higher bulk density than unaffected soils, but the underground galleries, etc., increase soil porosity and water infiltration (Lal, 1988). An important difference between termites and earthworms is that termites collect organic matter over a wide area and utilize it very efficiently, thus reducing the organic matter supply for other decomposers. Also, plant nutrients incorporated into termite mounds are withheld from circulation for a long time.

Other macroinvertebrates which can exert some mechanical effects on the soil include Coleoptera, molluscs, and dipterous larvae. Under infertile conditions where surface organic layers develop, the feces of smaller invertebrates, such as enchytraeids, Acari, Collembola, and dipterous larvae, contribute prominently to the soil fabric (Kubiena, 1953; Wood, 1966).

C. Invertebrates and Plants

The main soil invertebrates likely to damage plants are plant-feeding nematodes, molluscs, various dipterous, coleopterous and lepidopterous larvae, termites, and ants, while aerial plant structures may be attacked by a vast array of sapsuckers, defoliators, stem-borers, seed-feeders, etc. Pest attack is a recurring feature of crop ecosystems, but severely damaging pest outbreaks are less common in natural ecosystems where herbivore consumption usually falls within the range 1% to 10% of net primary production (Wiegert and Evans, 1967; Gibbs, 1976; Sinclair, 1975).

Herbivory influences plant growth rates and death rates, plant competitive ability, and botanical composition in many complex ways, and the outcome of herbivore/plant interactions is not easy to predict. Grazing is essential for the maintenance of much of the earth's seminatural grasslands, and moderate levels of grazing can increase primary production (McNaughton. 1979; Davidson, 1979). However, regrowth potential and tolerance for damage depend on many factors including the size and health of the plants and the degree to which they are already under stress.

Selective feeding on preferred species can alter botanical composition. Grasshopper feeding can suppress important ground cover plants in New Zealand tussock grassland (White, 1974), while control of frit fly and other pests can improve the establishment and persistence of ryegrass in grass leys (Henderson and Clements, 1979, 1981). Patchy damage by root-feeding species can provide opportunities for the colonization and establishment of new plants.

Invertebrates can influence plant distribution indirectly by disseminating disease organisms and pollen, while weed-feeding species may limit the prevalence of pasture weeds. Bees have a particularly important role in the pollination of legumes (Parker et al., 1987), while dramatic control of prickly pear cacti (*Opuntia* spp.) in Australia was achieved by the imported moth *Cactoblastis cactorum* Berg. Some herbivores, notably grasshoppers, destroy much more plant tissue than they consume, thereby influencing organic matter turnover (Andrzejewska and Wojcik, 1970; Mitchell and Pfadt, 1974). This effect is normally quantitatively small, but Rodell (1977) considers that it could have a significant influence on nutrient cycling in prairie grassland over a period of several years. The excreta of phytophagous invertebrates are rich in plant nutrients (Andrzejewska, 1979b), and could influence plant regeneration following severe pest outbreaks (Mattson and Addy, 1975). Some authors attribute an important role to invertebrate herbivores in influencing nutrient cycling and plant succession and consider that they can exert a regulatory function in ecosystem processes, but this is not universally accepted.

IV. Land Disturbance and Faunal Degradation

The extent of faunal degradation will depend on the nature of the original ecosystem and the degree of disturbance. For present purposes two situations will be considered—very severe disturbance associated with mining and industrial waste disposal, and less severe disturbance associated with land clearance and intensive agriculture.

A. Mining and Industrial Wastes

In general, extensive open cast mining completely removes biological communities and presents conditions which are extremely hostile for invertebrates. Features of newly restored mining and industrial waste sites likely to inhibit faunal establishment include lack of suitable food and adverse physicochemical conditions, particularly unfavorable moisture conditions and excessive fluctuations in surface temperatures. Extreme acidity resulting from the weathering of pyritic mine spoil, from ore extraction with acids, and from reduction of sulphides and other materials in tailings ponds, will prevent the establishment of all but the most acid-tolerant invertebrates. Earthworms such as *Lumbricus eiseni* Lev. and *Dendrobaena* spp. may be found in surface litter even in very acid soils, but most soil-dwelling species have a strong avoidance reaction to low pH (Satchell, 1955; Laverack, 1961).

Very alkaline wastes such as fresh pulverized fly ash (PFA), and wastes from bauxite refineries can also be toxic to soil fauna (Satchell and Stone, 1977; Southwell and Majer, 1982; Eijsackers et al., 1983). Fly ash toxicity declines with age, and once a surface organic mat develops, epigeic earthworm species can colonize. High salinity is probably the main reason for the toxicity of fresh fly ash to earthworms; this declines to harmless levels after weathering for 2 to 3 years (Townsend and Hodgson, 1973). High levels of salinity also occur in mine tailings from low grade copper and uranium mining in the western United States (Nielson and Peterson, 1973).

Metal toxicity can seriously impede rehabilitation of mine spoil, especially when the pH is low and high concentrations of metals are present in soluble form. Under these conditions revegetation of mine spoil may be inhibited although some metal-tolerant plants may be present (McNeilly, 1987). Marked depressions in microbial activity and litter decomposition rates and in microarthropod populations have been reported from heavily contaminated soil near metal smelters (Tyler, 1975; Strojan, 1978a, b). However, the toxicity of heavy metals is influenced by factors such as their chemical form, soil organic matter content, and pH. Earthworms can tolerate fairly high levels of most heavy metals (Ireland, 1983), although depressions in field population levels have been attributed to high levels of copper close to a copper refinery (Hunter and Johnson, 1982). Metals in ionic form pose greater risks than organically bound forms (Malecki et al., 1982), and adverse effects are more likely in acidic soils (Ma, 1988). Thus, liming and organic matter amendment of reclaimed soils are effective means of countering the effects of acidity and heavy metal toxicity.

Severe compaction of reclaimed sites due to natural settlement of very degraded substrate and to heavy machinery traffic also inhibits restoration, impeding plant rooting and earthworm penetration.

B. Land Clearance and Agricultural Management

Any form of human intervention will influence the soil fauna to a greater or lesser extent. As the proportion of net primary production channelled into commercially valuable species increases, invertebrate communities become increasingly simplified as plant diversity declines, and the surface litter layer disappears. The extent to which the organic cycle can be altered by management is apparent from a comparison of different systems of grassland utilization. Less than 5% of shoot production may be assimilated by cattle in semi-arid natural grassland (Coleman et al., 1976), up to 60% may be assimilated by sheep in temperate managed pasture (Hutchinson and King, 1980a), while over 90% may be harvested in mown grassland (Andrzejewska and Gyllenberg, 1980). However, high quality litter and root debris in managed grassland favor lumbricid earthworms, which become increasingly dominant under these conditions (Curry, 1983).

The most pronounced effects of management are seen under intensive annual cropping regimes, where the absence of plant cover for much of the year, low return of organic matter, periodic disruption of the soil by mechanical cultivation, and repeated use of pesticides can result in progressive depletion of soil organic matter, structural deterioration and compaction of soil, soil erosion and nutrient depletion, and marked reduction in the complexity and stability of the soil biological community.

The main practices which influence soil fauna are clearance of natural vegetation, fire, forage management, cultivation, irrigation and drainage, fertilizer and pesticide use. These are briefly considered below.

1. Clearing Natural Vegetation

The most striking consequence of clearing deciduous forest for agriculture is the disappearance of the litter layer with a consequent reduction in faunal diversity. However, many temperate forest species adapt well to grassland. In the case of earthworms, epigeic species decline in abundance with the disappearance of surface litter, but the anécique and endogeic (true soil) species increase in importance as soil fertility and food quality improve. The effects of deforestation in the tropics appear to be particularly marked. Epigeic species comprise most of the tropical earthworm fauna, and these, together with other litter-dwelling macroinvertebrates are destroyed by clearing and cropping. Lavelle and Pashanasi (1989) reported that soil macrofaunal biomass and population density in cultivated plots in the Peruvian Amazonia were reduced to 6% and 17% of those in primary forest (Fig. 1). The indigenous forest earthworm species largely disappeared, but when adapted species were available for recolonization, high earthworm population densities could be found under pasture. Lavelle and Pashanasi (1989) reported macroinvertebrate biomass of up to 160 g m^{-2} from pasture where the endogeic peregrine species *Pontoscolex corethrurus* Müller had become established.

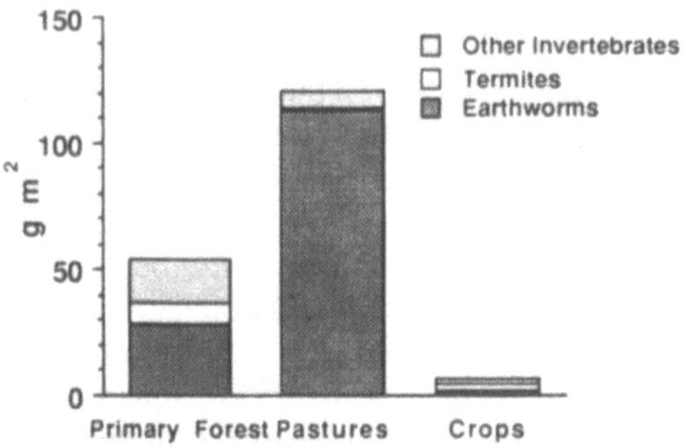

Figure 1. Macroinvertebrate biomass under different types of land use in the Peruvian Amazonia (after Lavelle and Pashanasi, 1989)

2. Fire

The total biomass of hemiedaphic (surface soil and litter-dwelling) invertebrates is usually drastically reduced by burning, the extent of the reduction depending on fire frequency, intensity, and duration (Athias, 1976; Lamotte, 1976; Edwards and Lofty, 1979; Majer, 1984). Populations of true soil dwellers may also decline in the longer term as organic matter is depleted by repeated burning.

3. Pasture Management

In general, increasing intensity of grassland utilization is accompanied by decreasing faunal diversity, with the simplification of the vegetation and the disappearance of the litter layer, while the dominance of earthworms, notably *Lumbricus terrestris*, increases as soil fertility and pasture productivity increase (Andrzejewska, 1979a; Curry, 1983).

Some faunal groups are favored by moderate levels of grazing (East and Pottinger, 1983; Seastedt et al., 1988), but heavy grazing depresses population densities of most hemiedaphic groups (King and Hutchinson, 1976; Huthinson and King, 1980b).

Mowing differs from grazing in being nonselective and in that the physical disturbance associated with stock trampling does not occur. The effect on invertebrates depends on the frequency and timing of mowing. Fewer invertebrate species are normally found in mown swards compared with-unmown, but judiciously timed annual cutting can be an effective means of conserving floral and faunal diversity (Morris, 1971, 1979; Wells, 1971). An important difference between grazing and mowing concerns the nature and extent of litter return to the soil: under both methods of herbage utilization over 90% of shoot production may be harvested, but in mown systems this material is removed, while under grazing regimes 60% may be returned to the soil in the form of dung (Andrzejewska and Gyllenberg, 1980; Hutchinson and King, 1980a).

4. Manures and Fertilizers

Variable soil invertebrate responses to mineral fertilizers have been reported. Positive responses to moderate fertilizer applications have been attributed to increased pH in acid soil (Huhta et al., 1986), and to increased litter quantity and quality (Edwards and Lofty, 1982a; Gerard and Hay, 1979; Lofs-Holmin, 1983). However, heavy applications of nitrogen can depress invertebrate populations (Nowak, 1976). By increasing soil acidity, sulfate of ammonia can be particularly toxic to earthworms in acidic soils (Satchell, 1955: Edwards, 1977, 1983).

Organic manures and crop residues can benefit soil invertebrates in various ways: by providing a food source for detritivores, by stimulating plant growth and litter return, and by contributing to the stabilization of the soil microclimate. Animal dung is a high quality food source which is exploited initially by coprophagous Diptera and dung beetles and later by a wide range of litter-dwelling invertebrates (Laurence, 1954; Valiela, 1974; Curry, 1979). Earthworms, and other soil invertebrates to a lesser extent, respond positively to moderate applications of farmyard manure, animal slurry and other organic wastes. However, very heavy applications can depress numbers of earthworms (Curry, 1976; Anderson, 1980; Curry and Cotton, 1980; see also Fig. 3) and micro-arthropods (Bolger and Curry, 1980); these effects are probably due to high concentrations of ammonia and organic salts.

5. Cultivation

Most groups of invertebrates, apart from Protozoa and Nematoda, are adversely affected by cultivation, although different species vary considerably in their ability to tolerate disturbance. Anécique earthworms such as *Lumbricus terrestris* and *Aporrectodea longa*, which feed on surface litter and have relatively permanent burrows, are severely affected, while species such as *Aporrectodea caliginosa* (Sav.) and *Allolobophora chlorotica* (Sav.), which do not occupy permanent burrows, and which benefit from ploughed-in plant residues, are less affected and usually become dominant in cultivated soils (Edwards, 1983). Earthworm populations in fragile tropical soils have little tolerance for cultivation (Lal, 1987a). Population densities of Acari and Collembola are often an order of magnitude lower in cultivated land than in undisturbed habitats, but very high population densities of some microbial-feeding species can sometimes be found associated with decaying crop residues (Andrén and Lagerlöf, 1983a; Emmanuel et al., 1985).

Most invertebrates, and especially deep-burrowing earthworms, are favored by minimum tillage and direct drilling, compared with conventional methods of cultivation (Gerard and Hay, 1979; Edwards and Lofty, 1982b; Hendrix et al., 1986). Some pest species may also be favored by direct drilling, but others are unaffected, and some may be less troublesome (Speight, 1983; House and Brust, 1989; Unger, 1990).

Crop residue management has important effects on soil properties and biological activity. Surface mulching enhances faunal activity in dry soils particularly (Lal, 1987b), and promotes earthworm burrowing to the soil surface (Zachman et al., 1987). Earthworms tend to be more abundant under crops such as cereals and fruit bushes where significant amounts of residues are left behind compared with root crops where most of the plant production is removed (Edwards, 1983).

6. Soil Water Management

Soil macroinvertebrates are scarcer in arid soils, and irrigation allows groups such as earthworms to become established where they were not previously present (Barley and Kleing, 1964; Reinecke and Visser, 1980). However, high salinity, soil wetness, and anaerobiosis due to excessive irrigation, canal seepage, etc., can limit biological activity and result in severe soil physical and chemical degradation (Fausey and Lal, 1990; Gupta and Abrol, 1990).

Biological activity is also severely limited in wet soils and invertebrate populations increase rapidly when habitats such as polders and peat are reclaimed (van Rhee, 1969a; Curry and Cotton, 1983; Curry and Momen, 1988; Meijer, 1989). Drainage of wet soils, however, can result in marked reductions in hydrophilic groups such as root-feeding *Tipula* spp.

7. Pesticides and Heavy Metals

Hundreds of different plant protection chemicals which are more or less toxic to non-target organisms are applied to cropland every year. Surface-dwelling invertebrates and surface-feeding earthworms such as *Lumbricus terrestris* which may ingest considerable quantities of contaminated crop residues are particularly at risk.

The older copper- and mercury-based fungicides were highly toxic to earthworms, while among the newer materials the most toxic are the substituted benzimidazoles such as benomyl, carbendazim, and thiophanate-methyl (Stringer and Wright, 1976; Lofs-Holmin, 1981). Soil fumigants such as DD, chloropicrin, methyl bromide, and carbon disulfide are highly toxic to most soil invertebrates.

Some organochlorine insecticides when used at high concentrations for the control of soil pests can reduce population densities of many soil invertebrates by more than 50% for several years (Edwards and Thompson, 1973; Brown, 1977). Only chlordane, heptachlor, and endrin have strongly adverse effects on earthworms at normal rates of application, but other substances may have sublethal effects (Lofs-Holmin, 1980; Reinecke and Venter, 1985). Also, residues may accumulate in earthworm tissues with adverse consequences for animals higher up the food chain (Edwards, 1973).

Organophosphates are less persistent and generally less injurious to soil invertebrates than organochlorines. Only phorate and, to a lesser extent parathion, are toxic to earthworms (Edwards 1980; Clements, 1982). Some carbamates including aldicarb, carbaryl, carbofuran, and methiocarb are quite toxic to various invertebrates including Acari, Collembola, and earthworms (Martin, 1975; Edwards, 1980, 1983; Clements et al., 1986; Lal, 1988). Synthetic pyrethroids have been shown to cause short-term reductions in numbers of polyphagous predators (Araneae, Carabidae, Staphylinidae) in winter cereals (Purvis et al., 1988).

Occasional pesticide use probably has little effect on ecosystem functioning, but repeated applications can reduce natural enemy impact on pest species and can affect decomposition and mineralization rates by eliminating earthworms. The development of a surface raw humus layer has been noted following earthworm suppression by copper fungicides in orchards (Raw, 1962; van Rhee, 1977b) and by repeated application of insecticides (mainly phorate) to suppress grassland pests (Clements, 1982).

Landspread municipal sludges and animal wastes may contain relatively high concentrations of heavy metals, but these are mainly organically bound and not considered to be toxic to earthworms (Hartenstein et al., 1980; Malecki et al., 1982; Neuhauser et al., 1984). However, copper toxicity has been cited as the likely cause of low earthworm populations in some sites heavily contaminated by pig slurry (van Rhee 1977b; Curry and Cotton, 1980; Ma, 1988).

V. Restoring Soil Fauna

A. General Considerations

The speed and extent of natural faunal colonization and establishment will depend on a number of factors relating to the nature of the disturbed site and the availability and attributes of potential colonizers. While the level of faunal restoration that is attainable will ultimately be constrained by climatic, and edaphic and management factors, the rates and extent of community development will be strongly influenced by the degree of degradation that the site has suffered and the extent of site rehabilitation. In the case of moderate degradation due to excessive cropping, once the land is taken out of cultivation the fauna recovers rapidly as the process of secondary succession gets under way. However, site conditions in newly reclaimed mining waste are often very hostile, and the site must undergo the slower and less predictable process of primary succession. Prerequisites for successful faunal establishment in such sites are the amelioration of severely limiting factors such as low pH, the stabilization of the physico-chemical environment, and the provision of a food base for the decomposer community in the form of organic waste or plant litter. Resoiling greatly facilitates restoration, although soil stored in large heaps for significant periods before spreading can undergo considerable degradation.

The size and shape of the reclaimed area and the proximity of suitable colonizers will influence the time scale of fauna restoration, particularly in the case of organisms with limited powers of dispersal. The nature and extent of revegetation has a major influence on the potential of reclaimed land to support fauna. Generally, studies on rehabilitated land show positive associations between faunal biomass, abundance, and diversity and vegetational parameters such as plant cover, floristic diversity, plant architecture, and the presence of habitats such as logs and litter (Majer, 1989c).

The early colonizers of disturbed habitats are opportunistic species; they are generally small and short-lived, with the capacity for rapid dispersal and rapid multiplication, characteristics associated with "r selected" species in the sense of Southwood (1977). Examples include many species of Acari and Collembola which are dispersed on other animals or by wind currents, and where there is sufficient organic matter, large numbers of dipterous larvae are soon found (Dunger, 1969*a*, *b*; Curry and Momen, 1988). As site conditions become more favorable and more predictable, larger-sized and more long-lived species with poorer powers of dispersal, with lower reproductive rates, and with longer generation times ("K selected" species) are able to become established; "r" and "K" selected types are two extremes of a continuum, and many of the species found in maturing sites will be intermediate in their characteristics between these two extremes. A third group ("A" selected) was proposed by Greenslade (1983) to categorize those species which can tolerate adverse environmental conditions.

B. The Process of Colonization and Succession

The classical view of succession as a highly deterministic process starting with a pioneer stage comprising a few species of early colonizers and proceeding through well-defined seral stages of increasing complexity towards a stable climax community (Odum, 1969) seems applicable in relatively few if any restoration cases. A diametrically opposed view considers succession as an essentially random process (Drury and Nisbet, 1973), while may studies (predominantly concerned with plant succession) point to the conclusion that there are several alternative mechanisms which can determine which species replace the early colonizers (Connell and Slatyer, 1977; MacIntosh, 1980; Begon et al., 1990).

The relationships between vegetational and faunal community development are most apparent in the above ground community, with maximal

Table 3. Colonization of mine spoils by selected groups, showing the importance of dispersal rate in succession (note that oribatids are an exception). Data refer to those species or taxa studied in each group, not to the group as a whole

Group	Tropic status	Dispersal rate	Age of site since rehabilitation[1–3] 2 years	>2 years	Source
Carabidae	Predatory	Rapid	+	−[1]	Strüve-Kusenberg (1981)
Chilpoda	Predatory	Slow	−	+[1]	Strüve-Kusenberg (1981)
Ispoda + Diplopoda	Detritivores	Slow	−	+[1]	Strüve-Kusenberg (1981)
Curculionidae	Herbivores	Rapid	+	−[1]	Strüve-Kusenberg (1981)
Oribatidae	Detritivores	Rapid	−	+[1,2]	Dunger (1989)
Collembola	Detritivores	Rapid	+	+[1]	Cross and Wilman (1982) Hutson (1980)
Protura	Mycorhizal and hyphal feeders	Slow	−	+[3]	Dunger (1989) Dunger (1989)

+, abundant; −, absent or uncommon
[1] 10–11 years
[2] 6 years
[3] 30 years

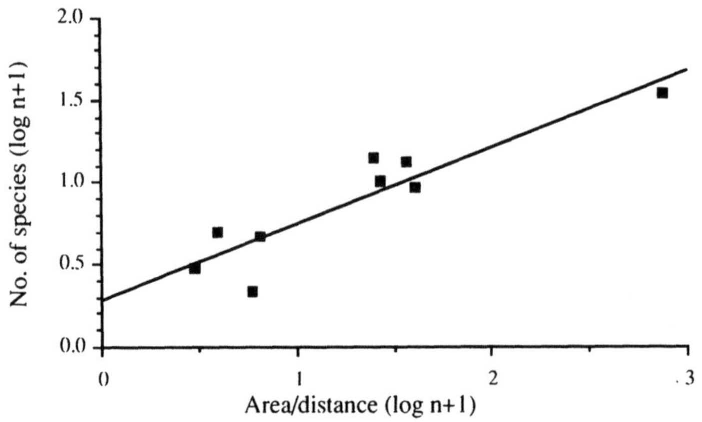

Figure 2. Regression of number of staphylinid beetle species (a rapidly dispersing group) in the Westmann Islands, against area/distance from mainland (Iceland) ($r^2 = 0.84$) (data from Lindroth et al., 1973). Fewer species occur in small islands at large distances from the source of colonists.

faunal diversity being associated with multilayered, multispecies plant communities (Andrzejewska, 1979a; Murdoch et al., 1972; Stinson and Brown, 1983). The parallels may not be so apparent in the case of the soil community, where factors such as organic matter accumulation and microclimatic stability may be of greater significance, to detritivores at least, than the species composition of the vegetation (Parr, 1978).

Current thinking on colonization owes much to ideas on island biogeography (MacArthur and Wilson, 1967). By analogy with oceanic islands, the equilibrium number of species in a restored area could be regarded as a function of the rate at which new species arrive (immigration rate) and the rate at which species leave (extinction rate). The equilibrium number of species for a given habitat will be determined by its size and structure and the range of resources available. The time taken to reach this equilibrium depends on factors such as the size of the pool of potential colonizing species, the distance from sources of colonizers, and the powers of dispersal and colonizing abilities of the individual species (Table 3). When habitats are similar, distance from source and size of area to be colonized are the major parameters for species with good dispersal abilities (Fig. 2).

Once vegetation cover is established faunal colonization of reclaimed land can proceed very rapidly. Hutson (1980) recorded a peak population density of 131×10^3 mites m^{-2} in mine spoil when reclamation was completed and grass-clover seed had germinated. Curry and Momen (1988) recorded 174 arthropod species and total population densities of 74×10^3 m^{-2}, comprising mainly Acari, Collembola, and Diptera within 2

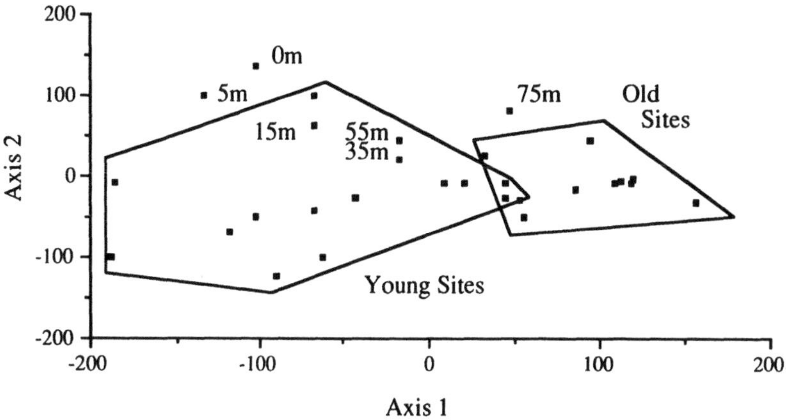

Figure 3. Detrended correspondence analysis ordination (using DECORANA program) of earthworm presence/absence data from Irish grassland sites on reclaimed peat (data from Curry and Cotton, 1983). Dots represent sites, and clusters represent young sites (1–24 years) and old sites (> 25 years). Axis 1 of the ordination (accounting for 43% of variance) shows a trend of succession from young to mature sites. Labelled dots represent samples from a different data set—a gradient of contamination by pig slurry overflow at a single grassland site (data from Curry and Cotton, 1980). Samples were taken 1 year after the overflow. Because the species composition was similar to that of the reclaimed sites they could be included in the same ordination (but because the sampling intensity was greater only the highest ranking six species were included). Their ordination positions show that the areas most contaminated (i.e., nearest source—0 m, 5 m, 15 m) contained an early successional fauna. If this site were monitored yearly, "movement" of the dots representing the heavily contaminated areas would be expected toward the right of the ordination, and recovery of the site could be regarded as complete when the dots occurred in the same region as the 'old sites' cluster

years of reclamation and seeding of cutaway peat. These early populations largely comprise r selected species, and it may take many years (possibly decades) before a balanced community with a good representation of K selected species is present. For example, significant earthworm population densities are often only recorded in reclaimed peat sites more than 25 years old (Fig. 3), although under the most favorable site conditions this time interval can be considerably less (Curry and Cotton, 1983).

C. Faunal Succession in Mine Spoil—a Case Study

While faunal succession will vary in many respects from site to site, many features of the process are well illustrated by data from rehabilitated coal mining dumps in the former German Democratic Republic summarized by Dunger (1989). Forty dumps were studied over a period of 25 years; most had been afforested with *Populus*, *Alnus*, and *Robinia* and seeded with clover and lupin. Some of the main conclusions are as follows:

1. The most important requirements for faunal rehabilitation of coal dumps were: (a) stabilization of the water regime; (b) nutrient build-up facilitated by rapid production of organic matter by N fixing plants; (c) detoxification, e.g., neutralization of mineral acids by liming.
2. Succession proceeds at different rates above ground and in the soil; the process is generally slower below ground.
3. Succession does not proceed uniformly across all groups with a steady increase in density and diversity. For some groups a "pioneer peak" may be followed by decreasing density and diversity.
4. Succession processes are modified by individual site factors and rehabilitation procedures. Moisture was the most important factor; the amount of dead organic input was of secondary importance.
5. Five stages of succession could be recognized:
   I. The pioneer stage characterized by hot, dry conditions and the absence of organic matter, and colonization by r strategists. This stage could last up to 30 years in harsh sites, but as little as 2 years under favorable conditions. Collembolans tend to be the most abundant micro-arthropods.
   II. The second stage is influenced by the method of rehabilitation and the degree of natural colonization by various organisms. Increasing litter accumulation from the herb layer occurs and high densities of dipterous larvae, Collembola, predatory Carabidae, and spiders are common. Minimal earthworm populations may be present. This stage may end at the 5th year, or may last until the 9th year or longer on acidic dumps.
   III. The third stage is characterized by a shrub layer without a closed canopy and a well-developed litter layer which was colonized and reduced by the epigeic earthworms *Dendrobaena* spp., and later, *Lumbricus rubellus* Hoff. This stage may last from about 5 years on better (Pleistocene) sites, to 10 or more years on Tertiary acid dumps.
   IV. This stage begins with the closure of the tree canopy. Earthworms dominate faunal activity and surface organic matter is incorporated into the soil by the activity of anécique species such as *Lumbricus terrestris*.
   V. During the transition to the fully developed woodland stage there is gradual immigration of further species, especially K strategists, and a marked increase in the activity of saprophagous fauna. Under optimum conditions the woodland stage may commence 20 to 25 years after rehabilitation. This stage coincides with the end of soil zoological development.

6. Many animal groups have adequate powers of dispersal to allow rapid colonization at the appropriate stage of reclamation, but this was not the case for certain earthworms, snails, isopods, and centipedes.

D. Promoting Faunal Restoration

1. Degraded Agricultural Soils

Since the main factors causing degradation of arable land are physical disturbance associated with cultivation, depletion of soil organic matter, reduced floral diversity, the absence of plant cover for part of the year, and frequent use of pesticides, faunal restoration will be facilitated by eliminating or alleviating the effects of these adverse practices.

Increasing concern about the adverse environmental and economic impacts of conventional, intensive, methods of crop production has stimulated interest in alternative (low input, sustainable) systems which seek to minimize fossil fuel inputs in the form of fertilizers, pesticides, and mechanical cultivations and to maintain crop yields through the use of renewable natural resources. Central to this concept is the need to promote and maintain the biological processes which sustain nutrient cycling and soil fertility and which promote natural pest control. Different approaches will be required for different soil and climatic conditions: potentially useful components of low input systems include the use of legumes, catch crops, living mulches, green manures, and organic wastes to provide nutrients and maintain soil organic matter and moisture levels; adoption of conservation tillage practices; agroforestry; use of intercropping, strip cropping, double row cropping, and ridge planting techniques; appropriate choice of cultivation equipment for seedbed preparation, planting, and mechanical weed control; timing of field operations for maximum effectiveness with minimum disturbance; and promotion of natural pest control through habitat manipulation (Lal, 1986, 1987b; Edwards, 1989; El Titi and Ipach, 1989; House and Brust, 1989; Pimentel et al., 1989; Unger, 1990).

The soil fauna responds positively to the favorable conditions obtaining under low input regimes, and although evidence is scarce, most accounts point to a positive influence on soil fertility and productivity. Edwards and Lofty (1978; 1980) demonstrated that soil animals, particularly earthworms, improve root growth and shoot biomass in direct drilled cereals. Through their interactions with organic matter and microflora, the fauna may contribute to decomposition and the regulation of nutrient cycling in a manner analogous to that in undisturbed ecosystems (House et al., 1984; Hendrix et al., 1986; House and Brust, 1989). Earthworm activity improves water movement and aeration in no-till soils (Wilkinson, 1975; Shipitalo and Protz, 1987), and can reduce surface crusting (Kladivko et al., 1986), although surface casting could contribute to soil erosion and crusting in areas exposed to raindrop impact (Sharpley et al., 1979; Shipitalo and Protz, 1988), while preferential flow of water in worm channels could result in loss of nutrients (and pesticide residues) to groundwater (Wild and Babiker, 1976; Bouma et al., 1982).

Termite activity can be encouraged by surface plant residues in compacted dry land soils, with beneficial consequences for soil properties (Lal, 1988). Minimum tillage methods and surface crop residues can promote the survival and activity of a range of crop pests including slugs and shoot flies (Edwards, 1975; Speight, 1983; Glen et al., 1984), but such conditions also favor polyphagous surface predators and the natural control of at least some pests (El Titi and Ipach, 1989; House and Brust, 1989). Pest control can be enhanced by a range of measures aimed at providing habitat for natural enemies; these include intercropping, strip cropping, uncultivated strips within the crop, uncultivated field margins, sowing of plants which are attractive to natural enemies adjacent to crop fields, etc. (Edwards, 1989, El Titi and Ipach, 1989; House and Brust, 1989). However, precise management is required as some of these measures can create new pest, disease, and weed problems.

2. Severely Disturbed Soils

The features of severely disturbed sites which most inhibit biological activity are likely to be lack of organic matter and suitable food, unfavorable microclimatic conditions, and chemical toxicity often related to low pH. Assuming that suitable colonizers are available in the area the process of faunal establishment can be greatly faciltated by site ameliorative measures such as the following:

1. Liming to counteract low pH and metal toxicity.
2. Return of topsoil to improve conditions for plant growth and soil fauna, especially earthworms. Lack of topsoil can be overcome by the use of suitable organic materials.
3. Organic amendment to provide substrate for decomposers, to stabilize soil moisture and temperature regimes, and to decrease heavy metal toxicity. Straw mulch, bark, animal manures, and sewage sludge may be suitable provided they do not contain unacceptably high levels of pollutants.
4. Establishment of plant cover greatly accelerates faunal return. Legumes are particularly beneficial in providing biologically fixed N and plant residues of high quality. Native species of legumes are generally preferable (Brenner et al., 1984).
5. Habitat diversity is important in determining the range of species which can become established. For example, the establishment of a rich ant, termite, and other invertebrate fauna on rehabilitated bauxite mine

sites in Australia was enhanced by the presence of a dense litter layer, large logs, and a diverse plant cover (Nichols et al., 1989).

6. Retention of patches or strips of undisturbed habitat within the area can greatly enhance faunal restoration (Recher, 1989).

In general, faunal recolonization can be left to chance. However, as the coal mine spoil example illustrates, natural colonization by some groups with poor powers of dispersal can be slow and there may be many situations in which deliberate introduction of key invertebrates (usually earthworms) can significantly accelerate the reclamation process.

E. Introducing Invertebrates

Resoiling can be an effective way of introducing a range of soil invertebrates, although redistribution of stored soil may be of limited value since most organisms perish during storage. The effectiveness of resoiling could be increased by distributing small quantities of fresh soil and litter containing a mixed inoculum of fauna from appropriate habitats. More specific introduction programs are required when the object is to introduce particular earthworm species or other key invertebrates. An example of the latter might be mound-building ants which are poor natural colonizers and which have an important role in soil cultivation and in promoting floral diversity in restored prairie (Baxter and Hole, 1967, cited by Kline and Howell, 1987).

Earthworms have been the subject of a number of introduction studies, and the consequent effects on soil properties have been well documented in a number of cases. Systematic earthworm introductions have been carried out in improved hill pasture in New Zealand using the sod-transplanting technique with impressive economic results (Stockdill, 1982). Among the effects noted following the establishment of *Aporrectodea caliginosa* were increased proportions of sown species in the sward, disappearance of the surface organic mat, improvement in soil structure, and an initial increase in dry matter production of 70% declining to a sustained level of about 30% (Stockdill, 1959; Stockdill and Cossens, 1966). Earthworm introductions into reclaimed Dutch polders have been described by van Rhee (1969a, b; 1977b), and by Hoogerkamp et al. (1983). The latter authors used infrared line scanning to monitor earthworm spread based on different patterns of heat exchange from land with and without a surface mat. The mat was found to be ingested and incorporated into the soil within 3 years of worm invasion, and improved soil structure, root growth, botanical composition, and grass yields were observed although areas with earthworms tended to be more prone to damage by cattle and soiling of grass in wet weather.

Dunger (1969a) reported successful introduction of *A. caliginosa* into reclaimed mine spoil where the surface had been mulched with leaf com-

Table 4. Effects of earthworms on grassland in reclaimed cutover peat in central Ireland (Curry and Bolger, 1985; Curry and Boyle, 1987; Boyle 1990)

Carrying capacity	100 g fresh mass m^{-2}
Litter consumption	365 g dry mass m^{-2} yr^{-1}
Soil consumption	1.3 kg m^{-2} yr^{-1} or 20 cm layer worked in 45 years
Soil properties affected	Bulk density
	Water infiltration
	Degree of humification
	Micromorphology
Increased grass/clover shoot production	25%–50% in field microplots receiving cattle slurry; 30% in glasshouse

post. Earthworm establishment was followed by greatly accelerated litter incorporation and a change in soil humus type from moder to mull (Dunger 1969a, b). Successful introductions of *Lumbricus terrestris* into reclaimed mine spoil in Ohio resulting in enhanced litter incorporation and improved soil structure were reported by Vimmerstedt and Finney (1973) and Hamilton and Vimmerstedt (1980).

Earthworm introductions into some reclaimed cutover peat sites in Ireland had beneficial consequences for soil development and grass growth (Table 4). However, under favorable site conditions deliberate introduction is probably unnecessary as significant populations can become established through natural colonization within a few years (Boyle 1990). Pilot introductions of *A. caliginosa* into limed coniferous forests in Finland have also been encouraging (Huhta, 1979).

Most earthworm introductions to date have been small in scale, using transplanted grass sods containing earthworm populations and coccoons, or releasing precollected live earthworms. The sod distribution technique has been successfully mechanized in New Zealand, but is of limited use for the introduction of valuable deep-working species such as *Lumbricus terrestris* and *Aporrectodea longa*.

VI. Faunal Indicators and Biological Monitoring of Soil Quality

A. Value of Faunal Indicators

Bioindicators and biological monitoring refer to assessments of the quality of the environment based on selected biological measurements, which can range from the molecular to the ecosystem level. Generally, for the restoration of sustainable soil productivity, physicochemical indicators (e.g., ^{137}Cesium concentration) (Piekarz, 1990) or microbial indicators (e.g., ATP concentration) (Sims, 1990) will suffice, but faunal indicators are useful in the rehabilitation of polluted sites, and also as indicators of ecosystem development.

Biological monitoring of pollution has several advantages compared with physicochemical monitoring: (1) monitoring is continuous rather than sporadic; (2) low concentrations may be toxic but undetectable by chemical monitoring; (3) toxic effects are being monitored under natural conditions—toxicity can vary widely under different environmental circumstances; (4) living organisms can monitor mixtures of chemicals which may have different effects together than separately; (5) impact on ecosystem components of particular value can be assessed (e.g., nitrogen cycling, pest predators). While the use of faunal indicators of pollution has been best developed in aquatic habitats, in the past decade there has been an increase in interest in the potential of soil animals as indicators of pollution due to acidification, pesticides, and heavy metals.

Soil fauna, as indicators of the rehabilitation of ecosystem quality, have the advantage that they can integrate information about other ecosystem values as well as soil quality, such as biocontrol potential, game food availability, and biodiversity. For example, insect monitoring in intensively managed English cereal fields has revealed declines in numbers of aphid natural enemies as well as insects important as food for partridge chicks (Potts, 1986). In rehabilitated amenity areas, earthworm species such as *Lumbricus terrestris* not only provide information on soil development, but are also an important food source for wildlife such as badgers, foxes, and thrushes (MacDonald, 1983). Due to this ability to integrate diverse information, faunal indicators have been claimed to have much potential as measures of ecosystem recovery on land reclaimed after mining or similar major disturbances (Majer, 1989b), and they also can suggest suitable management adjustments in the early period (Nichols et al., 1989). For example, Elkins et al., (1984) found that differences in the relative abundance of Acari and Collembola (at family level) reflected the superiority of bark-amended strip-mine spoils to top-soiled spoil and spoil with straw-mulch, and Majer et al. (1984) found certain seeding practices to result in improved restoration of Australian Jarrah forest after bauxite mining, using ant assemblages as indicators. However, to be effective, it is important that faunal monitoring is carefully designed in relation to the objectives of the rehabilitation program, and this will partly depend upon the criteria laid down by regulatory agencies.

Rehabilitation of toxic mine dumps and similar land with a potential toxic problem will often require biological monitoring of the developing ecosystem to estimate whether or not pollution is occurring. Furthermore, management for amelioration of degraded soils may itself sometimes increase certain types of pollution, for instance, during fertilizer and lime incorporation into acid forest soils (Bååth et al., 1980; Ulrich, 1987), and this also may require monitoring.

Various criteria have been proposed for biological indicators, some of which are debatable, and others which have general application. Much will depend upon the objectives of the monitoring program, and there will often be a trade-off between methodological difficulty and selection of sensitive and reliable indicators. It is necessary that some sort of baseline or standard is established in order to assess whether observed changes in indicator parameters are within normal fluctuations or not (Weinert, 1986; Eijsackers, 1987), although this can be difficult in ecosystems such as plantation forests which are continually changing. In the context of pollution it is necessary to define how much degradation due to pollution is acceptable (Logan, 1990), and to relate this to levels of change in indicator parameters. Generally, indicators need to be feasible in their use (Day, 1990), and they should be pollution or management related. However, many potential indicators can be difficult to assess for this latter criterion a priori, and their choice is often based on extrapolations from other studies, or on ecological generalities. There is a need for basic, analytical, and comparative research on soil fauna specifically addressing their indicator potential under different conditions.

B. Types of Faunal Indicator

Faunal indicators can be chosen to measure response at the individual, population, and community levels. Bioindicator species, which are either very sensitive or very tolerant of change, can be used to assess changes at individual or population levels, and various measures of community structure can be used to assess changes at the community level.

1. Bioindicator Species

At the individual level changes in behavior, biochemistry, morphology, physiology, and pathology, as well as in growth and reproduction, can be measured to determine impact of pollution or degradation. Such measures have the advantages of being sensitive (e.g., by indicating sublethal effects), and they possess good diagnostic potential. For example, Ma and Eijsackers (1989) conclude that reproductive success in earthworms is a good indicator of soil toxicity. However, it is important to choose the correct biological parameter. Christensen and Mather (1990) found that certain parameters (e.g., cocoon number) were considerably influenced by site conditions and management practices in a comparison of sites of differing soil quality.

At the population level, changes in relative abundance of a species are often cited as being more useful than measures of absolute density, but it is critical that sufficient evidence is available to show that population changes are in fact due to degradation and rehabilitation. An example is given by van Straalen et al. (1988), who found that changes in relative abundance of the oribatid mite, *Platynothrus peltifer* Koch, provided a good early warning indicator for monitoring decline in forest soil quality, when used in conjunction with chemical analysis. The species was found experimentally to avoid acid substrates, and to have a relatively high requirement for manganese. Comparison of 12 coniferous forests showed that the relative abundance of the species decreased in sites with low Mn concentration and poor tree vitality.

2. Community Structure

It is a widely held belief that the restoration of biological diversity will lead to increased stability. However, because of the many different meanings of both "diversity" and "stability" (see Pimm, 1984), this is often an oversimplification. For instance, measurement of "diversity" as arthropod species richness, and "stability" as fewer pest outbreaks, does not always support the diversity = stability relationship. In the context of agro-ecosystem rehabilitation, there is much evidence for diversity of vegetation and habitat structure leading to reduced pest problems (Altieri and Letourneau, 1982), but here again there are alternative arguments (see Sheehan, 1986), and Nordlund et al. (1984) are of the opinion that generalizations about natural enemies and plant diversity are not possible. In soil food webs, diversity of energetic pathways appears to be important, rather than species richness. An example is provided by the conclusions of Andrén et al. (1988), who proposed a hypothetical food web for barley straw-litter decomposition. The rate of mass loss was the same at high and low numbers of the detritivore springtail *Folsomia fimentaria* (L.), which could be explained by its role being taken over by other parallel trophic pathways in the food web.

Diversity indices have been popular as measures of change at the community level, and have the advantage that much information can be represented in a single index, but they have sometimes been found to be misleading, particularly in agro-ecosystems, which are already disturbed by cultivation, mowing, or grazing (Purvis and Curry, 1980; Dritschilo and Erwin, 1982; Siepel and van de Bund, 1988). Indices have the further disadvantage of not providing information on the behavior of the system, and they lack predictive value (Bernard, 1990). A review of these and other measures of community structure is given by Sheehan (1984).

Changes in species-abundance patterns (e.g., from a geometric series to a log-normal) have been found along pollution gradients, and theoretical models have been developed to account for these patterns, but such models and distributions are far from universal (Gray, 1987), and need to be interpreted with caution, especially in agro-ecosystems. However, a useful conclusion of comparisons of these patterns is that species of intermediate abundance show the most change under stress (Emmanuel et al., 1985; Gray, 1987), and indicator species can be found within this intermediate group (Pearson et al., 1984).

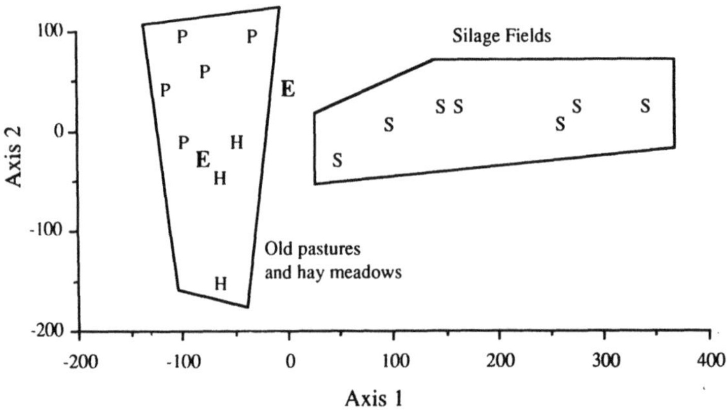

Figure 4. Detrended correspondence analysis ordination (using DECORANA program) of staphylinid beetle presence/absence data from Irish grassland sites (data from Good and Giller, 1990). Letters represent sites: S, silage; H, hay meadow; P, permanent lightly grazed pasture. Axis 1 of the ordination (accounting for 63% of variance) shows a separation of disturbed sites which have been cut for silage from hay meadows and lightly grazed pastures. Two sites are marked "E"; these are glacial esker grasslands, and of conservation interest. One of these, a site in a heritage zone designated for landscape conservation was represented outside the old field cluster and near to the disturbed field cluster, whereas the other esker site was within the old field cluster. The heritage zone site was grazed by cattle, and lighter grazing would be recommended to allow recovery of soil fauna

Multivariate analysis techniques such as detrended correspondence analysis and canonical correspondence analysis, are being increasingly used to detect changes in soil faunal assemblages due to environmental and management factors. Siepel and van de Bund (1988) identified several arthropod indicator species responding to nitrogen fertilization using canonical correspondence analysis. Meijer (1989), using the same technique, concluded that time and desalination were major factors influencing the faunal recolonization of reclaimed polders in The Netherlands, and Rushton et al. (1989) demonstrated changes in species composition of upland macro-arthropod communities due to the insecticide chlorpyrifos by detrended correspondence analysis. Examples of the application of this technique to evaluate the responses of different grassland and forest faunal groups to various kinds of disturbance are given in Figs. 3–5. Caution is necessary when comparing data from different regions and different seasons because of differences in species distributions and phenology—the technique requires species equivalence within data sets. As yet, it is not clear how similar species structure should be to predegraded or control sites to conclude that community structure is restored. Heatwole and Levins (1972) concluded from a study of faunal return to red mangrove

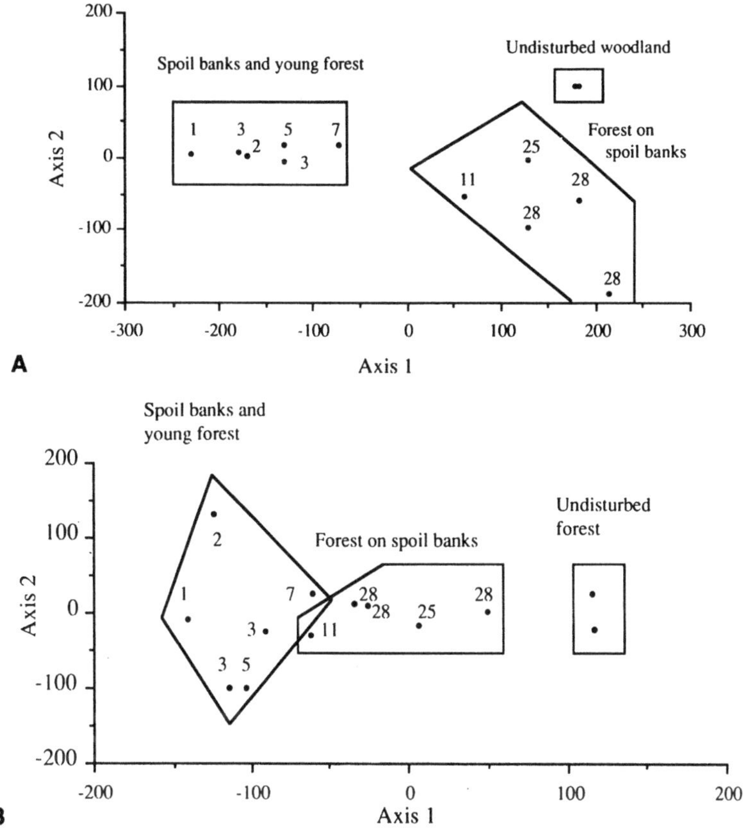

Figure 5. Detrended correspondence analysis ordination (using DECORANA program with downweighting of rare species) of carabid beetle (A) and diplopod and isopod (B) presence/absence data from German coal refuse dumps (data from Neumann, 1971; species comprising less than 2% of sample omitted). In both cases axis 1 shows a trend of faunal succession, and is significantly correlated with age of site (Spearman rank, carabids: $rs = 0.95$, $P < 0.01$; diplopods/isopods: $rs = 0.92$; $P < 0.01$). Axis 1 represents 71% of variation for carabids, and 37% for diplopods/isopods. Note that in both cases the oldest spoil banks are different from the natural woodland sites. However, a greater variety of undisturbed woodlands would need to be included to ascertain whether this is an important difference or not. The two plots show a difference in pattern, in that the carabids are more divergent in older sites, while the diplopods and isopods are more divergent in younger sites. A probable explanation for this is that rapidly colonizing carabids quickly form a distinct assemblage of species associated with disturbed habitats, but in later stages species with more specialized habitat requirements will invade, and subtle differences in habitat conditions in different sites will lead to different dominant species. In contrast, slowly dispersing diplopods and isopods are likely to colonize sites somewhat randomly in the early stages, and so large differences between sites in the early stages will be due to random recruitment of different species in different sites. Later on, the fauna will gradually add more species, but without the same degree of habitat specialization as in carabids

islands with intact vegetation (fauna removed by fumigation) that trophic structure was recovered within 1 year, whereas taxonomic structure was not. Parmenter and MacMahon (1987) found that for Coleoptera both species structure and trophic structure were different from control sites after early succession on rehabilitated strip-mined land. However, if a sufficiently large and variable data set is used to rate a site of interest, and the influence of rare or "tourist" species is reduced, this problem should be overcome. This can be seen in the examples in Figs. 3 and 4, where small but distinct differences in species composition due to site or season did not affect the overall ordination.

3. Vegetation Diversity and Soil Fauna

Recher (1989) states that plant species diversity can be used as an indicator of faunal habitat rehabilitation, based on the correlation between plant species richness and structural diversity. While this may hold true for vertebrates and plant-associated invertebrates, it has several limitations with regard to the soil fauna. In agro-ecosystems and forest plantations plant species diversity will obviously be a function of management, and even the presence of suitable faunal habitat may not indicate its occupancy, depending on the management practices used. In natural ecosystems, high plant diversity (e.g., in dry grazed calcareous grassland) will not indicate suitable litter/humus habitat which can support a well-developed fauna under ungrazed plant species-poor conditions. Nor will plant diversity tell much about faunal development in a forest cleared of dead wood. In rehabilitated soil, Dunger (1989) found a different succession pattern of hypogeal fauna when compared with plant community succession in coal-mine dumps.

C. Using Faunal Indicators

The kinds of indicators used in monitoring ecosystem rehabilitation are likely to differ from those used in soil pollution monitoring and surveillance, because there are different management objectives and the ecosystem components of interest will also differ.

1. Approaches to Monitoring Soil Pollution

It is well established that monitoring for toxic effects on ecosystem quality cannot be accomplished by use of a single method. Cairns (1981) criticized over-reliance on single species toxicity tests, and Cairns and van der Schalie (1980) conclude that a variety of dissimilar methods are required to provide sufficient information to adequately protect a potentially degradable ecosystem. This suite of dissimilar methods should ideally include linked biological and physicochemical monitoring, functional and structural indicators, descriptive and analytical approaches, modelling and direct observation, and community-level as well as bioindicator species approaches (Cairns, 1982). This would obviously be impractical on every potentially toxic site, but the arguments behind such combinations are important and need to be addressed by monitoring programs. It is important that biological changes can be shown to be due to the specific toxicant and not to some other factor, hence the need for simultaneous chemical monitoring. Several studies infer potential changes in processes based on observed structural changes (e.g., species richness, biomass). For instance, Hoy (1990) concluded that nutrient cycling may be affected by lindane application in a forest soil, based on reductions in densities of several oribatid mite species. However, where structural indicators are used to infer changes in processes such as microbial activity and nutrient cycling, it needs to be shown that such inferences are valid. Eijsackers (1987) stressed the need for both structural and functional measures on the premise that parameters such as soil respiration and species richness will not show true impact if resistant species have assumed the activities of sensitive species. Analytical aproaches involve hypothesis testing and often laboratory experimentation to demonstrate cause and effect, and have been emphasized in some pollution legislation (Herricks and Cairns, 1982). Such data can be incorporated into predictive models, and both analytical and modelling approaches allow much more precise location of components of the system that are most likely to be affected at an early stage, and hence assist the choice of the most useful indicators (Cairns, 1981). In considering single bioindicator species, one must accept that they are very unlikely to fully represent a complex ecosystem (Hellawell, 1977), and where they are used as the main monitoring method, some supplementary community measure is advisable.

2. Rate of Dispersal and Recolonization

It is important to understand the effect of dispersal ability on the reliability of indicators. The presence of many opportunist species and even foraging ants may not indicate site suitability (Ma and Eijsackers, 1989), and the uncritical use of presence/absence data should be avoided. Furthermore, rapidly colonizing groups such as carabids will show a different pattern of succession to slow dispersers like diplopods and isopods (Fig. 5, Table 3). Certain groups like oribatid mites can be absent from recently rehabilitated sites (Table 3) not because of poor dispersal ability, but because of the absence of accumulated litter (Dunger, 1989).

3. Choice of Indicator Groups

Virtually all diverse taxonomic groups of soil fauna have been proposed as being potential indicators. Given that most species-rich groups are likely to include species that will be sensitive to at least some form of soil pollution and rehabilitation management, e.g., acidophile and calciophile species occur in Enchytraeidae, Collembola, and Acari (Hågvar and Abrahamsen 1980), comparisons of different groups without regard to the specific objectives of monitoring, surveillance, or research are likely to be futile. Functional classifications may, however, be useful in determining how representative is an indicator group (Eijsackers, 1987). Arpin et al. (1984) found Collembola to be better indicators of differences in litter structure and Monochida (nematodes) better for chemical properties of the organic layers. Species identification will generally produce more reliable results than identification to higher taxonomic categories (e.g., family level): because species may replace each other without a change in numbers at family level, and, because of more precise ecological preferences, species data are likely to provide a more sensitive indicator of change (Good, 1988). However, groups which are taxonomically unstable, especially where species concepts are not fully worked out, are best avoided.

4. Sampling Methods

Indicators must be reliable, and the reliability of the sampling methods used must be known. Caution should be exercised in interpreting results using certain techniques such as pitfall traps (see Adis, 1979). Chiverton (1984) suggested that catches of large predatory carabid beetles could be increased on insecticide-treated plots due to prey depletion leading to increased locomotory activity of hungry individuals. This could cause misinterpretation of the impact of certain insecticides.

VII. Conclusions

The soil fauna has an important role in decomposition, nutrient cycling, maintaining soil structure, and influencing plant succession in healthy soils. Creating favorable conditions such as provision of organic matter, reduction in acidity, use of more selective insecticides, reduced cultivation, etc., will encourage faunal recolonization in rehabilitated soils. Recolonization depends upon dispersal ability, and, except for very large areas, most fauna will return on its own, but earthworms, isopods, and diplopods may sometimes need to be introduced. There is a clear succession of soil faunal communities following soil rehabilitation, which can take many years in difficult sites like mine spoils, but with judicious management this can be speeded up. In certain cases elements of the soil fauna can be useful as ecological indicators of soil pollution or ecosystem rehabilitation, because they can integrate information on various aspects of ecosystem structure. There is a need for more research on soil fauna in the context of specific management, monitoring, and modelling problems in soil and ecosystem rehabilitation.

References

Adis, J. 1979. Problems of interpreting arthropod sampling with pitfall traps. *Zool. Anz.* 202:177–184.

Altieri, M.A. and D.K. Letourneau. 1982. Vegetation management and biological control in agroecosystems. *Crop Prot.* 1:405–430.

Anderson, C. 1980. The influence of farmyard manure and slurry on the earthworm population (Lumbricidae) in arable soil. In: D.L. Dindal (ed.) *Soil Biology as Related to Land Use Practices*, p. 325–335. Proc. 7th Intern. Coll. Soil Zool. Washington, EPA.

Andrén, O. and J. Lagerlöf, 1983. Soil fauna (microarthropods, enchytraeids, nematodes) in Swedish agricultural cropping systems. *Acta Agric. Scand.* 33:33–52.

Andrén, O., K. Paustian, and T. Rosswall. 1988. Soil biotic interactions in the functioning of agroecosystems. *Agric. Ecosyst. Environ.* 24:57–67.

Andrén, O., T. Lindberg, K. Paustian, and T. Rosswall (eds.) 1990. Ecology of arable land—organisms, carbon and nitrogen cycling. *Ecol. Bull. (Copenhagen)* 40:1–222.

Andrzejewska, L. 1979a. Herbivorous fauna and its role in the economy of grassland ecosystems. I. Herbivores in natural and managed meadows. *Pol. Ecol. Stud.* 5:5–44.

Andrzejewska, L. 1979b. Herbivorous fauna and its role in the economy of grassland ecosystems. II. The role of herbivores in trophic relationships. *Pol. Ecol. Stud.* 5:45–76.

Andrzejewska, L. and G. Gyllenberg. 1980. Small herbivore subsystems. In: A.I. Breymeyer, and G.M. van Dyne (eds.) *Grasslands, Systems Analysis and Man*, pp. 201–267. Cambridge, Cambridge University Press.

Andrzejewska, L. and Z. Wojcik. 1970. The influence of Acridoidea on the primary production of a meadow (field experiment). *Ekol. Pol.* 18:89–109.

Arpin, P., J.-F. Ponge, B. Dabin, and A. Mori. 1984. Utilisation des nématodes Monochida et des Collemboles pour caractérises des phénomènes pédobiologiques. *Rev. Écol. Biol. Sol* 21:243–268.

Athias, F. 1976. Recherche sur les microarthropods du sol de la savane de Lamto (Côte d'Ivoirc). *Ann. Univ. Abidjan sér. E (Ecol.)* 9:193–271.

Athias, F., G. Josens, and P. Lavelle. 1974. Traite généraux du peuplement endogé: Le peuplement animal. In: F. Athias, G. Josens, P. Lavelle and R. Schaeter (eds.) Analyse d'un Ecosystème Tropicale Humidé: La Savane de Lamto (Côte d'Ivoire). Les Organismes Endogés de la Savane de Lamto. *Bull. de Liaison des Chercheurs de Lamto* (No. spécial) 5:45–54.

Bååth, E., B. Berg, U. Lohm, B. Lundgren, H. Lundkvist, T. Rosswall, B. Södersöm, and A. Wiren. 1980. Effects of experimental acidification and liming on soil organisms and decomposition in a Scots pine forest. *Pedobiologia* 20:85–100.

Barley, K.P. 1959. The influence of earthworms on soil fertility. II. Consumption of soil and organic matter by the earthworm *Allolobophora caliginosa* (Savigny). *Aust. J. Agric. Res.* 10:179–185.

Barley, K.P. and C.R. Kleing. 1964. The occupation of newly irrigated lands by earthworms. *Aust. J. Sci.* 26:290–291.

Begon, M., J.L. Harper, and C.R. Townsend. 1990. *Ecology: Individuals, Populations and Communities*, 2nd edn. Oxford, Blackwell.

Bernard, D. 1990. Community or ecosystem level indicators. *Environ. Monit. Assess.* 15:281–282.

Bolger, T. and J.P. Curry. 1980. Effects of cattle slurry on soil arthropods in grassland. *Pedobiologia* 20:246–253.

Bornebusch, C.H. 1930. The fauna of forest soil. Forst. ForsVaes. Danm. II, 1–224. Copenhagen.

Boström, U. 1988. *Ecology of Earthworms in Arable Land: Population Dynamics and Activity in Four Cropping Systems.* Department of Ecology and Environmental Research, Report 34. Uppsala, Swedish University of Agricultural Sciences.

Bouma, J., C.F.M. Belmans, and L.W. Dekker. 1982. Water infiltration and redistribution in a silt loam subsoil with vertical worm channels. *Soil Sci. Soc. Amer. J.* 46:917–921.

Boyle, K. 1990. *The Ecology of Earthworms (Lumbricidae) in Grasslands on Reclaimed Cutover Peatland and their Impact on Soil Physical Properties and Grass Yield.* Unpublished Ph.D. Thesis. Dublin, National University of Ireland.

Brenner, F.J., M. Werner, and J. Pike. 1984. Ecosystem development and natural succession in surface coal mine reclamation. *Minerals Environ.* 6:10–22.

Breymeyer, A. 1978. Analysis of the trophic structure of some grassland ecosystems. *Pol. Ecol. Stud.* 4:55–128.

Breymeyer, A.I. 1980. Trophic structure and relationships. In: A.I. Breymeyer, and G.M. van Dyne (eds.) *Grasslands, Systems Analysis and Man*, pp. 799–819. Cambridge, Cambridge University Press.

Brown, A.W.A. 1977. *Ecology of Pesticides.* New York, Wiley.

Bunnell, F.L., S.F. MacLean, and J. Brown. 1975. Barrow, Alaska, U.S.A. In: T. Rosswall, and O.W. Heal (eds.) Structure and Function of Tundra Ecosystems. *Ecol. Bull.* (*Stockholm*) 20:73–124.

Cairns, J., Jr. 1981. Biological monitoring. Part VI—Future needs. *Water Res.* 15:941–952.

Cairns, J., Jr. 1982. *Biological Monitoring in Water Pollution.* Oxford, Pergamon Press.

Cairns, J., Jr. and van der Schalie, W.H. 1980. Biological monitoring. Part I—Early warning systems. *Water Res.* 14:1179–1196.

Chernova, Yu. J., E.V. Dorogostaiskaya, T.V. Gerasimenko, I.V. Ignatenko, N.V. Matveyeva, O.M. Parinkina, T.G. Polozova, E.N. Romanova, V.F. Schamurin, N.V. Smirnova, I.V. Stepanova, B.A. Tomilin, A.A. Vinokurov, and O.V. Zalensky. 1975. Tareya, U.S.S.R. In: T. Rosswall and O.W. Heal (eds.) Structure and Function of Tundra Ecosystems. *Ecol. Bull.* (*Stockholm*) 20:159–181.

Chiverton, P.A. 1984. Pitfall trap catches of the carabid beetle *Pterostichus melanarius*, in relation to gut contents and prey densities, in insecticide treated and untreated spring barley. *Entomol. Exp. Appl.* 36:23–30.

Christensen, O. and J.G. Mather. 1990. Dynamics of lumbricid earthworm cocoons in relation to habitat conditions at three different arable sites. *Pedobiologia* 34:227–238.

Clarholm, M. 1984. Microbes as predators or prey. Heterotrophic, free-living Protozoa: Neglected microorganisms with an important task in regulating bacterial populations. In: M.J. Klug, and C.A. Reddy (eds.) *Current Perspectives on Microbial Ecology*, pp. 321–326. Washington, A.S.M.

Clements, R.O. 1982. Some consequences of large and frequent pesticide application to grassland. In: K.E. Lee (ed.) *Proc. 3rd Australas. Conf. Grassl. Invertebr. Ecol*, pp. 383–396. Adelaide, South Australian Government Printer.

Clements, R.O., B.R. Bentley, and C.A. Jackson. 1986. The impact of granular formulations of phorate, terbufos, carbofuran, carbosulfan and thiofanox on newly sown Italian ryegrass (*Lolium multiflorum*). *Crop Prot.* 6:389–394.

Coleman, D.C., R. Andrews, J.E. Ellis, and J.S. Singh. 1976. Energy flow and partitioning in selected man-managed and natural ecosystems. *Agro-ecosystems* 3:45–54.

Connell, J.H. and R.O. Slatyer. 1977. Mechanisms of succession in natural communities and their role in community stability and organization. *Amer. Nat.* 111:1119–1144.

Coulson, J.C. and J.B. Whittaker. 1978. Ecology of moorland animals. In: O.W. Heal, and D.F. Perkins (eds.) *The Ecology of some British Moors and Montane Grasslands*, pp. 52–93. Berlin, Springer-Verlag.

Cross, E.A. and J.M. Wilman. 1982. A preliminary study of the refaunation of alkaline shale coal surface mine spoil by soil arthropods. In: D.H. Graves (ed.) 1982 *Symposium on Surface Mining, Hydrology, Sedimentology and Reclamation*, pp. 353–358. Lexington, University of Kentucky.

Curry, J.P. 1969. The qualitative and quantitative composition of the fauna of an old grassland site at Celbridge, Co. Kildare. *Soil Biol. Biochem.* 1:219– 227.

Curry, J.P. 1976. Some effects of animal manures on earthworms in grassland. *Pedobiologia* 16:425–438.

Curry, J.P. 1979. The arthropod fauna associated with cattle manure applied as slurry to grassland. *Proc. R. Ir. Acad.* 79B:15–27.

Curry, J.P. 1983. Effects of management on soil decomposers and decomposition processes in grasslands and croplands. In: M.J. Mitchell, and J.P. Nakas (eds.) *Microfloral and Faunal Interactions in Natural and Agroecosystems*, pp. 349–398. Dordrecht, Martinius Nijhoff.

Curry, J.P. and T. Bolger. 1985. Growth, reproduction and litter and soil consumption by *Lumbicus terrestris* L. in reclaimed peat. *Soil Biol. Biochem.* 16:253–257.

Curry, J.P., and K.E. Boyle. 1987. Growth rates, establishment and effects on herbage yield of introduced earthworms in grassland on reclaimed cutover peat. *Biol. Fertil. Soils* 3:95–98.

Curry, J.P. and D.C.F. Cotton. 1980. Effects of heavy pig slurry contamination on earthworms in grassland. In: D.L. Dindal (ed.) *Soil Biology as Related to Land Use Practices*. Proc. 7th Intern. Coll. Soil Zool, pp. 336–343. Washington, EPA.

Curry, J.P. and D.C.F. Cotton. 1983. Earthworms and land reclamation. In: J.E. Satchell (ed.) *Earthworm Ecology—from Darwin to Vermiculture*, pp. 215–228. London, Chapman and Hall.

Curry, J.P. and F.M. Momen. 1988. The arthropod fauna of grassland on reclaimed cutaway peat in central Ireland. *Pedobiologia* 32:99–109.

Dangerfield, J.M. 1990. Abundance, biomass and diversity of soil macrofauna in savanna woodland and associated managed habitats. *Pedobiologia* 34:141–150.

Davidson, R.L. 1979. Effects of root feeding on foliar yield. In: T.K. Crosby, and R.P. Pottinger, (eds.) *Proc. 2nd Australas. Conf. Grassl. Invertebr. Ecol*, pp. 117–120. Wellington, Government Printer.

Day, K. 1990. Rapporteur's report of work group: indicators at the species and biochemical level. *Environ. Monit. Assess.* 15:277–279.
Dritschilo, W. and T.L. Erwin. 1982. Responses in abundance and diversity of cornfield carabid communities to differences in farm practices. *Ecology* 63:900–904.
Drury, W.H. and I.C.T. Nisbet. 1973. Succession. *J. Arnold Arbor.* 54:331–368.
Dunger, W. 1969a. Fragen der naturlichen und experimentellen Besiedlung kulturfeindlicher Boden durch Lumbriciden. *Pedobiologia* 9:146–151.
Dunger, W. 1969b. Über den Anteil der Arthropoden an der Umsetzung des Bestandesabfalles in Anfangs Bodenbildungen. *Pedobiologia* 9:366–371.
Dunger, W. 1989. The return of soil fauna to coal mined areas in the German Democratic Republic. In: J.D. Majer (ed.) *Animals in Primary Succession—the Role of Fauna in Reclaimed Lands*, pp. 307–337. Cambridge, Cambridge University Press.
East, R. and R.P. Pottinger. 1983. Use of grazing animals to control insect pests of pasture. *N.Z. Entomol.* 7:352–359.
Edwards, C.A. 1973. *Persistent Pesticides in the Environment*, 2nd edn. Cleveland, Ohio, CRC Press.
Edwards, C.A. 1975. Effects of direct drilling on the soil fauna. *Outlook Agric.* 8:243–244.
Edwards, C.A. 1977. Investigations into the influence of agricultural practice on soil invertebrates. *Ann. Appl. Biol.* 87:515–520.
Edwards, C.A. 1980. Interactions between agricultural practice and earthworms. In: D.L. Dindal (ed.) *Soil Biology as Related to Land Use Practices*. Proc. 7th Intern. Coll. Soil Zool, pp. 3–12. Washington, EPA.
Edwards, C.A. 1983. Earthworm ecology in cultivated soils. In: J.E. Satchell (ed.) *Earthworm Ecology—from Darwin to Vermiculture*, pp. 123–137. London, Chapman and Hall.
Edwards, C.A. 1989. The importance of integration in sustainable agricultural. systems. In: M.G. Paoletti, B.R. Stinner, and G.G. Lorenzoni (eds.) Agricultural Ecology and Environment. *Agric. Ecosyst. Environ.* 27:25–35.
Edwards, C.A. and J.R. Lofty. 1978. The influence of arthropods and earthworms upon root growth of direct drilled cereals. *J. Appl. Ecol.* 15:789–795.
Edwards, C.A. and J.R. Lofty. 1979. The effects of straw residues and their disposal on the soil fauna. In: E. Grossbard (ed.) *Straw Decay and its Effect on Disposal and Utilization*, pp. 37–44. Chichester, Wiley.
Edwards, C.A. and J.R. Lofty. 1980. Effects of earthworm inoculation upon the root growth of direct drilled cereals. *J. Appl. Ecol.* 17:533–543.
Edwards, C.A. and J.R. Lofty. 1982a. Nitrogenous fertilizers and earthworm populations in agricultural soils. *Soil Biol. Biochem.* 14:515–521.
Edwards, C.A. and J.R. Lofty. 1982b. The effect of direct drilling and minimal cultivation on earthworm populations. *J. Appl. Ecol.* 19:723–734.
Edwards, C.A. and A.R. Thompson. 1973. Pesticides and soil fauna. *Residue Rev.* 45:1–79.
Eijsackers, H. 1987. The impact of heavy metals on terrestrial ecosystems: biological adaptation through behavioural and physiological avoidance. In: O. Ravera (ed.) *Ecological Assessment of Environmental Degradation, Pollution and Recovery*, pp. 245–259. Amsterdam, Elsevier.
Eijsackers, H., N. Lourijsen, and J. Mentink. 1983. Effects of fly ash on soil fauna.

In: P. Lebrun, H.M. André, A. de Medts, C. Gregoire-Wibo, and G. Wauthy (eds.) *New Trends in Soil Biology*. Proc. VIII Intern. Coll. Soil Zoology, pp. 680–681. Ottignies-Louvain-la-Neuve, Dieu-Brichart.

Elkins, N.Z., L.W. Parker, E. Aldon, and W.G. Whitford. 1984. Responses of soil biota to organic amendments in stripmine spoils in northwestern New Mexico. *J. Environ. Qual.* 13:215–219.

Elliott, E.T. and D.C. Coleman. 1977. Soil protozoan dynamics of a Colorado shortgrass prairie. *Soil Biol. Biochem.* 9:113–118.

El Titi, A. and U. Ipach. 1989. Soil fauna in sustainable agriculture: results of an integrated farming system at Lautenbach, F.R.G. In: M.G. Paoletti, B.R. Stinner, and G.G. Lorenzoni (eds.) Agricultural Ecology and Environment. *Agric. Ecosyst. Environ.* 27:561–572.

Emmanuel, N., J.P. Curry, and G.O. Evans. 1985. The soil Acari of barley plots with different cultural treatments. *Exp. Appl. Acarol.* 1:101–113.

Fausey, N.R. and R. Lal. 1990. Soil wetness and anaerobiosis. In: R. Lal, and B.A. Stewart (eds.) Soil Degradation. *Adv. Soil Sci.* 11:173–186.

Ferrar, P. 1973. CSIRO dung beetle project. *Wool Technol. Sheep Breed.* 20:73–75.

Gerard, B.M. and R.K.M. Hay. 1979. The effect on earthworms of ploughing, tined cultivation, direct drilling and nitrogen in a barley monoculture system. *J. Agric. Sci.* 93:147–155.

Gibbs, G.W. 1976. The role of insects in natural terrestrial ecosystems. *N.Z. Entomol.* 6:113–121.

Glen, D.M., C.W. Wiltshire, and N.F. Milsom. 1984. Slugs and straw disposal in winter wheat. *Proc. 1984 Brit. Crop Prot. Conf.—Pests and Diseases* 1:136–143.

Good, J.A. 1988. The value and state of taxonomy of graminoid agroecosystem Staphylinidae (Coleoptera). In: C. Moriarty (ed.) *Taxonomy: Putting Plants and Animals in their Place*, pp. 14–30. Dublin, Royal Irish Academy.

Good, J.A. and P.S. Giller. 1990. Staphylinid beetles (Coleoptera) from cereal and grass fields in south-west Ireland. *Bull. Ir. Biogeog. Soc.* 13:2–22.

Gray, J.S. 1987. Species-abundance patterns. In: J.H.R. Gee, and P.S. Giller (eds.) *Organization of Communities Past and Present*. pp. 53–67. Oxford, Blackwell.

Greenslade, P.J.M. 1983. Adversity selection and the habitat templet. *Amer. Nat.* 122:352–365.

Gupta, R.K. and Abrol, L.P. 1990. Salt affected soils: their reclamation and management for crop production. In: R. Lal, and B.A. Stewart (eds.) Soil Degradation. *Adv. Soil Sci.* 11:223–288.

Gupta, S.R., R. Rajvanshj, and J.S. Singh. 1981. The role of the termite *Odontotermes gurdaspurensis* (Isopoda: Termitidae) in plant decomposition in a tropical grassland. *Pedobiologia* 22:254–261.

Hågvar, S. and G. Abrahamsen. 1980. Colonisation by Enchytraeidae, Collembola and Acari in sterile soil samples with adjusted pH levels. *Oikos* 34:215–258.

Hamilton, W.E. and J.P. Vimmerstedt. 1980. Earthworms on forested spoil banks. In: D.L. Dindal (ed.) *Soil Biology as related to Land Use Practices*. Proc. 7th Intern. Coll. Soil Zool, pp. 409–417. Washington, EPA.

Hartenstein, R., E.F. Neuhauser, and J. Collier. 1980. Accumulation of heavy metals in the earthworm *Eisenia foetida*. *J. Environ. Qual.* 9:23–26.

Heatwole, H. and R. Levins. 1972. Trophic structure stability and faunal change

during recolonization. *Ecology* 53:531–534.
Hellawell, J.M. 1977. Change in natural and managed ecosystems: detection, measurement and assessment. *Proc. R. Soc. Lond.* B 197:31–57.
Henderson, I.F. and R.O. Clements. 1979. Differential susceptibility to pest damage in agricultural grasses. *J. Agric. Sci., Cambridge* 73:465–472.
Henderson, I.F. and R.O. Clements. 1981. The effect of insecticide treatment on the establishment and growth of Italian ryegrass under different sowing conditions. *Grass For. Sci.* 35:235–241.
Hendrix, P.F., R.W. Parmelee, D.A. Crossley, Jr., D.C. Coleman, E.P. Odum, and P.M. Groffman. 1986. Detritus food webs in conventional and no-tillage agroecosystems. *Biosci.* 36:374–380.
Herricks, E.E. and J. Cairns, Jr. 1982. Biological monitoring. Part III—Receiving system methodology based on community structure. *Water Res.* 16:141–153.
Hoogerkamp, M., H. Rogäar, and H.J.P. Eijsackers 1983. Effect of earthworms on grassland on recently reclaimed polder soils in the Netherlands. In: J.E. Satchell (ed.) *Earthworm Ecology—from Darwin to Vermiculture*, pp. 85–105. London, Chapman and Hall.
Hopp, H. and C.S. Slater. 1948. Influence of earthworms on soil productivity. *Soil Sci.* 66:421–428.
House, G.J. and G.E. Brust. 1989. Ecology of low-input, no-tillage agroecosystems. In: M.G. Paoletti, B.R. Stinner, and G.G. Lorenzoni (eds.) Agricultural Ecology and Environment. *Agric. Ecosyst. Environ.* 27:331–345.
House, G.J., B.R. Stinner, D.A. Crossley, Jr., and E.P. Odum. 1984. Nitrogen cycling in conventional and no-tillage agroecosystems: analysis of pathways and processes. *J. Appl. Ecol.* 21:826–848.
Hoy, J.B. 1990. Long-term effects of lindane on the oribatid community in a pine plantation. *Environ. Entomol.* 19:898–900.
Hughes, R.D., M.D. Tindale-Biscoe, and 1. Walker. 1978. Effect of introduced dung beetles (Coleoptera: Scarabaeidae) on the breeding and abundance of the Australian bushfly, *Musca vetustissima* Walker (Diptera: Muscidae). *Bull. Entomol. Res.* 68:361–372.
Huhta, V. 1979. Effects of liming and deciduous litter on earthworm (Lumbricidae) populations of a spruce forest, with an inoculation experiment on *Allolobophora caliginosa*. *Pedobiologia* 19:340–345.
Huhta, V., R. Hyvönen, P. Kaaselainen, A. Koskenniemi, J. Muona, I. Mäkelä, M. Sulander, and P. Vilkamaa. 1986. Soil fauna of Finnish coniferous forests. *Ann. Zool. Fennici* 23:345–360.
Humphreys, G.S. 1981. The rate of ant mounding and earthworm casting near Sydney, New South Wales. *Search* 12:129–131.
Hunt, H.W., D.C. Coleman, E.R. Ingham, E.T. Elliot, J.C. Moore, S.L. Rose, C.P.P. Reid, and C.R. Morley. 1987. The detrital food web in a shortgrass prairie. *Biol. Fertil. Soil* 3:57–68.
Hunter, B.A. and M.S. Johnson. 1982. Food chain relationships of copper and cadmium in contaminated grassland ecosystems. *Oikos* 38:108–117.
Hutchinson, K.J. and K.L. King. 1980a Management impacts on structure and function of sown grasslands. In: A.I. Breymeyer, and G.M. van Dyne (eds.) *Grasslands, Systems Analysis and Man*, pp. 823–852. Cambridge, Cambridge University Press.
Hutchinson, K.J. and K.L. King. 1980b. The effects of sheep stocking level on

invertebrate abundance, biomass and energy utilization in a temperate, sown grassland. *J. Appl. Ecol.* 17:369–387.
Hutson, B.R. 1980. Colonization of industrial reclamation sites by Acari, Collembola and other invertebrates. *J. Appl. Ecol.* 17:255–275.
Ireland, M.P. 1983. Heavy metal uptake and tissue distribution in earthworms. In: J.E. Satchell (ed.) *Earthworm Ecology—from Darwin to Vermiculture*, pp. 247–265. London, Chapman and Hall.
Keogh, R.G. 1979. Lumbricid earthworm activities and nutrient cycling in pasture ecosystems. In: T.K. Crosby and R.P. Pottinger (eds.) *Proc. 2nd Australas. Conf. Grassl. Invertebr. Ecol*, pp. 49–51. Wellington, Government Printer.
King, K.L. and K.J. Hutchinson. 1976. The effects of sheep stocking intensity on the abundance and distribution of mesofauna in pastures. *J. Appl. Ecol.* 13:41–55.
Kladivko, E.J., A.D. Mackay, and J.M. Bradford. 1986. Earthworms as a factor in the reduction of soil crushing. *Soil Sci. Soc. Amer. J.* 50:191–196.
Kline, V.M. and E.A. Howell. 1987. Prairies. In: W.R. Jordan, III, M.E. Gilpin, and J.P. Aber (eds.) *Restoration Ecology—a Synthetic Approach to Ecological Research*, pp. 75–83. Cambridge, Cambridge University Press.
Kubiena, W.L. 1953. *The Soils of Europe*. London, Murby.
Lal, R. 1986. Soil surface management in the tropics for intensive land use and high arid sustained production. *Adv. Soil Sci.* 5:1–109.
Lal, R. 1987a. *Tropical Ecology and Edaphology*. New York, Wiley.
Lal, R. 1987b. Managing the soils of sub-saharan Africa. *Science* 236:1069–1076.
Lal, R. 1988. Effects of macrofauna on soil properties in tropical ecosystems. *Agric. Ecosyst. Environ.* 24:101–116.
Lamotte, M. 1976. The structure and function of a tropical savannah ecosystem. In: F.B. Golley, and E. Medina (eds.) *Ecological Studies. II. Tropical Ecological System: Analysis and Synthesis*, pp. 179–222. New York, Springer-Verlag.
Laurence, B.R. 1954. The larval inhabitants of cow pats. *J. Anim. Ecol.* 23:234–260.
Lavelle, P. 1974. Les vers de terre de la savane de Lamto. In: F. Athias, G. Josens, P. Lavelle and R. Schaefer (eds.) Analyse d'un Écosystème Tropicale Humidé: La Savane de Lamto (Côte d'Ivoire). Les Organismes Endogés de la Savane de Lamto. *Bull. de Liaison des Chercheurs de Lamto* (No. spécial) 5:133–166.
Lavelle, P. 1978. Les vers de terre de la savane de Lamto (Côte d'Ivoire). Peuplements, populations et fonctions de l'écosystème. *Publ. Lab. Zool. E.N.S.* 12:1–301.
Lavelle, P. and B. Pashanasi. 1989. Soil macrofauna and land management in Peruvian Amazonia (Yurimaguas, Loreto). *Pedobiologia* 33:283–291.
Laverack, M.S. 1961. Tactile and chemical perception in earthworms. II. Response to acid pH solutions. *Comp. Biochem. Physiol.* B 2:22–34.
Lee, K.E. 1985. *Earthworms: Their Ecology and Relationships with Soils and Land Use*. London, Academic Press.
Lee, K.E. and T.G. Wood. 1971a. Physical and chemical effects on soils of some Australian termites and their pedological significance. *Pedobiologia* 11:376–409.
Lee, K.E. and T.G. Wood. 1971b. *Termites and Soils*. London, Academic Press.
Lepage, M. 1972. Recherches écologiques sur une savane Sahelienne du Ferlo septentrional, Sénégal: données préliminaires sur l'écologie des termites. *Terre Vie* 26:383–409.

Lindroth, C.H., H. Andersson, H. Bödvarsson, and S.H. Richter. 1973. Surtsey, Iceland. The development of a new fauna, 1963–1970. Terrestrial invertebrates. *Ent. Scand.*, Suppl. 5:1–280.

Lofs-Holmin, A. 1980. Measuring growth of earthworms as a method of testing sublethal toxicity of pesticides. *Swedish J. Agric. Res.* 10:25–32.

Lofs-Holmin, A. 1981. Influence in field experiments of benomyl and carbendazim on earthworms (Lumbricidae) in relation to soil texture. *Swedish J. Agric. Res.* 11:141–147.

Lofs-Holmin, A. 1983. Influence of agricultural practices on earthworms. *Acta Agric. Scand.* 33:225–256.

Logan, T.J. 1990. Chemical degradation of soil. In: R. Lal, and B.A. Stewart (eds.) Soil Degradation. *Adv. Soil Sci.* 11:187–221.

Ma, W.-C. 1988. Toxicity of copper to lumbricid earthworms in sandy agricultural soils amended with Cu-enriched organic waste materials. *Ecol. Bull.* (Copenhagen) 39:53–56.

Ma, W.-C. and H. Eijsackers. 1989. The influence of substrate toxicity on soil macrofauna return in reclaimed land. In: J.D. Majer (ed.) *Animals in Primary Succession—the Role of Fauna in Reclaimed Lands*, pp. 223–244. Cambridge, Cambridge University Press.

MacArthur, R.H., and E.O. Wilson. 1967. *The Theory of Island Biogeography*. Princeton, Princeton University Press.

MacDonald, D.W. 1983. Predation on earthworms by terrestrial vertebrates. In: Satchell, J.E. (ed.) *Earthworm Ecology from Darwin to Vermiculture*, pp. 393–414. London, Chapman and Hall.

MacIntosh, R.P. 1980. The relationship between succession and the recovery process in systems. In: J. Cairns, Jr. (ed.) *The Recovery Process in Damaged Ecosystems*, pp. 11–62. Ann Arbor, Ann Arbor Science Publications.

Majer, J.D. 1984. Short term responses of soil and litter invertebrates to a cool autumn burn in Jarrah (*Eucalyptus marginata*) forest in Western Australia. *Pedobiologia* 26:229–247.

Majer, J.D. (ed.) 1989a. *Animals in Primary Succession: the Role of Fauna in Reclaimed Lands*. Cambridge, Cambridge University Press.

Majer, J.D. 1989b. Fauna studies and land reclamation technology—a review of the history and need for such studies. In: J.D. Majer (ed.) *Animals in Primary Succession—the Role of Fauna in Reclaimed Lands*, pp. 5–33. Cambridge, Cambridge University Press.

Majer, J.D. 1989c. Long-term colonization of fauna in reclaimed land. In J.D. Majer (ed.) *Animals in Primary Succession—the Role of Fauna in Reclaimed Lands*, pp. 143–177. Cambridge, Cambridge University Press.

Majer, J.D., J.E. Day, E.D. Kabay, and W.S. Perriman. 1984. Recolonization by ants in bauxite mines rehabilitated by a number of different methods. *J. Appl. Ecol.* 21:355–375.

Malecki, M.R., E.F. Neuhauser, and R. Loehr. 1982. The effect of metals on the growth and reproduction of *Eisenia foetida* (Oligochaeta, Lumbricidae). *Pedobiologia* 24:129–137.

Martin, N.A. 1975. Effect of four insecticides on the pasture ecosystem. II. The fauna collected in pit-traps. *N.Z.J. Agric. Res.* 18:179–182.

Mattson, W.J. and N.D. Addy. 1975. Phytophagous insects as regulators of forest primary production. *Science* 190:515–522.

McNaughton, S.J. 1979. Grassland-herbivore dynamics. In: A.R.E. Sinclair, and M. Norton-Griffiths (eds.) *Serengeti: Dynamics of an Ecosystem*, pp. 46–81. Chicago, Chicago University Press.

McNeilly, T. 1987. Evolutionary lesson from degraded ecosystems. In: W.R. Jordan, III, M.E. Gilpin, and T.P. Aber (eds.) *Restoration Ecology—a Synthetic Approach to Ecological Research*, pp. 271–286. Cambridge, Cambridge University Press.

Meijer, J. 1989. Sixteen years of faunal invasion and succession in the Lauwerszeepolder. In: J.D. Majer (ed.) *Animals in Primary Succession—the Role of Fauna in Reclaimed Lands*, pp. 339–369. Cambridge, Cambridge University Press.

Mitchell, J.E. and R.E. Pfadt. 1974. A role of grasshoppers in a shortgrass prairie ecosystem. *Environ. Entomol.* 3:358–360.

Morris, M.G. 1971. The management of grassland for the conservation of invertebrate animals. In: E. Duffey and A.S. Watt (eds.) *The Scientific Management of Animal and Plant Comunities for Conservation*, pp. 527–552. Oxford, Blackwell.

Morris, M.G. 1979. Responses of grassland invertebrates to management by cutting. II. Heteroptera. *J. Appl. Ecol.* 16:417–432.

Murdoch, W.W., F.C. Evans, and C.H. Peterson. 1972. Diversity and pattern in plants and insects. *Ecology* 53:819–828.

Neuhauser, E.F., M.R. Malecki, and R.C. Loehr. 1984. Growth and reproduction of the earthworm *Eisenia fetida* after exposure to sublethal concentrations of metals. *Pedobiologia* 27:89–97.

Neumann, U. 1971. Die Sukzession der Bodenfauna (Carabidae (Coleoptera), Diplopoda und Isopoda) in den forstlich rekultivierten Gebieten des Rheinischen Braunkohlenreviers. *Pedobiologia* 11:193–226.

Nichols, O., B.J. Wykes, and J.D. Majer. 1989. The return of vertebrate and invertebrate fauna to bauxite mined areas in south-western Australia. In: J.D. Majer (ed.) *Animals in Primary Succession—the Role of Fauna in Reclaimed Lands*, pp. 397–422. Cambridge, Cambridge University Press.

Nielson, R.F. and H.B. Peterson. 1973. Establishing vegetation on mine tailings waste. In: R.J. Hutnik and G. Davis (eds.) *Ecology and Reclamation of Devastated Land*, vol. 2, pp. 103–115. New York, Gordon and Breach.

Nordlund, D.A., R.B. Chalfant, and W.J. Lewis. 1984. Arthropod populations, yield and damage in monocultures and polycultures of corn, beans and tomatoes. *Agric. Ecosyst. Environ.* 11:353–367.

Nowak, E. 1976. The effect of fertilization on earthworms and other soil microfauna. *Pol. Ecol. Stud.* 2:195–207.

Odum, E.P. 1969. The strategy of ecosystem development. *Science* 164:262–270.

Parker, F.D., S.W.T. Batra, and V.T. Tepedino. 1987. New pollinators for our crops. *Agric. Zool. Rev.* 2:279–304.

Parmenter, R.R. and J.A. MacMahon. 1987. Early successional patterns of arthropod recolonization on reclaimed strip mines in southwestern Wyoming: the ground-dwelling beetle fauna (Coleoptera). *Environ. Entomol.* 16:168–177.

Parr, T.W. 1978. An analysis of soil micro-arthropod succession. *Scient. Proc. R. Dubl. Soc.* 6A:185–196.

Paustian, K., O. Andrén, M. Clarholm, A.-C. Hansson, G. Johansson, J. Lagerlöf, T. Lindberg, R. Petersson, and B. Sohlenius. 1990. Carbon and nitrogen budgets of four agro-ecosystems with annual and perennial crops, with and without N fertilization. *J. Appl. Ecol.* 27:60–84.

Pearson, T.H., J.S. Gray, and P. Johannessen. 1984. Objective selection of sensitive species indicative of pollution-induced changes in benthic communities. 2. Data analyses. *Marine Ecol. Prog. Ser.* 12:237–255.

Persson, T. 1983. Influence of soil animals on nitrogen mineralization in a northern Scots pine forest. In: Ph. Lebrun, H.M. André, A. De Medts, C. Grégoire-Wibo and G. Wauthy (eds.) *New Trends in Soil Biology*. Proc. VIII Intern. Coll. Soil Zool, pp. 117–126. Ottignies-Louvain-la-Neuve, Dieu-Brichart.

Persson, T., E. Bååth, M. Clarholm, H. Lundkvist, B.E. Söderström, and B. Sohlenius. 1980. Trophic structure, biomass dynamics and carbon metabolism of soil animals in a Scots pine forest. In: T. Persson (ed.) Structure and Function of Northern Coniferous Forests—an Ecosystem Study. *Ecol. Bull. (Stockholm)* 32:419–459.

Persson, T. and H. Lohm. 1977. Energetical significance of the annelids and arthropods in a Swedish grassland soil. *Ecol. Bull. (Stockholm)* 23:1–211.

Piekarz, D. 1990. Rapporteur's report of work group: indicators and assessment of agricultural sustainability. *Environ. Monit. Assess.* 15:307–308.

Pimentel, D., T.W. Culliney, I.W. Buttler, D.J. Reinemann, and K.B. Beckman. 1989. Low-input sustainable agriculture using ecological management practices. In: M.G. Paoletti, B.R. Stinner, and G.G. Lorenzoni (eds.) Agricultural Ecology and Environment. *Agric. Ecosyst. Environ.* 27:3–24.

Pimm, S.L. 1984. The complexity and stability of ecosystems. *Nature* 307:321–326.

Potts, G.R. 1986. *The Partridge*. London, Collins.

Purvis, G. and J.P. Curry. 1980. Successional changes in the arthropod fauna of a new ley pasture established on previously cultivated arable land. *J. Appl. Ecol.* 17:309–321.

Purvis, G., N. Carter, and W. Powell. 1988. Observations on the effects of an autumn application of a pyrethroid insecticide on non-target predatory species in winter cereals. In: R. Cavalloro and K.D. Sunderland (eds.) *Integrated Crop Protection in Cereals*, pp. 153–166. Rotterdam, A.A. Balkema.

Raw, F. 1962. Studies of earthworm populations in orchards. 1. Leaf burial in apple orchards. *Ann. Appl. Biol.* 50:389–404.

Recher, H.F. 1989. Colonization of reclaimed land by animals: an ecologist's overview. In: J.D. Majer (ed.) *Animals in Primary Succession: the Role of Fauna in Reclaimed Lands*, pp. 441–448. Cambridge, Cambridge University Press.

Reinecke, A.J. and F.A. Visser. 1980. The influence of agricultural land use practices on the population densities of *Allolobophora trapezoides* and *Eisenia rosea* (Oligochaeta) in Southern Africa. In: D.L. Dindal (ed.) *Soil Biology as related to Land Use Practices*. Proc. 7th Intern. Coll. Soil Zool, pp. 310–324. Washington, EPA.

Reinecke, A.J. and J.M. Venter. 1985. Influence of dieldrin on the reproduction of the earthworm *Eisenia foetida* (Oligochaeta). *Biol. Fertil. Soils* 1:39–44.

Rodell, C.F. 1977. A grasshopper model for a grassland ecosystem. *Ecology* 58:227–245.

Rushton, S.P., M.L. Luff, and M.D. Eyre. 1989. Effects of pasture improvement and management on the ground beetle and spider communities of upland grasslands. *J. Appl. Ecol.* 26:489–503.

Satchell, J.E. 1955. Some aspects of earthworm ecology. In: D.K. McE. Kevan (ed.) *Soil Zoology*, pp. 180–201. London, Butterworths.

Satchell, J.E. 1963. Nitrogen turnover by a woodland population of *Lumbricus terrestris*. In: J. Doeksen and J. van der Drift (eds.) *Soil Organisms*, pp. 60–66. Amsterdam, North-Holland.
Satchell, J.E. 1967. Lumbricidae. In: A. Burges and F. Raw (eds.) *Soil Biology*, pp. 180–202. London, Butterworths.
Satchell, J.E. and D.A. Stone. 1977. Colonization of pulverized fuel ash sites by earthworms. *P. Cent. Pir. Biol. Exp.* 9:59–74.
Seastedt, T.R. 1984. The role of microarthropods in decomposition and mineralization processes. *Annu. Rev. Entomol.* 29:25–46.
Seastedt, T.R., R.A. Ramundo, and D.C. Hayes. 1988. Maximization of densities of soil animals by foliage herbivory: empirical evidence, graphical and conceptual models. *Oikos* 51:243–248.
Sharpley, A.N., J.K. Syers, and J.A. Springett. 1979. Effect of surface-casting earthworms on the transport of phosphorus and nitrogen in surface runoff from pasture. *Soil Biol. Biochem.* 11:459–462.
Sheehan, P.J. 1984. Effects on community and ecosystem structure and dynamics. In: P.J. Sheehan, D.R. Miller, G.C. Butler, and Ph. Bourdeau (eds.) *Effects of Pollutants at the Ecosystem Level*, pp. 51–99. New York, Wiley.
Sheehan, W. 1986. Response by specialist and generalist natural enemies to agroecosystem diversification: a selective review. *Environ. Ent.* 15:456–461.
Shipitalo, M.J. and R. Protz. 1987. Comparison of morphology and porosity of a soil under conventional and zero tillage. *Can J. Soil Sci.* 67:445–456.
Shipitalo, M.J. and R. Protz. 1988. Factors influencing the dispersability of clay in worm casts. *Soil Sci. Soc. Amer. J.* 52:764–769.
Siepel, H. and C.F. van de Bund. 1988. The influence of management practices on the microarthropod community of grassland. *Pedobiologia* 31:339–354.
Sims, G.K. 1990. Biological degradation of soil. *Adv. Soil Sci.* 11:289–330.
Sinclair, A.R.E. 1975. The resource limitation of trophic levels in tropical grassland ecosystems. *J. Anim. Ecol.* 44:497–520.
Sohlenius, B. 1980. Abundance, biomass and contribution to energy flow by soil nematodes in terrestrial ecosystems. *Oikos* 34:186–194.
Southwell, L.T. and Majer, J.D. 1982. The survival and growth of the earthworm *Eisenia foetida* (Lumbricidae: Oligochaeta) in alkaline residues associated with the bauxite refining process. *Pedobiologia* 23:42–52.
Southwood, T.R.E. 1977. Habitat, the templet for ecological strategies? *J. Anim. Ecol.* 46:337–365.
Speight, M.R. 1983. The potential of ecosystem management for pest control. *Agric. Ecosyst. Environ.* 10:183–199.
Springett, J.A. 1983. Effect of five species of earthworm on some soil properties. *J. Appl. Ecol.* 20:865–872.
Stinson, C.S.A. and V.K. Brown. 1983. Seasonal changes in the architecture of natural plant communities and its relevance to insect herbivores. *Oecologia* 56:67–69.
Stockdill, S.M.J. 1959. Earthworms improve pasture growth. *N.Z.J. Agric.* 98:227–233.
Stockdill, S.M.J. 1982. Effects of introduced earthworms on the productivity of New Zealand pastures. *Pedobiologia* 24:29–35.
Stockdill, S.M.J. and G.G. Cossens. 1966. The role of earthworms in pasture production and moisture conservation. *Proc. N.Z. Grassl. Assoc.* 28:168–183.

Stout, J. and O.W. Heal. 1967. Protozoa In: A. Burges and F. Raw (eds.) *Soil Biology*, pp. 149–195. London, Academic Press.

Stringer, A. and M.A. Wright. 1976. The toxicity of benomyl and some related 2-substituted benzimidazoles to the earthworm *Lumbricus terrestris*. *Pesticide Sci.* 7:459–464.

Strojan, C.L. 1978a. Forest leaf litter decomposition in the vicinity of a zinc smelter. *Oecologia* 32:203–212.

Strojan, C.L. 1978b. The impact of zinc smelter emissions on forest litter arthropods. *Oikos* 31:41–46.

Strüve-Kusenberg, R. 1981. Succession and trophic structure of the soil fauna of fallow land. *Pedobiologia* 21:132–141.

Svensson, B.H., U.-L. Boström, and L. Klemedtsson. 1986. Potential for higher rates of denitrification in earthworm casts than in the surrounding soil. *Biol. Fertil. Soils* 2:147–149.

Syers, J.K. and J.A. Springett. 1983. Earthworm ecology in grassland soils. In: J.E. Satchell (ed.) *Earthworm Ecology—from Darwin to Vermiculture*, pp. 67–83. London, Chapman and Hall.

Townsend, W.N. and D.R. Hodgson. 1973. Edaphological problems associated with deposits of pulverized fuel ash. In: R.J. Hutnik and G. Davis (eds.) *Ecology and Reclamation of Devastated Land*, vol. 1, pp. 45–56. New York, Gordon and Breach.

Tyler, G. 1975. Heavy metal pollution and mineralisation of nitrogen in forest soils. *Nature* 255:701–702.

Ulrich, B. 1987. Air pollution effects in terrestrial ecosystems and their restoration. In: O. Ravera (ed.) *Ecological Assessment of Environmental Degradation, Pollution and Recovery*, pp. 275–290. Amsterdam, Elsevier.

Unger, P.W. 1990. Conservation tillage systems. In: R.P. Singh, J.F. Parr, and B.A. Stewart (eds.) Dryland Agriculture: Strategies for Sustainability. *Adv. Soil Sci.* 13:27–68.

Valiela, I. 1974. Composition, food webs and population limitation in dung arthropod communities during invasion and succession. *Amer. Midl. Nat.* 92:370–385.

van Rhee, J.A. 1965. Earthworm activity and plant growth in artificial cultures. *Plant and Soil* 22:45–48.

van Rhee, J.A. 1969a. Inoculation of earthworms in a newly drained polder. *Pedobiologia* 9:128–132.

van Rhee, J.A. 1969b. Development of earthworm populations in polder soils. *Pedobiologia* 9:133–140.

van Rhee, J.A. 1977a. A study of the effect of earthworms on orchard productivity. *Pedobiologia* 17:107–114.

van Rhee, J.A. 1977b. Effects of soil pollution on earthworms. *Pedobiologia* 17:201–208.

van Straalen, N.M., M.H.S. Kraak, and C.A.J. Denneman. 1988. Soil microarthropods as indicators of soil acidification and forest decline in the Veluwe area, the Netherlands. *Pedobiologia* 32:47–55.

Vimmerstedt, J.P. and J.H. Finney. 1973. Impact of earthworm introduction on litter burial and nutrient distribution in Ohio strip-mine spoil banks. *Proc. Soil Sci. Soc. Amer.* 37:388–391.

Wallwork, J.A. 1970. *Ecology of Soil Animals*. London, McGraw-Hill.

Wasilewska, L. 1979. The structure and function of soil nematode communities in natural ecosystems and agrocenoses. *Pol. Ecol. Stud.* 5:97–145.
Weinert, E. 1986. Selected methods of bioindication in terrestrial ecosystems. In: J. Salanki (ed.) *Biological Monitoring of the State of the Environment: Bioindicators*, pp. 67–70. Oxford, IRL Press.
Wells, T.C.E. 1971. A comparison of the effects of sheep grazing and mechanical cutting on the structure and botanical composition of chalk grassland. In: E. Duffey, and A.S. Watt (eds.) *The Scientific Management of Animal and Plant Communities for Conservation*, pp. 497–515. Oxford, Blackwell.
White, E.G. 1974. Grazing pressures of grasshoppers in an alpine tussock grassland. *N.Z.J. Agric. Res.* 17:357–372.
Wiegert, R.G. and F.C. Evans. 1967. Investigations of secondary productivity in grasslands. In: K. Petrusewicz (ed.) *Secondary Productivity of Terrestrial Ecosystems*, vol. 2, pp. 499–518. Warsaw, Polish Academy of Sciences.
Wild, A. and I.A. Babiker. 1976. The asymmetric leaching pattern of nitrate and chloride in a loamy sand under field conditions. *J. Soil Sci.* 27:460–466.
Wilkinson, G.E. 1975. Effect of grass fallow rotations on the infiltration of water into a savanna zone soil in northern Nigeria. *Trop. Agric.* 52:97–103.
Wood, T.G. 1966. The fauna of grassland soils with special reference to Acari and Collembola. *Proc. N.Z. Ecol. Soc.* 13:79–85.
Zachman, J.E., D.R. Linden, and C.E. Clapp. 1987. Macroporous infiltration and redistribution as affected by earthworms, tillage and residue. *Soil Sci. Soc. Amer. J.* 51:1580–1586.

Creation and Restoration of Wetlands: Some Design Consideration for Ecological Engineering

W.J. Mitsch and J.K. Cronk

I. Introduction

Wetlands are estimated to cover 860 million hectares or 6% of the land surface of the world (Maltby and Turner, 1983; see Fig. 1). They have been described as the kidneys of the landscape for the biogeochemical and hydrologic roles that they provide (Mitsch and Gosselink, 1986). They prevent floods, cleanse waters, protect shorelines, and recharge groundwater aquifers. Just as important to some, wetlands provide haven for a wide variety of flora and fauna and offer a unique habitat for many rare and endangered species. But like the species that they harbor, wetlands themselves have

Advances in Soil Science, Volume 17

A

B

C

D

E

Figure 1a–e. Examples of some major wetlands types: (a) salt marsh near Charleston, South Carolina; (b) freshwater marsh, Everglades, Florida; (c) forested bog near Pinsk, Byelorussia, USSR; (d) wet meadow near Glumsø, Denmark; (e) managed freshwater marsh, Lake Erie coast, Ohio (photos by Mitsch)

been considered by some to be on the endangered list in the United States. While the extent of wetlands in presettlement United States is difficult to determine, our wetlands in the coterminous United States are estimated to have decreased, from approximately 60 to 75 million hectares to about 42 million hectares today (OTA, 1984; Mitsch and Gosselink, 1986). In midwestern states such as Ohio and Illinois, over 90% of the wetlands were drained, partially in response to The Swamp Lands Acts of 1849, 1950, and 1860. The riparian and coastal wetlands of the midwestern United States, which once were connected to all the streams, rivers, and Great Lakes of the glaciated Midwest are, for all intents and purposes, gone from the landscape. With their loss, our rivers and streams have lost their ability to cleanse themselves and the Great Lakes are no longer buffered from upland regions.

A. No Net Loss of Wetlands

In 1987, a National Wetlands Policy Forum was convened by the Conservation Foundation at the request of the U.S. Environmental Protection Agency to investigate the issue of wetland management in the United States (NWPF, 1988; Davis, 1989). This distinguished group of 20 members, which included three governors, a state legislator, state and local agency heads, chief executive officers of environmental groups and businesses, farmers, ranchers, and academic experts, published a report (NWPF, 1988) which set significant goals for the nation's remaining wetlands. The Forum recommended a policy: "to achieve no overall net loss of the nation's remaining wetlands base and to restore and create wetlands, where feasible, to increase the quantity and quality of the nation's wetland resource base" (NWPF, 1988).

It was recommended as an interim goal that the holdings of wetlands in the United States should decrease no further and as a long-term goal that the wetlands and their quality should increase (NWPF, 1988). President Bush, in his 1990 budget address to Congress, echoed the "no net loss" concept as a national goal (Davis, 1989), shifting the activities of a great number of agencies such as the Department of Interior, the U.S. Environmental Protection Agency, the U.S. Army Corps of Engineers, and the Department of Agriculture to leadership toward a unified and seemingly simple goal. It was not anticipated that there would be a complete halt of wetland drainage in the United States when economic or political reasons dictated otherwise, so implied in the "no net loss" concept is an increase in wetlands through wetland restoration and creation. The ultimate increase in wetland holdings implies even more wetland restoration and creation.

B. Ecological Engineering

We propose that the design and construction of wetlands needs to be done in an ecologically sound and predictable way and have proposed "ecological engineering" as the proper approach (Mitsch, 1988, in press; Mitsch et al., 1989). Ecological engineering was defined in the early 1960s by H.T. Odum as "those cases in which the energy supplied by man is small relative to the natural sources, but sufficient to produce large effects in the resulting patterns and processes" (Odum, 1962) and as "environmental manipulation by man using small amounts of supplementary energy to control systems in which the main energy drives are still coming from natural sources" (Odum et al., 1963). Ecological engineering and ecotechnology have been

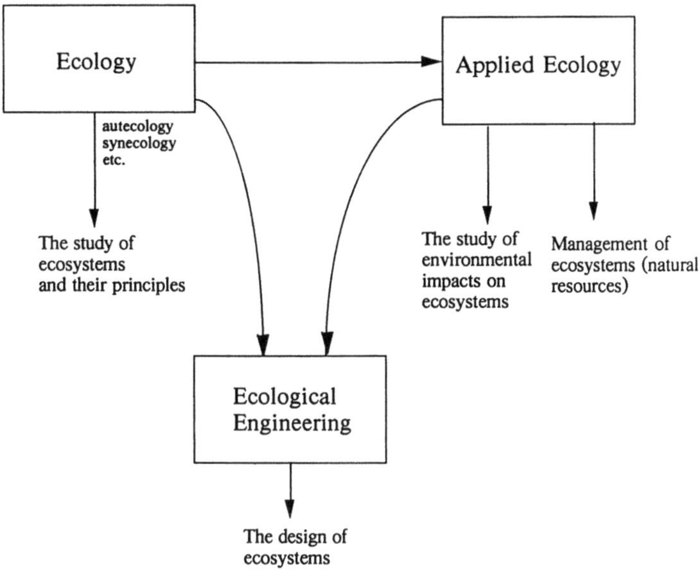

Figure 2. Ecological engineering relationship to basic ecology and applied ecology. As an applied field, ecological engineering has a basis in both ecology and traditional applied ecology, but is distinct from both. The construction of wetlands, if done with ecological principles, is a good example of ecological engineering (modified from Mitsch and Jørgensen, 1989)

recently defined as "the design of human society with its natural environment for the benefit of both" (Mitsch and Jørgensen, 1989) and as "the techniques of designing and operating the economy with nature" (Odum, 1989). Odum (1989) has further suggested that while we have several fields for the study of systems of humanity and its environment (e.g., landscape ecology, ecological economics, human ecology), study is not enough. Both basic and applied ecology provide important fundamentals to ecological engineering but do not define it completely (Fig. 2). Ecological engineering is the "prescriptive" discipline of ecology. Ecological engineering has its roots close to ecology, just as chemical engineering is close to chemistry and and biochemical engineering is close to biochemistry. It should remain a branch of ecology. Ecological engineering is not the same as environmental engineering or biotechnology (see Mitsch and Jørgensen, 1989, for discussion of these distinctions). Although not proposed as an additional field for traditional engineering, ecological engineering shares the concept of design with more traditional engineering in that design of ecosystems is involved. As stated in Mitsch and Jørgensen (1989): "[Ecological engineering] is engineering in the sense that it involves the design of this natural environment using quantitative approaches and basing our approaches on basic science. It is technology with the primary tool being self-designing ecosystems. The components are all of the biological species of the world."

C. Constructed and Restored Wetlands

Ecological engineering can apply to either wetland restoration or wetland construction. "Wetland restoration" usually refers to the rehabilitation of wetlands that may be degraded or hydrologically altered. "Wetland creation" refers to the construction of wetlands where they did not exist before. These "created wetlands" are also called "constructed wetlands" or "artificial wetlands," although the last term is not preferred by many wetland scientists. No one has estimated the number of such wetlands in the United States, but it is probably in the thousands. Sometimes wetland construction or restoration can be combined with other improvements in the landscape to optimize their effectiveness. For example, reduced tillage combined with wetland restoration along streams may have a significant benefit to water quality of downstream systems (Loucks, 1989). Wetland restoration and construction are among the talents that will be required of modem landscape management in the future.

When ecological engineering is applied to the building or restoration of wetlands, it is done for a number of reasons or objectives. The three most popular reasons for wetland construction in the United States have been for waste-water treatment, for coal mine drainage control, and for replacement of wetland loss elsewhere.

1. Wastewater Wetlands

The use of wetlands for waste-water treatment was stimulated by a number of studies that have indicated the ability of these wetlands to remove nutrients, particularly nitrogen and phosphorus from domestic waste water (see e.g., Nichols, 1983; Godfrey et al., 1985; Knight et al., 1987). Early ecosystem-level experiments which set many of the standards for today's use of wetlands for domestic waste-water treatment were with forested cypress swamps in Gainesville, Florida (Odum et al., 1977; Ewel and Odum, 1984; see Fig. 3a) and a rich fen system in Houghton Lake, Michigan (Kadlec and Kadlec, 1979). This ability of wetlands to adsorb nutrients is related to several factors, notably: (1) the shallow nature of the system which allows maximum contact between sediments and overlying waters; (2) the presence of both aerobic and anaerobic processes; (3) the general high primary productivity of these systems; and (4) the accumulation of organic matter or peat in many wetlands (Mitsch and Gosselink, 1986). Studies of wetlands that have received high nutrient waste water show some cases where wetlands are still effective in retaining nutrients (Boyt et al., 1977), while wetlands can lose the ability to retain nutrients in other cases (Kadlec, 1985).

A

B

Figure 3a, b. Examples of natural and constructed wetlands used to improve water quality: (a) cypress dome near Gainesville, Florida, used for domestic wastewater treatment (photo by Mitsch); and (b) constructed cattail marsh near Coshocton, Ohio used for coal mine drainage control (photo by Fennessy)

2. Coal Mine Drainage Wetlands

The use of wetlands for mine drainage control was probably first considered by observing volunteer *Typha* wetlands near acid seeps in a harsh environment where no other vegetation could grow. In the past decade, perhaps 100 or more wetlands have been constructed in the Appalachian region of the Eastern United States to control mine drainage (e.g., Wieder and Lang 1984; Brodie et al., 1988; Fennessy and Mitsch, 1989a, b; see Fig. 3b). The goal of these systems is usually the removal of iron from the water column to avoid its discharge downstream, but sulfate reductions and recovery of pH from extremely acidic conditions have also been goals of these systems.

3. Mitigation Wetlands

Strict enforcement of Section 404 of the Clean Water Act by the U.S. Army Corps of Engineers has led to the common practice of requiring that a wetland system be built to replace any wetland that is lost in a development such as highway construction, coastal drainage and filling, or commercial development (Larson and Neill, 1987; Kusler et al., 1988). There has been little follow-up of these "mitigation wetlands," and there are few methods available to determine the "success" of a mitigation wetland in replacing the functions lost with the original wetland.

Other Objectives

Wetlands have also been preserved, but rarely built solely for flood control. More recently, the control of nonpoint source pollution, sediment retention, and river floodplain restoration have been proposed as valid applications of ecological engineering of wetlands (Livingston, 1989; Hey et al., 1989; see Fig. 4). Wildlife enhancement is an obvious benefit of many of these constructed wetlands and is the primary goal of constructed wetlands in some cases. In fact, construction of wetlands for the combination of waste-water treatment and wildlife enhancement objectives is often possible and solves two national problems—cleaning up our nation's waterways and adding to our nation's wetlands.

II. Wetland Design

A. Defining Objectives

The design of a proper wetland or series of wetlands, whether for the control of nonpoint source pollution, for a wildlife habitat, or for waste-water treatment, starts with the overall objectives of the wetland. One view is that the wetlands should be designed so as to maximize systems longevity

A

B

C

Figure 4a–c. Constructed experimental wetlands at Des Plaines River Wetland Demonstration Project, Lake County, Illinois: (a) during construction, April, 1987; (b) after construction, June, 1990; (c) outflow structure for flow measurement (photos by Mitsch). These experimental wetlands are used to investigate the control of nonpoint source pollution and river restoration by wetlands and efficiency and minimize cost (Girts and Knight, 1989). The most important aspect of designing a wetland is to define the goal of the wetland project (Willard et al., 1989). Among the possible goals for wetland construction are the following:

1. Flood control
2. Waste-water treatment (e.g., domestic waste-water or acid mine drainage)
3. Storm water or nonpoint source pollution control
4. Ambient water quality improvement (e.g., instream system)
5. Wildlife enhancement
6. Fisheries enhancement
7. Replacement of similar habitat (mitigation wetlands)
8. Research wetland

This objective, or a series of desired objectives, needs to be determined before a specific site is chosen or a wetland is designed. If several objectives are desired, one must be identified as the primary objective.

B. Preliminary Design Considerations

Ecological engineering dictates that we take advantage of our ever increasing knowledge of ecology and its principles (e.g., succession, energy flow, self-design) to design a system that will be as close to a natural feature of the landscape as possible and will require a minimum amount of maintenance. This means resisting the ever-present temptation to over-engineer, to channel energies that cannot be channeled, to impose species that the design does not support. We agree with Boule (1988) who recommends that the design of wetlands should be kept simple without reliance on complex technological approaches that invite failure: "Simple systems tend to be self-regulating and self-maintaining" (Boule, 1988).

Some of the principles of ecological engineering of wetlands are outlined below:

1. Design the system for minimum maintenance. Instead, the system of plants, animals, microbes, substrate, and water flows should be developed for self-maintenance and self-design (Mitsch and Jørgensen, 1989; Odum, 1989).
2. Design a system that utilizes natural energies such as potential energy of streams as natural subsidies to the system. Pulsing streams in midwestern winters provide great quantities of nutrients in relatively short periods.
3. Design the system with the landscape, not against it. Floods and droughts are to be expected, not feared. Outbreak of plant diseases or invasion of alien species are often symptomatic of other stresses and may indicate faulty design rather than ecosystem failure.

4. Design the system to be multi-objective but identify at least one major objective and several secondary objectives.
5. Design the system as an ecotone. This means including a buffer strip around the site but also means that the wetland site itself needs to be viewed as a buffer system between upland and aquatic systems.
6. Give the system time. Wetlands are not functional overnight and several years may elapse before nutrient retention or wildlife enhancement are optimum. Strategies that try to short-circuit ecological succession or over-manage it are often doomed to failure.
7. Design the system for function, not form. If initial plantings and animal introductions fail, but the overall function of the wetland, based on its initial objectives, is intact, then the wetland has not failed. Expect the unexpected.
8. Do not over-engineer wetland design with rectangular basins, rigid structures and channels, and regular morphology. Ecological engineering recognizes that natural systems should be mimicked, not simplified, to accommodate biological systems (Brooks, 1989).

C. Site Selection

Several important aspects relate to site selection. When the objective is defined, the proper site should allow for a maximum probability that the objective can be met, that it can be done at a reasonable cost, that the system will perform in a generally predictable way, and that the long-term maintenance of the system be kept to a minimum. Among the ways in which this can be reasonably assured for wetlands are the following (Brodie, 1989; Willard et al., 1989):

1. Find a site where wetlands previously existed or where nearby wetlands still do exist. This assures that the proper substrate may also be present, that seed sources may be on site or nearby, and that the proper hydrologic conditions may exist. A historical meander of a stream that has been abandoned or channelized makes an excellent potential site for restoring a wetland.
2. Take into account the surrounding land use and the future plans for the land. Future land use plans, such as the abandoning of agricultural fields to become old-field ecosystems may obviate the need for a wetland to control nonpoint runoff.
3. Hydrologic conditions are paramount. Without water for at least part of the growing season, a wetland is impossible. A detailed hydrologic study of the site should be undertaken. This should include a determination of the potential interaction of groundwater with the wetland.
4. Find a site where natural flooding is frequent and the flooding water has somewhere to go. Sites should be inspected during flood season

and heavy rains and the annual and extreme event flooding history of the site should be determined as closely as possible.

5. The soils should be inspected and characterized in some detail, not only to determine their permeability and depth but also to determine their chemical content. Highly permeable soils are not likely to support a wetland unless water inflow rates are excessive.
6. Quality of groundwater, surface flows, and flooding streams and rivers that may influence the site water quality need to be determined. Even if the wetland is being built primarily for wildlife enhancement, chemicals in the water may be significant either to wetland productivity or bioaccumulation of toxic materials.
7. Site and nearby seed banks should be evaluated as to their viability and response to hydrologic conditions.
8. The availability of necessary fill material, seed and plant stocks, and access to infrastructure (e.g., roads, electricity) should be ascertained. This is particularly important for the construction phase.
9. Ownership of the land, and hence the price, are often overriding considerations. Additional lands may need to be purchased in the future to provide a buffer zone to the site.
10. For wildlife and fisheries enhancement, it needs to be determined if the wetland site is along ecological corridors such as migratory flyways or spawning runs.
11. Site access by the public will eventually need to be controlled to avoid vandalism and personal injury. A remote site is also preferable to an urban one for reasons of potential mosquito complaints, changing property values, or other social impacts.
12. An adequate amount of land should be available to meet the objectives. If "aging" of a wetland, defined as an impairment of wetland function after several years of perturbation (Kadlec, 1985), is anticipated due to loading of sediments, nutrients, or other materials, then larger land parcels to build additional wetlands in the future should be considered.
13. Riparian wetlands present a particular problem because flooding also causes scouring, sediment shifts, erosion, and deposition. Convex sides of river channels may be preferable to concave sides because of higher erosive forces on the latter.
14. The advantages of locating several small wetlands in the upper reaches of a watershed (but not in the streams themselves) versus fewer larger wetlands in the lower reaches should be considered. Loucks (1989) argues that a better strategy for wetlands to survive extreme events is to locate a larger number of low-cost wetlands in the upper reaches of a watershed rather than to build fewer high-cost wetlands in the lower reaches. A modelling effort on flood control by Ogawa and Male (1983) suggested the opposite; the usefulness of wetlands in decreasing flooding increases with the distance the wetland is downstream.

D. Hydrology

Hydrology is the most important variable in wetland design. If the proper hydrologic conditions are developed, the chemical and biological conditions will, to a degree, respond accordingly (Mitsch and Gosselink, 1986). The hydrologic conditions, in turn, are dependent on climate, seasonal patterns of streamflow and runoff, and possible groundwater influences. It is improper hydrology that leads to the failure of many created wetlands (D'Avanzo, 1989). Improper hydrologic conditions will not always correct themselves as will the more forgiving biological components of the system such as vegetation and animals which can self-correct initial mistakes. Ultimately the loading rate, or amount of water entering the system per unit area of wetland per unit time, determines much of the wetland function. This is functionally tied to the hydroperiod and retention time of the wetland.

1. Hydroperiod and Depth

One of the most basic design parameters for constructed wetlands is the pattern of depth over time, called the hydroperiod (Mitsch and Gosselink, 1986). Included in this parameter is not only the depth but the frequency of flooding and the seasonal patterns. Typical hydroperiods to be expected for riparian wetlands and others are shown in Fig. 5. Bottomland forested wetlands frequently have extended periods of dry conditions during the growing season. If herbaceous rather than woody vegetation is desired, more frequent flooding is necessary. Seasonal patterns of river flooding are part of the hydroperiod and offer the most natural way to hydrologically replenish a riparian wetland (Mitsch et al., 1979; Niering, 1989). After start-up, a variable hydroperiod, with dry periods interspersed with flooding, is a natural cycle in midwestern wetlands (Weller, 1981) and fluctuating water levels should be accepted if not encouraged (Willard et al., 1989). A fluctuating water level can often provide needed oxidation of organic sediments and can, in some cases, rejuvenate a system to higher levels of chemical retention (Kobriger et al., 1983; Faulkner and Richardson, 1989).

Wetlands with variable depths have the most potential for developing a diversity of plant and animal species. Deep-water areas, devoid of emergent vegetation, offer habitat for fish (i.e., *Gambusia affinis* that control mosquitoes), can enhance nitrification as a prelude to later denitrification if nitrogen removal is desired, and can provide low velocity areas where water flow can be redistributed (Steiner and Freeman, 1989). Allen et al. (1989) describe four general vegetation zones by depth although no recommendation is given as to their distribution:

1. Deep zone, 91 to 152 cm
2. Midzone, 15 to 91 cm

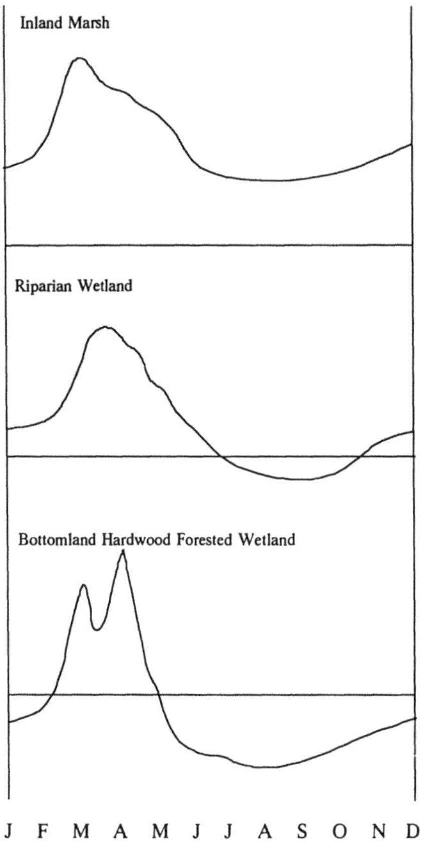

Figure 5. Typical hydroperiods (water level versus time) for inland marsh, riparian marsh, and bottomland hardwood forested wetland

3. Shallow zone, 15 cm below to 15 cm above normal water level
4. Transitional zone, 15 to 45 cm above normal water level

Several states have developed guidelines for bottom profiles of wetlands. Florida regulations for the Orlando area (described by Palmer and Hunt, 1989) require a littoral shelf with gently sloping sides of 6:1 or flatter out of a point 60 to 77 cm (2 to 2.5 feet) below the water surface. They further recommend less than 70% open water, with the rest of the wetland established with aquatic vegetation. A mean depth of 1 to 3 m (3 to 10 feet) is recommended for permanent pools. For Maryland, specific water depth requirements for wetlands for the use of storm-water runoff treatment are presented by Athanas (1987): 75% of the wetland should have a depth under 30 cm (50% less than or equal to 15 cm and 25% 15 to 20 cm), 25% should have depths ranging from 60 cm to 100 cm.

Water levels need to be controlled through inflow and outflow structures to allow the proper water depths for planting and controlling undesirable plants. During the start-up period of constructed wetlands, lower water levels are needed to avoid flooding new emergent plants while complete flooding is necessary for floating leaved and submerged plants (Allen et al., 1989). Specific periods of drawdown may be necessary if wetland plants are to be started from germinating seeds. Start-up periods for the establishment of plants may take up to 2 to 3 years and an adequate litter-sediment compartment may take another 2 to 3 years (Kadlec, 1989).

2. Loading Rates

Most of what is known about loading rates (volume of water applied per unit area per unit time) to wetlands comes from studies done on design of wetlands to treat waste-water. Watson et al. (1989) reviewed loading rates to wetlands for waste-water treatment from small municipalities and reported surface loading rates ranging from 1.4 to 22 cm/day while subsurface rates varied between 1.3 and 26 cm/day. Their analysis suggests that surface flow systems generally have lower loading rates than do subsurface systems. Wile et al. (1985) have recommended 2 cm/day for waste-water wetlands as optimum. Brown (1987) designed a mosaic of wetland "cells" amid forested flood-plain wetlands and estimated a hydraulic loading rate of waste-water of 2.2 cm/day. Adamus (1990) reports an EPA guideline for loading rates of less than 5 cm/week (<0.7 cm/day) for wetlands receiving waste-water. This limitation was developed from State of Florida guidelines for forested wetlands which, in turn, were originally developed from experiments with cypress domes in Gainesville, Florida (Odum et al., 1977).

Hydraulic loading rates as high as 29 cm/day have been suggested for wetlands designed for acid mine drainage (Pesavento, unpublished data as cited in Watson et al., 1989) although Fennessy and Mitsch (1989b) recommend 5 cm/day as a conservative loading rate for this type of wetland.

The above loading rate limitations are probably too restrictive for wetlands used for control of nonpoint pollution, but few studies have been undertaken to see the optimum design rates for wetlands used for nonpoint pollution control or river restoration projects. The Des Plaines River Demonstration Project in northeastern Illinois (Hey et al., 1989) has initial experiments designed for loading rates of river water from 1 to 8 cm/day with pumps delivering the water from the Des Plaines River. However, these rates were estimated from rates for comparable waste-water wetlands and may be too low for riparian wetlands receiving flood waters.

3. Seasonal Pulses

Inflows to riparian wetlands may be primarily due to flooding streams and rivers. Most of the loading of phosphorus-laden sediments to riverine wetlands occurs during the inter and spring (Livingston and Loucks, 1978; Mitsch et al., 1979, 1989), and a good wetland design will take advantage of these pulses if phosphorus retention is the primary objective. Loucks

(1989) points out that these seasonal pulses, in which the magnitude of flooding increases with its higher return period, are themselves not constant but vary in intensity from year to year. The rarer events, say the 100-year flood, may be the stream flow for which the wetland is designed. Infrequent and nonperiodic flooding and droughts are also important for dispersing biological species to the wetland and adjusting resident species composition.

4. Retention Time

Watson and Hobson (1989) present the wetland residence time as given by:

$$t = LWnd/Q$$

where L represents the length of the system, W the width of the system, n the porosity (which is 1.0 for surface flow without vegetation and < 1.0 for subsurface flow or surface flow with vegetation), d the average depth, and Q the average flow.

They suggest that porosity (n) of surface flow wetlands is actually slightly less than 1 because of the volume occupied by the vegetation. They estimate the effective porosity of wetlands of various species to be 0.86 to 0.98.

As with loading rate, most of the experience in retention time of wetlands is based on wetlands designed to treat waste-water. The optimum detention time has been suggested to be 5 to 14 days for treatment of municipal waste-water (Wile et al., 1985; Watson and Hobson, 1989). Florida regulations on wetlands (cited in Palmer and Hunt, 1989) require that the volume in the permanent pools of the wetland must provide for a residence time of at least 14 days. Brown (1987) suggested a retention time of a riparian wetland system in Florida of 21 days in the dry season and more than 7 days in the wet season. Fennessy and Mitsch (1989b) recommend a minimum retention time of 1 day for acid mine drainage wetlands with more effective iron removal at much longer periods.

5. Flow Characteristics

Several aspects related to the morphology of the basin need to be considered when designing wetlands, especially riparian wetlands. Wetland bottom slopes of less than 1% are recommended for wetlands built to control runoff and for wetlands in riparian settings (Bell, 1981; Kobriger et al., 1983, cited by Willard et al., 1989), while Steiner and Freeman (1989) suggest a substrate slope, from inlet to outlet, of 0.5% or less for surface flow wetlands used to control waste water. Flow conditions should be designed so that the entire wetland is effective in nutrient and sediment retention, if these are desired objectives. This may necessitate several inflow locations and a wetland configuration to avoid short-circuiting of flows. Steiner and Freeman (1989) suggest a length to width ratio (L/W) of at least 10 if water is purposely introduced to the system. In riparian settings, the wetland should probably be designed longitudinally to minimize flooding effcts. Dikes built perpendicular to the flow on the flood plain are more susceptible to erosion.

In some cases, individual wetland cells, placed in series or parallel, offer the most effective design. Sometimes wetlands are built in several "cells" or compartments to create different habitats or establish different functions (Steiner and Freeman, 1989). In some cases, cells can be parallel so that alternate drawdowns can be accomplished for mosquito control or redox enhancement. In other cases, cells in series can be used to enhance biological processes. Wetlands in Ohio (Fennessy and Mitsch, 1989a, b), California (Metz, 1987) and Florida (Redmond, 1981) have utilized separate cells. In other cases, wetlands are designed with cells that actively receive polluted water amid passive wetlands that receive overflow from the active cells (Best, 1987; Brown, 1987).

E. Chemical Inputs

When water flows into wetlands it brings with it chemicals that may be either beneficial or possibly detrimental to the functioning of a wetland. In an agricultural watershed, this inflow will include nutrients such as nitrogen and phosphorus as well as sediments and trace amounts of various pesticides. If the wetland is designed to retain nutrients, then it is desirable to know how well that occurs for various nutrient inflows. Some work, mainly from the compilation of data from several wetland sites, has given some indication of the nutrient retention of wetlands (Knight et al., 1984; Richardson and Nichols, 1985; Athanas, 1988; Faulkner and Richardson, 1989). Figure 6 illustrates the percent removal of nitrogen and phosphorus for various loading rates in waste-water wetlands, and Fig. 7 shows the same variables as a function of wetland size and volume for wetlands designed to control storm-water runoff. Maristany and Bartel (1989) used an empirical model for reservoirs to estimate the sediment and phosphorus retention in a Florida wetland (Fig. 8). They found that removal efficiencies increase rapidly with increasing wetland size (decreasing chemical load) to a point but level off quickly after that. For example, suspended solids retention increases rapidly as a function of wetland size until the wetland is approximately 1% of the watershed area, above which the removal efficiencies improve much more slowly (Fig. 8). Empirical models such as these can be used as first estimates of the potential nutrient and sediment retention in freshwater wetlands (Mitsch et al., 1989).

With the exception of a significant literature on acid mine drainage wetlands (e.g., Fennessy and Mitsch, 1989a, b), few evaluations are found on the efficiency of wetlands to remove chemicals other than nutrients. McArthur (1989) reports on a case study in Florida where lead was reduced by

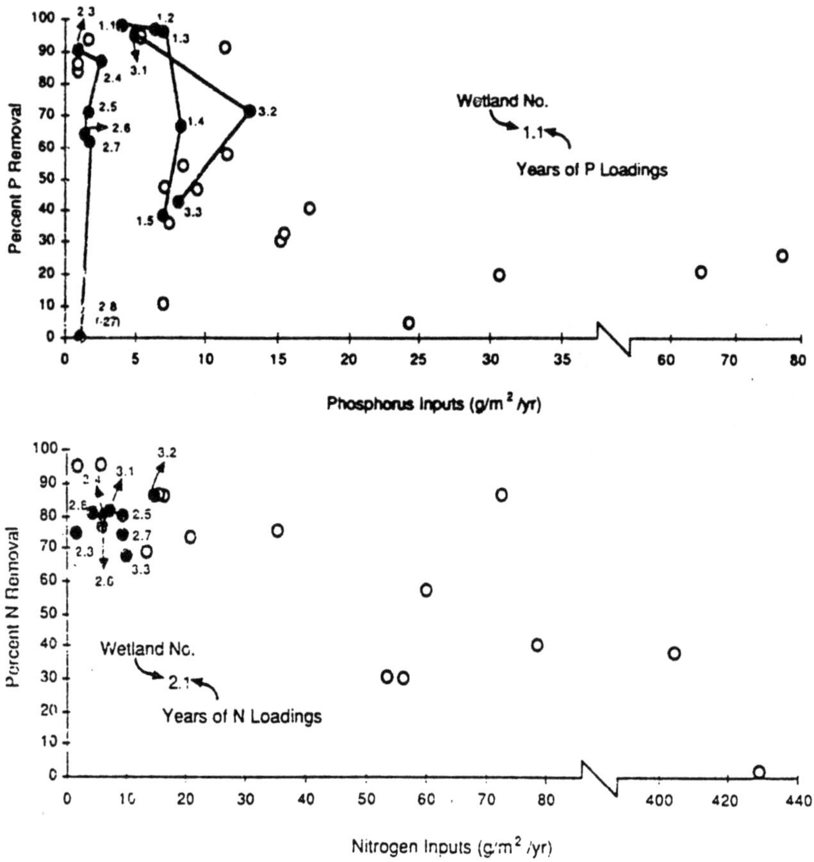

Figure 6. Percent removal graphs for wetlands receiving wastewater for (a) phosphorus and (b) nitrogen (Faulkner and Richardson, 1989). From *Constructed Wetlands for Wastewater Treatment: Municipal, Industrial and Agricultural*. Donald A. Hammer, Ed. Copyright 1989. Lewis Publishers, Inc., Chelsea, MI. Used with permission.

83%, zinc by 70% and total solids by 55% as storm water passed through a 0.4 ha (0.9 acre) pond-wetland complex.

F. Substrate/Soils

The substrate is important to the overall function of a constructed wetland and is the primary medium supporting rooted vegetation (Willard et al., 1989). If a wetland is being used to improve water quality, the substrate or soils of a wetland provide a significant role in the ability of a wetland to retain certain chemicals. The sediments provide a unique role in retaining

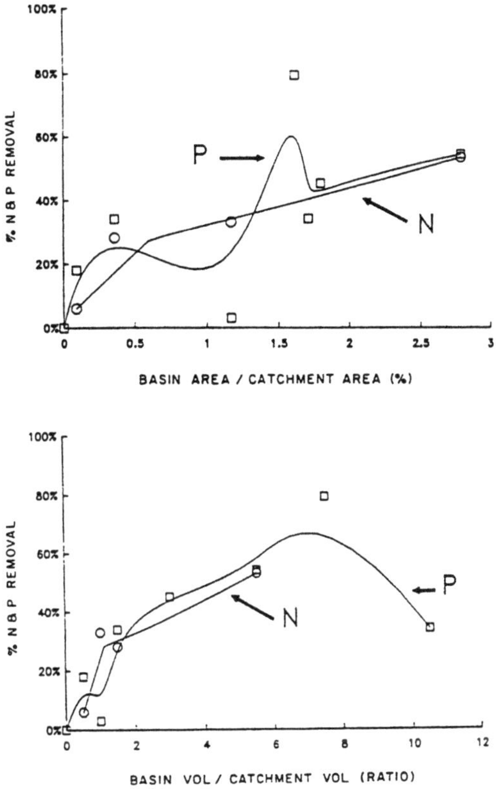

Figure 7. Percent phosphorus and nitrogen removal versus size for stormwater wetlands for (a) wetland surface area/watershed area ratio (×100) and (b) wetland volume/mean runoff ratio (×100) (Athanas, 1988). From *Wetlands: Increasing our Wetland Resources.* National Wildlife Federation, Washington D.C. Used with permission.

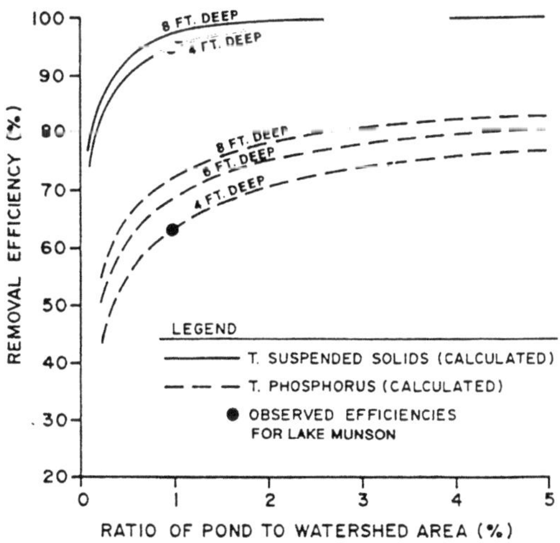

Figure 8. Percent removal of suspended solids and total phosphorus for shallow lakes in Florida (Maristany and Bartel, 1989). From *Wetlands Concerns and Successes* Copyright American Work Resources Association, Bethesda, MD. Used with permission.

certain chemicals and in providing the habitat for micro- and macroflora and fauna that are involved in several chemical transformations. The substrate is not as significant for retention of suspended organics and solids (along with the chemicals partitioned to sediment particles), as their retention is based primarily on the net rate of sedimentation (sedimentation-resuspension) that results from the low velocities characteristic of wetlands (Steiner and Freeman, 1989).

Generally, constructed wetlands are designed for either surface flow over the substrate or subsurface flow through the substrate (Watson et al., 1989; Steiner and Freeman, 1989). The surface flow system is generally less effective in removing some pollutants at first but is closer in design to natural wetlands.

The characteristics of substrate that do play a role in the design of wetlands are reviewed here.

1. Organic Content

The general organic content of soils has some significance on the retention of chemicals in a wetland. Mineral soils generally have lower cation exchange capacity than do organic soils, with the former dominated by various metal cations and the latter by the hydrogen ion. Organic soils can therefore remove some contaminants (e.g., certain metals) through ion exchange and can enhance nitrogen removal by providing an energy source and anaerobic conditions appropriate for denitrification. Wetland soils generally vary between 15% and 75% organic matter (Faulkner and Richardson, 1989) with higher concentrations in peat-building systems such as bogs and fens and lower concentrations in open wetlands such as riparian bottomland hardwood wetlands subject to mineral sedimentation or erosion. When wetlands are constructed, especially subsurface flow wetlands, organic matter such as mushroom compost, peat, or detritus is often added in one of the layers. For many wetlands, organic soils are not preferred because they are low in nutrients, can cause low pH, and often provide inadequate support for rooted aquatic plants (Allen et al., 1989).

2. Soil Texture

Clay material, while often favored for surface wetlands to prevent percolation of water to groundwater, may also limit root and rhizome penetration and may be impermeable to water for plant roots. In this case, loam soils are preferable. Sandy soils, while generally low in nutrients, do anchor plants and allow water to readily reach the plant roots (Allen et al., 1989). The use of local soils, underlain with impermeable clay to prevent downward percolation, is often the best design. If on-site top soils are to be returned to the wetland, adequate temporary storage should be provided (Willard et al., 1989). If clay is not available on site, it may be advisable to add a layer to slow percolation.

3. Subsurface Flow and Gravel

Wieder et al. (1989) suggest that subsurface flow through artificial wetlands can be through soil media (root-zone method) or through rocks or sand (rock-reed filters) with the flow in both cases 15 to 30 cm below the surface. Gravel has been added to the substrate of artificial wetlands to provide a relatively high permeability that allows water to percolate into the root zone of the plants where microbial activity is high (Gersberg et al., 1983, 1984, 1986, 1989). In a survey of several hundred wetlands built in Europe for sewage treatment in rural settings, Cooper and Hobson (1989) report that gravel is used in combination with soil, but that the substrate retains the greatest uncertainty in artificial reed (*Phragmites*) wetlands in Europe. Gravel can be silica-based or limestone based, with the former having much less capacity for phosphorus retention (Cooper and Hobson, 1989). More recent evaluation of the European-design subsurface wetlands indicates that they decrease in hydrologic conductivity after a few years and become essentially overloaded surface flow wetlands (Steiner and Freeman, 1989). One study, which investigated six different substrates for their effectiveness in controlling acid mine drainage, concluded that all performed basically the same (Brodie et al., 1988).

4. Depth and Layering of Substrate

The depth of substrate has been an important design consideration for waste-water and mine drainage wetlands, particularly those which use subsurface flow. Depth of substrate is of less concern for surface water wetlands than for subsurface flow wetlands. All wetlands should have an adequate depth of clay materials if downward percolation is not desired. Meyer (1985) discusses a layered substrate in wetlands to control stormwater runoff as having the following materials: 60 cm of 1.9 cm limestone; 30 cm of 2 mm crushed limestone to raise pH and aid in precipitation of dissolved heavy metals and phosphate; 60 cm of coarse to medium sand as filter, and 50 cm organic soil. A common depth of substrate for subsurface flow wetlands is 60 cm (Steiner and Freeman, 1989). The depth of suitable substrate should be adequate to support and hold vegetation roots (see discussion below).

5. Nutrients

While exact specifications of nutrient conditions in substrate necessary to support aquatic plants are not well known (Allen et al., 1989), low nutrient levels characteristic of either organic, clay, or sandy soils can cause problems in initial plant growth. While fertilization may be necessary in some cases to establish plants and enhance growth, it should be avoided if possible in wetlands which eventually will be used as a sink for macronutrients. When fertilization is required, slow release granular and tablet fertilizers are often useful (Allen et al., 1989).

7. Iron and Aluminum

When soils are submerged and anoxic conditions result, iron is reduced from ferric (Fe^{3++}) to ferrous (Fe^{2+}) forms, releasing phosphorus that was previously held as ferric phosphate compounds. The Fe-P compound is referred to operationally as reductant-soluble phosphorus (RS-P) and can be a significant source of phosphorus to overlying and interstitial waters (Faulkner and Richardson, 1989) after flooding and anaerobic conditions occur. Phosphorus can also be retained by wetlands by oxides and hydroxides of iron and aluminum (Richardson, 1985). There is apparent disagreement in the literature as to whether the phosphorus is retained by ligand exchange or by precipitation (Faulkner and Richardson, 1989). Nevertheless, the iron and aluminum content of soils has significant implications of the ability of a wetland, whether constructed or natural, to retain nutrients such as phosphorus.

G. Vegetation

1. Types

Vegetation types, of course, depend on the region and climate of the constructed wetland as well as the design characteristics described above. Common emergent plants used in the U.S. include soft-stem bulrush (*Scirpus validus*), cattails (*Typha latifolia* and *Typha angustifolia*), and sedges (*Carex* spp.), and floating-leaved aquatic plants such as white water lily (*Nymphaea odorata*) and spatterdock (*Nuphar luteum*). Submerged plants are not as common in wetland design, and their propagation is often hampered by turbidity. Table 1 summarizes some of the species investigated for wetlands constructed for waste-water treatment.

In some cases, certain plants are viewed as desirable or undesirable for reasons such as their value to wildlife or their esthetics. Reed grass (*Phragmites australis*) is often favored in constructed wetlands in Europe (Cooper and Hobson, 1989) but is not generally used in the United States. Some plants are considered undesirable in wetlands because they are aggressive alien species. In the southern United States, the floating aquatic plant, water hyacinth (*Eichhornia crassipes*), and in the northern United States, the emergent purple loostrife (*Lythrum Salicaria*), are considered undesirable alien plants in wetlands, although the former has frequently been evaluated for its role in nutrient retention (e.g., Mitsch, 1977; Ma and Yan, 1989). Throughout the United States, cattail (*Typha* spp.) is championed by some and disdained by others as it is a rapid colonizer but of limited wildlife value (Odum, 1987). The list of possible plants for constructed wetlands is large and can be found elsewhere (e.g., Midwestern Wetland Flora, Field Office Guide to Plant Species, USDA, Lincoln, NE) and an almost infinite number of combinations of plants exist.

Table 1. Some plant species tested for use in constructed wetlands for stormwater and wastewater treatment

Emergent	**Submerged**	**Woody Plants**
Scirpus robustus	*Ceratophyllum demersum*	*Cornus stolonifera*
Scirpus lacustris	*Elodea nuttallii*	*Celtis occidentalis*
Schoenoplectus lacustris	*Myriophyllum aquaticum*	*Cephalanthus occidentalis*
Phragmites australis	**Floating**	*Salix* spp.
Phalaris arundinacea	*Lagorosiphon major*	*Alnus* spp.
Typha domingensis	*Salvinia rotundifolia*	*Populus deltoides*
Typha angustifolia	*Canna flaccida*	*Acer saccharinum*
Typha latifolia	*Pistia stratiotes*	*Quercus* spp.
Iris pseudacorus	*Lemna minor*	
Scirpus americanus	*Eichhornia crassipes*	
Scirpus validus	*Wolffia arrhiza*	
Scirpus pungens	*Azolla caroliniana*	
Glyceria maxima	*Hydrocotyle umbellata*	
Eleocharis dulcis	*Lemna gibba*	
Eleocharis sphacelata	*Lemna* spp.	
Typha orientalis	*Spirodela polyrhiza*	
Zantedeschia aethiopica		
Colocasia esculenta		
Leersia oryzoides		
Panicum virgatum		

Adapted from Guntenspergen et al. (1989) and Willard et al. (1989)

An important general consideration of wetland design is whether plant material is going to be allowed to develop naturally with some initial seeding and planting, or whether continuous horticultural selection for desired plants will be imposed. We refer to the wetland in the former condition as a self-design wetland, the latter a designer wetland. To develop a wetland that is ultimately a low-maintenance one, natural successional processes need to be allowed to proceed. Often, this means some initial period of invasion by undesirable species, but if proper hydrologic conditions are imposed, those invasions may be temporary. The best strategy is to introduce, by seeding and planting, as many choices as possible to allow self-design to sort out the species and communities in a timely fashion (Odum, 1989; Dunn, 1989). Some help to this selection process, e.g., selective weeding, may be necessary in the beginning, but ultimately the system needs to survive with its own successional patterns unless significant labor-intensive management is used. Odum (1987) distinguishes freshwater wetland succession from coastal saltwater wetland succession by stating that "in many freshwater wetland sites it may be an expensive waste of time to plant species which are of greater value to wildlife . . . . It may be wiser to simply accept the establishment of disturbance species as a cheaper although somewhat less attractive solution."

2. Planting Techniques

Plants in the wetland can be introduced by transplanting roots, rhizomes, tubers, or whole plants, by broadcasting seeds obtained commercially or from other sites, by importing substrate with their seed bank from nearby wetlands, or by relying completely on the seed bank of the original site. If planting stocks rather than site seed banks are used, it is most desirable to have plants chosen from wild stock rather than nurseries because the former are generally better adapted to the environmental conditions that they will face in constructed wetlands. The plants should come from nearby if possible and should be planted within 36 h of collection. If nursery plants are used, they should be from the same general climatic conditions and should be shipped by express service to minimize losses (Allen et. al., 1989). Esry and Cairns (1989) describe three cells of a marsh designed for storm-water runoff which were planted differently to determine effectiveness and hardiness. The first cell was planted with locally obtained sawgrass (*Cladium*) and the second cell was planted with bulrush (*Scirpus*) obtained in nurseries. Both of these species were hand-planted in rows. The third cell was planted with pickerelweed (*Pontederia*) which was gathered in a nearby farm pond and spread with heavy machinery. The plants of the first and the third cells have successfully taken hold and thrived. The second cell still contains some bulrushes, but their survival has been limited. There has been rapid volunteer growth of duckweed (*Lemna*) and other emerged plants.

For emergent plants, the use of planting materials with at least 20- to 30-cm stems is recommended (Tomljanovich and Perez, 1989) and whole plants, rhizomes, or tubers rather than seeds have been most successful. In temperate climates, both fall and spring planting times are possible for certain species, with spring plantings generally more successful (Herricks et al., 1982 in Willard et al., 1989; Allen et al., 1989). Garbisch (1989) suggests that spring planting is desirable to minimize the destructive grazing of plants in the winter by migratory animals and to reduce the uprooting of plants by ice. Transplanting plugs or cores (8 to 10 cm in diameter) from existing wetlands is another technique that has been used with success as it brings seeds, shoots, and roots of a variety of wetland plants to the newly constructed wetland (Kobriger et al., 1983; Allen et al., 1989). Brown (1987) gives planting recommendations for a Floridian site used for renovation of treated effluent. He suggests that marshes be planted at densities to insure seed source, rapid colonization, and effective competition with *Typha* spp. Specifically, this could mean from 2000 to 5000 plants/ha (800 to 2000 plants/acre). Brown (1987) and Willard et al. (1989) found that a varied bed form adds to diversity of the vegetation.

Athanas (1987) and Livingston (1989) present guidelines for wetland construction for the state of Maryland for storm-water management. Very specific recommendations are given for plant types, with planting to in-

Table 2. Recommended plant species for stormwater managements in Maryland (Athanas 1987 and Livingston 1989)

Primary species	*Leersia oryzoides*
Sagittaria latifolia	*Nuphar luteum*
Scirpus americanus	*Peltandra virginica*
Scirpus validus	*Pontederia cordata*
	Saururus cernuus
Secondary species	
Acorus Calamus	**Undesirable species**
Cephalanthus occidentalis	*Typha latifolia*
Hibiscus Moscheutos	*Typha angustifolia*
Hibiscus laevis	*Phragmites australis*

clude at least two aggressive wetland species (primary species) and three secondary species planted in smaller numbers than the primary species (Table 2). *Typha* spp. and *Phragmites australis* are considered aggressive plants but without much value to wildlife. For this reason, they are not included on the list of desirable species. *Peltandra virginica* is an example of a secondary species which is less aggressive (probably because it depends more on seed germination than on vegetative propagation) but which seems to have good wildlife value.

The planting guidelines given are:

1. Primary species should cover 30% of the shallow zone and be spaced at 1 meter (3 feet) intervals.
2. These species should be in four monospecific areas.
3. One hundred clumps per hectare (40 clumps/acre) should be distributed over the rest of the wetland.
4. For secondary species, plant 125 individuals per hectare (50 per acre) and, for each species, plant 25 clumps of five individuals per hectare (10 per acre) close to edge of wetland yet with clumps as far apart as possible to segregate species.

3. Seeding Techniques

If seeds and seed banks are used for wetland vegetation, several precautions must be taken. The seed bank should be evaluated as to seed viability and species present (van der Valk, 1981). The use of seed banks from other nearby sites can be an effective way to develop wetland plants in a constructed wetland if the hydrologic conditions in the new wetland are similar. Weller (1981) suggests success of seed bank transplants with many different species including sedges (*Carex* spp.), *Sagittaria* sp., *Scirpus acutus*, *S. validus*, and *Typha* spp. Disruption of the wetland site where the seed bank is obtained must also be considered.

When seeds are used directly to vegetate a wetland, they must be collected when they are ripe and stratified if necessary (Willard et al., 1989). If commercial stocks are used, the purity of the seed stock must be determined (Garbisch, 1989; Willard et al., 1989). The seeds can be added with commercial drills or by broadcasting from the ground, watercraft, or aircraft. If broadcast seeding is used, it is usually desirable for it to occur when there is little to no standing water in the wetland.

H. Management after Construction

1. Plant Harvesting

Harvesting of plants generally does not result in a great quantity of chemicals, e.g., nutrients, being removed from the system unless the material is harvested several times in a growing season. There are some cases when plant harvesting may be advisable as part of the routine maintenance of the constructed wetland. Wieder et al. (1989) suggest plant harvesting as a mechanism to alter the effect that plants have on the system, generally by putting them back into an earlier stage of succession when net growth may be greater. Plant harvesting may also be necessary to control mosquitoes, reduce congestion in the water, increase the retention time of the basin, and allow for greater plant diversity. Yan Jingsong (personal communication, Nanjing, China, 1989) has suggested that the practice in China is to continually harvest emergent plants, thereby increasing the size, strengths, and number of shoots, and encouraging vegetative reproduction. Suzuki et al. (1989) suggest that harvesting *Phragmites* twice during the growing season, once at peak nutrient content and the second at the end of plant growth, leads to a maximum removal of nitrogen and phosphorus.

Sometimes other plant management techniques such as drawdowns followed by burning are used as a means of controlling the invasion of woody vegetation if that invasion is considered undesirable (Warners, 1987). When controlled burning of wetlands is used, the wetland manager needs to consider the impact on wildlife (Willard et al., 1989) as well as the potential reintroduction of inorganic nutrients to the water column from the sediments when water is again added.

2. Wildlife Management

While the development of wildlife is a welcomed and often desired aspect of created wetlands, beaver or muskrat burrows in dikes or obstructions to flow can cause problems (Tomljanovich and Perez, 1989); sand, gravel, or wire screening can be used to discourage the burrowing. In other cases, animals grazing on newly planted perennial herbs and seedlings is particularly destructive to newly planted material (if planting techniques are used). Garbisch (1989) suggests that the timing of planting is important, especially when migratory animals are involved in destructive grazing in the winter. Likewise, deeper wetlands often become havens for undesirable fish such as carp which can cause excessive turbidity and uproot vegetation. Weller (1981) recommends 50% open water in midwestern basin wetlands to encourage wildlife, while Brinson et al. (1981) suggest creating diverse habitats with live and dead vegetation, islands, and floating structures.

In many cases of wetland construction, wildlife enhancement begins within a few years following construction. At a constructed wetland at Pintail Lake in Arizona, the area's waterfowl population has increased dramatically. By the second year of use, duck nest density increased 97% over the first year (Wilhelm et al., 1989). Considerable wildlife activity has already been noted at the Des Plaines River wetlands in northeastern Illinois approximately 1 to 2 years after their construction.

3. Mosquitoes

Mosquitoes are controlled in California constructed wetlands by changing the conditions of the wetland (e.g., hydrology) to inhibit mosquito larvae development or by using chemical or biological control (Martin and Eldridge 1989). Wieder et al. (1989) and others have suggested mosquito control by fish, especially with the air-gulping mosquito fish (*Gambusia affinis*). Apparently little is known about the role of water quality on the control of mosquitoes (Martin and Eldridge, 1989). Bacterial insecticides (e.g., *Bacillus sphaericus*) and the fungus *Lagenidium giganteum* are known pathogens of mosquito larvae, but they have not been extensively tested (Martin and Eldridge, 1989).

4. Perturbation Control

The average conditions of a wetland used in its design do not reflect the actual conditions, where seasonal and less frequent perturbations are uncertain in frequency and magnitude and yet may require some response (Willard and Hiller, 1989; Brooks, 1989; Girts and Knight, 1989). Willard and Hiller (1989) suggest that we design and plan wetlands for the worst case conditions of perturbations but that we maintain a balance among form, function, and persistence. Tables 3 and 4 list some of the seasonal and unpredictable disturbances that occur in managed wetlands and the possible responses to those disturbances. It is useful to remember that wetland design is an inexact science and perturbations may change the original design (e.g., selected plant species) to something else. If wetland functions, particularly those which are related to the objectives of the wetland, remain intact, changes in species and forms are not as significant.

5. Sediment Dredging

This is an optional management technique in constructed wetlands. Its use depends on whether the wetlands are filling in with sediments, thus

Table 3. Possible solutions to predictable system disturbances which may occur throughout the lifetime of a constructed wetland (Girts and Knight, 1989). From Constructed Wetlands for Wastewater Treatment: Municipal, Industrial and Agricultural, Donald A, Hammer, Ed. Copyright 1989, Lewis Publishers, Inc. Chelsea, MI. Used with permission.

Disturbance	Features	Symptoms	Operations Modifications
Start-up	Vegetation establishment Microbial flora colonization Technical system debugging	Widely fluctuationg inflow-outflow chemistries, after taking flow velocity changes into account	Control loading rates, i.e., water flow rate, chemical concentrations * water inflow rates * fresh water source Control water depths-critical for vegetation establishment and development of conditions suitable for target microbial population Dilution/recirculation Chemical additions
Seasonal	Extreme precipitation	High loading rates	Control loading rates, i.e., water flow rates, chemical concentration * water inflow rates * dilution * recirculation
		Decreased storage capacity	Increase storage capacity * stormwater diversion * detention pond * increase water depth
		Insufficient residence time; Channeling	Control outflow rates Installation of baffles
	Extreme low temperatures	Insufficient flow	Freshwater source, recirculation, parallel cells

	Freezing; sheet flow over ice surface	Recirulation, aeration, control water depth Preheated water
Végetation growth/decay	Flushing of chemicals, nutrients, and microbes from decaying vegetation, sediment	Secondary treatment pond for treatment Vegetation harvest or burning Recirculation to increase nutrient retintion Drawdown and oxidation
Population composition changes (microbial, algal)	Gradual change in treatment efficiency	Secondary treatment pond Recirculation

Table 4. Possible solutions to unpredictable system disturbances throughout the lifetime of a constructed wetland (Girts and Knight, 1989). From Constructed Wetlands for Wastewater Trestment: Municipal Inductrial and Agricultural, Donald A. Hammer, Ed. copyright 1989. Lewis Publishers, Inc., Chelsea, MI. Used with permission.

Disturbance	Symptoms	Water inflow	Water outflow	Water depth
Record storm event	Hight hydraulic loading rates	X		
	Decreased storage capacity	X		
	Insufficient residence time	X	X	X
	Channeling	X		
	High sediment loads			X
	High chemical loads		X	
Change in chemical constituents and concentrations	High chemical loads	X		
	Increased toxicity (vegetation, wildlife)			
	Release of chemicals from sediments/vegetation		X	X
	Change in chemical form			X
Vegetation damage	Increased debris, flow hindrance	X		X
	Elemental release from vegetation		X	X
	Change in conditions for replanting	X		X
Pests (beavers, mosquitoes, etc.)	Complaints from neighbors			X
	Reduced flow and water level control			X
Malfunctions/construction failures	Reduced flow and water level control	X	X	X
	Inability to respond to need for changes in operation[b]			
Design flaw	Limited treatment capacity[b]			
	Limited lifespan[b]			

[a] New species
[b] All operation modifications may need to be considered

Operation modifications						
Dilution	Recirculation	Pretreatment pond	Chemical addition	Vegetation harvest	Replant	Predator control
	X	X				
		X				
X	X	X	X			
		X				
		X				
X	X	X	X			
X		X	X			
X				X	X	
	X			X	X	
X			X			
				X	X	
	X			X	X	
		X		X	X[a]	
			X			X
X	X	X	X			

shortening their effective lives, or whether organic sediment accumulation is viewed as an undesirable feature relative to the objectives of the wetland. Dredging is generally a very expensive operation and one that should not be attempted frequently in the life of a constructed wetland. Dredging not only carries out sediments, but also removes seed bank and rooted plants themselves. The very process of dredging sediments from constructed wetlands may require a regulatory permit, even if it is done to "improve" wetland function. The best approach, unless dredging is unavoidable, is to "accept the [sediment] accumulation as a natural part of wetland dynamics" (Willard et al., 1989).

I. Economics

1. Construction Costs

The construction of a new wetland involves careful consideration of a number of criteria, including a realistic look at cost. The amount of funding available, the period of time for which it is available, and the limits and rules concerning its expenditure are questions to be dealt with early in the planning stages of a constructed wetland (Newling, 1982). Tomljanovich and Perez (1989) suggest that an estimate of the cost of a new wetland's development will necessarily include the following items:

1. Engineering plan
2. Preconstruction site preparation
3. Construction (labor, equipment, materials, supervision, indirect and overhead charges)

The cost of wetland construction varies widely and depends on location, type and objectives of the wetland, as well as the maintenance required (Table 5). Factors that add to the cost variation include access to the site, substrate characteristics, cost of protective structures, local labor rates, and the availability of equipment (Newling, 1982). Compared with many other systems for water quality improvement, wetlands are relatively inexpensive to build. As Table 5 indicates, some wetlands which require human and technological intervention, such as the Santee Marsh in California and the Iselin treatment plant in Pennsylvania, are more costly to construct. The Pintail Lake and Jacques Marsh sites in Arizona were fairly inexpensive to build since they were constructed in pre-existing, but dry, lake basins. Digging and basin formation were not necessary at these sites and the natural formations helped to bring down the construction costs (Wilhelm et al., 1989).

2. Operational Costs

Operating and maintenance costs vary according to the wetland's use and to the amount and complexity of mechanical parts and plumbing that the

Table 5. Costs of constructed wetlands for various uses in the United States

Wetland	State	Use	Area (ha)	Cost ($/m²)	Source
Ballon a Wetland	California	Habitat, recreation	87.4	$7.01	Metz (1987)
Greenwood Urban Wetland	Florida	Stormwater runoff	11.0	$5.15	Palmer and Hunt (1989)
Lake Jackson Restoration	Florida	Urban runoff	4.0	$19.95[a]	Esry and Cairns (1989)
Santee Marsh	California	Wastewater treatment	0.1	$182.00[b]	Gersberg et al. (1989)
Iselin Marsh/Pond/Meadow	Pennsylvania	Wastewater treatment	0.2	$208.00[b]	Conway and Murtha (1989)
Pintail Lake	Arizona	Wastewater treatment	20.2	$7.38	Wilhelm et al. (1989)
Jacques Marsh	Arizona	Wastewater treatment	18.0	$7.53	Wilhelm et al. (1989)
Kash Creek (Impoundment 3)	Alabama	Acid mine drainage	0.4	$8.42	Brodie et al. (1989a)
Peabody	Ohio	Acid mine drainage	0.2	$48.00	Kolbash and Murphy (1987)
Widows Creek Steam Plant	Alabama	Ash pond seepage	0.5	$6.98	Brodie et al. (1989b)
Kingston	Alabama	Ash pond seepage	0.9	$14.21	Brodie et al. (1989b)
Bolivar Peninsula	Texas	Disposal site for dredge	8.0	$3.41	Newling (1982)
Windmill Point	Virginia	Disposal site for dredge	8.0	$2.53	Newling (1982)
Blue River Reclamation Project	Colorado	Riparian restoration	12.0	$4.13	Roesser (1988)
		Average:		$37.48	

[a] Includes area on which an impoundment and filter were built
[b] Cost reflects entirely artificial wetland and includes a good deal of plumbing

wetland contains. Fewer data are available on these operational costs. A pump, filter, impoundment tank, and piping add considerably to both the construction and maintenance costs of a wetland. Wetlands fed by natural runoff or by water that enters the site from adjacent waterways using only the force of gravity are far less expensive to maintain than highly mechanized wetlands.

3. Benefits

Benefits derived from a constructed wetland may be presented in terms of the savings that a wetland can provide over other technological approaches. For example, Kash Creek wetland in Alabama was built on a TVA site that had been in chronic violation of water quality standards due to acid mine drainage. The switch from chemical to natural wetland processes brought about water quality in compliance with standards and sent the yearly cost of treatment plummeting from $50 000 per year to only $3700 (Brodie et al., 1989a). The treatment of domestic waste water by wetlands is often cheaper than traditional chemical means. Odum et al. (1977) estimated a decrease in cost of waste-water treatment by half when the waste water passed through cypress domes rather than through chemical treatment plants. Boyt et al. (1977) estimated a savings of nearly $2 million over 25 years through the use of swampland for tertiary treatment instead of a waste-water treatment plant.

Wetland benefits are not limited to the amount of savings the users enjoy. Examples of benefits derived from constructed wetlands that were not originally a part of the design are many. Kreutzwiser (1981) assessed the recreational value of Long Point Marsh on the Canadian shore of Lake Erie. In 1978, 17 000 people visited the 750-hectare marsh for a variety of recreational purposes (fishing, photography, canoeing, hunting), and the monetary value of their use was estimated at over $200 000. The users also spent approximately $225 000 in surrounding businesses. This report emphasizes the importance of considering the value of enjoyment and recreation derived from a wetland.

4. Wetland Valuation

A wetland's value may include a number of parameters that are difficult to quantify. The debate over wetland valuation started with estimates of dollar values of salt marshes by Gosselink et al. (1974). Those authors considered commercial and sports fisheries, aquaculture, and waste treatment in their evaluation of a salt marsh. Their estimates range from $128 000/ha ($52 000 per acre) for moderately used marshland to $205 000/ha ($83 000 per acre) for highly productive areas. These values do not include nonconsumptive benefits such as added wildlife habitat, esthetic values, and flood protection which would bring the dollar value of a wetland even higher (Landin, 1982). Costanza et al. (1989) have estimated that the present value (1983 $US) of an average Louisiana coastal salt marsh is $6000 to $42 000 per hectare ($2429 to $17 000 per acre), depending on the discount rate and the method used for the estimate (e.g., willingness to pay or energy analysis).

A wetland may be given a value according to the products that can be harvested (utilitarian value) such as fish and wildlife harvest or esthetics, or according to its indirect role in maintaining ecosystem processes which support direct values such as water quality improvement, flood control, or supporting food webs and nutrient cycling (contributory value) (Costanza et al., 1989). Contributory values may far outweigh utilitarian values of wetlands. An estimate of the value of Canadian wetlands states that nonconsumptive values may be as much as 40% higher than the profits gained from consumptive use (Rubec, 1987).

More and better data on wetland function, along with standardized criteria, would aid in the valuation of wetlands and help make more realistic economic estimates (King et al., 1978; Jaworski, 1981; Kusler, 1985; Shabman, 1985). Many of the methods for valuation of wetlands are discussed by Mitsch and Gosselink (1986). Wetland valuation does impose an anthropocentric viewpoint on a natural structure and reflects the value of an ecosystem only in terms of its worth to humans (Mitsch and Gosselink, 1986). Nevertheless, economic valuation provides a recognizable gauge with which to communicate the importance and benefits of a wetland.

III. Summary

Policies such as "no net loss" of wetlands and relatively recent recognition of wetland values have stimulated restoration and creation of these systems. Restored or created wetlands have specific objectives such as waste-water treatment, mine drainage control, storm-water retention and improvement, mitigation of unavoidable wetland losses, or wildlife enhancement. Many of these constructed and restored wetlands have been successful in providing the desired results. Likewise, there have been some cases of "failure" of constructed or restored wetlands; generally the reasons relate to a lack of the proper hydrologic conditions. Ecological engineering of wetlands implies that their designs have reliance on self-design and a minimum of human maintenance. Among the hydrologic design parameters to be considered for constructing wetlands are hydroperiod, loading rates, seasonal pulses, flow patterns, and retention times. Wetland managers in the past have used mainly water depth to control the functioning of wetlands; a more comprehensive management of the flow-through characteristics of the wetland is needed. Chemical loading rates are important for wetlands being designed for water pollution control. Some guidelines on chemical loading are available for nitrogen and phosphorus but not for many other chemicals. Substrate plays an important role in plant development and chemical processes with substrate characteristics such as organic content, texture, nutrients, iron, and aluminum playing important roles in wetland design and construction. A wide variety of vegetation types and planting and seeding techniques are available for wetland construction. Vegetation success should be measured more by the success of the original objective of the wetland than by the success of individual species. Management after wetland construction and restoration can involve plant harvesting, wildlife management, mosquito control, perturbation control, and sediment dredging but ecological engineering suggests that these management options be kept to a minimum. Costs estimates for wetland construction are available and tend to be quite site specific; maintenance costs and economic benefits of constructed wetlands are even more difficult to estimate.

Acknowledgments

This paper was supported in part by a contract "Wetlands for the Control of Nonpoint Source Pollution" from the Ohio Environmental Protection Agency, P.O. Number 607107. Hugh Crowell was the OEPA Project Officer. Some salaries and research support were provided by State and Federal Funds appropriated to the Ohio Agricultural Research and Development Center, The Ohio State University. Manuscript number 210-90.

References

Adamus, P.R. 1990. *Wetlands and Water Quality: EPA's Research and Monitoring Implementation Plan for the Years 1989–1994.* Environmental Research Laboratory, Office of Research and Development, USEPA, Corvallis, Oregon. USEPA Office of Wetlands Protection, USEPA Environmental Research Laboratory, and Roy F. Weston, Inc.

Allen, H.H., G.J. Pierce, and R. Van Wormer. 1989. Considerations and techniques for vegetation establishment in constructed wetlands. In: D.A. Hammer (ed.) *Constructed Wetlands for Wastewater Treatment*, pp. 405–416. Lewis Publishers, Chelsea, MI.

Athanas, C. 1987. *Guidelines for Constructing Wetland Stormwater Basins*. Maryland Department of Natural Resources.

Athanas, C. 1988. Wetlands creation for stormwater treatment. In: J. Zelazny and J.S. Feierabend (eds.) *Wetlands: Increasing our Wetland Resources*. pp. 61–66. Proceedings of the Conference Wetlands: Increasing our Wetland Resources, Washington, D.C., October 4–7, 1987. Corporate Conservation Council, National Wildlife Federation, Washington, D.C.

Bell, H.E. 1981. *Illinois Wetlands: Their Value and Management*. State of Illinois Institute of Natural Resources. Document Number 81/33, Springfield, Illinois.

Best, R.G. 1987. Natural wetlands-southern environment: wastewater to wetlands, where do we go from here? In: K.R. Reddy and W.H. Smith (eds.) *Aquatic Plants for Water Treatment and Resource Recovery*, pp. 99–119. Proceedings of the Conference Research and Application of Aquatic Plants for Water Treatment and Resource Recovery, Orlando, Florida. Magnolia Publishing, Orlando, FL.

Boule, M.E. 1988. Wetland creation and enhancement in the Pacific Northwest. In: J. Zelazny and J.S. Feierabend (eds.) *Wetlands: Increasing our Wetland Resources*, pp. 130–136. Proceedings of the Conference Wetlands: Increasing our Wetland Resources, Washington D.C. Corporate Conservation Council, National Wildlife Federation, Washington, D.C.

Boyt, F.L., S.E. Bayley, and J. Zoltek. 1977. Removal of nutrients from treated wastewater by wetland vegetation. *J. Water Pollution Control Federation* 49: 789–799.

Brinson, M.M., B.L. Swift, R.C. Plantico, and J.S. Barclay. 1981. Riparian ecoysystems: their ecology and status. United States Fish and Wildlife Service, Kearneysville.

Brodie, G.A. 1989. Selection and evaluation of sites for constructed wastewater treatment wetlands. In: D.A. Hammer (ed.) *Constructed Wetlands for Wastewater Treatment*, pp. 307–318. Lewis Publishers, Chelsea, MI.

Brodie, G.A., D.A. Hammer, and D.A. Tomljanovich. 1988. An evaluation of substrate types in constructed wetlands acid drainage treatment systems. In: *Mine Drainage and Surface Mine Reclamation. Vol. I: Mine Water and Mine Waste*, pp. 389–398. United States Department of the Interior, Pittsburgh, PA.

Brodie, G.A., D.A. Hammer, and D.A. Tomljanovich. 1989a. Treatment of acid drainage with a constructed wetland at the Tennessee Valley Authority 950 coal mine. In: D.A. Hammer (ed.) *Constructed Wetlands for Wastewater Treatment*, pp. 201–210. Lewis Publishers, Chelsea, MI.

Brodie, G.A., D.A. Hammer, and D.A. Tomljanovich. 1989b. Constructed wetlands for treatment of ash pond seepage. In: D.A. Hammer (ed.) *Constructed Wetlands for Wastewater Treatment*, pp. 211–220. Lewis Publishers, Chelsea, MI.

Brooks, R.P. 1989. Wetland and waterbody restoration and creation associated with mining. In: J.A. Kusler and M.E. Kentula (eds.) *Wetland Creation and Restoration: The Status of the Science*, pp. 117–136. USEPA, Corvallis, OR.

Brown, M.T. 1987. Conceptual design for a constructed wetlands system for the renovation of treated effluent. Report from the Center for Wetlands, University of Florida.

Conway, T.E. and J.M. Murtha. 1989. The Iselin Marsh Pond Meadow. In: D.A. Hammer (ed.) *Constructed Wetlands for Wastewater Treatment*, pp. 139–144. Lewis Publishers, Chelsea, MI.

Cooper, P.F. and J.A. Hobson. 1989. Sewage treatment by reed bed systems: the present situation in the United Kingdom. In: D.A. Hammer (ed.) *Constructed Wetlands for Wastewater Treatment*, pp. 153–172. Lewis Publishers, Chelsea, MI.

Costanza, R., S.C. Farber, and J. Maxwell. 1989. Valuation and management of wetland ecosystems. *Ecological Economics* 1:335–361.

D'Avanzo, C. 1989. Long-term evaluation of wetland creation projects. In: J.A. Kusler and M.E. Kentula (eds.) *Wetland Creation and Restoration: The Status of the Science*, pp. 75–84. USEPA, Corvallis, OR.

Davis, D.G. 1989. No net loss of the nation's wetlands: a goal and a challenge. *Water Environment and Technology* 4:513–514.

Dunn, W.J. 1989. Wetland succession-what is the appropriate paradigm? In: D.W. Fisk (ed.) *Wetlands Concerns and Successes*, pp. 473–488. Proceedings of the Conference American Water Resources Association: Wetlands Concerns and Successes, Tampa, Florida. American Water Resources Association, Tampa, FL.

Esry, D.H. and D.J. Cairns. 1989. Overview of the Lake Jackson Restoration Project with artificially created wetlands for treatment of urban runoff. In: D.W. Fisk (ed.) *Wetlands Concerns and Successes*, pp. 247–257. Proceedings of the Conference American Water Resources Association: Wetlands Concerns and Successes, Tampa, Florida. American Water Resources Association, Tampa, FL.

Ewel, K.C. and H.T. Odum (eds.) 1984. *Cypress Swamps*. University Presses of Florida, Gainesville.

Faulkner, S.P. and C.J. Richardson. 1989. Physical and chemical characteristics of freshwater wetlands. In: D.A. Hammer (ed.) *Constructed Wetlands for Wastewater Treatment*, pp. 41–72. Lewis Publishers, Chelsea, MI.

Fennessy, M.S. and W.J. Mitsch. 1989a. Design and use of wetlands for renovation of drainage from coal mines. In: W.J. Mitsch and S.E. Jorgensen (eds.) *Ecological Engineering: An Introduction to Ecotechnology*, pp. 231–253. Wiley, New York.

Fennessy, M.S. and W.J. Mitsch. 1989b. Treating coal mine drainage with an artificial wetland. *Research Journal Water Pollution Control Federation* 61 (11/12):1691–1701.

Garbisch, E.W. 1989. Wetland enhancement, restoration, and construction. In: S.K. Majumdar, R.P. Brooks, F.J. Brenner, and J.R.W. Tiner (eds.) *Wetlands Ecology and Conservation: Emphasis in Pennsylvania*, pp. 261–275. Pennsylvania Academy of Science. Easton, PA.

Gersberg, R.M., B.V. Elkins, and C.R. Goldman. 1983. Nitrogen removal in artificial wetlands. *Water Resources* 17 (9):1009–1014.

Gersberg, R.M., B.V. Elkins, and C.R. Goldman. 1984. Use of artificial wetlands to remove nitrogen from wastewater. *J. Water Pollution Control Federation* 56 (2):152–156.

Gersberg, R.M., B.V. Elkins, S.R. Lyon, and C.R. Goldman. 1986. Role of aquatic plants in wastewater treatment by artificial wetlands. *Water Resources* 20 (3):363–368.

Gersberg, R.M., S.R. Lyon, R. Brenner, and B.V. Elkins. 1989. Integrated wastewater treatment using artificial wetlands: a gravel marsh case study. In: D.A. Hammer. (ed.) *Constructed Wetlands for Wastewater Treatment*, pp. 145–152. Lewis Publishers, Chelsea, MI.

Girts, M.A. and R.M. Knight. 1989. Operations optimization. In: D.A. Hammer (ed.) *Constructed Wetlands for Wastewater Treatment*, pp. 417– 430. Lewis Publishers, Chelsea, MI.

Godfrey, P.J., E.R. Kaynor, S. Pelczarski, and E.J. Benforado. 1985. *Ecological Considerations in Wetland Treatment of Municipal Wastewaters*. Van Nostrand Reinhold, New York.

Gosselink, J.G., E.P. Odum, and R.M. Pope. 1974. *The Value of the Tidal Marsh*. Louisiana State University. LSU-SG-74-03.

Guntenspergen, G.R., F. Stearns, and J.A. Kadlec. 1989. Wetland vegetation. In: D.A. Hammer (ed.) *Constructed Wetlands for Wastewater Treatment*, pp. 73–88.

Lewis Publishers, Chelsea, MI.
Herricks, E.E., A.J. Krzysik, R.E. Szafoni, and D.J. Tazik. 1982. *Best Current Practices for Fish and Wildlife on Surface-Mined Lands in the Eastern Interior Coal Region*. United States Fish and Wildlife Service. FWS/OBS-80/68.
Hey, D.L., M.A. Cardamone, J.H. Sather, and W.J. Mitsch. 1989. Restoration of riverine wetlands: The Des Plaines River wetlands demonstration project. In: W.J. Mitsch and S.E. Jorgensen (eds.) *Ecological Engineering: An Introduction to Ecotechnology*, pp. 159–183. Wiley, New York.
Jaworski, E. 1981. The economics of wetland protection. In: A. Champagne (ed.) *Ontario Wetlands*, pp. 58–62. Proceedings of the Conference Ontario Wetlands, Toronto, Ontario, Federation of Ontario Naturalists and the Department of Applied Geography, Ryerson Polytechnical Institute, Don Mills, Ontario.
Kadlec, R.H. 1985. Aging phenomenon in wastewater wetlands. In: P.J. Godfrey, E.R. Kaynor, S. Pelczarski, and J. Benforado, (eds.) *Ecological Considerations in Wetland Treatment of Municipal Wastewaters*, pp. 338–350. Van Nostrand Reinhold. New York.
Kadlec, R.H. 1989. Hydrologic factors in wetland water treatment. In: D.A. Hammer (ed.) *Constructed Wetlands for Wastewater Treatment*, pp. 21–40. Lewis Publishers, Chelsea, MI.
Kadlec, R.H. and Kadlec, J.A. 1979. Wetlands and water quality. In: P.E. Greeson, J.R. Clark, and J.E. Clark (eds.) *Wetland Functions and Values: The State of Our Understanding*, pp. 436–456. American Water Resources Association, Minneapolis, Minnesota.
King, W., M. Hay, and J. Charbonneau. 1978. Valuation of riparian habitats. In: R.R. Johnson and J.F. McCormick (eds.) *Strategies for Protection and Management of Floodplain Wetlands and Other Riparian Ecosystems*, pp. 161–165. Proceedings of the Conference Strategies for Protection and Management of Floodplain Wetlands and Other Riparian Ecosystems, Callaway Gardens, Georgia. U.S. Department of Agriculture Forest Service General Technical Report WO-12 Washington, D.C.
Knight, R.L., B.H. Winchester, and J.C. Higman. 1984. Carolina bays-feasibility for effluent advanced treatment and disposal. *Wetlands* 4:177–204.
Knight, R.L., T.W. McKim, and H.R. Kohl. 1987. Performance of a national wetland treatment system for wastewater management. *J. Water Pollution Control Federation* 59:746–754.
Kobriger, N.P., T.V. Dupuis, W.A. Kreutzberger, F. Steams, G. Guntenspergen, and J.R. Keough. 1983. *Guidelines for the Management of Highway Runoff on Wetlands*. National Research Council Transportation Research Board, Washington D.C. National Cooperative Highway Research Program Report 264.
Kolbash, R.L. and E. Murphy. 1987. Simco/Peabody Wetland: An Overview. Paper presented to the Mine Drainage Task Force, Coshocton, Ohio.
Kreutzwiser, R. 1981. The economic significance of the Long Point Marsh, Lake Erie, as a recreational resource. *J. Great Lakes Research* 7 (2):105–110.
Kusler, J.A. 1985. Introduction, wetland assessment: the regulator's perspective. In: J.A. Kusler and P. Riexinger (eds.) *The National Wetland Assessment Symposium*, pp. 2–8. Proceedings of the Conference The National Wetland Assessment Symposium, Portland, Maine, Association of State Wetland Managers, Omnipress. Madison, WI.
Kusler, J.A., M.L. Quammen, and G. Brooks (eds). 1988. *Mitigation of impacts*

and losses. Proceedings of the Conference The National Wetland Symposium: Mitigation of Impacts and Losses, New Orleans, Louisiana. Omnipress. Madison, WI.

Landin, M.C. 1982. Brunswick, Georgia harbor deepening project: habitat development and marsh valuation. In: M.C. Landin, (ed.) *Habitat Development at Eight Corps of Engineers Sites: Feasibility and Assessment*, pp. 85–101. U.S. Army Engineer Waterways Experiment Station Environmental Laboratory. Vicksburg, MI.

Larson, J.S. and C. Neill (eds.) 1987. *Mitigating Freshwater Wetland Alterations in the Glaciated Northeastern United States: An Assessment of the Science Base*. Proceedings of the Conference. University of Massachusetts, Amherst.

Livingston, E.H. 1989. Use of wetlands for urban stormwater management. In: D.A. Hammer (ed.) *Constructed Wetlands for Wastewater Treatment*, pp. 253–264. Lewis Publishers, Chelsea, MI.

Livingston, R.J. and O.L. Loucks. 1978. Productivity, trophic interactions, food-web relationships in wetlands and associated systems. In: P.E. Greeson, J.R. Clark and J.E. Clark (eds.) *Proceedings of the Conference Wetland Functions and Values: the State of Our Understanding*, pp. 101–119. American Water Resources Association. Minneapolis, MN.

Loucks, O.L. 1989. Restoration of the pulse control function of wetlands and its relationship to water quality objectives. In: J.A. Kusler and M.E. Kentula (eds.) *Wetland Creation and Restoration: The Status of the Science*, pp. 55–74. USEPA, Corvallis, OR.

Ma, S. and J. Yan. 1989. Ecological engineering for treatment and utilization of wastewater. In: W.J. Mitsch and S.E. Jorgensen (eds.) *Ecological Engineering: An Introduction to Ecotechnology*, pp. 185–218. Wiley, New York.

Maltby, E. and R.E. Turner. 1983. Wetlands of the world. *Geog. Mag*. 55:12–17.

Maristany, A.E. and R.L. Bartel. 1989. Wetlands and stormwater management: a case study of Lake Munson Part I: long-term treatment efficiencies. In: D.W. Fisk (ed.) *Wetlands Concerns and Successes*, pp. 215–230. Proceedings of the Conference American Water Resources Association: Wetlands Concerns and Successes, Tampa, Florida. American Water Resources Association. Tampa, FL.

Martin, C.V. and B.F. Eldridge. 1989. California's experience with mosquitoes in aquatic wastewater treatment systems. In: D.A. Hammer (ed.) *Constructed Wetlands for Wastewater Treatment*, pp. 393–398. Lewis Publishers, Chelsea, MI.

McArthur, B.H. 1989. The use of isolated wetlands in florida for stormwater treatment. In: D.W. Fisk (ed.) *Wetlands Concerns and Successes*, pp. 185–194. Proceedings of the Conference American Water Resources Association: Wetlands Concerns and Successes, Tampa, Florida. American Water Resources Association, Tampa, FL.

Metz, E.D. 1987. Guidelines for planning and designing a major wetlands restoration project: Ballona Wetland Case study. In: J. Zelazny and J.S. Feierabend (eds.) *Wetlands: Increasing our Wetland Resources*, pp. 80–87. Proceedings of the Conference Wetlands: Increasing our Wetland Resources, Washington D.C., Corporate Conservation Council, National Wildlife Federation, Washington, D.C.

Meyer, J.L. 1985. A detention basin/artificial wetland treatment system to renovate stormwater runoff from urban, highway, and industrial areas. *Wetlands* 5:135–145.

Mitsch, W.J. 1977. Water hyacinth (*Eichhornia crassipes*) nutrient uptake and metabolism in a north-central Florida marsh. Archiv. *für Hydrobiologia* 81:188–210.

Mitsch, W.J. 1988. Ecological engineering and ecotechnology with wetlands: applications of systems approaches. In: A. Marani (ed.) *Advances in Environmental Modelling*, pp. 565–580. Elsevier, Amsterdam.

Mitsch, W.J. in press. Applications of ecotechnology to the creation and rehabilitation of temperate wetlands. In: M. Wali (ed.) *Environmental Rehabilitation: Preamble to Sustainable Development*, vol. 2. SPB Academic Publishing, The Hague, The Netherlands. (In press).

Mitsch, W.J. and J.G. Gosselink. 1986. *Wetlands*. Van Nostrand Reinhold, New York.

Mitsch, W.J. and S.E. Jørgensen. 1989. Introduction to ecological engineering. In: W.J. Mitsch and S.E. Jorgensen (eds.) *Ecological Engineering: An Introduction to Ecotechnology*, pp. 3–12. Wiley, New York.

Mitsch, W.J., C.L. Dorge, and J.W. Wiemhoff. 1979. Ecosystem dynamics and a phosphorus budget of an alluvial cypress swamp in southern Illinois. *Ecology* 60:1116–1124.

Mitsch, W.J., B.C. Reeder, and D.M. Klarer. 1989. The role of wetlands in the control of nutrients with a case study of western Lake Erie. In: W.J. Mitsch and S.E. Jorgensen (eds.) *Ecological Engineering: An Introduction to Ecotechnology*, pp. 129–158. Wiley, New York.

National Wetlands Policy Forum. 1988. *Protecting America's wetlands: an action agenda*. Conservation Foundation, Washington, D.C.

Newling, C.J. 1982. Feasibility report on a Santa Ana River marsh restoration and habitat development project. In: M.C. Landin (ed.) *Habitat Development at Eight Corps of Engineers Sites: Feasibility and Assessment*, pp. 45–84. U.S. Army Engineer Waterways Experiment Station Environmental Laboratory, Vicksburg, MI.

Nichols, D.S. 1983. Capacity of natural wetlands to remove nutrients from wastewater. *J. Water Pollution Control Federation* 55 (5):495–505.

Niering, W.A. 1989. Wetland vegetation development. In: S.K. Majumdar, R.P. Brooks, F.J. Brenner, and J.R.W. Tiner (eds.) *Wetlands Ecology and Conservation: Emphasis in Pennsylvania*, pp. 103–113. The Pennsylvania Academy of Science. Easton, PA.

Odum, H.T. 1962. Man in the ecosystem. *Bull. Conn. Agric. Station*. 652:57–75.

Odum, H.T., W.L. Siler, R.J. Beyers, and N. Armstrong. 1963. Experiments with engineering of marine ecosystems. *Publications of the Institute of Marine Science University of Texas* 9:374–403.

Odum, H.T. 1989. Ecological engineering and self-organization. In: W.J. Mitsch and S.E. Jorgensen (eds.) *Ecological Engineering: An Introduction to Ecotechnology*, pp. 79–101. Wiley, New York.

Odum, H.T., K.C. Ewel, W.J. Mitsch, and J.W. Ordway. 1977. Recycling treated sewage through cypress wetlands in Florida. In: F.M. D'Itri (eds.) *Wastewater Renovation and Reuse*, pp. 35–67. Marcel Dekker, New York.

Odum, W.E. 1987. Predicting ecosystem development following creation and restoration of wetlands. In: J. Zelazny and J.S. Feierabend (eds.) *Wetlands: Increasing Our Wetland Resources*, pp. 67–70. Proceedings of the Conference Wetlands: Increasing our Wetland Resources, Washington D.C., Corporate Conservation Council, National Wildlife Federation, Washington, D.C.

Office of Technology Assessment. 1984. *Wetlands: Their Use and Regulation*. OTA, U.S. Congress. OTA-O-206, Washington, D.C. 208 p.

Ogawa, H. and J.W. Male. 1983. *The Flood Mitigation Potential of Inland Wetlands*. University of Massachusetts, Amherst. Water Resources Research Center Publication No. 138.

Palmer, C.N. and J.D. Hunt. 1989. Greenwood urban wetland a manmade stormwater treatment facility. In: D.W. Fisk (ed.) *Wetlands Concerns and Successes*, pp. 1–10. Proceedings of the Conference American Water Resources Association: Wetlands Concerns and Successes, Tampa, Florida. American Water Resources Association. Tampa, FL.

Redmond, A.M. 1981. Considerations for design of an artificial marsh for use in stormwater renovation. In: R.H. Stovall (ed.) *Wetlands Restoration and Creation*, pp. 189–199. Proceedings of the eighth annual conference on wetlands restoration and creation, Hillsborough Community College, Tampa, FL.

Richardson, C.J. 1985. Mechanisms controlling phosphorus retention capacity in freshwater wetlands. *Science* 228:1424–1427.

Richardson, C.J. and D.S. Nichols. 1985. Ecological analysis of wastewater management criteria in wetland ecosystems. In: P.J. Godfrey, E.R. Kaynor, S. Pelczarski, and J. Benforado (eds.) *Ecological Considerations in Wetlands Treatment of Municipal Wastewaters*, pp. 351–391. Van Nostrand Reinhold, New York.

Roesser, J.C. 1988. The Blue River Reclamation Project. In: K.M. Mutz, D.J. Cooper, M.L. Scott and L.K. Miller (eds.) *Proceedings of the Conference Restoration, Creation and Management of Wetland and Riparian Ecosystems in the American West*, pp. 94–101. Denver, Colorado, PIC Technologies, Denver, CO.

Rubec, C.D.A. 1987. Canada's $10 billion gamble-can our wetlands conservation dollar be expanded? In: K.M. Mutz and L.C. Lee (eds.) *Proceedings of the Eighth Annual Meeting of the Society of Wetland Scientists: Wetland and Riparian Ecosystems of the American West*, pp. 16–21. Seattle, WA.

Shabman, L. 1985. The contribution of economics to wetlands valuation and management. In: J.A. Kusler and P. Riexinger (eds.) *Proceedings of the Conference, The National Wetland Assessment Symposium*, pp. 9–13. Portland, Maine. Association of State Wetland Managers. Omnipress, Madison, WI.

Steiner, G.R. and R.J. Freeman Jr. 1989. Configuration and substrate design considerations for constructed wetlands for wastewater treatment. In: D.A. Hammer (ed.) *Constructed Wetlands for Wastewater Treatment*, pp. 363–378. Lewis Publishers, Chelsea, MI.

Suzuki, T., W.G.A. Nissanka, and Y. Kurihara. 1989. Amplification of total dry matter, nitrogen and phosphorus removal from stands of *Phragmites australis* by harvesting and reharvesting regenerated shoots. In: D.A. Hammer (ed.) *Constructed Wetlands for Wastewater Treatment*, pp. 530–535. Lewis Publishers, Chelsea, MI.

Tomljanovich, D.A. and O. Perez. 1989. Constructing the wastewater treatment wetland-some factors to consider. In: D.A. Hammer (ed.) *Constructed Wetlands for Wastewater Treatment*, pp. 399–404. Lewis Publishers, Chelsea, MI.

van der Valk, A.G. 1981. Succession in wetlands: a Gleasonian approach. *Ecology*. 62 (3):688–696.
Warners, D.P. 1987. Effects of burning on sedge meadow studied. *Restoration and Management Notes* 5 (2):90–91.
Watson, J.T., S.C. Reed, R.H. Kadlec, R.L. Knight, and A.E. Whitehouse. 1989. Performance expectations and loading rates for constructed wetlands. In: D.A. Hammer (ed.) *Constructed Wetlands for Wastewater Treatment*, pp. 319–361. Lewis Publishers, Chelsea, MI.
Watson, J.T. and J.A. Hobson. 1989. Hydraulic design considerations and control structures for constructed wetlands for wastewater treatment. In: D.A. Hammer (ed.) *Constructed Wetlands for Wastewater Treatment*, pp. 379–392. Lewis Publishers, Chelsea, MI.
Weller, M.W. 1981. Freshwater marshes. University of Minnnesota, Minneapolis, MI.
Wieder, R.K. and G.E. Lang. 1984. Influence of wetlands and coal mining on stream water chemistry. *Water. Air. and Soil Pollution* 23:381–396.
Wieder, R.K., G. Tchobanoglous, and R.W. Tuttle. 1989. Preliminary considerations regarding constructed wetlands for wastewater treatment. In: D.A. Hammer (ed.) *Constructed Wetlands for Wastewater Treatment*, pp. 297–306. Lewis Publishers, Chelsea, MI.
Wile, I., G. Miller, and S. Black. 1985. Design and use of artificial wetlands. In: P.J. Godfrey, E.R. Kaynor, S. Pelczarski, and J. Benforado (eds.) *Ecological Considerations in Wetland Treatment of Municipal Wastewaters*, pp. 26–37. Van Nostrand Reinhold, New York.
Wilhelm, M., S.R. Lawry, and D.D. Hardy. 1989. Creation and management of wetlands using municipal wastewater in northern Arizona: a status report. In: D.A. Hammer (ed.) *Constructed Wetlands for Wastewater Treatment*, pp. 179–185. Lewis Publishers, Chelsea, MI.
Willard, D.E. and A.K. Hiller. 1989. Wetland dynamics: considerations for restored and created wetlands. In: J.A. Kusler and M.E. Kentula (eds.) *Wetland Creation and Restoration: The Status of the Science*, pp. 47–54. USEPA. Corvallis, OR.
Willard, D.E., V.M. Finn, D.A. Levine, and J.E. Klarquist. 1989. Creation and restoration of riparian wetlands in the agricultural midwest. In: J.A. Kusler and M.E. Kentula (eds.) *Wetland Creation and Restoration: The Status of the Science*, pp. 333–358. USEPA. Corvallis, OR.

Bioremediation of Soils Contaminated with Selenium

E.T. Thompson-Eagle and W.T. Frankenberger, Jr.

Advances in Soil Science, Volume 17

I. Introduction

Selenium (Se) has been blamed for the extinction of dinosaurs, problems encountered by Marco Polo during his thirteenth century adventures, and inducing cancer (Koch, 1967; Nelson et al., 1943; Rosenfeld and Beath, 1964). Recently, public attention has been drawn to Se because, in several areas of the world, it has been discovered to be an environmental threat. Selenium is a widespread contaminant in the United States in areas including Arizona, California, Colorado, Montana, Nevada, New Mexico (Williams et al., 1940), South Dakota, Texas (Clark et al., 1980), Utah, and Wyoming (Boon, 1989). This chapter will focus on the efforts that have been made to develop a bioremediation approach to deselenify the environment of Se.

II. Geochemistry of Selenium

Selenium was discovered in 1817 by Berzelius and Gahn when working with Se-bearing pyrites. Selenium is classified as a metalloid having properties of both a metal and a nonmetal. It is markedly similar to sulfur in its chemistry, with its primary oxidation states being $^{+}6$, $^{+}4$, 0, and $^{-}2$. Elemental Se (Se^{o}) exhibits a zero-valence state and is often associated with sulfur in compounds such as selenium sulfide (Se_2S_2) and polysulfides. Selenate (SeO_4^{2-}), and selenite (SeO_3^{2-}) are common ions in natural waters and soils. Reduced Se compounds include volatile methylated species such as dimethylselenide (DMSe, $[CH_3]_2Se$), dimethyl diselenide (DMDSe, $[CH_3]_2Se_2$) and dimethyl selenone ($[CH_3]_2SeO_2$), and sulfur substitution in amino acids including selenomethionine, selenocysteine, and selenocystine. Inorganic reduced Se forms include mineral selenides and hydrogen selenide (H_2Se).

Among the elements, Se ranks seventieth in order of abundance and is widely dispersed in the earth's crust at low concentrations (Berrow and Ure, 1989; Crystal, 1973). The principal sources of Se for commercial applications are copper-bearing ores and sulfur deposits. Selenium is a semiconductor with a low carrier mobility and exhibits photoconductivity. It is therefore used in photocell devices as well as in xerography. Other uses include the manufacture of batteries, glass, electronic equipment, antidandruff products, veterinary therapeutic agents, feed additives, and fertilizers (Mayland et al., 1989).

A. Cycling of Selenium

Selenium is distributed throughout the environment by processes such as volcanic activity and hot springs, sea salt spray, forest wildfires, combustion of fossil fuels, incineration of municipal waste, weathering of rocks and soils, dust, soil leaching, copper/nickel production, lead and zinc smelting, iron and steel production, crop-fallow and irrigation practices, fertilizers, groundwater transport, plant and animal uptake and release, adsorption and desorption, chemical and biological redox reactions and mineral formation (Mayland et al., 1989; McNeal and Balistrieri, 1989; Nriagu, 1989). Estimated Se fluxes indicate that the natural sources of Se emission are as important as anthropogenic emissions. These natural sources are responsible for a worldwide atmospheric flux of 8400 t/year, while the anthropogenic emission of particulate Se is estimated to be 6302 t/year (Nriagu, 1989). Nearly 95% of the total natural emission can be attributed to biogenic processes in both the terrestrial and aquatic environment. It is suggested that the biological release of DMSe into the atmosphere may be an important factor in the global cycling and distribution of this element in the same manner as gaseous sulfur compounds participate in the sulfur cycle (Doran and Alexander, 1977; Francis et al., 1974). Evidence supporting this view comes from the similarity of behavior between aerosol Se and non-sea salt sulfate as a function of latitude in the North and South Pacific Ocean (Mosher and Duce, 1989) and the work of Läg and Steinnes (1978) who showed that much of the Se supplied to Norwegian soils is through precipitation. Selenium has been detected in remote regions of the world such as Antarctica (Zoller et al., 1974) and the ice sheets of Greenland (Weiss et al., 1974). Because Se is highly susceptible to biomethylation in a diverse range of environments, this reaction may be an important transformation in its global cycling.

1. Terrestrial Systems

The concentration and distribution of Se in terrestrial systems (Table 1) have been extensively reviewed by Berrow and Ure (1989) and Mayland et al. (1989). Most soils contain between 0.1 and 2 mg kg^{-1} (Elrashidi et al.,

Table 1. Selenium in terrestrial systems

Matrix	Se concentration range (mg kg^{-1})
Calcareous rocks	01–24
Magmatic rocks	0.05
Sulfide minerals	<10–<94000
Coals and oil	<0.1–11
Meteorites	0.0016–34
Soils	0.1–2
Seleniferous U.S. soils	4.5 (max. 80)
California agricultural drainage sediments	<1–>500

1989; Mayland et al., 1989), but elevated concentrations of Se are associated with soils of marine sedimentary parent material, coal, and petroleum by-products (including coal fly-ash wastes), metal refining operations, and mine tailings (Gruebel et al., 1988). The Se concentration in soil depends upon the parent material, climate, topography, age of the soil, and agricultural or industrial utilization. The rate and extent to which Se is mobilized depends upon its chemical speciation and partitioning in soils and sediments.

a. Forms of Selenium

The concentration and speciation of Se in soil depends on the pH, redox potential, solubility, complexing ability of soluble and solid ligands, biological interactions, and reaction kinetics (Barrow and Whelan, 1989; McNeal and Balistrieri, 1989). Under acidic, reducing conditions in soils which may be waterlogged and rich in organic matter, elemental Se and selenides are the predominant species (McNeal and Balistrieri, 1989). Since metal selenides, Se-sulfides, and elemental Se are insoluble, they are less biologically available for uptake. From pH 4 to 8, stable adsorption complexes or co-precipitates with sesquioxides are prevalent (Ullrey, 1981) and at moderate redox potentials, either $HSeO_3^-$ or SeO_3^{2-} is the predominant species in soil solution. Although soils with high Se concentrations are found in Hawaii and Puerto Rico (6 to 15 mg kg^{-1}), they are nonhazardous because of the low pH (4.5 to 6.5), high iron content, and humid climate. At a high redox, in well-aerated, alkaline soils the highly soluble SeO_4^{2-} is the predominant species (Elrashidi et al., 1989). Selenate ions do not form stable adsorption complexes or co-precipitates with sesquioxides (Ullrey, 1981). Selenate is stable in oxidized environments and is the Se form most readily taken up by plants (Gissel-Nielson and Bisbjerg, 1970; Eisler, 1985). Therefore, under most pH and redox conditions, SeO_3^{2-} and SeO_4^{2-} are the dominant forms of Se found in soils.

b. Movement of Selenium

Selenate is soluble and therefore mobile in soil at most pH values because it is weakly adsorbed by soil particles (Arlichs and Hossner, 1987). Since SeO_4^{2-} is the principal Se species at high pH and redox values, it is easily leached from alkaline soils (Arlichs and Hossner, 1987). Most SeO_3^{2-} salts are less soluble than the corresponding SeO_4^{2-} salts (Elrashidi et al., 1987). Selenite has a high affinity for ions of variable charge while SeO_4^{2-} has a low affinity for these ions. Selenite is rapidly sorbed at all pH values to soil particles such as clays (Bar-Yosef and Meek, 1987) and in particular to Fe oxyhydroxides (Balistrieri and Chao, 1987). The adsorption of Se to sesquioxides in soils is a common phenomenon and may control the distribution of Se between solid surfaces and the solution phase. Because the solubility data of Se minerals is still incomplete, it is not clear whether Se in soil solution is controlled by an adsorption-desorption mechanism or by precipitation-dissolution reactions although the former concept is more commonly accepted (Elrashidi et al., 1989). The solubility and hence biological availability of Se is therefore rarely related to the total Se concentration.

Column studies with alkaline soils have shown that the amount of Se leached out of soil is independent of the total Se inventory (Calderone et al., 1990). A considerable pool of Se is unavailable for leaching even at a soil pH of >8.0. Temperature can affect Se mobility since the amount of Se subject to leaching doubled in soil columns when the temperature was increased from 25° to 35° C (Calderone et al., 1990). Organic matter can also have an effect on the mobilization of Se. Korte et al. (1976) first suggested that organic amendments could influence Se speciation. The addition of organic materials (gluten, orange peel, casein, and manure) to seleniferous soils (7.5 to 40.7 mg kg^{-} soil) decreased the amount of Se leached out of the soil with a concurrent increase in Se volatilization (Calderone et al., 1990). This is a similar finding to that of Abu-Eirreish et al. (1968) and Zieve and Peterson (1981) who directly correlated a decrease in water-soluble Se with an increase in Se volatilization.

Polysulfides and thiols in mildly alkaline, sulfidic pore water can increase Se solubility (Weres et al., 1989). Some areas of Kesterson Reservoir (Merced Co., CA), highly contaminated with Se, have thick organic deposits and were subject to strongly reducing conditions. Selenium penetration in the sediment at Kesterson Reservoir has occurred to a depth of 66 cm (Weres et al., 1989). This suggests that Se can be mobile even under strongly reducing conditions, possibly as a result of increased Se solubility in the presence of hydrogen sulfide.

Microbial processes can affect the mobilization of Se in soil (Doran, 1982). Selenium can be immobilized as Se° from SeO_3^{2-} and SeO_4^{2-} (Rosenfeld and Beath, 1964; Vokal-Borek, 1979). It can also be converted into organic and inorganic selenides (Chau et al. 1976; Cutter, 1982; Cutter and Bruland, 1984). Dimethylselenide can be anaerobically metabolized into CH_4 and CO_2 (Oremland and Zehr, 1986), and Se^0 may be solubilized by oxidation into SeO_3^{2-} and SeO_4^{2-} (Sarathchandra and Watkinson, 1981). Anaerobic bacteria can transform SeO_4^{2-} and seleno-amino acids into more volatile Se compounds such as H_2Se and DMSe (Doran, 1982). However, in contrast to H_2S, the formation of H_2Se contributes little to the global cycling of Se because it is readily oxidized to elemental Se (Lakin, 1961). Bacteria and fungi volatilize SeO_3^{2-}, SeO_4^{2-}, elemental Se, and organoselenides into mobile methylated derivatives such as DMSe and DMDSe (Barkes and Fleming, 1974; Challenger, 1951; Chau et al., 1976; Cox and Alexander, 1974; Doran and Alexander, 1975, 1977; Francis et al., 1974; Frankenberger and Karlson, 1990; Karlson and Frankenberger, 1988a, b, 1989, 1990a, b; Reamer and Zoller, 1980). Because of the high vapor pressure of DMSe (Frankenberger and Karlson, 1988), it has been suggested that microbial Se methylation might play an important role in the cycling of Se in the same manner as gaseous sulfur compounds participate in the sulfur cycle.

2. Aquatic Systems

Most natural waters have low concentrations of Se, <0.01 mg L^{-1} (McNeal and Balistrieri, 1989). However, some evaporation ponds in the California San Joaquin Valley, California are reaching levels in excess of 1000 μg L^{-1} (Thompson-Eagle et al., 1989; Tanji and Grismer, 1988). The California State Water Resources Control Board (1987) stipulates that the Se content in the San Joaquin Valley drainage water must not exceed 5 μg L^{-1}. The upper limit of Se in drinking water set by the United States Department of Health, Education and Welfare (1962) is 10 μg Se L^{-1}. As with soils, under most pH and redox conditions, SeO_3^{2-} and SeO_4^{2-}, are the dominant forms of Se in water, with several forms of Se^{2-} also being present (Cutter and Bruland, 1984). This is also true for saline, alkaline evaporation pond water in the San Joaquin Valley, California (Izbecki, 1984; Presser and Barnes, 1984). Marine studies have shown that both SeO_3^{2-} and SeO_4^{2-} are present in seawater, with higher concentrations in deep waters than at the surface (Sugimura et al., 1977). In the surface waters of the North and South Pacific Oceans, it is thought that organic Se^{2-} makes up about 80% of the total dissolved Se (Cutter and Bruland, 1984). However, these conclusions are based on sequential extraction procedures and need further confirmation. Gruebel et al. (1988) believe that these techniques lead to an overestimate of the amount of Se associated with organic material. Cutter and Bruland (1984) propose that Se is cycled in the marine environment, whereby Se is selectively taken up, reductively incorporated into biogenic material, submerged into the deep sea as particulate organic selenide via sinking detritus, and regenerated as soluble Se through a multistep oxidation process (Cutter and Bruland, 1984). Similarly, in freshwater systems, Se can rapidly be lost from the water column to fine-grained, highly organic sediments which can result in a reservoir of Se which may be remobilized by biological activity.

3. Atmosphere

Submicrometer particulate Se is the dominant phase in the atmosphere with the remaining 25% being in the vapor phase (Mosher and Duce, 1989). The global distribution of Se is relatively uniform with concentrations ranging from 5 ng m^{-3} in urban areas to 0.05 to 0.1 ng m^{-3} in remote marine and continental areas (Mosher and Duce, 1989). Little is known about the speciation of Se in the atmosphere because of problems associated with sampling, fractionation, and measuring Se concentrations that are on the border of detectability. Atmospheric Se is usually reported in terms of total concentration. Most attempts to speciate atmospheric Se

Table 2. Rate constants for the gas phase reaction of DMSe and DMS with OH and NO_3 radicals and O_3 (Atkinson et al., 1990. Reprinted with permission from *Environ. Sci. Tech.*, vol. 24, p. 1330. Copyright 1990 American Chemical Society)

Reactant species	Rate constant (cm^3 molecule^{-1} s^{-1})		$(CH_3)_2Se$ Lifetime
	$(CH_3)_2Se$	$(CH_3)_2S$	
OH	$(6.78 \pm 1.7) \times 10^{-11}$	6.3×10^{-12}	2.7 h[a]
NO_3	1.4×10^{-11}[b]	1.0×10^{-12}	5 min[c]
O_3	$(6.8 \pm 0.72) \times 10^{-17}$	$>1 \times 10^{-18}$	5.8 h[d]

[a] At a 12-h average daytime OH radical concentration of 1.5×10^6 molecule cm^{-3}
[b] Extrapolation to zero NO_3 concentration
[c] At a 12-h average nighttime NO_3 radical concentration of 2.4×10^8 molecule cm^{-3} (10 parts per trillion)
[d] At an average O_3 concentration of 7×10^{11} molecule cm^{-3} (30 parts per trillion)

have been made by measuring the immediate headspace above the source being surveyed rather than the troposphere. Jiang et al. (1983) identified three different organoselenium species; DMSe, DMDSe, and dimethylselenone in the atmospheric vapor phase in the immediate vicinity of a sewage treatment plant, a coal-fired plant, a smelter, and some lakes. Reamer and Zoller (1980), Frankenberger and Karlson (1988, 1989b, 1990) and Thompson-Eagle and Frankenberger (1990a) detected DMSe in the vicinity of sewage digestion tanks, moist seleniferous soils, and seleniferous pond waters, respectively. Andren et al. (1975) indicated that the vapor-phase Se produced by coal-fired power plants was in the elemental form. Jiang et al. (1983) reported that DMSe and dimethylselenone were present in the atmosphere 1.5 km downstream and downwind of a coal-fired plant while no detectable quantities of organoselenium compounds were found 3 km upwind.

Although these measurements of Se species in the boundary layer give useful information on the production of volatile Se compounds, they do not provide information on the fate of volatile Se in the atmosphere. It has been suggested that a substantial portion of particulate Se in remote, marine regions may result from the gas-to-particle conversion of biologically produced gaseous species (Mosher and Duce, 1989). The first attempt to determine the fate of a volatile Se compound, DMSe, in the atmosphere was by Atkinson et al. (1990) who found that DMSe reacted with OH and NO_3 radicals, and ozone (O_3). The rate constants at room temperature and atmospheric pressure with respect to the gas-phase reaction with OH and NO_3 radicals, and O_3 and DMSe lifetime are shown in Table 2. The rate constants are also shown for dimethylsulfide (DMS) for comparison. No evidence of photolysis of DMSe was observed. A comparison between the two alkylated compounds (DMS and DMSe) in terms of their fate in the atmosphere indicates that DMSe is slightly more reactive than DMS.

III. Deficiencies and Toxicity of Selenium

Until recently, Se was thought to be a requirement for only a few microorganisms such as methanogens, where it plays a role in the formation of Se-dependent formate dehydrogenases (Jones and Stadtman, 1977; Krishnan et al., 1989). These enzymes have since been found in *Escherichia coli*, several clostridia, and in *Methanococcus vannielii* (Stadtman, 1980). At least four other selenoenzymes are now known to occur in bacteria including glycine reductase, nicotinic acid hydroxylase, xanthine dehydrogenase, and thiolase as well as some seleno-tRNAs whose biochemical role is as yet unknown (Stadtman, 1980).

Although Se has not been proven to be essential for plant growth, it is an essential element for animals. The essentiality of Se for animals was first demonstrated in 1957 by Schwarz and Foltz. Since then, considerable work has been devoted to this trace element in relation to animal nutrition. The clinical effect of Se deficiency (white muscle disease) has been observed in grazing animals occupying regions with low soil Se concentrations including the Pacific Northwest and the eastern third of the U.S.A. (Mayland, 1985), Finland (Yläranta, 1983), Western Australia (Gardiner, 1961), and Canada. In order to satisfy the nutritional requirements of grazing animals, Se needs to be present in pastures at a concentration of at least 50–100 μg kg^{-1} and should not exceed 2–5 mg kg^{-1} (Gissel-Nielsen et al., 1984). Selenium deficiency in animals may be overcome by injections of Se pellets, and shoulder or rumen inplants (Jenkins and Hidiroglou, 1972; Whelan, 1989). Alternatively, Se can be added as a fertilizer to pasture. Whelan et al. (1989) recommend that *Selcote* which is a proprietary Se fertilizer in the form of pelleted selenate plus an inert filler be applied to pasture at 10 g Se ha^{-1}. In Se deficient areas of the eastern San Joaquin Valley, ranchers commonly supplement their cattlefeed with Se in the form of sodium selenite at 0.3 mg kg^{-1} or 3 mg Se per head. Intraruminal boluses (*Permasel*) consisting of 90% iron, and 10% elemental Se have also been applied (Ben Norman, personal communication, Extension Veternarian, Veterinary Medicine Extension, School of Veterinary Medicine, U.C. Davis, CA 95616).

Evidence indicating that there is a human requirement for Se was established by the findings that Se is a component of the hydrogen peroxide-degrading enzyme, glutathione peroxidase (Rotruck et al., 1971) and from the discovery by Chinese scientists in the 1970s that Se deficiency causes Keshan disease, an endemic heart disease (Burau, 1985). Glutathione peroxidase acts as an antioxidant to prevent free radical damage to tissues (Underwood, 1981). Selenium may also be important in the prevention of certain diseases such as some forms of cancer, cardiovascular diseases, and degenerative problems. New studies show that Se can be beneficial to man, particularly during cancer therapy.

The threshold between deficiency and toxicity is very narrow for Se. Animals require 0.05 to 0.1 mg Se kg^{-1} in their diets to prevent Se deficiency but suffer Se toxicosis when dietary levels exceed 5 to 15 mg Se kg^{-1} (Mayland et al., 1989). The concentration of Se in animal tissues is dependent on the chemical form as well as the quantity of Se in the food. This subject is reviewed in detail by Thomassen and Aaseth (1989) and Shamberger (1983). Organic Se is retained longer than inorganic Se in the body, but because it may be absorbed much more slowly (Levander et al., 1983) it is not necessarily more toxic than inorganic Se. Selenium toxicosis is the cause of alkali disease in farm livestock (Mayland et al., 1989) and embryonic mortality and multiple developmental abnormalities in aquatic birds as well as reduced growth and mortality of adult birds (Ohlendorf, 1989). Skorupa et al. (1990) have found that teratogenesis (embryonic deformities) in waterfowl was consistently associated in populations that had a mean egg Se concentration > 20 mg kg^{-1}. In the Tulare Lake basin, California, ponded drain water with as little as 15 μg Se L^{-1} has been associated with substantial bioaccumulation of Se in waterfowl eggs (Skorupa et al., 1990). Selenium toxicity to fish and waterfowl is also dependent on its chemical form. In general, organic forms of Se in the diet appear to be more toxic to fish and wildlife. When ^{75}Se was introduced as SeO_3^{2-} and selenomethionine into experimental ponds, Se-methionine was found to be assimilated by the biota more rapidly than SeO_3^{2-} (Graham et al., 1990). Primary producers rapidly incorporate Se in their tissue with the higher trophic levels accumulating Se via the food chain (Graham et al., 1990). However, inorganic Se has been shown to be more toxic to fish than organic forms in some exposure tests. The life stage is also important as demonstrated by studies with chinook salmon (Johns and Watkins, 1989).

A. Interaction with Other Elements

Selenium exhibits a number of antagonistic and synergistic relationships with other elements in microorganisms, plants, and animals. The response may vary from organism to organism. Sulfur compounds are almost always antagonistic to Se toxicity in animals (Shamberger, 1983). Ammonium sulfate and elemental sulfur markedly decreased Se uptake by perennial ryegrass and red clover (Williams and Thornton, 1972). Sulfate also interferes with SeO_4^{2-} uptake by the alga, *Chlorella vulgaris* (Shrift, 1954), while SeO_4^{2-} blocks the uptake of SO_4^{2-} by the fungus *Penicillium chrysogenum* (Yamamoto and Segal, 1966). The growth inhibition and Se accumulation by tall fescue cultivars were reversed by the addition of SO_4^{2-} (Wu et al., 1990). In contrast, methylation of Se by aquatic microorganisms is unaffected by the presence of high SO_4^{2-} levels (Thompson-Eagle and Frankenberger, 1991a). Competitive antagonism between SeO_3^{2-} and sulfur compounds has never been demonstrated. Selenate is a competitive inhibitor of sulfate respiration in bacteria (Banat and Nedwell, 1984; Peck, 1959; Postgate, 1952), but at high SO_4^{2-} concentrations (1 m*M*) noncompetitive inhibition may occur (Zehr and Oremland, 1987). Antagonism has also been demonstrated between volatile sulfur and Se substrates for demethylation (Oremland and Zehr, 1986). The addition of DMS to sediments or a pure culture of an obligate methylotropic bacterium (DMS-grown) competitively inhibited the anaerobic degradation of DMSe into methane and carbon dioxide (Oremland and Zehr, 1986). These results suggest that there is a common enzyme system for DMS and DMSe metabolism.

IV. Vegetation Uptake

The accumulation of Se in plants is highly variable and can favorably or adversely affect their growth, survival, and reproduction (Trelease, 1945; Eisler, 1985). The extent of Se accumulation depends on the plant, speciation of Se, pH, salinity, and the calcium carbonate content of the soil (Gissel-Nielsen et al., 1984; McNeal and Balistrieri, 1989). Soils of Hawaii may contain 6–15 mg Se (+4) kg^{-1} yet do not yield seleniferous vegetation, while plants growing on alkaline soils containing <1 mg Se(+6) kg^{-1} may accumulate Se to potentially toxic concentrations (Lakin, 1973; U.S. Environmental Protection Agency, 1975). The type of soils which give rise to high concentrations of Se in plants—hence potential animal toxicosis—are usually alkaline and contain free calcium carbonate (Lakin, 1961; Rosenfeld and Beath, 1964). Seleniferous vegetation has been detected in Australia, Ireland, Israel, South Africa, U.S.A., and Venezuela. Based on their relationship to Se uptake, plants may be divided into three groups: primary Se accumulators containing Se concentrations in excess of 100 mg per kg dry weight (e.g., *Astralagus*, *Haplopappus*, *Machaeranthera*, and *Stanleya*,) while secondary Se absorbers rarely concentrate more than 50 to 100 mg kg^{-1} (e.g., *Aster*, *Astragalus*, *Atriplex*, *Castilleja*, *Grindelia*, and *Gutierrezia*). Plants that usually do not absorb more than 50 mg kg^{-1} (group 3) growing on seleniferous soil include grasses (Rosenfeld and Beath, 1964; Shrift, 1973). In addition to Se-accumulating plants there are some plants that take up surprisingly little Se considering the high concentration that is present in the soil (e.g., *Trifolium repens* L). Aquatic plants such as algae may also absorb enormous quantities of Se from water. Some plants appear to be able to absorb Se from the atmosphere and transport it to the roots (Zieve and Peterson, 1987). Studies with higher plants have shown that some have the ability to volatilize Se with DMSe being the principal product of a nonaccumulator species of cabbage and DMDSe the product of *Astragalus*, an accumulator species (Evans et al., 1968; Lewis et al., 1974).

V. Microbial Transformations

A. Reduction

In living systems, Se tends to be reduced rather than oxidized (Shamberger, 1983). Anaerobic cell extracts of *Micrococcus lactilyticus*, *Clostridium pasteurianum*, and *Desulfovibrio desulfuricans* have been shown to use molecular hydrogen to reduce SeO_3^{2-} but not SeO_4^{2-} to Se^{2-} (Woolfolk and Whitely, 1962). However, Zehr and Oremland (1987) reported that washed cells of *D. desulfuricans* subsp. *aestuarii* were capable of reducing SeO_4^{2-} to Se^{2-}. Upon an acetate-SO_4^{2-} enrichment of a seleniferous soil, Macy et al. (1989) were able to isolate two anaerobic bacteria; one which reduced SeO_3^{2-} to Se° and a *Pseudomonas* sp. which reduced SeO_4^{2-} to SeO_3^{2-}. It is likely that the SeO_4^{2-} reduction in this organism is not carried out by NO_3^- reducing enzymes (Macy and Rech, 1990). *Pseudomonas maltophilia* O-2, an isolate from a toxic waste site, reduced SeO_3^{2-} as well as salts of Hg, Pb, Cd, Ag, Au, Cr, and Sn to their elemental forms (Latinwo et al. 1990). Bacterial resistance to Hg^{2+}, SeO_3^{2-}, and Pb^{2+} was found to be associated with a single large plasmid (Latinwo et al., 1990). Selenite is reduced to Se° by aerobically grown *Salmonella heidelberg* (McCready et al., 1966) as well as by cell-free extracts of *Streptococcus faecalis* and *S. faecium* (Tilton et al. 1967). Oremland et al. (1989) demonstrated *in situ* (anoxic sediments) and in pure culture, the anaerobic bacterial conversion of seleno-oxyanions to Se°. Their results indicate that dissimilatory SeO_4^{2-} reduction to Se° may be a major biological transformation for reduction of Se oxyanions in anoxic sediments.

B. Oxidation

Although there are several reports of Se reduction, there are few well-documented cases of Se oxidation. Reduced forms of Se such as elemental Se (Bisbjerg, 1972; Geering et al., 1968; Lipman and Waksman, 1923; Sarathchandra and Watkinson, 1981), SeO_3^{2-} (Bisbjerg, 1972; Geering et al., 1968), and copper selenide (Torma and Habashi, 1972) have been reported to be oxidized by laboratory bacterial cultures and soils. *Aspergillus niger* has also been reported to oxidize SeO_3^{2-} to SeO_4^{2-} (Bird et al., 1948).

C. Demethylation

Little work has been conducted on the demethylation of organoselenides. It has been demonstrated that both air-dry and moist soils have the capacity to sorb substantial amounts of DMSe (Zieve and Peterson, 1985). There is some confusion in the literature as to whether methanogenic bacteria are active methylators or demethylators of certain metals and metalloids in the environment. Oremland et al. (1989) put forward the hypothesis that hydrogen-oxidizing methanogens such as *Methanobacterium omelianskii* may be involved in reductive methylation, while methylotrophic bacteria carry out demethylation. Dimethylselenide can be demethylated in anoxic sediments as well as anaerobically by an obligate methylotroph similar to *Methanococcoides methylutens* in pure culture (Oremland and Zehr, 1986). A soil pseudomonad converted trimethylselenonium chloride to DMSe, apparently by cleaving the C-Se bond (Doran and Alexander, 1977).

D. Methylation

Volatilization through methylation is thought to be a protective mechanism used by microorganisms to detoxify their surrounding environment. The process permanently removes Se from soil and water. The predominant groups of Se methylating organisms isolated from soils and sediments are bacteria and fungi (Abu-Eirreish, 1968; Barkes and Fleming, 1974; Challenger, 1945; Doran, 1982; Karlson and Frankenberger, 1988b), while in water, bacteria are thought to play a more dominant role (Table 3) (Thompson-Eagle and Frankenberger 1991a). Biomethylation of toxic Se species including SeO_3^{2-}, SeO_4^{2-}, Se^O, and various organoselenium compounds into a less toxic, volatile form (DMSe) is apparently a widespread transformation in seleniferous environments (Chau et al., 1976; Doran, 1982; Francis et al., 1974). Dimethylselenide is the major metabolite of Se volatilization (Barkes and Fleming, 1974; Doran and Alexander, 1975; Doran, 1982; Francis et al., 1974; Karlson and Frankenberger, 1988b, 1989; Thompson-Eagle et al, 1989), although other Se compounds such as DMDSe, methaneselenone ($[CH_3]_2SeO_2$), methane selenol (CH_3SeH), and dimethyl selenenyl sulfide (CH_3SeSCH_3) may also be produced (Chasteen et al., 1990; Chau et al., 1976; Reamer and Zoller, 1980; Shrift, 1973).

1. Historical Facts

Historically, the methylation of various trace elements (e.g., Hg, As, Pb) has been of environmental concern, with emphasis being placed on the lethal aspects of alkylation. Approximately 150 years ago, several cases of arsenical poisoning occurred in Germany due to the inhalation of arsenical gases produced from moldy, As-containing pigments used in domestic wallpapers (Challenger, 1935). Scheele's green and Schweinfürter green, were fairly common arsenical wallpaper pigments used at that time. During the 1930s, a similar case arose in Britain, causing the death of a child (Challenger, 1935). The combination of damp, moldy walls containing arsenic and an available carbon source (wallpaper and wallpaper paste) led to the fungal methylation of As. Upon close examination of the family's living quarters, a garlic odor was often detected. This is a characteristic of volatile methylated As compounds as well as Se and tellurium (Te) methylated gases. It took several years of investigations by various workers, summarized by Challenger (1951) before the methylated As gas was finally identified as being trimethylarsine.

In contrast, most reports on Se methylation are more benevolent and are viewed as a mechanism in which Se could be safely eliminated from the organism's environment. The first report of gaseous emissions of Se from higher organisms was by Japha (see Challenger and North, 1934) who administered inorganic Se to humans and dogs and noticed the production of a garlic odor in their breath within a few hours. In 1902, Maassen (see Challenger and North, 1934) also experimented with Se-injected dogs and *Penicillium brevicaule* (*Scopulariopsis brevicaulis*) cultivated on Se-spiked media and observed the evolution of odiferous Se compounds. It was Challenger and North in 1934 who showed that the volatile Se product evolved by the fungus, *Scopulariopsis brevicaulis* grown on sterile breadcrumbs containing sodium selenite or selenate was DMSe. Since then a large number of microorganisms, plants, and animals have been described in the literature as having the ability to volatilize Se. Frankenberger and Karlson (1989a) were the first to suggest that Se volatilization could be used beneficially in the detoxification of seleniferous soil.

2. Biochemical Pathway

Little progress has been made on the microbial Se methylation pathway since Doran's review in 1982. Although methylation of inorganic Se is known to involve a reduction and a methylation step, the order of these reactions is still unknown. Challenger (1951) proposed a pathway consisting of successive methylation and reduction steps resulting in the production of dimethylselenone as an intermediate:

$$HSeO_3^- \longrightarrow CH_3SeO_3H \longrightarrow CH_3SeO_2^- \longrightarrow (CH_3)_2SeO_2 \longrightarrow (CH_3)_2Se$$

selenite	methane selenonic acid	ion of methane-seleninic	dimethyl selenone	dimethyl selenide

Unfortunately he did not identify and confirm these intermediates from a culture solution. This pathway has been given some credence by the tentative identification of dimethylselenone in the headspace above soil, and sewage sludge matrices (Reamer and Zoller, 1980). However, there is no mention of DMDSe in this proposed pathway. Reamer and Zoller (1980) suggested that Challenger's proposed methylation pathway could be modified to include a concentration-dependent branch at the dimethylselenone intermediate whereby reduction would result in the formation of either

Table 3. Microorganisms that volatilize selenium

Microorganisms	Source	Se substrate	Se concentration (ppm)	Aerobic	Anaerobic	Se product	References
Bacteria							
Aeromonas sp. *Flavobacterium* sp. *Pseudomonas* sp.	Lake sediment	SeO_3	5	+	–	$(CH_3)_2Se$ $(CH_3)_2Se_2$ Unknown volatile Se	Chau et al. (1976)
Corynebacterium sp.	Seleniferous soil	SeO_3, SeO_4, Se°	—	+	–	$(CH_3)_2Se$	Doran and Alexander (1975)
Pseudomonas fluorescens	Evaporation pond sediment	SeO_4	0.8	–	+	$(CH_3)_2Se$ $(CH_3)_2Se_2$ CH_3SeSCH_3	Chasteen et al. (1991)
Unidentified sp.	Evaporation pond water	Mainly SeO_4	1.2	+	–	$(CH_3)_2Se$	Unpublished data, Thompson-Eagle and Frankenberger
Fungi							
Cephalosporium sp. *Fusarium* sp. *Penicillium* sp. *Scopulariopsis* sp.	Garden soil	SeO_3, SeO_4	457 418	+	–	$(CH_3)_2Se$	Barkes and Fleming (1974)
Scopulariopsis brevicaulis	Unspecified	SeO_3, SeO_4	15	+	–	$(CH_3)_2Se$	Challenger and North (1934)

Schizopyllum commune	Wood	SeO_4	366	+	–	$(CH_3)_2Se$	Challenger and Charlton (1947)
Aspergillus niger	Unspecified	SeO_4	—	+	–	$(CH_3)_2Se$	Challenger (1951)
Candida humicola	Sewage	SeO_3, SeO_4	46	+	–	$(CH_3)_2Se$	Cox and Alexander (1974)
		SeO_3, SeO_4	418	+	–	$(CH_3)_2Se$	
Acremonium falciforme *Penicillium citrinum* *Ulocladium tuberculatum*	Evaporation pond sediment	$^{75}SeO_3$	100	+	–	$(CH_3)_2Se$	Karlson and Frankenberger (1988b)
Acremonium falciforme *Penicillium citrinum*	Evaporation pond sediment	SeO_4	0.79	+	–	$(CH_3)_2Se$ $(CH_3)_2Se_2$	Chasteen et al. (1991)
Penicillium sp.	Sewage	SeO_3	457	+	–	$(CH_3)_2Se$	Fleming and Alexander (1972)
Alternaria alternata	Evaporation pond water	SeO_3, SeO_4	1 100	+	–	$(CH_3)_2Se$	Thompson-Eagle et al. (1989)

CH_3SeOH or CH_3SeH which would rapidly yield $(CH_3)_2Se_2$. They suggested that when the Se concentration is sufficiently high, less DMSe and more DMDSe is produced. The formation of DMDSe is likely to be energetically favorable over the formation of DMSe. An alternative theory is that DMSe-producing organisms are inhibited by high Se concentrations while DMDSe-producing organisms are more Se-tolerant.

Doran (1982) challenged Challenger's proposed pathway and suggested that reduction to the selenide form occurs before methylation as follows:

$$\underset{\text{selenite}}{SeO_3^{2-}} \longrightarrow \underset{\text{elemental Se}}{Se^\circ} \longrightarrow \underset{\text{selenide}}{HSeX} \longrightarrow \underset{\text{methane selenol}}{CH_3SeH} \longrightarrow \underset{\text{dimethyl selenide}}{(CH_3)_2Se}$$

While Doran did not test for selenides and methane selenol as intermediates, trace concentrations of methane selenol have been detected in active methylating fungal cultures (Bird and Challenger, 1942) as well as in the headspace of denitrifying bacteria which methylate Se (R. Fall, personal communication, Department of Chemistry and Biochemistry, and Cooperative Institute for Research in Environmental Sciences, University of Colorado, Boulder, CO 80309). Additional support for this pathway comes from various mammalian studies which have led other researchers to propose methane selenol as an intermediate in the methylation of Se in animal tissues (Doran, 1982). In addition to the Se methylation products, DMSe and DMDSe, Chasteen et al. (1990) have shown that an anaerobically incubated culture of *Pseudomonas fluorescens* K27 grown under denitrifying conditions evolved dimethyl selenenyl sulfide (CH_3SeSCH_3).

Little work has been conducted on the enzymology of Se methylation in microorganisms. One study with the ciliate, *Tetrahymena thermophila* suggested that a selenide methyltransferase is involved in the S-adenosylmethionine-dependent methylation of sodium selenide with the end product being methane selenol (Drotar et al., 1987). It is possible that bacteria and fungi also possess selenide methyltransferases in addition to sulfide methyltransferases.

3. Fate of Methylated Gas

Once the Se is methylated, it is released into the atmosphere, diluted and dispersed by air currents directly away from the contaminated source. DMSe reacts with OH and NO_3 radicals and O_3 within a few hours to yield products which are as yet unknown (Atkinson et al., 1990). However, it is likely that these oxidized products may be scavenged onto aerosols or sorbed onto particulates which have a relatively long residence time (7 to 9 days) in the atmosphere (Mosher and Duce, 1989), and can travel considerable distances (Mayland et al., 1989). The changes in speciation that occur in the atmosphere are as yet unknown due to the problems of collection and analysis. At the present time, the fate of DMSe in the atmosphere is subject to much debate among Se researchers. The controversy is based on the concentration of Se in remote regions of the world, DMSe emission rates, and interactions in relation to the sulfur cycle.

4. Toxicity of the Methylated Products

Dimethylselenide is 500 to 700 times less toxic to rats than aqueous SeO_3^{2-} and SeO_4^{2-} ions (Franke and Moxon, 1936; McConnell and Portman, 1952; Frankenberger and Karlson, 1988). Recently an acute toxicity study by O.G. Raabe and M.A. Al-Bayati (see Frankenberger and Karlson, 1988) was conducted on the inhalation of DMSe by rats. This study consisted of 85 adult rats exposed to four concentrations of DMSe (0, 1607, 4499, and 8034 ppm) for 1 h. Not a single animal was killed by gaseous DMSe. After exposure, the animals were observed for a 1-week period for clinical abnormalities and all appeared normal. The exposed and control rats were killed and their major tissues and organs were examined. The effect of DMSe was one of irritation rather than injury since there was a slight increase in the lung weight after 1 day of exposure, a small injury to the spleen at the highest concentration tested, and elevated Se levels in the lungs and serum. Within 7 days, all affected organs exhibited complete recovery. The half-life of DMSe appears to be very short and the compound was eliminated mainly via the lung. The data indicate that inhaled DMSe vapor is nontoxic to the rat at concentrations of up to 8034 ppm or 34000 mg m^{-3} (Frankenberger and Karlson, 1988).

5. Factors Enhancing Biomethylation of Selenium

Characterization of this naturally occurring microbial Se transformation and removal process has led to the discovery that biomethylation can be accelerated to the point where there is a significant decline in the initial Se inventory within a relatively short time. In order to effectively utilize this novel biotechnology to bioremediate seleniferous sediments and water, it is important to determine the factors that affect volatilization of Se in both soil and water.

a. Microrganisms

Selenium volatilization is microbially mediated (Table 3). Sterilization of seleniferous soil and water by autoclaving completely eliminates the reaction (Abu-Eirresh et al., 1968; Ganje and Whitehead, 1958; Karlson and Frankenberger, 1989; Reamer and Zoller, 1980; Thompson-Eagle and Frankenberger, 1991a). The addition of the bactericide, chloramphenicol, to soil reduces Se volatilization rates which indicates that both bacteria and fungi are important in this process (Zieve and Peterson, 1981). Adding a fungal innoculum of 2.8×10^7 cells of *Candida humicola* to soil caused Se evolution to double (Zieve and Peterson, 1981). However, Karlson and

Frankenberger (1989) found no increased rate of DMSe production when an active methylating inoculum was added to seleniferous sediments, probably because there was a sufficient population of microflora capable of producing gaseous Se. Selenium-methylating fungi and bacteria have been isolated from seleniferous soils and agricultural drainage waters, respectively, of the western San Joaquin Valley, in California. These microbes are able to withstand extreme osmotic stress produced by fluctuating saline conditions (Karlson and Frankenberger, 1989; Thompson-Eagle and Frankenberger, 1991a). There is often no need to add a microbial innoculum to seleniferous soils or water during a bioremediation program.

b. Nutrients

Soil alkylselenide production is carbon-limited. In general, the rate of Se evolution from soils, sediments, and water increases with the addition of certain organic materials (Tables 4 and 5). It is possible to achieve more than a 10-fold increase in volatile Se evolution with the addition of organic amendments to soil. While stimulation of volatilization may vary between different soil types (Table 4), no single parameter solely governs Se volatilization rates because the physical, chemical, and biological properties in soils all govern the potential for volatilization. Stimulation of volatilization is dependent on the form of organic amendment. Short-term studies conducted in our laboratory with naturally seleniferous sediments indicated that Se biomethylation is accelerated through the provision of saccharides, amino acids, and especially proteins (Karlson and Frankenberger, 1988b). The best treatments for accelerating methylation of Se from seleniferous soils are gluten, casein, pectin, and orange peel (Frankenberger and Karlson, 1990; Karlson and Frankenberger, 1990b; Calderone et al., 1990). In some soils, nitrogen may be a limiting factor. The optimum C/N ratio in Los Banos clay loam was found to be 20:1 (Karlson and Frankenberger, 1988b).

Unlike soils, the addition of mono-, poly-, and acidic saccharides, alcohol, amino acids, fats, and oils had little effect on biomethylation in water (Thompson-Eagle and Frankenberger, 1990a, 1991a). However, proteins such as casein, and albumen dramatically stimulated biomethylation (Table 5). All protein and peptide sources appeared to be stimulatory to the deselenification process. Manufacturer's by-products which stimulated Se biomethylation in water included cheese whey, whey protein, cottonseed and soybean meals, and yeast sludge (Thompson-Eagle and Frankenberger, 1991b).

c. Selenium Concentration

Although the Se biomethylation capacity of a soil is dependent on the Se concentration (Karlson and Frankenberger, 1988b), it is the level of available or water-soluble Se that directly governs the process. Zieve and Peter-

Table 4. Enhancement of selenium volatilization from soils and dewatered sediments by organic amendments

Matrix	Organic amendment	Amendment application rates (g kg^{-1})	Native Se concentration (mg kg^{-1})	Se spike concentration (mg kg^{-1})	Species	% Se volatilized: Unamended	% Se volatilized: Amended	Incubation time (days)	References
Soils									
Lima loam	Wheat	20	0.9			0	0.67	25	Doran and Alexander (1977)
	Wheat	20	0.9	50	(IV)	0.4	3.1		
Sansac clay	Wheat	20	36.0			0.01	0.03		
	Wheat	20	36.0	50	(IV)	0.16	2.24		
Pierre Formation, South Dakota									
0–12″	Wheat	50	6.6			0.05	0.05	25	Abu-Eirreish et al. (1968)
12–24″	Wheat	50	6.9			0.08	0.46		
24–36″	Wheat	50	9.1			0.05	0.62		
Muck soil	Wheat	25	0.42	0.04	$(IV)^{75}$	$0.68Se^{75}$	$0.51Se^{75}$	60	Hamdy and Gissel-Nielson (1976)
Clay loam	Wheat	25	0.35	0.04	$(IV)^{75}$	$1.41Se^{75}$	$2.32Se^{75}$		
Sandy soil	Wheat	25	0.14	0.04	$(IV)^{75}$	$2.12Se^{75}$	$0.70Se^{75}$		
Los Banos	Galacturonic acid	2 g C	0.22	100.0	$(IV)^{75}$	$2.90Se^{75}$	$6.87Se^{75}$	13	Karlson and Frankenberger (1988a)

Table 4 (Cont.)

Matrix	Organic amendment	Amendment application rates (g kg^{-1})	Native Se concentration	Se spike concentration (mg kg^{-1})	Species	% Se volatilized		Incubation time (days)	References
						Unamended	Amended		
Clay loam	Pectin	2 g C	0.22	100.0	$(IV)^{75}$	$2.90Se^{75}$	$8.77Se^{75}$		
	Cellulose	2 g C	0.22	100.0	$(IV)^{75}$	$2.90Se^{75}$	$3.27Se^{75}$		
	Sewage sludge	2 g C	0.22	100.0	$(IV)^{75}$	$2.90Se^{75}$	$2.50Se^{75}$		
	Corn	2 g C	0.22	100.0	$(IV)^{75}$	$2.90Se^{75}$	$6.53Se^{75}$		
	Cowpea	2 g C	0.22	100.0	$(IV)^{75}$	$2.90Se^{75}$	$7.35Se^{75}$		
	Manure	2 g C	0.22	100.0	$(IV)^{75}$	$2.90Se^{75}$	$3.53Se^{75}$		
S. Dakota clay	Glucose	10	30			0	0	45	Francis et al. (1974)
Seleniferous shale	Oat straw	10	0.4	12.3	$(IV)^{75}$	$0.41Se^{75}$	$5.69Se^{75}$	60	Ganje and Whitehead (1985)
	Astralagus	10	0.4 + 0.6	12.8	$(IV)^{75}$ +org Se		$70.31Se^{75}$		
Seleniferous sediments (dewatered)									
Kesterson Pond 1	Casein	7.5	7.5			2.40	3.47	140	Calderone et al. (1990)
	Gluten	7.5	7.5				7.73		
	Manure	7.5	7.5				3.33		
	Orange peel	7.5	7.5				7.47		

San Luis Drain	Casein	7.5	17.1	2.11	4.85		
	Gluten	7.5	17.1		3.57		
	Cattle manure	7.5	17.1		2.69		
	Orange peel	7.5	17.1		6.37		
Kesterson Pond 4	Casein	2 g C	60.7	0.56	9.00	5	Karlson and Frankenberger (1988b)
	Albumin	2 g	60.7		10.2		
	Gluten	2 g C	60.7		1.40		
	Methionine	2 g C	60.7		1.83		
Kesterson Salt-grass vegetation	Serine	20	7.3	6.85	6.85	35	Weres et al. (1989)
	Methionine	20	7.3		31.5		
	Cysteine	20	7.3		28.7		

Table 5. Enhancement of selenium volatilization from evaporation pond water by organic amendments (Thompson-Eagle and Frankenberger, 1990a)

Matrix	Organic amendment	Amendment application rates ($g\ L^{-1}$)	Native Se concentration ($mg\ L^{-1}$)	% Se volatilized		Incubation time (days)
				Unamended	Amended	
Seleniferous waters (Evaporation pond water)						
Sumner Peck	Glucose	2 g C	1.2	0.81	1.21	21
Ranch	Maltose	2 g C			1.21	
	Casein	2 g C			17.5	
	Methionine	2 g C	1.1	0.09	4.06	15
	Serine	2 g C			0.90	
	Albumin	20 g C	1.1	0.45	66.8	43
	Gluten	2 g C			13.9	
	Casein	2 g C			64.4	

son (1981) were able to correlate a decrease in water-soluble Se as volatilization increased. Calderone et al. (1990) found an interaction ($r = 0.81$) between the decrease in the cumulative soluble Se collected upon leaching soil columns and an increase in the cumulative volatilization upon amendments with organic materials. Experiments with Los Banos clay loam, a fine, montmorillonitic, thermic Typic Haploxeralf, showed that as the Se inventory (SeO_3^{2-} or SeO_4^{2-}) increased above 20 mg kg^{-1}, the percentage of Se volatilization decreased (Table 4). This trend was reversed for SeO_3^{2-} and to a lesser extent for SeO_4^{2-} in the presence of an organic amendment. The Se concentration of soil not only affects the volatilization capability but can also change the ratio of volatile Se species evolved. Reamer and Zoller (1980) demonstrated that the relative abundance of volatile Se species evolved from SeO_3^{2-}-amended sewage sludge was directly dependent on the Se concentration. The major volatile Se species evolved with an initial SeO_3^{2-} concentration between 1 and 10 mg kg^{-1} was DMSe while at concentrations of ≥ 100 mg kg^{-1}, DMSe decreased markedly and the relative concentrations of DMDSe and dimethylselenone increased. The major product of Se volatilization from naturally seleniferous sediments and pond water is DMSe (Karlson and Frankenberger, 1989; Thompson-Eagle and Frankenberger, 1990a,b; 1991a). However, when organically amended pond water was spiked with SeO_4^{2-} at concentrations of Se greater than 500 μg kg^{-1}, a small quantity of DMDSe was evolved (Thompson-Eagle and Frankenberger, unpublished data) and in seleniferous soil when the concentration is above 100 mg kg^{-1} (Frankenberger, 1986).

d. Selenium Species

Selenium in certain organic forms is more readily transformed into volatile derivatives than inorganic Se (Table 6). Doran and Alexander (1977) found that the transformation into volatile Se was an order of magnitude greater for TMSe, selenomethionine, and selenocysteine than for SeO_3^{2-}, SeO_4^{2-} and elemental Se (Table 6). They also found that biomethylation of these organic Se substrates was not consistant from soil to soil with 87%, 28%, and 7% of TMSe, selenomethionine and selenocystine, respectively being volatilized from Lima loam, a fine-loamy, mixed, mesic Glossoboric Hapludalf, while in Sansarc clay, a montmorillonitic (calcareous), mesic, shallow Typic Ustorthent, 24%, 37%, and 6.2% were volatilized. Frankenberger and Karlson (1990) found that seleniferous soils methylated organic Se compounds in the following order: selenomethionine > selenocysteine = selenoguanosine = selenoinosine > selenoethionine = selenopurine > selenourea. Chau et al. (1976) also observed the production of DMSe, DMDSe, and an unknown product from both soil and sediment samples enriched with inorganic and organic Se species including sodium selenite, sodium selenate, selenocystine, selenourea, and seleno-DL-

Table 6. Effect of selenium speciation on volatilization from soils, sediments, sludge and water

Matrix	Native Se concentration ($mg\ kg^{-1}$)	Se spike concentration ($mg\ kg^{-1}$)	Species	% Se volatilized	Organic amendment	Incubation time (days)	Reference
Soils							
Lima loam	0.9	50	(IV)	3.08	−	25	Doran and Alexander (1977)
	0.9	0.6	(org)	2.0	+	25	
	0.9	50	(org + IV)	2.08	+	25	
	0.9	250	Se°	0.01	−	17	
	0.9	50	(IV)	0.59	−	17	
	0.9	50	(VI)	0.10	−	17	
	0.9	5.0	Secysteine	7.0	+	32	
	0.9	5.0	Semethionine	28.0	+	32	
	0.9	5.0	TMSe	87.0	+	32	
Sansarc clay	36.0	50	(IV)	2.2	+	25	Doran and Alexander (1977)
	36.0	0.6	org	0.19	+	25	
	36.0	50	(org + IV)	1.05	+	25	
Seleniferous shale	0.4	12.3	$(IV)^{75}$	5.69^{75}	+	60	Ganje and Whitehead (1958)
	0.4	3.0	$(IV)^{75}$	100.00^{75}	+	60	
	0.4 + 0.6 org	12.8	$(IV)^{75}$	70.31^{75}	+	60	
	0.4 + 0.6 org	12.8	$(IV)^{75}$	53.33^{75}	+	60	
Los Banos clay loam	0.22	5	$(IV)^{75}$	10.4^{75}	−	29	Karlson and Frankenberger (1989)
	0.22	20	$(IV)^{75}$	9.0^{75}	−	29	
	0.22	100	$(IV)^{75}$	4.8^{75}	−	29	
	0.22	5	$(VI)^{75}$	4.0^{75}	−	29	

	0.22	25	(VI)^{75}	2.3^{75}	–	29	
	0.22	150	(VI)^{75}	1.5^{75}	–	29	
	0.22	5	(VI)^{75}	43.1^{75}	+	29	
	0.22	25	(VI)^{75}	51.1^{75}	+	29	
	0.22	5	(IV)^{75}	41.6^{75}	+	29	
	0.22	25	(IV)^{75}	31.0^{75}	+	29	
Dewatered seleniferous sediments							
Kesterson Pond 4	60.7	219	Secysteine	0.2	–	5	Frankenberger and Karlson (1990)
		219	Seethionine	0.1	–	5	
		263	Semethionine	0.6	–	5	
		132	Seguanosine	0.2	–	5	
		263	Sepurine	0.1	–	5	
		1317	Seurea	0.01	–	5	
		132	Seinosine	0.2	–	5	

methionine, but no substrate concentrations were given, nor did they indicate any differences in volatilization of the different Se species. Not all microorganisms are able to transform organoselenium compounds as easily as some inorganic compounds. An *Alternaria alternata* species was found to methylate various Se species in pure culture in the following order of magnitude; SeO_4^{2-} > SeO_3^{2-} >> selenoinosine > selenomethionine >> selenopurine > selenium sulfide (Thompson-Eagle et al., 1989). When organoselenium compounds in the form of plant material are added to soil, biomethylation appears to be less efficient compared with noncomplexed selenoamino acids or inorganic Se species (Table 6). However, comparisons are difficult to make because of the differences in substrate concentrations and incubation conditions carried out by different researchers.

Although laboratory experiments with Se-amended soils and sewage sludge show that Se° is poorly methylated compared with other inorganic Se species, probably as a result of its low solubility (Reamer and Zoller, 1980), subsequent experiments which monitored the change in total Se, water-soluble Se, and Se° concentrations over time in a naturally seleniferous soil, indicate that Se° is readily available for microbial Se methylation (Weres et al., 1989). A substantial percentage of elemental Se (29%) was removed from a nutrient-amended soil system in a 5-week incubation period.

According to Doran's hypothesis (1982), the pathway of Se biomethylation requires reduction to the Se^{2-} species and subsequent methylation to form DMSe. It should therefore be more energetically favorable to methylate SeO_3^{2-} rather than SeO_4^{2-}. Experimental evidence confirms this hypothesis (Table 6), however because the availability of SeO_3^{2-} can be limited by its ability to bind to Fe-oxides and clays (see section on mobility of Se in terrestrial systems) particularly at lower pHs, the microbiological uptake and therefore biomethylation of this compound may be limited in some matrices. Doran and Alexander (1977) found that the formation of volatile Se compounds was the most rapid with 50 mg kg^{-1} Se-SeO_3^{2-} followed by 50 mg kg^{-1} Se-SeO_4^{2-}. Karlson and Frankenberger (1988b) found that without a carbon amendment, volatilization rates were an order of magnitude higher when SeO_3^{2-} was the substrate compared with SeO_4^{2-}. The addition of an organic amendment largely cancelled out this difference possibly because there was more energy available for the reduction of SeO_4^{2-} to SeO_3^{2-}. Barkes and Fleming (1974) found that out of 11 pure isolates of soil fungi, all were capable of producing DMSe from SeO_3^{2-} while only six organisms were capable of producing DMSe from SeO_4^{2-}. Interestingly, none of the bacterial isolates which had the ability to reduce Se to red amorphous Se° were capable of methylating Se (Barkes and Fleming, 1974). In contrast, Ganje and Whitehead (1958) and Thompson-Eagle and Frankenberger (1990b) discovered that seleniferous shale and evaporation pond water, respectively, methylated SeO_4^{2-} more efficiently than SeO_3^{2-}, regardless of the presence of organic matter.

e. Temperature

Selenium volatilization is temperature-dependent. The maximum release of DMSe from lake sediments occurred at 20° C (Chau et al. 1976), in California evaporation pond water at 35° C (Thompson-Eagle and Frankenberger, 1990a), from a British loamy soil at 20° C (Zieve and Peterson, 1981), and from a California sandy-textured soil at 35° C (Frankenberger and Karlson, 1990). However, in each case, maximum DMSe emission occurred at the maximum temperature tested, therefore the optimum temperature for biomethylation may not have been reached. Soil temperatures in the field seasonally vary between 4° and 50° C in the Central Valley of California (Frankenberger and Karlson, 1988). During the winter months Se emission is relatively low but increases during the spring and summer months. There is a diurnal peak of volatile Se emission during the midday to midafternoon which correlates with soil temperature (Frankenberger and Karlson, 1988; Weres et al., 1989).

f. Moisture

Being a biological process, Se biomethylation requires the presence of water. Air-drying the soil severely inhibits methylation (Zieve and Peterson, 1981) while water-saturating it to a 1:1 or 1:3 soil:water paste causes anaerobiosis, decreasing the transfer of volatile Se from soil to air and biomethylation is less efficient (Frankenberger and Karlson, 1990). Field studies have also shown that Se emission rates are much lower in dry sites than in corresponding damp or wet conditions (Frankenberger and Karlson, 1988, 1989b, 1990; Weres et al., 1989). Maximal Se biomethylation in seleniferous dewatered sediments occurred at 70% of the water-holding capacity (field-moist soil) (Frankenberger and Karlson, 1990), while in a heavy clay soil between 18% and 25% moisture was optimal (Abu-Eirreish et al., 1968). In a loam soil, Zieve and Peterson (1981) found that 28% was optimal for volatilization, while 16% and 40% moisture gave rise to considerably less volatile Se. Fluctuations in the soil water content appear to stimulate Se volatilization since sequential drying and rewetting of soil promotes the release of volatile Se (Hamdy and Gissel-Nielsen, 1976). This trend may be explained by the fact that the decomposition of organic matter in soil is also directly related to repeated drying-rewetting cycles and hence nutrients may become more available for the soil microflora and increase their metabolic activity under these conditions (Sørensen, 1974).

g. pH

Because Se volatilization is a biologically mediated process, the optimum pH is likely to be at the optimal pH for growth of the methylating microorganisms. Another factor to consider is the increasing solubility and hence availability of Se with increasing soil pH. The optimum pH for biomethylation in seleniferous Kesterson sediments (pH 7.7) was 8.0 (Karlson and Frankenberger 1989). The addition of lime to a sandy soil increased the pH (6 to 7) and Se volatilization 1.2-fold (Hamdy and Gissel-Nielson, 1976). Horne (1990) has suggested that it may be possible to reduce the recycling and ecosystem toxicity of Se in water bodies contaminated with Se by reducing the pH of the sediment.

h. Aeration

Most studies show that greater quantities of volatile Se are evolved aerobically than anaerobically regardless of the matrix. Saturation of a Pierre formation soil with N_2 gas almost completely eliminated Se evolution during a 26-day incubation period (Abu-Eirresh et al., 1968). Francis et al. (1974) found that glucose- and $Na_2SeO_3^{2-}$-amended seleniferous clay soil but not a silt loam evolved trace quantities of DMSe under argon. Under air, the same soils methylated > 83- and > 64-fold more DMSe. Soil, sewage sludge, and seleniferous pond water samples exposed to air produced larger quantities of volatile Se than corresponding samples exposed to N_2 (Reamer and Zoller, 1980; Thompson-Eagle and Frankenberger, 1991a). In contrast, Doran and Alexander (1977) and Frankenberger (unpublished data) measured substantial evolution of volatile Se from soil even under anaerobiosis.

i. Cofactors

Since it has been proposed that Se volatilization consists of both reduction and methylation reactions (Challenger, 1951; Doran, 1982), it is therefore likely that reducing agents, methyl donors, and prosthetic groups are involved in volatilization. Methyl donors include S-adenosylmethionine (SAM) and its precursors or derivatives, such as homocysteine and methionine. Methionine has been found to be stimulatory to Se volatilization in soil, dewatered sediments, water (Frankenberger and Karlson, 1990; Thompson-Eagle and Frankenberger, 1990a; Weres et al., 1989), and pure cultures of fungi such as *Penicillium* sp. (Fleming and Alexander, 1972) and *Alternaria alternata* (Thompson-Eagle et al., 1989). However, it is not clear if the stimulatory action of this amino acid is as a methyl donor or serves a nutritional role in microbial metabolism. The following methyl donors (1 mg kg^{-1}) were stimulatory to Se volatilization in dewatered sediments: S-adenosyl-L-methionine chloride (1.7-fold stimulation), S-adenosyl-L-homocysteine (1.6-fold stimulation), methionine sulfone (2.3-fold stimulation), and methionine sulfoxide (4-fold stimulation) (Frankenberger, unpublished data) The biological methyl donors, SAM and methylcobalamine had little effect on the methylation of Se in pond water, but cell-free studies are needed to determine if Se volatilization is a transmethylation reaction (Thompson-Eagle and Frankenberger, 1991a).

The alkyl-transfer to Tl (III), As (III), and Hg (II) as well as the synthesis of monomethyl and dimethylmercury (II) are known to require a cobalamine-containing cofactor, methylcobalamin, (Halpern, 1982; Wood et al., 1968). Although the addition of methylcobalamine (10 μM) to pond water did not stimulate Se volatilization (Thompson-Eagle and Frankenberger, 1991a), Se methylation in seleniferous soils increased up to 2.5-fold with the addition of relatively high levels (25 mmol kg^{-1}) of cobalt (Co) (Karlson and Frankenberger, 1988b). The addition of a metal activator as a trace fertilizer may be a feasible way of stimulating Se volatilization in the field.

Thiols are reducing agents and the ability of glutathione, cysteine residues, and coenzyme A to form Se complexes (selenotrisulfides) is known to be an important pathway by which inorganic Se is initially incorporated into living systems (Ganther, 1986; Garberg and Hogberg, 1986; Kice, 1981).

$$4\,RSH + H_2SeO_3 \longrightarrow RS\text{-}Se\text{-}SR + RSSR + 3H_2O$$

It has also been demonstrated that thiols, thiol derivatives, polysulfides, and H_2S increase the solubility of Se° in water and soil systems (Weres et al., 1989). Reduced glutathione and homocysteine, (10 μM) stimulated Se volatilization in unamended pond water 27- and 71-fold, respectively as well as in peptone-amended pond water, 5- and 14-fold, respectively (Thompson-Eagle and Frankenberger, 1991a). Glutathione (0.05 mg kg^{-1}) and homocysteine (1 mg kg^{-1}) stimulated volatilization of Se in dewatered sediments 1.2- and 2.8-fold, respectively (Frankenberger, unpublished data). In contrast, homocysteine and homocystine had no effect on Se methylation by *Penicillium* sp. (Fleming and Alexander, 1972). Glutathione, cysteine, and a propionic acid-thiol derivative were also found to be stimulatory to biomethylation in dewatered sediments by Weres et al. (1989). The effect of adding thiols to soil or water systems may therefore increase the availability and microbial uptake of Se.

6. Factors Inhibitory to Biomethylation of Selenium

a. Heavy Metals

There are few studies on the effect of heavy metals on biomethylation of Se. Karlson and Frankenberger (1988b) found that the addition of 5 mmol kg^{-1} molybdenum (Mo), mercury (Hg), chromium (Cr), and lead (Pb) to seleniferous soils greatly inhibited Se volatilization, while arsenic (As), boron (B), and manganese (Mn) had little effect. The addition of Co, zinc (Zn), and nickel (Ni) to seleniferous sediments stimulated volatilization of Se. Karlson and Frankenberger (1988b) postulate that Zn and Ni may inhibit the utilization of a readily available organic source by the non-methylating microbial population, thus making C more available to the Se methylating microbiota.

b. Salinity

Some Se methylating organisms are extremely tolerant to saline conditions. The methylation of SeO_3^{2-} by *Candida humicola* in a glucose salts medium was not affected by 1000 mg L^{-1} KH_2PO_4, NaH_2AsO_4, or Na_2TeO_4 (Cox and Alexander, 1974), but 1000 mg L^{-1} SO_4 decreased methylation by 25%. Although it is generally recognized that in alkaline environments, species diversity decreases as salinity and alkalinity increases (Grant and Tindall, 1986; Reed 1986), Frankenberger and Karlson (1990) and Thompson-Eagle and Frankenberger (1990a) have demonstrated considerable Se volatilization in soils and water with EC_e values as high as 22 dS m^{-1} and 10 to 30 dS m^{-1}, respectively. It therefore appears that some methylating microflora have adapted to these saline conditions and can tolerate extreme fluctuations in the environment. The addition of salts to unamended, saline evaporation pond water did not appear to affect Se biomethylation unless protein-amended and then only at high concentrations. The addition of NaH_2PO_4 (1 to 1000 m*M*) had no effect on Se biomethylation in water, while 0.1, 1 *M* Na_2SO_4 increased methylation 2.6- and 1.6-fold, respectively, and 0.1, 1 *M* $CaCl_2$ increased methylation 2- and 8-fold, respectively (Thompson-Eagle and Frankenberger, 1990a, 1991a). Protein-mediated biomethylation was inhibited by 100 m*M* PO_4 and 1000 m*M* $CaCl_2$, while SO_4^{2-} had no effect. Fleming and Alexander (1972) found that SeO_3^{2-} alkylation increased with increasing SO_4^{2-} concentrations. The addition of low concentrations of Na_2SO_4 and $CaCl_2$ were found to stimulate Se volatilization from a nonsaline Panhill soil, while at 20 dS m^{-1}, Na_2SO_4, NaCl and $CaCl_2$ reduced Se volatilization rate coefficients by an average of 16.8%, 18.3%, and 24.3%, respectively (Karlson and Frankenberger, 1990a). Soil microbial production of volatile Se was slightly more sensitive to Cl^- than to SO_4^{2-} ions and more sensitive to Na^+ than to Ca^{2+} ions (Karlson and Frankenberger, 1990a).

c. Nitrates

The presence of high levels of NO_3^- and NO_2^- inhibits Se biomethylation. Methylation of Se in evaporation pond water was inhibited by NO_3^- and NO_2^- ions at concentrations of 0.1 *M* and above (Thompson-Eagle and Frankenberger, 1990a,b). The application of KNO_3 to seleniferous soil in combination with galacturonic acid inhibited methylation by 11.8% when added to yield a C/N equal to 5 (Karlson and Frankenberger, 1988b). Nitrates have also been found to inhibit anaerobic Se transformations (Oremland et al., 1989).

d. Antibiotics

The addition of a fungicide (200 mg L^{-1} cycloheximide) to protein-amended evaporation pond water slightly stimulated biomethylation of Se or else had no effect on volatilization (200 mg L^{-1} nystatin), while a number of bactericides tested (100 mg L^{-1} penicillin G, 100 mg L^{-1} polymyxin B sulfate, 10 mg L^{-1} crystal violet) were extremely inhibitory to biomethylation (Thompson-Eagle and Frankenberger, 1991a). The methylating bacteria present in pond water exhibited resistance to the more broad spectrum bactericides such as 100 mg L^{-1} chlortetracycline, and 100 mg L^{-1} streptomycin sulfate (Thompson-Eagle and Frankenberger, 1991a) as did some dewatered sediment microflora to 40 mg kg^{-1} of penicillin-G and streptomycin (Weres et al., 1989) and 100 mg kg^{-1} chloramphenicol, penicillin-G, and streptomycin sulfate (Frankenberger, unpublished data). Streptomycin actually stimulated volatilization of Se from soil by 3.5-fold. In contrast, microbial volatilization of Se from a seleniferous British loam soil was inhibited by concentrations $\geq$ 14 mg kg^{-1} of chlortetracycline (Zieve and Peterson, 1981).

VI. Bioremediation of Selenium-Contaminated Soils: San Joaquin Valley, California—A Case History

A. Geology

The San Joaquin Valley was once covered by a large inland sea. The western side was uplifted and formed the Coast Range of California about 60 million years ago. With further uplifting, the sea evaporated leaving behind sediments. One of these sediments is the Corcoran clay which is an impervious subsurface layer that interferes with natural drainage in some parts of the western San Joaquin Valley. The Panoche Fan is an alluvial outwash of Panoche Creek, which originates in the Diablo Range. Soils in the Panoche Fan contain elevated levels of Se, which have remained in high concentrations because of the semi-arid climate (Moore, 1989). Shallow groundwater affects about 120000 ha of agricultural land in the western San Joaquin Valley (U.S. Bureau of Reclamation, 1984). Using an R-mode factor analysis, Deveral et al. (1984) found that Se concentrations in the seleniferous soil between Panoche Creek and Cantua Creek alluvial fans derived from the western Coast Range are associated with high levels of sulfur and sodium.

B. Historical Background

Throughout history, irrigated agriculture has been plagued by shallow groundwater and salinity problems. The San Joaquin Valley of California has an arid climate, receiving less than 20 cm of rain annually (Westlands Water District, 1989). It was not until the 1930s that government-sponsored irrigation projects brought in irrigation water from Northern California. Saline soils and saline drainage water have caused water management problems for agriculture in the San Joaquin Valley since the late 1800s (Fujii and Deveral, 1989). The management and disposal of this agricultural drainage water has developed into an issue of national concern because of the toxic effects of trace elements on wildlife. California is one of the most productive agricultural regions in the world, despite the drainage problems in the Central Valley, because the climate is mild and the soils are fertile. Farmers are able to grow some 35 commercial crops and contribute in excess of 70% of the nation's broccoli, cauliflower, lettuce, tomatoes, almonds, apricots, avocados, dates, figs, grapes, kiwifruit, lemons, nectarines, olives, pistachios, plums, pomegranates, prunes, strawberries, and walnuts (Land Preservation Association, 1990a). In 1988, crops grown in the Westlands Water District, an area of nearly 1000 square miles (600 000 acres), were valued at almost $668 million.

A shallow water table containing high levels of salts throughout the west side of the valley can lower the yield of crops by saturating the root zone unless drainage is improved. These salts must be removed from the crop root zone. Over 200000 ha of irrigated farmland in the San Joaquin Valley are affected by salinity (Mikkelson et al., 1986). Many growers have installed subsurface drainage systems to collect the excess, saline drainage water. Disposal of the drainage water is and continues to be a major problem because, in some districts, trace elements including Cr, Hg, and Se exceed the water quality criteria for freshwater aquatic life and therefore cannot be disposed of in river systems (Deveral et al., 1984).

A master drain (San Luis Drain) was originally designed in the 1950s to carry subsurface drainage water from farms in the San Joaquin Valley north to a discharge point in the San Francisco Bay-Delta region. However, due to economic, environmental, and political constraints, only 85 miles of the San Luis Drain (SLD) were built and terminated approximately 10 miles north of Los Banos, California. In 1978, twelve shallow evaporation ponds of Kesterson Reservoir became the terminus for drainage water carried by the SLD. Kesterson Reservoir was originally intended as a regulatory system for the SLD, with a secondary role serving as a wildlife habitat. Because Kesterson lies within the Pacific migratory bird flyway, it was designated as a National Wildlife Refuge. In 1978, the SLD was mainly composed of tail water (runoff) with an increasing amount of drainage water going into the Reservoir. By 1981, the agricultural drainage water collected from $<$ 42000 acres of irrigated agricultural land in the Westlands Water District was the sole source of water entering into the Reservoir. In 1982, the U.S. Fish and Wildlife Service (USFWS) biologists began monitoring elevated levels of Se in fish collected from the SLD and in the Reservoir (Saiki, 1985). Kesterson Reservoir became the focus of national attention in 1983 when a high incidence of deaths and deformities was noted among several species of birds and elevated Se concentrations in fish, invertebrates, and plants exposed to the water (U.S. Bureau of Reclamation, 1984). In 1986, the SLD was closed and the State Water Resources Control Board (SWRCB) demanded a clean-up plan by the Department of the

Interior, Bureau of Reclamation. The SWRCB adopted the last option proposed by the Bureau and that was to scrape up the seleniferous sediment and dump it into an on-site landfill at a cost of $48.2 million. The subcommittee on Energy and Water Development of the U.S. House Appropriation Committee asked the Bureau of Reclamation to persuade the State of California to reconsider the toxic dump plan. The state delayed the clean-up order and supported further research into a permanent clean-up plan. Selenium volatilization is currently being considered as a bioremediation approach at Kesterson Reservoir. Meanwhile, growers nearby have to dispose of their drainage water in on-farm evaporation ponds. In time these are likely to become mini-Kestersons of the future, since problems with the wildlife are already starting to surface. Although it has not conclusively been proven that all of the severe impacts to the wildlife observed at Kesterson Reservoir and now at the on-farm evaporation ponds are caused solely by Se, it has been clearly demonstrated through controlled laboratory experiments that Se, at the same concentrations present within evaporation ponds, leads to the same devastating biological effects (Moore, 1989). There are currently 21 active evaporation ponds in the San Joaquin Valley (Wass, 1990) which cover a total area of 6650 acres (Westcot et al., 1988; Dennis Westcot, personal communication, California Regional Water Quality Control Board, Sacramento). Approximately 20000 more acres are in the planning and construction stages. Approximately 10 to 15 acres of cropland are taken out of production in farmland to provide space for these evaporation ponds. It is estimated that "between 1,200,000 and 1,500,000 acres will need drainage during the next 100 years in order to sustain intensively managed irrigated agriculture and associated high levels of crop productivity" (Moore, 1989). The San Joaquin Valley Drainage Program is a specially created interagency group with the responsibility for developing action plans for short- and long-term management of irrigation return flows in problem areas of the San Joaquin Valley (Quinn, 1989). As yet, no remediation scheme has been implemented with the exception of encouraging the growers to use irrigation water more conservatively. Without successful treatment and disposal technologies, the San Joaquin Valley will suffer continued loss of agricultural productivity, degradation of water quality, destruction of fish and wildlife resources, and potentially adverse impacts on public health (Moore, 1989).

C. Selenium Composition of Drainage Water

Most of the soluble, inorganic Se occurring in alkaline, oxidized soils of the San Joaquin Valley exists in the SeO_4^{2-} form (Fujii and Deveral, 1989) which is very mobile and moves with drainage water. The average Se composition of the drainage water in the western San Joaquin Valley is 98% SeO_4^{2-} (Deveral and Millard, 1988). Fujii and Deveral (1989) fractionated two soils and found that the adsorbed Se was partitioned equally between

SeO_4^{2-} (51%) and SeO_3^{2-} (49%). In contrast, the soluble Se consisted almost entirely of SeO_4^{2-} (>97%). Selenium concentrations in the drain water discharged from 1984 to 1986 in the SLD averaged about 300 $\mu g\ L^{-1}$ (Fujii and Deveral, 1989), while dewatered sediments at Kesterson Reservoir currently range between <1 to over 700 mg kg^{-1} (Frankenberger, 1989; Frankenberger and Karlson, 1988). The on-farm evaporation pond facilities used by some growers in the San Joaquin Valley are rich in sodium and sulfate ions with concentrations in excess of 1 g L^{-1}. Selenium concentrations in the water vary between <30 and >2000 $\mu g\ L^{-1}$ (Tanji and Grismer, 1988; Thompson-Eagle and Frankenberger, 1990a,b, 1991a). Although substantial progress has been made in the development of treatment technologies to remove hazardous constituents from drainage water, none so far have been able to remove Se at a cost affordable by the agricultural community (Moore, 1989).

D. Utility of Evaporation Ponds

In the absence of alternative remediation methods, agricultural drainage water is impounded in evaporation ponds. One advantage of evaporation ponds is that they can be used to dispose of large volumes of drainage water. Potential problems of concern with these facilities include: (1) accumulation of toxic constituents to hazardous levels; (2) harmful effects on waterfowl and other wildlife; (3) seepage to groundwater and adjacent lands; and (4) the ultimate disposal of precipitated salts that may contain toxic constituents (State Water Resources Control Board, 1988). Hydrogeological assessment reports are required for evaporation ponds by the Regional Water Quality Control Board (RWQCB) under the Toxic Pits Act. Under California State law, surface impoundments of any liquid must not degrade groundwater. When Se levels exceed 1000 $\mu g\ L^{-1}$, the water is considered to be hazardous waste and the grower is required to convert the pond into a Class 1 hazardous waste site with double linings and leachate collector systems (Subchapter 15 of Title 23, California SWRQCB). This can greatly increase the cost of disposal. According to Hall et al. (1989), an evaporation pond can be constructed for about $1000/acre plus the cost of the land, however, a similar facility designed to meet the strict containment requirements of a Class 1 site can cost $200,000/acre or more. In addition to regulation by the RWQCB, the Department of Health and Services (DHS) imposes strict requirements over waste containment (Hall et al., 1989).

VII. Remediation Of Seleniferous Sediments and Water

In 1980, the U.S. Environmental Protection Agency (EPA) established a drinking water standard for dissolved Se at 10 $\mu g\ L^{-1}$. The EPA guideline for protection of aquatic life is currently set at 45 $\mu g\ L^{-1}$. SWRCB is having difficulties in establishing a water quality standard for freshwater systems. Currently the SWRCB (1988) stipulates that the Se content in the Central Valley drainage water must not exceed 5 $\mu g\ L^{-1}$ before discharge into surface water. An effort is being made by a number of state and government agencies to investigate alternative water management strategies and remediation technology. Meanwhile growers are being encouraged to use irrigation water more conservatively (Gates and Grismer, 1989).

A. Chemical Treatment

The drainage water could be chemically treated before it reaches the evaporation ponds to prevent accumulation of Se. This process is hampered by the alkalinity (pH 7.5 to 9.3) of the drainage water and the ratio of sulfur to Se ion concentrations ($>$10000 mg S L^{-1}) (James Montgomery, Consulting Engineers, 1985). In general, alum and ferric sulfate coagulation, lime softening, activated alumina and ion exchange with strong-base resins are more effective at removing SeO_3^{2-} than SeO_4^{2-} ions, which are the dominant Se species in the San Joaquin Valley drainage water (Cutter, 1988; Cooke and Bruland, 1987). Desalination is a very expensive process (Hall et al., 1989). Cogeneration is another process which converts drainage water into pure distilled water, salt, and electricity (Land Preservation Association, 1990b). A pilot scale cogeneration project is being initiated by the Santina Water Company (SWC) near Mendota, California.

B. Deep Well Injection

Deep well injection of the drainage water is extremely expensive due to the high costs of drilling wells 1 mile deep. This is obviously not an environmentally sound solution because the water may not remain immobilized for long periods of time. Other problems include digging enough wells to cope with the ongoing problem and finding suitable land to drill. The low capacity of the wells obviously is a disadvantage in disposal. These wells merely serve as storage basins of wastewater. A deep well prototype experiment in California was temporarily stopped in November, 1989, because the injection rate was 86% less efficient than expected (Westlands Water District, 1990). Tests showed that the injection formation was a far less porous sandstone than expected. During the 18-month field trial, the Westlands Water District has spent about $1.7 million to test the effectiveness of this disposal technique (Land Preservation Association, 1989).

C. Soil Washing

Soil washing can only be used to treat dewatered evaporation ponds. Soil washing is inefficient since it removes only the water-soluble fraction of the soil Se inventory. An undesirable by-product of the process would be the generation of huge quantities of seleniferous water which would then require disposal.

D. Deep Plowing

Deep plowing is a means of diluting the Se concentration by subsurface burial. This technology obviously does not involve permanent removal of Se from soil.

E. Containment into Landfills

This is a form of immobilization which is dependent on the permanency of the landfill itself. Liners currently being used for containment are known not to last indefinitely and often rip or tear. Capping a clay-lined landfill with fill material is not always effective. This method is therefore not a permanent solution, but merely buries the problem for a finite period of time.

F. Vegetative Uptake

The accumulation and removal of Se from drainage water by vegetation may be a viable bioremediation approach. Uptake is generally restricted to the soluble fraction of Se. In order to be effective, the vegetation would have to rely on the soil microflora to mineralize the complexed, insoluble organic Se into a soluble form. Vegetative uptake holds some promise because it is a permanent removal process. However, the seleniferous plant material would have to be disposed of. Recently tall fescue cultivation was found to reduce soil Se by 50% in seleniferous areas of California in one growing season (Wu et al., 1990), however volatilization was not accounted for in this study. Some of the San Joaquin Valley water districts are supporting funding for an agroforestry project to receive drainage water (Westlands Water District, 1990). The project, entitled "Los Arboles" calls for the planting of 600 acres of eucalyptus trees and salt-tolerant shrubs, which would serve to concentrate about 2000 acre-feet of drainage water each year. Westlands Water District plans to have about 5000 trees in the ground by 1991. Further work is being carried out to determine the tolerance of eucalyptus trees and shrubs to the extreme salinities and high concentrations of boron found in the drainage waier and their ability to selectively take up SeO_4^{2-} over such large quantities of SO_4^{2-}. Competitive plant uptake has been demonstrated between these two oxyanions.

G. Volatilization

Selenium volatilization permanently removes Se from contaminated soil and water. It therefore has potential applications for the successful maintenance and operation of evaporation pond disposal sites as well as drainage water itself. Research in our laboratory has shown that microorganisms naturally present in the saline, alkaline drainage water, and soil methylate Se into DMSe. This naturally occurring transformation can be dramatically accelerated by the addition of specific amendments. The advantages of biostimulation of the indigenous microflora over bioaugmentation (adding preselected micro-organisms) are twofold; first, survival of alien organisms is not required and second, the indigenous microflora have adapted to their surrounding environment, and are capable of multiplying rapidly and making a substantial contribution to deselenification in a relatively short period of time. When the Se levels decline below the acceptable level, amendments are no longer necessary and the microbial population quickly dies back to its original level.

1. Landfarming Seleniferous Sediments

Microbial volatilization of Se is being considered as a bioremediation technique to remove toxic levels of Se at Kesterson Reservoir. A field investigation was initiated in July of 1987 with the goals of identifying the most effective practices for accelerating volatilization and to obtain information necessary for the determination of time and factors affecting this technology (Frankenberger, 1989; Frankenberger and Karlson, 1989b; Frankenberger et al., 1990). This field study was conducted on cattail-enriched sediments in Pond 4 representing one of the more contaminated areas containing Se concentrations ranging from 10 to 209 mg kg^{-1} (median = 39 mg kg^{-1}). The treatments consisted of the application of water alone, or with cattail straw, cattle manure, citrus (orange) peel, and protein sources (casein and gluten). Some plots were also treated with fertilizers such as ammonium nitrate and zinc sulfate. All subplots were sampled for gaseous Se with an inverted box and an alkali peroxide trap (Fig. 1). Soil samples were collected at monthly intervals in a 5-point pattern at 0 to 6 inches depth in order to account for soil depletion with time. Seasonal variation of gaseous Se emission was evident with the highest emission recorded in the late spring and summer months. The biomethylation rates were correlated with soil temperature (Fig. 2). Less volatile Se was released in the fall and winter months. The greatest emission of gaseous Se with all treatments occurred at the initiation of the project when the Se inventory was high. As time proceeds, the Se inventory available for methylation appears to be decreasing with each season of warmer temperature.

The emission flux of gaseous Se varied with each of the treatments (Frankenberger, 1989). Irrigation with tillage alone resulted in an average volatile Se emission of 16 μg Se m^{-2} h^{-1} during this 2-year field study. The most stimulatory soil treatment for Se volatilization was citrus peel + N + Zn. The highest emission rate recorded with this treatment was 808 μg Se m^{-2} h^{-1} which is approximately 42-fold greater than the background level. The application of casein, a milk protein, also promoted methylation of Se with an average emission rate of 50 μg Se m^{-2} h^{-1}.

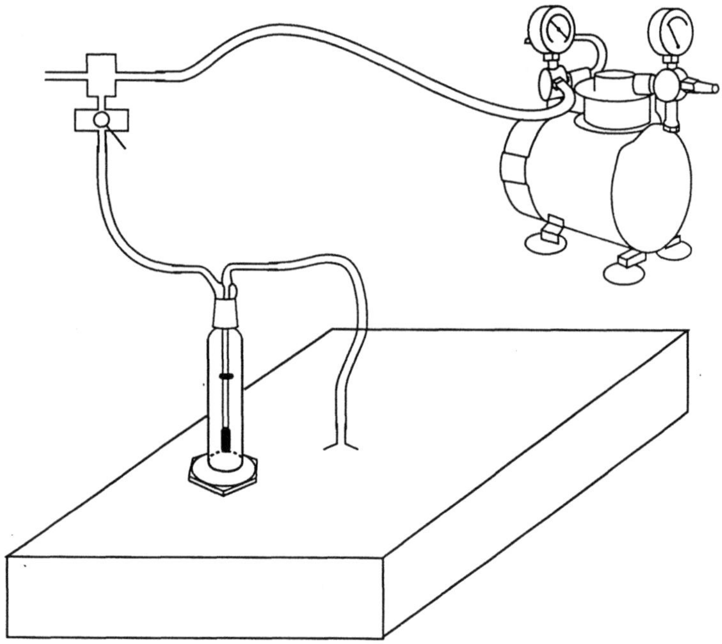

Figure 1. Apparatus used to monitor alkylselenide production in the field

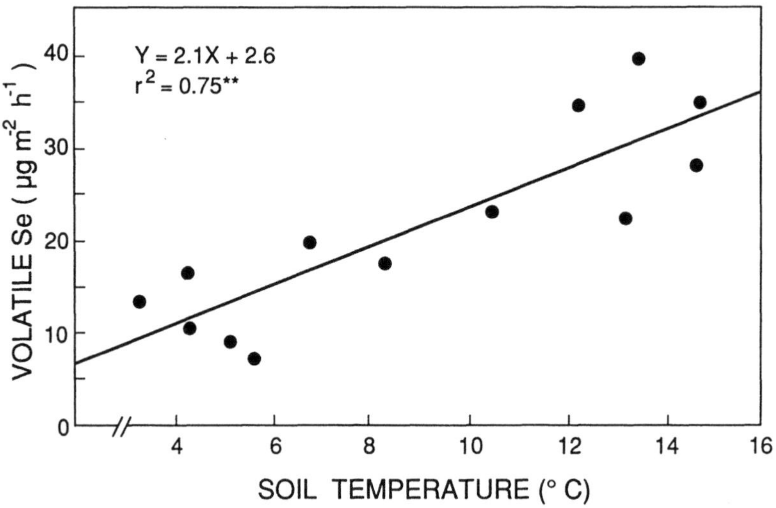

Figure 2. Linear regression analysis of alkylselenide production in the field and soil temperature

Figure 3. Soil selenium depletion in response to specific amendments added to Kesterson Pond 4 sediments

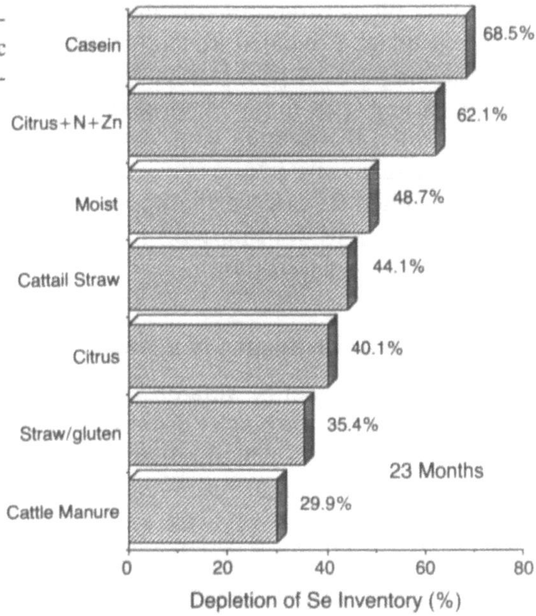

Soil samples collected at different depths in Pond 4 indicated that 90% of the total Se inventory was within the upper 6 inches of the soil profile. The decline in residual soil Se revealed that cattle manure was the least effective treatment with 30% Se removal in 23 months, while citrus peel + N + Zn and casein were the most effective treatments with 62% and 69% of the initial Se inventory removed in the Ap layer after 2 years of investigation, respectively (Fig. 3). (Frankenberger, 1989).

Another field experiment was initiated at the Sumner Peck Ranch (near Fresno, CA) in October of 1987 to assess microbial volatilization from a sediment comprised mainly of clay (Frankenberger et al., 1990). The dewatered sediment was rototilled and plots were staked out and amended with moisture, citrus peel, cattle manure, barley straw, and grape pomace. Some subplots were fertilized with N (ammonium sulfate) and Zn (zinc sulfate). All plots were rototilled to approximately 6 inches in depth and sprinkler irrigation was applied to keep the sediments moist. Emanation of volatile Se indicated that there was seasonal variation in the quantity of gaseous Se released from the sediment. Overall the greatest emission of volatile Se was recorded in the summer months with the lowest emission occurring during the winter. The most effective amendment was cattle manure. After 21 months of study, the application of water removed approximately 32% of the Se inventory while treatment with cattle manure and straw removed 57.8%.

In summary, the parameters which enhance volatilization of Se in the field are an available carbon source, aeration, moisture, and high temperatures. Rototilling promotes volatilization as long as the soil is kept moist. Frequent tillage is needed to support the aerobic methylating organisms, enhance soil porosity (this facilitates the diffusion of the alkylselenide gas) and break any crust that may form as a result of sprinkler irrigation. Also recommended is irrigation with wetting and drying cycles in order to release the organically bound Se to methylating organisms. Water should only be applied to moisten the upper few inches of soil otherwise water-soluble Se may be transported out of the surface layer, thus making it unavailable for volatilization.

H. Atmospheric Dissipation of Selenium

Air quality impacts of Se volatilization have been assessed using WYND-valley, a numerical air quality dispersion model (U.S. Bureau of Reclamation, 1988). Selenium volatilization rates from ongoing field trials at Kesterson Reservoir were incorporated into a model simulating different wind conditions including a worst-case scenario of a severely stagnant wind episode. The model shows that the atmospheric Se released as DMSe becomes diluted and dispersed by air currents away from the contaminated source (U.S. Bureau of Reclamation, 1988). The highest 24-h average air Se concentration simulated was about 250 ng m^{-3}. Acceptable ambient air Se concentrations for states that have standards range from 2700 ng m^{-3} over 24 h to 5000 ng m^{-3} over 8 h. No simulated air Se concentrations exceeded these standards (U.S. Bureau of Reclamation, 1988). The impact of deposition of volatilized Se on lands surrounding Kesterson is expected to be minimal. The estimated deposition fluxes of Se range from 4.5 g ha^{-1} $year^{-1}$ near the Reservoir to 0.9 g at a distance of 10 km and 0.009 g at a distance of 100 km. If 4.5 g Se ha^{-1} were mixed with the upper 10 cm of soil, the soil Se concentration would be increased by approximately 0.005 mg kg^{-1} This report concludes that enhanced volatilization of Se from Kesterson Reservoir is not expected to substantially alter the regional Se distribution. Accelerated biomethylation of Se on the west side of the San Joaquin Valley occurs at optimal levels during the warm spring and summer months. This is a period of very little precipitation, and the prevailing winds blow mainly in a southeasterly direction, toward the Se-deficient areas on the east side of the Valley.

I. Implementation and Economics

The bioremediation technology of Se volatilization could be operated in two different ways, either through series evaporation pond management or through a primary pond operation (Frankenberger et al., 1990). Although the two systems differ in the way they are managed, the bioremediation principle is the same. Ponds are used to evaporate the incoming drainage water to dryness and the process is repeated until the sediment approaches 100 mg Se kg^{-1}. At this stage, volatilization is optimized as a treatment process until the Se inventory decreases significantly to an acceptable concentration, and then the pond is placed back into operation. In order to study the economics of Se bioremediation, the following factors must be taken into consideration: (1) selenium volatilization rate, (2) irrigation flow rate, (3) income generated from the land, (4) selenium concentration in the drainage water, (5) land requirement for evaporation ponds (amount removed from production), (6) pond construction costs, (7) water depth and evaporation rate, (8) costs of Se volatilization as a treatment process including nutrient amendments, water, labor, and equipment needed to implement sediment aeration, irrigation, and land amendments.

The economics of implementing this process are reported in a previous publication (Thompson-Eagle and Frankenberger, 1991b). Selenium volatilization is projected to cost between $45 and $151 per productive acre for 0.25 to 0.5 acre-ft/acre-year drainage water containing Se concentrations of between 50 and 100 μg Se L^{-1} (Frankenberger et al., 1990). Other analyses have been conducted on the economics of Se remediation of drainage water. According to Johns and Watkins (1989), the 1987 costs of treating seleniferous drainage water to achieve a final Se concentration of 5 μg L^{-1} were estimated to range from $53 to $77 per acre, with a total capital cost ranging from $52,400,000 to $64,400,000. No details of these treatment technologies have been reported, but feedback from the San Joaquin Valley Drainage Program indicates that these estimates are considered to be low.

VIII. Conclusions

At the present time, field projects involving volatilization of Se from sediments at Kesterson Reservoir and other evaporation ponds (e.g., Sumner Peck Ranch) are showing a rapid decline in the soil Se inventory. Volatilization looks extremely promising and future research will focus on generating the necessary information to successfully initiate a full-scale Se volatilization operation. Future studies on aquatic biomethylation will focus on the development of a deselenification water treatment process which can be used on-site to treat incoming drainage water. Bioremediation of seleniferous environments is likely to be an important operation in sustaining high crop productivity on the west side of the San Joaquin Valley.

References

Abu-Eirreish, G.M., E.I. Whitehead, and O.E. Olson. 1968. Evolution of volatile selenium from soils. *Soil Science* 106:415–420.

Andren, A.W., D.H. Klein, and Y. Talmi. 1975. Selenium in coal-fired steam plant emissions. *Environ. Sci. Technol.* 9:856–858.

Arlichs, J.R. and L.R. Hossner. 1987. Selenate and selenite mobility in overburden by saturated flow. *J. Environ. Qual.* 16:95–98.

Atkinson, R., S.M. Aschmann, D. Hasegawa, E.T. Thompson-Eagle, and W.T. Frankenberger, Jr. 1990. Kinetics of the atmospherically-important reactions of dimethyl selenide. *Environ. Sci. Technol.* 24:1326–1332.

Banat, I.M. and D.B. Nedwell. 1984. Inhibition of sulfate reduction in anoxic marine sediment by group VI anions. *Estuarine Coastal Shelf Sci.* 18:361–366.

Barkes, L. and R. W. Fleming. 1974. Production of dimethylselenide gas from inorganic selenium by eleven fungi. *Bull. Environ. Contam. Toxicol.* 12:308–311.

Barrow, N.J. and B.R. Whelan. 1989. Testing a mechanistic model. VIII. The effects of time and temperature of incubation on the sorption and subsequent desorption of selenite and selenate by a soil. *J. Soil Sci.* 40:29–37.

Balistrieri, L.S. and T.T. Chao. 1987. Selenium adsorption by goethite. *Soil Sci. Soc. Am. J.* 51:1145–1151.

Bar-Yosef, B. and D. Meek. 1987. Selenium adsorption by kaolinite and montmorillonite. *Soil Sci.* 144:11–19.

Berrow, M.L. and A.M. Ure. 1989. Geological materials and soils. In: M. Ihnat (ed.) *Occurrence and Distribution of Selenium*, pp. 295–325. CRC Press, Boca Raton, FL.

Bird, M.L. and F. Challenger. 1942. Studies in biological methylation. Part IX. The action of *Scopulariopsis brevicaulis* and certain penicillia on salts of aliphatic seleninic and selenonic acids. *J. Chem. Soc.* pp. 574–577.

Bird, M.L., F. Challenger, P.T. Charlton, and J.O. Smith. 1948. Studies on biological methylation. XI. The action of molds on inorganic and organic compounds of arsenic. *Biochem. J.* 43:78–83.

Bisbjerg, B. 1972. *Studies on Selenium in Plants and Soil.* Danish Atomic Energy Commission, Riso Report No. 200. Copenhagen.

Boon, D.Y. 1989. Potential selenium problems in Great Plains soils. In: L.W. Jacobs (ed.) *Selenium in Agriculture and the Environment*, pp. 107–131. SSSA special publication No. 23. New Orleans, LA.

Burau, R.G. 1985. Environmental chemistry of selenium. *Calif. Agric.* July–August, pp. 16–18.

Calderone, S.J., W.T. Frankenberger, Jr., U. Karlson, and D.R. Parker. 1990. Influence of temperature and organic amendments on the mobilization of selenium in soil. *Soil Biol. Biochem.* 22:615–620.

California State Water Resources Control Board. 1987. *Regulation of agricultural drainage to the San Joaquin River.* Tech. Comm. Rep., SWRCB, Sacramento, CA.

Challenger, F. 1935. The biological methylation of compounds of arsenic and selenium. *Chem. Indus.*, pp. 657–662.

Challenger, F. 1945. Biological methylation. *Chem. Rev.* 36:315–316.

Challenger, F. 1951. Biological methylation. *Adv. Enzymol.* 12:429–491.

Challenger, F. and P.T. Charlton. 1947. Studies on biological methylation. Part X. The fission of the mono- and di-sulphide links by moulds. *Quarterly Rev. Chemical Soc.*, London, pp. 424–429.

Challenger, F. and H.E. North. 1934. The production of organo-metalloidal compounds by microorganisms. Part II. Dimethyl selenide. *J. Chem. Soc.*, pp. 68–71.

Chasteen, T.G., G.M. Silver, J.W. Birks, and R. Fall. 1990. Fluorine-induced chemiluminescence detection of biologically methylated tellurium, selenium and sulfur compounds. *Chromatographia* 30:181–185.

Chau, Y.K., P.T.S. Wong, B.A. Silverberg, P.L. Luxon, and G.A. Bengert 1976. Methylation of selenium in the aquatic environment. *Science* 912:1130–1131.

Clark, P.J, R.A. Zingarro, K.J. Irgolic, and A.N. McGinley. 1980. Arsenic and selenium in Texas lignite. *Int. J. Environ. Anal. Chem.* 7:295–314.

Cooke, T.D. and K.W. Bruland. 1987. Aquatic chemistry of selenium: evidence of biomethylation. *Environ. Sci. Technol.* 21:1214–1219.

Cox, D.P. and M. Alexander. 1974. Factors affecting trimethylarsine and dimethylselenide formation by *Candida humicola. Microb. Ecol.* 1:136–144.

Crystal, R.G. 1973. Elemental selenium: structure and properties. In: D.L. Klayman and W.H.H. Gunther (ed.) *Organic Selenium Compounds: Their Chemistry and Biology*, pp. 13–27. Wiley Interscience, New York.

Cutter, G.A. 1982. Selenium in reducing waters. *Science* 217:829–831.

Cutter, G.A. 1988. Selenium in freshwater systems. In: M. Inhat (ed.) *Occurrence and Distribution of Selenium*. pp. 243–262. CRC Press, Boca Raton, Florida.

Cutter, G.A. and K.W. Bruland. 1984. The marine biogeochemistry of selenium: a re-evaluation. *Limnol. Oceanogr.* 29:1179–1192.

Deveral, S.J., R.J. Gilliom, R. Fujii, J.A. Izbecki, and J.C. Fields. 1984. *Areal Distribution of Selenium and Other Inorganic Constituents in Shallow Ground Water of the San Luis Drain Service Area, San Joaquin Valley, California: a Preliminary Study*. U.S. Geol. Surv. Water Resour. Invest. Rep. 84-4319.

Deveral, S.J. and S.P. Millard. 1988. Distribution and mobility of selenium and other trace elements in shallow ground water of the western San Joaquin Valley, California. *Environ. Sci. Technol.* 22:697–702.

Doran, J.W. 1982. Microorganisms and the biological cycling of selenium. *Adv. Microb. Ecol.* 6:1–32.

Doran, J.W. and M. Alexander. 1975. Microbial formation of dimethylselenide and the biological cycling of selenium. *Abstracts of the Annual meeting of the America Society for Microbiology*, N22, p. 188.

Doran, J.W. and M. Alexander. 1977. Microbial formation of volatile selenium compounds in soil. *Soil Sci. Soc. Am. J.* 41:70–73.

Drotar, A., L.R. Fall, E.A. Mishalanie, J.E. Tavernier, and R. Fall. 1987. Enzymatic methylation of sulfide, selenide, and organic thiols by *Tetrahymena thermophila. Appl. Environ. Microbiol.* 53:2111–2118.

Eisler, R. 1985. Selenium hazards to fish and wildlife. *Serv. Biol. Rep.* 85(1.5). U.S. Gov. Print. Office, Washington, D.C.

Elrashidi, M.A., D.C. Adriano, S.M. Workman, and W.L. Lindsay. 1987. Chemical equilibria of selenium in soils: a theoretical development. *Soil Sci.* 144: 141–152.

Elrashidi, M.A., D.C. Adriano, and W.L. Lindsay. 1989. Solubility, speciation and transformation of selenium in soils. In: L.W. Jacobs (ed.) *Selenium in Agriculture and the Environment*, pp. 51–63. SSSA special publ. No. 23. Madison, WI.

Evans, C.S., C.J. Asher, and C.M. Johnson. 1968. Isolation of dimethyl diselenide and other volatile selenium compounds from *Astralagus racemosus* (pursh). *Aust. J. Biol. Sci.* 21:13–20.

Fleming, R.W. and M. Alexander. 1972. Dimethylselenide and dimethyltelluride formation by a strain of *Penicillium*. *Appl. Microbiol.* 24:424–429.

Francis, A.J., J.M. Duxbury, and M. Alexander. 1974. Evolution of dimethylselenide from soils. *Appl. Microbiol.* 28:248–250.

Franke, K.W. and A.L. Moxon. 1936. A comparison of the minimum fatal doses of selenium, tellurium, arsenic and vanadium. *J. Pharmacol. Exptl. Ther.* 58:454–459.

Frankenberger, W.T. Jr. 1986. *Microbial Transformation in Seleniferous Soils.* 1985–86 technical progress report, UC Salinity/Drainage Task Force, Division of Agriculture and Natural Resources, University of California, September.

Frankenberger, W.T. Jr. 1989. *Dissipation of Soil Selenium by Microbial Volatilization at Kesterson Reservoir.* Prepared for the U.S. Dept. of Interior, Bureau of Reclamation, December. Contract No. 7-FC-20-05240.

Frankenberger, W.T. Jr. and U. Karlson. 1988. *Dissipation of Soil Selenium by Microbial Volatilization at Kesterson Reservoir.* December. Prepared for the U.S. Dept. of Interior, Bureau of Reclamation, Contract No. 7-FC-20-05240.

Frankenberger, W.T. Jr. and U. Karlson. 1989a. Land treatment to detoxify soil of selenium. U.S. Patent and Trademark Office. Patent. 4,861, 482.

Frankenberger, W.T. Jr. and U. Karlson. 1989b. *Dissipation of Soil Selenium by Microbial Volatilization at Kesterson Reservoir.* December 1989. Prepared for the U.S. Dept. of Interior, Bureau of Reclamation, Contract No. 7-FC-20-05240.

Frankenberger, W.T. Jr. and U. Karlson. 1990. Environmental factors affecting microbial production of dimethylselenide in a selenium-contaminated sediment. *Soil Sci. Soc. Am. J.* 53:1435–1442.

Frankenberger, W.T. Jr., U. Karlson, and K.E. Longley. 1990. *Microbial Volatilization of Selenium from Sediments of Agicultural Evaporation Ponds.* Final draft report prepared for the State Water Resources Control Board under Interagency agreement No. 7-125-250-1.

Fujii, R. and S.J. Deveral. 1989. Mobility and distribution of selenium and salinity in groundwater and soil of drained agricultural fields, western San Joaquin Valley of California In: L.W. Jacobs (ed.) *Selenium in Agriculture and the Environment*, pp. 195–212. SSSA special publ. No. 23. Madison, WI.

Ganje, T.J. and E.I. Whitehead. 1958. Selenium uptake by plants as affected by the forms of selenium in the soil. *Proc. South Dakota Acad. Sci.* 37:85–88.

Ganther, H.E. 1986. Pathways of selenium metabolism including respiratory excretory products. *J. Am. College of Toxicol.* 5:1–5.

Garberg, P. and J. Hogberg. 1986. The role of selenium-oxygen interactions in selenium metabolism. *Ambio* 14:354–355.

Gardiner, M.R. 1961. White muscle disease of sheep. *J. Dept. Agric. West. Austral.* 2:497–500.

Gates, T.K., and M.E. Grismer. 1989. Irrigation and drainage strategies in salinity-affected regions. *J. Irrig. Drain. Eng.* 115:255–284.

Geering, H.R., E.E. Cary, L.H.P. Jones, and W.H. Allaway. 1968. Solubility and redox criteria for the possible forms of selenium in soils. *Soil Sci. Soc. Am. Proc.* 32:35–40.

Gissel-Nielson, G. and B. Bisbjerg. 1970. The uptake of applied selenium from soils by plants. 2. The utilization of various selenium compounds. *Plant Soil* 32:382–396.

Gissel-Nielson, G., U.C. Gupta, M. Lamand, and T. Westermarck. 1984. Selenium in soils and plants and its importance in livestock and human nutrition. *Adv.*

Agron. 37:397–460.
Graham, R.V., B.G. Blaylock, F.O. Hoffman, and M. Frank. 1990. Comparison of selenomethionine and selenite cycling in freshwater experimental ponds. Abstract. International Conference on metals in soils, waters, plants and animals, April 30–May 3, 1990, Orlando, Florida, p. 272.
Grant, W.D. and B.J. Tindall. 1986. The alkaline saline environment. In: R.A. Herbert and G.A. Codd (eds.) *Microbes in Extreme Environments*, pp. 25–54. Special publications of the Society for General Microbiology, vol. 17. Academic Press. London.
Gruebel, K.A., J.A. Davis, and J.O. Leckie. 1988. The feasibility of using sequential extraction techniques for arsenic and selenium in soils and sediments. *Soil Sci. Soc. Am. J.* 52:390–397.
Hall, S.K., W.R. Johnston, and W. J. Miller. 1989. Agricultural drainage water-how should it be regulated in California. *J. Irrig. Drain. Eng.* 115:3–8.
Halpern, J. 1982. Chemistry and significance of vitamin B_{12} model system. In: D. Dolphin (ed.) B_{12}, vol. 1, pp. 501–541. Wiley Interscience, New York.
Hamdy, A.A. and G. Gissel-Nielsen. 1976. Volatilization of selenium from soils. *Z. für Pflanzen. Boden.* 6:671–678.
Horne, A.J. 1990. Selenium detoxicification in wetlands by permanent flooding: a long-term, large-scale mesocosm experiment at Kesterson Reservoir, California. Abstract. International Conference on metals in soils, waters, plants and animals, April 30–May 3, 1990, Orlando, Florida, p. 229.
Izbecki, J.A. 1984. *Chemical Quality of Water at 14 Sites near Kesterson National Wildlife Refuge, Fresno and Merced Counties, California.* Open File Report 84-582. U.S. Geological Survey, Washington, D.C.
Jenkins, K.J. and M. Hidiroglou. 1972. A review of selenium/vitamin E responsive problems in livestock: A case for selenium as a feed additive in Canada. *Can. J. Anim. Sci.* 52:591–620.
Jiang, S., H. Robberecht, and F. Adams. 1983. Identification and determination of alkylselenide compounds in environmental air. *Atmo. Environ.* 17:111–114.
Johns, G.E. and D.A. Watkins. 1989. Regulation of agricultural drainage to San Joaquin River. *J. Irrig. Drain. Eng.* 115:29–41.
Jones, J.B. and T.C. Stadtman. 1977. *Methanococcus vannielii*: culture and effects of selenium and tungsten on growth. *J. Bacteriol.* 130:1404–1406.
Karlson, U. and W.T. Frankenberger. 1988a. Determination of gaseous selenium-75 evolved from soil. *Soil Sci. Soc. Amer. J.* 52:678–681.
Karlson, U. and W.T. Frankenberger, Jr. 1988b. Effects of carbon and trace element addition on alkylselenide production by soil. *Soil Sci. Soc. Amer.* 52:1640–1644.
Karlson, U. and W.T. Frankenberger. 1989. Accelerated rates of selenium volatilization from California soils. *Soil Sci. Soc. Amer.* 53:749–753.
Karlson, U. and W.T. Frankenberger. 1990a. Alkylselenide production in salinized soil. *Soil Sci.* 149:56–61.
Karlson, U. and W.T. Frankenberger. 1990b. Volatilization of selenium from agricultural evaporation pond sediments. *Sci. Total Environ.* 92:41–54.
Kice, J.L. 1981. The mechanism of the reaction of thiols with selenite and other Se (IV) species, In: J.E. Spallholz, J.L. Martin, and H.E. Ganther (eds.) *Selenium in Biology and Medicine.* pp. 17–32. AVI Publishing Comp., Westport, Conn.
Koch, N.C. 1967. Disappearance of the dinosaurs. *J. Palaeontol.* 41:970–972.

Korte, N.E., J. Skopp, W.H. Fuller, E.E. Niebla, and B.A. Aleshi. 1976. Trace element movement in soils: influence of soil physical and chemical properties. *Soil Sci.* 122:350–359.

Krishnan, S., N. Vasanthy and K. Lalitha. 1989. Effect of selenium on the metabolism of a novel methanogen. In: J. Neve, and A. Favier (eds.) *Selenium in Medicine and Biology*, Proceedings of the Second International Congress on Trace Elements in Medicine and Biology, pp. 385–388. Walter de Gruyter, New York.

Läg, J. and E. Steinnes. 1978. Regional distribution of selenium and arsenic in humus layers of Norwegian forest soils. *Geoderma* 20:3–14.

Lakin, H.W. 1961. Geochemistry of selenium in relation to agriculture. In: M.S. Anderson et al. (ed.) *Selenium in Agriculture*, pp. 3–12. USDA Handb. 200. U.S. Gov. Print. Office, Wahington, D.C.

Lakin, H.W. 1973. Selenium in our environment. *Adv. Chem.* 123:96–111.

Land Preservation Association. 1989. Deep-well project hits snag. *West Valley J.* December, p. 2.

Land Preservation Association. 1990a. California's share of major crops. *West Valley J.* July, p. 1.

Land Preservation Association. 1990b. Cogeneration plant moves closer to reality. *West Valley J.* July, p. 3.

Latinwo, L., L. Overbye, J. Caguiat, R. Blake, N. Revis, and J.H. Jackson. 1990. Reduction of selenium, mercury and lead salts encoded by a single plasmid in a *Pseudomonas* strain from a toxic waste site. Abstract. 90th Annual Meeting of the American Society for Microbiology, Anaheim, May. 1990. Q-244, p. 329.

Levander, O.A., G. Alfthan, H. Arvilommi, C.G. Gref, J.K. Huttunen, M. Kataja, P. Koivistoinen, and J. Pikarainen. 1983. Bioavailability of selenium to Finnish men as assessed by platelet glutathione peroxidase activity and other blood parameters. *Am. J. Clin. Nutr.* 37:887–897.

Lewis, B.G., C.M. Johnson, and T.C. Broyer. 1974. Volatile selenium in higher plants. The production of dimethyl selenide in cabbage leaves by enzymatic cleavage of Se-methyl selenomethionine selenonium salt. *Plant Soil* 40:107–118.

Lipman, J.G. and S.A. Waksman. 1923. The oxidation of selenium by a new group of autotrophic microorganisms. *Science* 57:60.

Macy, J.M. and S. Rech. 1990. The selenate reductase and nitrate reductase of a selenate- and nitrate-respiring *Pseudomonas* sp. are different enzymes. Abstract. 90th Annual Meeting of the American Society for Microbiology, Anaheim, May. 1990. K-131, p. 241.

Macy, J.M., T.A. Michel, and D.G. Kirsch. 1989. Selenium reduction by a *Pseudomonas* species: a new mode of anaerobic respiration. *FEMS Microbiol. Letts.* 61:195–198.

Mayland, H.F. 1985. Selenium in soils and plants. In: J. Maas (ed.) *Selenium Responsive Diseases in Food Animals*, pp. 5–10. Western States Veterinary Conf. Symp., Las Vegas, NV. 18 Feb. Veterinary Learning Systems Co., Princeton Junction. N.J.

Mayland, H.F., L.F. James, K.E. Panter, and J.L. Sondregger. 1989. Selenium in seleniferous environments. In: L.W. Jacobs (ed.) *Selenium in Agriculture and the Environment*, pp. 15–50. SSSA special publ. No. 23. Madison, WI.

McConnell, K.P. and O.W. Portman. 1952. Toxicity of dimethylselenide in the rat and mouse. *Proc. Soc. Exp. Biol. Med.* 79:230–231.

McCready, R.G.L., J.N. Campbell. and J.I. Payne. 1966. Selenite reduction by

Salmonella heidelberg. Can. J. Microbiol. 1:703–714.
McNeal, J.M. and L.S. Balistrieri. 1989. Geochemistry and occurrence of selenium: an overview. In: L.W. Jacobs (ed.) *Selenium in Agriculture and the Environment*, pp. 1–13. SSSA special publ. No. 23. Madison, WI.
Mikkelsen, R.L., A.L. Page, and F.T. Bingham. 1986. Geochemistry and health in California: recent experiences with selenium. In: D.D. Hemphill (ed.) *Trace Substances in Environmental Health*, vol. 2, pp. 413–423. A symposium. University of Missouri, CO.
Moore, S.B. 1989. Selenium in agricultural drainage: essential nutrient or toxic threat? *J. Irrig. Drain. Eng.* 115:21–28.
Mosher, B.W. and R.A. Duce. 1989. The atmosphere. In: M. Ihnat (ed.) *Occurrence and Distribution of Selenium*, pp. 295–325. CRC Press, Boca Raton, FL.
Nelson, A.A., O.G. Fitghugh, and H.O. Calvery. 1943. Liver tumors following cirrhosis caused by Se in rats. *Cancer Res.* 3:230–236.
Nriagu, J.O. 1989. Global cycling of selenium. In: M. Inhat (ed.) *Occurrence and Distribution of Selenium*, pp. 327–340. CRC Press, Boca Raton, FL.
Ohlendorf, H.M. 1989. Bioaccumulation and effects of selenium in wildlife. In: L.W. Jacobs (ed.) *Selenium in Agriculture and the Environment*, pp. 133–177. SSSA special publ. No. 23. Madison, WI.
Oremland, R.S. and J.P. Zehr. 1986. Formation of methane and carbon dioxide from dimethylselenide in anoxic sediments and by a methanogenic bacterium. *Appl. Environ. Microbiol.* 52:1031–1036.
Oremland, R.S., J.T. Hollibaugh, A.S. Maest, T.S. Presser, L.G. Miller, and C.W. Culbertson. 1989. Selenate reduction to elemental selenium by anaerobic bacteria in sediments and culture: biogeochemical significance of a novel, sulfate-independent respiration. *Appl. Environ. Microbiol.* 55:2333–2343.
Peck, H.D. Jr. 1959. The ATP-dependent reduction of sulfate with hydrogen in extracts of *Desulfovibrio desulfuricans. Proc. Natl. Acad. Sci. USA* 45:701–708.
Postgate, J.R. 1952. Competitive and non-competitive inhibitors of bacterial sulfate reduction. J. *Gen. Microbiol.* 6:128–142.
Presser, T.S. and I. Barnes. 1984. *Selenium Concentrations in Waters Tributary to and in the Vicinity of the Kesterson National Wildlife Refuge, Fresno and Merced Counties, California.* Report 84-4122, 1-25. U.S. Geological Survey, Washington, D.C.
Quinn, N.W.T. 1989. A screening study of management alternatives for agriculture for selenium drainage reduction in the San Joaquin Valley of California. In: L.W. Jacobs (ed.) *Selenium in Agriculture and the Environment*, pp. 213–233. SSSA special publ. No. 23. Madison, WI.
Reamer, D.C. and W.H. Zoller. 1980. Selenium biomethylation products from soil and sewage sludge. *Science* 208:500–502.
Reed, R.H. 1986. Halotolerant and halophilic microbes. In: R.A. Herbert, and G.A. Codd (eds.) *Microbes in Extreme Environments*, pp. 55–81. Special publication of the Society for General Microbiology, 17. Academic Press, London.
Rosenfeld, I. and O.A. Beath. 1964. *Selenium. Geobotany, Biochemistry, Toxicity and Nutrition.* Academic Press, New York.
Rotruck, J.T., W.G. Hoekstra, and A.L. Pope. 1971. Glucose-dependent protection by dietary selenium against haemolysis of rat erythrocytes in vitro. *Nature New Biol.* 231:223–224.
Saiki, M.K. 1985. Concentrations of selenium in aquatic food-chain organisms and

fish exposed to agricultural tile drainage water in selenium and agricultural drainage: implications for San Francisco Bay and the California environment. Proc. of the Second Selenium Symposium, Berkely, Calif., pp. 25–33.

Sarathchandra, S.U. and J.H. Watkinson, 1981. Oxidation of elemental selenium to selenite by *Bacillus megaterium*. *Science* 21:600–601.

Schwarz, K. and C.M. Foltz. 1957. Selenium as an integral part of Factor 3 against dietary necrotic liver degeneration. *J. Am. Chem. Soc.* 79:3292–3293.

Shamberger, R.J. 1983. *Biochemistry of Selenium.* Plenum Press, New York.

Shrift, A. 1954. Sulfur-selenium antagonism. I. Antimetabolite action of selenate on the growth of *Chlorella vulgaris*. *Am. J. Bot.* 41:223–230.

Shrift, A. 1973. Metabolism of selenium by plants and microorganisms. pp. 763–814. In: D.L. Klayman and W.H.H. Gunther (eds.) *Organic Selenium Compounds: Their Chemistry and Biology*. Wiley, New York.

Skorupa, J.P., D.L. Roster, W.L. Hohman, R.G. Stein, and D. Welsh. 1990. Assessment of drainwater contaminant impacts on breeding waterbirds in the Tulare basin, California. Agricultural evaporation ponds. Abstracts. March 1990 UC Salinity/Drainage Task Force Annual Research Conference, Sacramento, CA.

Sørensen, L.H. 1974. Rate of decomposition of organic matter in soil as influenced by repeated air drying-rewetting and repeated additions of organic material. *Soil Biol. Biochem.* 6:287–292.

Stadtman, T.C. 1980. Bacterial selenoenzymes and seleno-tRNAs. In: J.E. Spallholz, J.L. Martin, H.E. Ganther. (eds.) *Selenium in Biology and Medicine*, pp. 203–208. AVI Publishing Co., Westport, CT.

State Water Resources Control Board. 1988. *Evaporation Ponds for the Disposal of Agricultural Waste Water.* Final report. Submitted under Contract Number WRCB 5-190-150-0 to the SWRCB and in fulfillment of Contract Number C-060000-23-0 to the U.S. Environmental Protection Agency. June, 1988.

Sugimura, Y., Y. Suzuki, and Y. Miyake. 1977. The content of selenium and its chemical form in seawater. *J. Oceanogr. Soc. Jap.* 32:235–241.

Tanji, K. and M. Grismer. 1988. *Evaporation Ponds for the Disposal of Agricultural Waste Water.* Submitted to S.W.R.C.B., Contract No. 5-190-150-0 and U.S. E.P.A., Contract No. C-060000-23-0. Dept. of Land, Air and Water Resources, University of California, Davis.

Thomassen, Y. and J. Aaseth. 1989. Human tissues. In: M. Ihnat. (ed.) *Occurrence and Distribution of Selenium*, pp. 169–212. CRC Press, Boca Raton, Florida.

Thompson-Eagle, E.T. and W.T. Frankenberger, Jr. 1990a. Volatilization of selenium from agricultural evaporation pond water. *J. Environ. Qual.* 19:125–131.

Thompson-Eagle, E.T. and W.T. Frankenberger, Jr. 1990b. Protein-mediated selenium biomethylation in evaporation pond water. *Environ. Toxicol. Chem.* 9:1453–1462.

Thompson-Eagle, E.T. and W.T. Frankenberger, Jr. 1991a. Selenium biomethylation in an alkaline, saline environment. *Water Research.* 25:231–240.

Thompson-Eagle, E.T., W.T. Frankenberger, Jr., and K.E. Longley. 1991b. Removal of selenium from agricultural drainage water through soil transformations. pp. 169–186. In: A. Dinar and D. Zilberman (eds) *The Economics and Management of Water and Drainage in Agriculture.* Kluwer Academic Publishers, Norwell, MA.

Thompson-Eagle, E.T. and W.T. Frankenberger, Jr, and U. Karlson. 1989. Volatilization of selenium by *Alternaria alternata*. *Appl. Environ. Microbiol.* 55:1406–1413.

Tilton, R.C., H.B. Gunner, and W. Litsky. 1967. Physiology of selenite reduction by enterococci. I. Influence of environmental variables. *Can. J. Microbiol.* 13:1175–1185.

Torma, A.E. and F. Habashi. 1972. Oxidation of copper (II) selenide by *Thiobacillus ferrooxidans*. *Can. J. Microbiol.* 18:1780–1781.

Trelease, S.F. 1945. Selenium in soils, plants and animals. *Soil Sci.* 60:125–130.

Ullrey, D.E. 1981. Selenium in the soil-plant food chain. In: J.E. Spallholz et al. (eds.) *Selenium in Biology and Medicine*, pp. 176–191. AVI Publishing Co., Westport, CN.

Underwood, E.J. 1981. *The Mineral Nutrition of Livestock*. CAB Books, London. England.

U.S. Bureau of Reclamation. 1984. *Kesterson Reservoir and Waterfowl*. U.S. Bur. Reclam. Inf. Bull. 2.

U.S. Bureau of Reclamation. 1988. *Air Quality Impacts of Enhanced Selenium Volatilization at Kesterson Reservoir*. Prepared by CH2M Hill, May. Sacramento, CA.

U.S. Department of Health, Education and Welfare. 1962. *National Water Quality Network Annual Compilation of Data*. Public Health Service Publication 956, Washington D.C.

U.S. Environmental Protection Agency. 1975. *Federal Register*, vol. 40, no. 248, pp. 59566–59588. Dec. 24. National Interim Primary Drinking Water Standards Administrator: Russel E. Train.

Vokal-Borek, H. 1979. *Selenium*. Univ. of Stockholm Inst. of Physics Rep. 79-16. Stockholm, Sweden.

Wass, L. 1990. Location and physical characteristics of subsurface agricultural drainage evaporation basins. Agricultural evaporation ponds. Abstracts. March 1990 UC Salinity/Drainage Task Force Annual Research Conference, Sacramento, CA.

Weiss, H.V., M. Koide, and E.D. Goldberg. 1974. Selenium and sulfur in a Greenland ice sheet: relative to fossil fuel combustion. *Science* 172:261–263.

Weres, O., A. Jaouni and L. Tsao. 1989. The distribution, speciation and geochemical cycling of selenium in a sedimentary environment, Kesterson Reservoir, California, U.S.A. *Appl. Geochem.* 4:543–563.

Westcot, D., S. Rosebaum, B. Grewell, and K. Belden. 1988. *Water and Sediment Quality in Evaporation Basins Used for the Disposal of Agricultural Subsurface Drainage Water in the San Joaquin Valley, California*. Regional Water Quality Control Board Report, July 1988. Sacramento, CA.

Westlands Water District. 1989. *Water Conservation and Drainage Reduction Programs, 1987–1988*. Westlands Water District, 3130 Fresno St., P.O. Box 6056, Fresno, CA 93703.

Westlands Water District. 1990. Drainage update. *The Westside Water Rep.* 1:3.

Whelan, B.R. 1989. Uptake of selenite fertiliser by subterranean clover pasture in Western Australia. *Austral. J. Exper. Agric.* 29:517–522.

Whelan, B.R., D.W. Peter, N.J. Barrow, P. Young, D.J. Buscall, and V.C. Cox. 1989. The effect of selenium fertilizer on the selenium status of sheep grazing annual pasture in south-west western Australia. XVI International Grassland Congress, Nice, France, 1989, pp. 761–762.

Williams C. and Thornton I. 1972. The effect of soil additives on the uptake of molybdenum and selenium from soils from different environments. *Plant Soil* 36:395–406.

Williams, K.T., H.W. Lakin, and H.G. Byers. 1940. Selenium occurrence in certain soils in the United States, with a discussion of related topics: fourth report. *U.S. Dept. Agric. Tech. Bull.* 702:1–60.

Wood, J.M., F.S. Kennedy, and C.G. Rosen. 1968. Synthesis of methyl-mercury compounds by extracts of a methanogenic bacterium. *Nature* 220:173–174.

Woolfolk, C.A. and H.R. Whitely. 1962. Reduction of inorganic compounds with molecular hydrogen by *Micrococcus lactilyticus. J. Bacteriol.* 84:647–658.

Wu, L., R.M. Sachs, R.G. Burau, and E. Epstein. 1990. Study of effects of chloride and sulfate salinity on selenium accumulation by Se and salt tolerant genotypes of forage and native halophyte grass species. Abstract 88-1. UC Salinity/Drainage Task Force Annual Research Conference, March.

Yamamoto, L.A. and I.H. Segal. 1966. The inorganic sulfate transport system of *Penicillium chrysogenum. Arch. Biochem. Biophys.* 114:523–538.

Yläranta, T. 1983. Selenium in Finnish agricultural soils. *Ann. Agric. Fenn.* 22:122–136.

Zehr, J.P. and R.S. Oremland. 1987. Reduction of selenate to selenide by sulfate-respiring bacteria: experiments with cell suspensions and estuarine sediments. *Appl. Environ. Micobiol.* 53:1365–1369.

Zieve, R. and P.J. Peterson. 1981. Factors influencing the volatilization of selenium from soil. *Sci. Tot. Environ.* 19:277–284.

Zieve, R. and P.J. Peterson. 1985. Sorption of dimethylselenide by soils. *Soil Biol. Biochem.* 17:105–107.

Zieve, R. and P.J. Peterson. 1987. Selenium in plants: Soil versus atmosphere origin. In: G.F. Combs, Jr., O.A. Levander, J. E. Spallholz, and J.E. Oldfield (ed.) *Selenium in biology and medicine,* Part A, pp. 548–555. Van Nost and Reinhold, New York.

Zoller, W.H., E.S. Gladney, and R.A. Duce. 1974. Atmospheric concentrations and sources of trace metals at the South Pole. *Science* 183:198–200.

Reclamation of Mine Tailings

L.R. Hossner and F.M. Hons

I. Introduction

A variety of environmental problems, including air, land, and water pollution arise from mining activities. The detrimental impact of mining has been a topic of concern for many years. Abandoned mine tailings from a wide variety of industries are found around the world. The chemical com-

Advances in Soil Science, Volume 17

position of tailings depends on the minerals mined and the extraction technique. For example, cyanide compounds have been used in some gold (Au) operations. Concentrated sodium hydroxide is used in the extraction of aluminum (Al) from bauxite ore. These processes add to the potential reclamation difficulties at these sites. Chemicals reaching tailings ponds may undergo further reaction over an extended period of time changing their character. Alleviation of the chemical and physical limitations of tailings followed by reclamation of the tailings and the associated landscapes is a challenging task.

Tailings are defined here as the waste materials generated by the grinding and processing of ores and other materials containing economically retrievable minerals. Many processing methods involve grinding of rock and ores, chemical and/or physical removal of the desired commodity, and transportation of the wastes, often as slurry, to a tailings, or retention pond. These impoundments may range from a few to thousands of hectares in size. More than 99% of the original material may finally become tailings when utilizing low quality ores (Gemmell, 1977). Early evaluation of the potential chemical and physical problems associated with a particular site is essential to the timely and successful reclamation of that site.

Reviews on various aspects of reclamation of mine tailings have been offered including modeling of water quality (Rogowski et al., 1977), a review of the results of major research efforts in disturbed land reclamation in the southwest United States (Thames, 1977), a reclamation guide for planners and engineers (Bradshaw and Chadwick, 1980), a review of international literature as it relates to reclamation of mineland and mine tailings in Alberta, Canada (Sims et al., 1984), and a review of the chemistry and biology of solid wastes, dredge materials, and mine tailings (Salomons and Forstner, 1988).

II. Distribution of Tailings

The area of land disturbed by mining was estimated to be approximately 386000 ha per year in 1986 and by 2000 A.D., is projected to increase to 924000 ha per year (Soni and Vasistha, 1986). According to a 1966 estimate by the U.S. Bureau of Mines, more that 16000 inactive and/or abandoned underground mines with their associated tailings were scattered throughout the western United States (Harwood, 1979). Recent estimates have suggested that this list is incomplete and the actual count may be more than twice as high. For example, Colorado has more than 10 000 abandoned prospect sites. The total accumulated mineral waste in the United States in 1976, including overburden, submarginal ores, milling wastes, and stripmine spoils was estimated to be greater than 23 billion metric tons and to cover 2 million hectares of land (Donovan et al., 1976). Although tailings comprise only a fraction of the total reclamation effort, reclamation of tailings is often difficult because of adverse physical and chemical characteristics. Many of these locations contribute visual disruption of scenic landscapes and continuing environmental damage in the form of sediments and acid drainage into streams, lakes, and underground water supplies (Harwood, 1979). Abandoned mine workings and other industrial wastes associated with the exploitation of nonferrous metal ores are ubiquitous. The extent of derelict land affected by waste, mainly colliery waste, metalliferous smelter waste, mine spoil, chemical wastes, and china clay waste in the United Kingdom is estimated to be 100 000 ha (Gemmell, 1977). Derelict buildings and mineral spoil heaps are typical features of the landscape in areas of Central/North Wales, southwest England, and the Central/North Pennines (Johnson et al., 1977). In Wales alone there are some 2000 ha of metal contaminated derelict land comprising mainly waste products discarded during the separation of valuable metals, chiefly lead (Pb) and zinc (Zn). In Peninsular Malaysia, approximately 2% of the land (200 000 ha) is covered with abandoned tin tailings (Kho, 1970). The total area of land being mined in India is equivalent to one-third of that under agricultural production (Soni and Vasistha, 1986). Reclamation of orphaned tailings and planning for reclamation of new tailing materials will occupy the efforts of planners, scientists, and reclamationists for many years.

III. Environmental Consequences

Abandoned mine tailings have extremely diverse physical, chemical, and ecological conditions (Berg et al., 1975; Gemmell, 1973; Hunter and Whiteman, 1974; Ludeke, 1977; Shamshuddin et al., 1986). The tailings are normally variable in physical composition with depth and low in organic matter and essential plant nutrients, particularly nitrogen (N), phosphorus (P), and potassium (K). Acid drainage due to sulfide oxidation may be a consideration. Some tailings may have elevated levels of heavy metals or other toxic materials (Whitby and Hutchinson, 1974; Hutchinson and Whitby, 1974). Plant uptake of potentially toxic chemicals or heavy metals and their incorporation into the food chain are real concerns. Some tailings contain radioactive nuclides which can pose long-term health considerations. Erosion by wind and water with the associated environmental degradation is a universal concern associated with tailing materials (Johnson and Eaton, 1980; Sheppard et at., 1984).

IV. Limitations to Tailings Reclamation

There are several potential soil limitations to plant establishment and growth on mine tailings. Each site must be evaluated separately to identify adverse substrate characteristics prior to preparation for revegetation. At many sites there is evidence of two or more adverse factors and it is often the interaction of these factors that determines successful reclamation as measured by plant establishment and vegetative growth.

A. Acidity

Mining often exposes sulfide-bearing minerals (pyrite, pyrrhotite, chalcopyrite, arsenopyrite, cobaltite) to the atmosphere. Overburdens adjacent to mineral concentrations often contain low concentrations of pyritic minerals that are not economically retrievable. Metal processing also does not remove all pyritic minerals and tailings, therefore, they often have significant sulfide concentrations (Berg et al., 1975; Fuller and Lanspa, 1975; Sorensen et al., 1980). Oxidation of pyrites and production of acid can be illustrated using iron (Fe) sulfide as an example in the following general equations:

$$2FeS_2 + 2H_2O + 7O_2 = 2FeSO_4 + 2H_2SO_4 \quad (1)$$

$$2FeSO_4 + H_2SO_4 + \frac{1}{2}O_2 = Fe_2(SO_4)_3 + H_2O \quad (2)$$

$$Fe_2(SO_4)_3 + 6H_2O = 2Fe(OH)_3 + 3H_2SO_4 \quad (3)$$

$$2FeS_2 + \frac{15}{2}O_2 + 7H_2O = 2Fe(OH)_3 + 4H_2SO_4 \quad \text{(Summary)}$$

Under conditions found in acid mine tailings, oxidation of ferrous iron by oxygen (Reaction 1) is much slower than the oxidation of iron disulfide by ferric iron (Reaction 4).

$$FeS_2 + 14Fe^{3+} + 8H_2O = 15Fe^{2+} + 2SO_4^{2-} + 16H^+ \quad (4)$$

This reaction is catalyzed at pH values of approximately 3 by iron oxidizing *Thiobaccillus ferroxidans* bacteria, which greatly speed the oxidation of ferrous (Fe^{2+}) to ferric (Fe^{3+}) iron thereby regenerating the supply of ferric ions in solution (Nordstrom, 1982).

The amount of acidity produced by a given sample of FeS_2 over time is a function of crystal structure, surface area, temperature, oxygen concentration, water partial pressure, pH, ferrous/ferric adsorption ratio, and total Fe concentration, bacteria, adsorbed impurities, and flushing frequency (Caruccio et al., 1988; Pugh et al., 1981, 1984; Arora et al., 1978).

Formation of sulfuric acid decreases the pH of the tailing environment and results in increased solubility of metals and minerals that may be present. An additional process that may account for acidic tailings is the direct addition of acid in the processing of ore for product extraction.

A baseline study in southern Arizona on abandoned mines which were worked from the 1600s to 1949 for silver (Ag), Pb, and Zn revealed pH measurements of tailings drainage ranging from 2.9 to 4.2 directly below three sampled mines (Dean and Fogel, 1982). Drainage from the tailings contained maximum concentrations of 4200 mg sulfate L^{-1}, 1860 mg Fe L^{-1}, and 286 mg Mn L^{-1}.

Selection of tolerant plant species, and even genotypes within species, is frequently needed to revegetate strongly acidic mine tailings. Aluminum toxicity is the most important growth limiting factor for plants in many acid soils and mine spoils (Foy, 1974, 1983). A tolerant plant such as limpograss, *Hermarthria altissima*, (poir) Stapf & C.D. Hubb (PI 364344), was found to be exceptionally tolerant to Al in an Al-toxic Tatum subsoil (pH 4.1), in an acid mine spoil (pH 4.0), and in nutrient solutions containing 0 to 24 ppm Al added at an initial pH 4.5 or 4.0 (Foy and Oakes, 1984).

Traditional guidelines used for applying limestone to neutralize soil acidity in agricultural soils often do not apply to highly acidic sulfidic tailings (Sorensen et al. 1980; Carrucio et al., 1988). Hydrolysis of Al species is chiefly responsible for acidity in agricultural soils. Methods, such as the SMP buffer method (Shoemaker et al., 1961) were developed for lime requirement determinations in soils based primarily on Al hydrolysis. These methods tend to grossly underestimate potenial acidity where sulfide oxidation contributes substantial acidity. Estimates of potential acidity, from the oxidation of pyritic minerals, and neutralization potential must also be included to evaluate the residual acid potential of the tailing (O'Shay et al., 1990; Sobek et al., 1978; Sorensen et al., 1980). Oxidation of pyritic minerals can also contribute significant soluble salts to tailings.

B. Salinity and/or Sodicity

Excess concentrations of soluble salts are present in many tailing materials. In some instances, the salt may consist largely of sodium (Na) which may introduce additional problems in reclamation of the tailings. These salts commonly accumulate as the ore body is processed and are concentrated by recycling of water. In many cases, leaching of at least a portion of the salt from the material must be accomplished before plants can be established and grown.

Use of saturation percentage and saturation extracts for evaluating soil, overburden, and mine spoils in mined-land studies has been recommended by a number of authors (Dollhopf et al., 1980; Merrill et al., 1980; Merrill et al., 1983; U.S. Department of Agriculture, 1979; Schafer, 1979). Saturation percentage is a useful parameter for detecting low water-holding capacity tailings and estimating the degree of sodic hazard as indicated by the sodium adsorption ratio (SAR). A saturation percentage of 25 has been used as an indicator point for low water-holding capacity and 80% to 90% saturation has been used as an indicator for swelling tendency associated with sodic hazard (Merrill et al., 1987).

C. Nutrient Deficiencies

Tailings are almost universally deficient in N. Many tailings are deficient in P and are commonly deficient in K. In addition, secondary and micronutrients are sometimes deficient. Diagnosis of potential nutrient deficiencies and application of adequate quantities of essential elements are necessary before revegetation can be successful. The addition of a complete fertilizer at normal agricultural rates enormously improves the growth of both native and introduced plants on mine tailings. In many cases, native tolerant plants can be made to grow on previously bare mine soils only with the addition of a complete fertilizer (Antonovics et al., 1971). Warman (1988) reported that high levels of N-P-K fertilizer should be applied and maintained for successful revegetation of Pb-Zn wastes. These results were consistent in both pot and field experiments. Numerous studies have shown that if the major nutrients are at high levels, the harmful effects of metals can be partially eliminated (Johnson et al., 1977; Smith and Bradshaw, 1979). An element that appears to be important in reclamation is calcium (Ca), either by virtue of its direct effect, or by its influence on soil pH. Observations that vegetation types and metal uptake are correlated with the Ca content of the soil have led some to conclude that soil pH and Ca can be more important than the heavy metal concentration. Since Ca may alleviate the uptake and toxic effects of Zn, the Ca/Zn ratio has been proposed as a possible measure of Zn availability (Antonovics et al., 1971).

D. Toxic Ions

Toxic ions are frequently present in tailings in sufficient concentrations to prevent revegetation unless considerable amelioration is undertaken. Heavy metals decrease root respiration, water and nutrient uptake, and inhibit cell mitosis in root meristematic regions (Gemmell, 1977). Heavy metals reduce enzymatic activity and the microbial and microfaunal populations in soils so that the effects of heavy metals (i.e., copper, iron, lead, manganese, nickel, and zinc) on higher plants could be due partly to their inhibition of soil enzymes (Clark and Clark, 1981).

Toxic ions may also contaminate soils and waters adjacent to smelters (Rutherford and Bray, 1979) or tailings through seepage, runoff waters, and eroded sediments (Johnson and Eaton, 1980; Sheppard et al., 1984). Soils contaminated with heavy metals differ from each other and from normal soils in many ways other than metal content. The nutrient status, organic matter content, and texture may all affect the number and types of plants growing on the soil. Metal toxicities or imbalances are common in tailings. Copper (Cu), Pb, and Zn are the most extensively mined metals in the world and often occur in waste materials at concentrations toxic to plants (Jennett and Linneman, 1977; Wong, 1986). These metals are naturally present in the ore and extraction techniques commonly increase the acidity or alkalinity of the tailings and increase the concentration of metal ions. Some of the modern waste materials are less toxic that in the past. The total toxic metal concentration rarely exceeds 1% by weight because of improvements in mining procedures and processing methods (Wong, 1986).

Table 1. Total concentrations of copper (Cu), lead (Pb), zinc (Zn), nickel (Ni), and manganese (Mn) in mine tailings compared with the normal concentration range of these elements in soil

	Element					
	Cu	Pb	Zn	Ni	Mn	Reference
			($mg\ kg^{-1}$)			
			Mine tailing			
Asbestos	4	—	70	462	—	Ellery and Walker (1986)
Lead	113	20 530	1 600	—	—	Simon (1978)
Lead	42 000	1 200	1 250	—	—	Nunn and Riches (1978)
Lead-zinc	—	12 300	37 940	—	—	Morrey et al. (1984)
Lead-zinc	625	14 000	34 000	219	—	Johnson et al. (1977)
Copper	2 060	327	124	—	—	McNeilly and Johnson (1978)
Copper	1 740	—	125	95	—	Lopez and Lee (1977)
Copper	890	7	73	33	490	Ruppin (1987)
Copper	15 400	29 900	20 200	—	—	Simon (1978)
Copper-nickel	440	—	43	330	250	Crowder et al. (1982)
Copper-lead-zinc	99	1 125	255	—	1 325	Dean and Fogel (1982)
Zinc	57	12 930	22 700	—	—	Simon (1978)
Metal	1 329	—	101	1 323	334	Hogan et al. (1977)
Metal	—	21 320	1 273	—	—	Williams et al. (1977)
Metal	—	35 833	19 376	—	—	Simon (1978)
			Normal soil range			
Soil	2–100	2–200	10–300	5–500	20–3 000	Lindsay (1979)

There is a paucity of data available in the literature on toxicity limits to various plants and practically no information on interacting toxicities. Data presented in Table 1 indicate concentrations of *total* metal in a number of metalliferous mine tailings and compare these values to normal soil values reported by Lindsay (1979). Limited research and a lack of agreement appears to exist for methods to determine available metal concentrations of mine tailings. Availability of metals to plants will depend on such properties as pH, redox potential, organic content, and crystalline properties of metal compounds.

Two general approaches have been used in the revegetation of tailings with toxic levels of heavy metals. One approach is to add sewage sludge, compost, or similar wastes to the surface of the tailings site (Goodman et al., 1973; Harwood et al., 1987). Included are application of inert substrates (Johnson et al., 1977), topsoil treatment (Day and Ludeke, 1980a, 1982), and amendments such as sewage sludge and fly ash (Bergholm and Steen, 1989; Brierly, 1956; Gadgil, 1969; Goodman et al., 1973; Ruschena et al., 1974; Street and Goodman, 1966; Wong, 1986). The second approach is to seed genotypes of grass collected from old metal mine sites with a genetic tolerance to heavy metal contamination (Bradshaw et al., 1978; Drew and Reilly, 1972; Goodman et al., 1973; Hogan et al., 1977; Morrey et al., 1984; Palaniappan, 1974; Simon, 1978; Smith and Bradshaw, 1979; Wong, 1986). Stabilization of toxic materials by vegetation is generally superior to alternative techniques (Williamson and Johnson, 1984). Limited plant rooting zones and the tendency for plant roots to penetrate impermeable barriers and grow into underlying toxic materials have been significant problems with this approach, however.

Selection of plant species for tailings reclamation is based on several criteria: (1) chemical and physical properties of the tailings, (2) geographic location and climatic characteristics, (3) elevation, (4) season of seeding, (5) compatibility with other vegetation, (6) topographic exposure, and (7) land use objectives. If selected plant species are not compatible with one or more of the above criteria, then revegetation is likely to fail.

Soil properties which affect the proportion of metal which is either in solution or exchangeable and which therefore may determine the degree of toxicity of a soil for plant growth include clay and organic matter contents, cation exchange capacity (CEC), pH, and the concentrations of Ca and P. Alvarez et al. (1974) showed that on the spoil heaps of an area of abandoned Pb, Zn, and Ag mines in Utah, the amounts of vegetation and the density of species were correlated with levels of P and that the species were also distributed in relation to levels of Ca. Plant distribution on heavy metal minetips and outcrops was found to be related to levels of K, P, or Ca in a number of studies (Smith and Bradshaw, 1979).

Physical barriers to prevent leaching of soluble metal constituents have been used to reduce surface- and groundwater pollution. Covering tailings from Cu, Ag, and Au smelters with clay layers to prevent leaching, combined with a rock layer to prevent capillary rise and a cap of soil for revegetation, have successfully reduced pollution of river water at the Captains Flat Mine in New South Wales, Australia (Craze, 1977a, b). Experimental embankments on this site consisted of dump material covered with a 15-cm layer of compacted clay to act as an impervious layer and covered with 45 cm of rockfill (shale) and 30 cm of soil to support plant growth (Keane and Craze, 1978). Mini-embankments on 1-in-3 and low slopes were used successfully to establish vegetation on the reshaped and covered mine dumps. Attempts to eliminate or reduce acidic, heavily mineralized water containing considerable quantities of soluble heavy metals (Cu, Fe, Zn) in drainage from the Sheldon Mine Complex in Arizona by landform modification were largely unsuccessful. Landform reclamation and revegetation, however, have reduced erosion of the mine spoils and have esthetically improved the area (Lampkin and Sommerfeld, 1981). Failure of the reclamation to reduce heavy metal pollution may be due, in part, to the ability of the reclaimed spoils to retain water on-site. Rapid surface runoff for erosion control was prevented by channels and furrows designed to slow water movement. During prolonged rainfall and/or snowmelt, infiltration may increase and the spoils may become saturated, resulting in an increase in seepage through the wastes rather than being channeled away on the surface.

E. Physical Limitations

Tailings generally possess one or more chemical and/or physical properties that limit successful revegetation. The texture of tailings ranges from sand to clay depending on the compostion of the original material, stratification, and method of slurry entry into the tailings pond. Coarser fractions generally are poorly buffered, devoid of organic matter, deficient in nutrients, structureless, prone to crusting, and have a low water-holding capacity (Verma et al., 1977).

Tailings exposed to direct solar radiation can have temperatures of 55° to 65° C at a 1- to 2-cm depth when air temperatures are 35° to 38° C (Shetron, 1983). High potential evapotranspiration and low water-holding capacity suggest that water deficits limit revegetation of coarser tailings, especially in arid regions. Finer textured tailings usually exhibit greater water-holding capacity, but water infiltration is often limited by poor structural characteristics.

Tailings are commonly handled as a water slurry. They are transported and deposited into dammed artificial ponds where the particles settle out on the basis of size (Murray, 1977). Stratification based on particle size frequently occurs within the tailings pond. Larger, sand-sized particles settle near the slurry inlet, while clay-sized particles are located at the outlet end where water is ponded. Crusting, cracking, and a general lack of structure are common characteristics of mine tailings brought about by differences in texture, lack of organic matter and variable mineralogy. Extremely high clay or sand contents are commonly encountered and add to the physical limitations of the tailings. The fines (slimes) have been found to be relatively impervious to water, difficult to leach, and have poor physical properties. In contrast, the coarser fractions of tailings have a low available water capacity and are commonly deficient in plant nutrients. Physical constraints to revegetation of tailings must be modified or removed prior to successful revegetation.

Bulk densities of mine tailings are sometimes elevated due to compaction. Root penetration and moisture stress due to limited rooting volumes generally becomes a problem with dry bulk densities above 1.5 Mg m^{-3} (Craze, 1977a). However, Crews (1984) reported that compaction of a loam mine-soil to a density of 2.0 Mg m^{-3} may not be detrimental to the growth and yield of KY-31 tall fescue (*Festuca arundinacea* Schreb), but such compaction may drastically reduce the growth of sericea lespedeza (*Lespedeza cuneata*). Roots of the tall fescue were largely confined to the upper 3 cm of compacted soil. Shetron and Spindler (1983) found that alfalfa (*Medicago sativa* L.) roots concentrated in clay strata of mixed tailings, presumably because of greater available water. Sand-textured tailings in this study had greater bulk densities than clay-sized tailings.

Although tailings ponds may have been abandoned for years and appear dry on the surface, suspended colloids may still be present at depth, resulting in the submergence of heavy equipment when driven onto the surface. Trenches or drainage pipes may be required to remove excess water from tailings (Harwood, 1979; Ward, 1987).

F. Dust

An almost universal problem associated with unreclaimed tailings is blowing dust. These wind-blown silicious particles pollute air, streams, and lakes, abrade vegetation, decrease esthetics, and create health problems for communities near these wastes. Approximately 40% of mineral waste is fine-sized material that requires some kind of stabilization if air and water pollution are to be controlled (Day and Ludeke, 1973). Blowing dust has been reported as a problem for asbestos dumps (Ellery and Walker, 1986), red muds, and sands from bauxite processing (Ward, 1987; Coffey et al., 1986), Cu and Ni tailings (Peters, 1984; McNeilly and Johnson, 1978; Peterson and Nielson, 1973; Day and Ludeke, 1973), Fe tailings basins (Jordan and Dewar, 1985), china and bentonite clay wastes (Jeffries et al., 1981; Smith et al., 1985), and for Au, Ag, and Zn tailings (Harwood, 1979; Creswell, 1973). These wastes normally are susceptible to both wind and water erosion. Methods of dust abatement include suppressants such as bitumen or paper mulches, rock mulching, irrigation, vegetative ground cover, and windbreaks (Ward, 1987). Coffey et al. (1986) proposed a model to relate wind speed-up, as influenced by the disposal area profiles, to an erosion ratio which can predict the relative rates of wind erosion. This allows a strategy for minimizing erosion of mine tailings through shape selection as a part of the overall environmental plan.

V. Mine Tailings Reclamation

Reclamation of specific mine tailings will be discussed in the following sections. The discussion was limited to tailings where adequate information was available in the literature to allow some interpretation of the major problems and proposed reclamation strategies. Some tailings are discussed under one heading (i.e., Pb-Zn) since the metals commonly occur in the same ore body and are discussed together in the literature.

A. Asbestos

Chrysotile, or fibrous serpentine, is the mineral normally mined for asbestos production. The combination of this mineral's fibrous morphology, low heat conductivity, high electrical resistance, and chemical inertness have contributed to its wide industrial application. The Thetford Mines, Black Lake, and Asbestos districts of southeastern Quebec are the largest producers of chrysotile in the world. The main problems associated with vegetation establishment in asbestos tailings are extremely alkaline conditions, low nutrient concentrations, and surface crusting. The first two problems can be counteracted by gypsum and fertilizer and the latter problem by topsoiling to improve water-holding capacity and microbial activity.

Blowing dust has been reported as a problem for asbestos dumps (Ellery and Walker, 1986) as it has for many types of tailings. Because of health problems associated with prolonged asbestos exposure, dust from chrysotile tailings can have considerable impact on nearby communities and should be abated. Although asbestos tailings in South Africa had pH values of 9.6, available heavy metal concentrations in the tailings were still greater than the limits normally tolerated by plants (Ellery and Walker, 1986). Decreasing the pH to 8.0 further exacerbated the problem. These same authors demonstrated significant plant dry matter increases to N, P, and K application on these tailings. Nutrient imbalances may have been present because magnesium (Mg) was in extremely high concentration compared with Ca and K (Proctor and Woodwell, 1975). Applications of gypsum ($CaSO_4$) have been recommended to balance the Ca:Mg ratio and to increase the long-term success of revegetation (Moore and Zimmerman, 1977; Meyer, 1980). Magnesium is the principal metallic constituent of chrysotile. *Sporobolus spicatus* (Vahl) Kunth exhibited the most potential as a pioneer species for vegetation of asbestos tailings because of rapid stolon production and less dependence on fertilizers for growth (Ellery and Walker, 1986).

B. Bauxite

Bauxite is a heterogeneous mixture of aluminum oxides and various impurities of iron oxides, silica, aluminum silicates, and titanium oxides. Because of its versatility, Al consumption worldwide is exceeded only by Fe (McCawley and Baumgardner, 1985). Bauxite mining and processing can create several challenges for reclamation. Overburden spoils from bauxite mining are often acidic, while residue sands and red muds from bauxite processing are highly alkaline. Phillips and Spooner (1983) reported that bauxite spoils with pH 3.0 in Arkansas required a limestone rate of 48.9 t ha^{-1} for acid neutralization. The low pH was caused by iron sulfide oxidation. Soluble salts were also high. The spoil was low in fertility and microbial population, coarse-textured, and droughty. Pearl millet [*Pennisetum typhoides* (Burm.) Stapf and C.E. Hubb.] produced the most dry matter, followed by common bermudagrass (*Cynodon dactylon* L. Pers.), and several legumes. Bauxite spoils are highly erodible and a minimum ground cover of 80% is recommended.

Red muds and residue sands are common waste products of bauxite pro-

Table 2. Chemical characteristics of residue sand and red mud from bauxite processing[a] (after Meecham and Bell, 1977a)

Parameter	Residue sand	Red mud
pH in water (1:2)	9.75	11.2
Electrical conductivity (dSm^{-1})	30.2	37.7
Extractable cations ($cmol\ kg^{-1}$)		
Na^+	40.3	93.5
Ca^{+2}	4.6	0.4
Mg^{+2}	1.0	2.4
K^+	5.3	10.4
Cation exchange capacity ($cmol\ kg^{-1}$)	11.1	17.3
Sodium adsorption ratio	52.8	28.3
Exchangeable sodium percentage	88	81

[a] Extracted with 1 *M* NH_4OAC

cessing. The Bayer process, the most common method of bauxite processing, utilizes NaOH to dissolve alumina and form a sodium aluminate solution. Insoluble impurities are filtered and transported to retention ponds as slurries. Depending on texture, these impurities are referred to as residue sands and red muds. Both wastes are highly alkaline (pH often greater than 10) and are also often saline (electrical conductivity commonly 30 dSm^{-1} or greater) and sodic (Meecham and Bell, 1977a; Fuller et al., 1982; Fuller and Richardson, 1986; Ward, 1987) (Table 2). Both Meecham and Bell (1977a) and Ward (1987) reported residue sands with pH values of almost 10 and red muds with $pH > 11$. Red mud from the former study had an electrical conductivity (EC) of 30.2 dSm^{-1} and an exchangeable sodium percentage (ESP) of 81, while red mud from the latter study exhibited an EC of 36 dSm^{-1} and an ESP of 96. Residue sands generally have low water-holding capacities, while red muds have low infiltration and hydraulic conductivity because of fine texture and high Na content. Both residue sands and red muds are also commonly reported to be low in plant available N, P, and several micronutrients. Potassium can also be moderately limiting (Ward, 1983).

Leaching of residue sands can sufficiently reduce pH, salinity, and sodicity so that vegetation can be supported (Meecham and Bell, 1977a; Ward, 1987) (Table 3). Leaching of red muds is more difficult because of dispersion and poor hydraulic conductivity. Barrow (1982) showed that gypsum addition and subsequent leaching can reduce pH, salinity, and sodicity and increase flocculation of red mud. Barrow (1982) further reported decreased P adsorption and increased P availability with gypsum addition to red mud.

Ward (1983) and Barrow (1982) found that red mud–gypsum mixtures added to sandy soils in Australia improved the growth of annual legumes

Table 3. Effect of leaching on chemical characteristics of residue sands from bauxite processing (after Meecham and Bell, 1977a)

Parameter	Prior to leaching	After leaching
pH	9.8	8.8
Electrical conductivity (dSm^{-1})	47.6	2.0
Sodium adsorption ratio	55.8	6.0
Exchangeable sodium percentage	49.0	5.5

compared with soil alone. Ward (1987) recommended that a 1-m layer of well-drained residue sand be applied onto red muds to improve revegetation success. Vegetation was essentially nonexistent where red mud surfaces were exposed. Fuller et al. (1982) demonstrated that sewage sludge was a suitable amendment for the revegetation of red muds in Alabama. These authors attributed the benefits of sewage sludge addition to reduced pH, increased nutrient availability, and increased Al complexation. Other organic amendments such as wheat (*Triticum aestivum* L.) straw, paper pulp waste, glucose, and pine (*Pinus taeda* L.) needles and complete nutrient additions failed to produce consistent vegetation cover.

Fuller et al. (1982) used the alkali-tolerant grasses, desert saltgrass (*Distichlis spicata* var. stricta), alkali sacaton (*Sporobolus airoides*), and tall

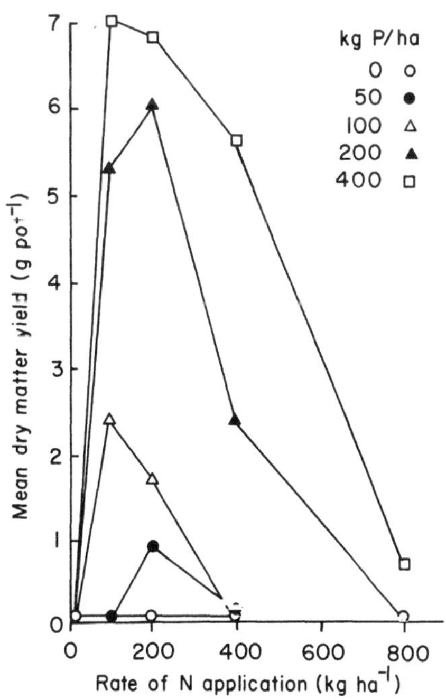

Figure 1. Rhodes grass response on red sand to nitrogen application at various levels of phosphorus (after Meecham and Bell, 1977b)

wheatgrass (*Agropyron elongatum*) as successful revegetating species. Meecham and Bell (1977b) reported excellent responses of Rhodes grass (*Chloris gayana* Kunth.) to added P and good responses to added N on residue sands in Australia (Fig. 1).

C. Clay

Although clay tailings and wastes do pose problems for reclamation, these difficulties are often innocuous compared with most metal tailings. The primary clays mined on a significant scale are bentonite and kaolinite (kaolin). Bentonite is a 2:1 expanding clay formed from the alteration of volcanic ash deposited in shallow marine seas. The United States is the world's largest producer of bentonitic clay, with Wyoming accounting for almost 70% of its production. Kaolinite is a 1:1 silicate clay mineral formed in acid environments. The soluble Al and Si released in such environments can recrystallize to form kaolinite. Kaolinite minerals are mined worldwide. The inner Coastal Plain of the southeastern United States contains one of the world's largest kaolin reserves as does southwest England (Bradshaw and Chadwick 1980).

Alkalinity, salinity, and sodicity can be problems in certain clay tailings. Sieg et al. (1983) showed that natural revegetation of nontopsoiled bentonite clay tailings in the northern High Plains of the U.S. was extremely slow. Schuman and Sedbrook (1984) demonstrated that wood residue could increase revegetation of such sites, primarily by reducing spoil density and increasing water infiltration and storage. Smith et al. (1985) worked with bentonite tailings in Wyoming with electrical conductivity of 13.4 dSm^{-1} and sufficient Na to be classified as sodic. The treatment receiving no wood residue had only 8% cover and a dry matter yield of 80 kg ha^{-1}. A uniform application of 90 kg P ha^{-1} combined with 90 Mg ha^{-1} wood residue and 2.5 kg N Mg^{-1} residue gave the greatest increase in seeded plant density, canopy cover and above-ground biomass (Smith et al., 1985). Second-year canopy covers of seeded perennial grasses were significantly increased for both native and introduced species by increasing rates of wood residue and fertilizer N (Fig. 2). Dollhopf and Bauman (1981) demonstrated that topsoiling increased plant survival, but because of shallow surface soil and other considerations, topsoiling was not practical for many situations. Bentonite also may contain high Pb, titanium (Ti), Mn, and other trace elements.

China clay is composed chiefly of the silicate clay, kaolinite. Compared with metal tailings, china clay wastes normally do not contain heavy metals or large quantities of oxidizable sulfides (Gemmell, 1977). Deficiencies of plant nutrients, especially N, P, and K, are often the principal limiting constraints to revegetation of such wastes. Nutrient deficiencies are often intensified by the coarse texture and low cation exchange capacity which allows rapid leaching. If china clay wastes become acidic, their low buffering capacity results in low lime requirements. Jeffries et al. (1981) reported

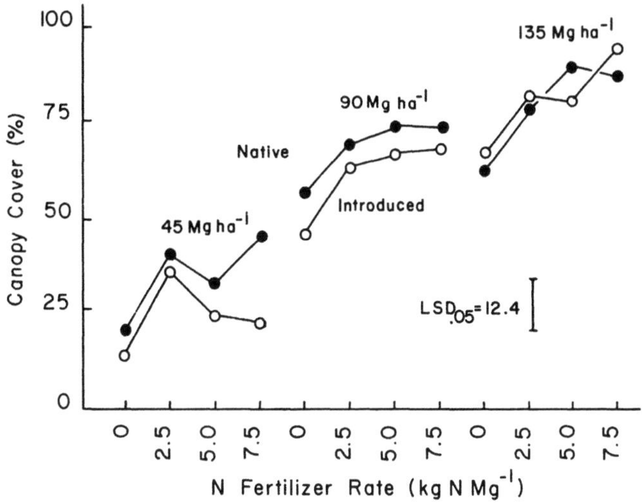

Figure 2. Second-year canopy cover of seeded perennial grasses (naive and introduced mixtures) in relation to N fertilization (kg N Mg^{-1} of wood residue) and species mixture on spoils amended at 45-, 90-, and 135-Mg ha^{-1} wood residue (after Smith et al., 1985)

on a china clay sand waste with pH 4.2 and low total concentrations of N, P, K, Ca, Mg, and Fe. The waste was amended with 1000 kg lime ha^{-1} and 30, 40, and 40 kg ha^{-1} of N, P, and K, respectively. *Lupinis perennis* was the highest yielding legume species on the amended tailings, while *Cytisus scoparius* and *Ulex europaeus* were the best adapted shrubs. Nitrogen transfer from the legume to associated grasses occurred within 2 years of sowing, with as much as 76 kg N ha^{-1} being transferred. Dwyer (1986) found that 10 cm of topsoil over refractory clay (kaolinite) tailings was necessary for rapid establishment of legume ground cover in New South Wales, Australia. The most suitable legume species were *Lolium perenne* and *Vicia dasycarpa*, while the best adapted trees were *Eucalyptus albens*, *E. pilligaensis*, *E. viridis*, *E. woollsiana*, and *E. torquata*. Tree species tended to establish better on nontopsoiled sites. Loblolly pine (*Pinus taeda* L.) height, stem diameter, and fresh weight were significantly increased by the addition of sewage sludge to kaolin spoil in Georgia (Berry and Marx, 1977). Sludge addition also increased mycorrhizal development and availability of all measured nutrients. As with other tailings, nonvegetated clay tailings are susceptible to severe wind and water erosion.

D. Copper, Gold, Nickel, and Silver

Silver (Ag), gold (Au), and copper (Cu) were mined extensively along Lynx Creek (near Prescott, Arizona) resulting in numerous waste dumps and mill tailing ponds that were abandoned after the ore was exhausted.

Table 4. Chemical characteristics of water saturation extracts of copper tailings (after Peterson and Nielson, 1973)

Sample location	Electrical conductivity	pH	B	Cu	Soluble ions Fe	Mn	Zn	Pb
	(dSm^{-1})				(mg L^{-1})			
New Mexico	9.5	2.2	10	>600	3000	24	21	>5
Montana	3.2	5.2	0.4	195	0.6	154	2.4	>5
Montana	2.1	7.6	0.2	<1	<0.5	0.6	0.2	<5
Arizona	3.4	7.4	0.1	<1	<0.5	<0.2	0.1	<5
Arizona	3.1	7.3	0.1	<1	<0.5	<0.2	0.2	<5
Utah	9.5	7.0	0.4	<1	0.6	<0.2	<0.6	<5

Drainage from the orphaned mine sites contributes toxic mineral and sediment pollution into Lynx Creek and eventually into Lynx Lake (Verma and Felix, 1977). Acid mine drainage and elevated levels of Cu, Fe, Mn, Zn, and SO_4 in the creek were the primary problem areas identified for treatment. Liming, topsoiling, fertilization, and hydromulch seeding resulted in successful revegetation over much of the area.

Unreclaimed Cu, Au, and Ag tailings are susceptible to both wind and water erosion (Creswell, 1973; Day and Ludeke, 1973, 1981; Harwood, 1979; McNeilly and Johnson, 1978; Peters, 1984; Peterson and Nielson, 1973). Heavy metal tailings frequently are or become highly acidic over time because of the presence of sulfide-containing minerals. Overburdens adjacent to mineral concentrations often contain low concentrations of iron sulfide minerals that are not economically retrievable or metal processing does not remove all iron sulfide minerals from the tailings. Therefore, they often exhibit significant sulfide concentrations (Berg et al., 1975; Sorensen et al., 1980). Additional acidity may be contributed by acid extraction of these ores during processing.

Metal ions are frequently present in sufficient concentrations in acidic Cu and Au tailings to inhibit or terminate plant growth (Wong, 1986) (Table 4). Heavy metals in tailings decrease plant root respiration, water and nutrient uptake, and inhibit cell mitosis in root meristematic regions (Gemmell, 1977). Toxic ions may also contaminate soils and waters adjacent to tailings through seepage, runoff waters, and eroded sediments. Peterson and Nielson (1973) reported that Cu tailings in the western United States contained up to 5700, 875, 420, and 140 mg kg^{-1} of water soluble Cu, Fe, Mn, and Zn, respectively. A portion of these tailings also exhibited electrical conductivities as high as 150 dSm^{-1}. Soil reactions ranged from 2.2 to 7.1, but most were pH 4.0 or lower. Leaching of tailings is often recommended to decrease soluble salts, acidity, and soluble toxic ions. Shetron (1983) reported available Cu, Fe, Mn, and Zn concentrations to be as high as 2444, 445, 668, and 32 mg kg^{-1} in Cu tailings from Michigan. This au-

Table 5. Total (dissolved + suspended) mean element concentrations of waters entering Lynx Creek (after Lampkin and Sommerfeld, 1981)

Element	Sheldon Mine pond seep	Spruce Creek	Rich Gulch Creek
		($mg\ L^{-1}$)	
P	0.4	0.1	0.2
Ca	176.4	35.4	90.1
Mg	53.1	12.2	26.6
Na	13.4	9.6	14.6
K	1.5	1.6	3.2
Cu	21.0	1.2	6.5
Fe	109.1	3.8	6.9
Mn	16.1	0.3	2.3
Zn	115.6	2.0	24.3

thor also suggested the difficulty in equating sufficiency and toxicity levels in tailings with those derived for soils. Lopez and Lee (1977) found that Cu could be approaching concentrations toxic to fish and aquatic invertebrates in a lake receiving Cu tailings drainage and sediments. McNeilly and Johnson (1978) reported available Cu and Pb concentrations of 2060 and 327 mg kg^{-1} for metalliferous tailings in Great Britain. Crowder et al. (1982) reported higher concentrations of Fe, Cu, and nickel (Ni) in tailings and vegetation growing on tailings than for soils. Metal concentrations were generally greatest after summer storms because of the flushing effect of rainfall on tailings that became well-oxidized during the summer. Lampkin and Sommerfeld (1981) reported on metal contaminated drainage from abandoned Au tailings in the Bradshaw Mountains of Arizona (Table 5). Drainage from the Sheldon Mine tailings pond contained average Cu, Fe, Mn, and Zn concentrations of 21, 109, 16, and 116 mg 1^{-1}, respectively, and was implicated in the failure of a trout fishery located in the same watershed. Iron sulfide-bearing tailings in Canada still contained DTPA-extractable concentrations of Cu, Fe, Mn, and Zn as high as 310, 364, 8, and 40 987 mg kg^{-1}, respectively, after cropping in the greenhouse (MacLean and Dekker, 1976). The pH of these tailings was 2.5. Sheets et al. (1982) recommended that Ni ore processed tailings be washed to reduce soluble salts and NH_4-N prior to revegetation, and that native soil be added to the surface to reduce crusting.

Besides inhibiting revegetation, plants grown on metal mine waste may contain metal concentrations high enough to retard organic matter decomposition, decrease humus formation, reduce soil urease activity, and support smaller microbial and microfaunal populations (Williams et al., 1977; Tyler, 1975). Microbes in metal contaminated sites have been reported to develop an increased tolerance to heavy metals (Griffiths et al., 1974).

Active acidity in metal tailings can normally be neutralized or even leached below the plant rooting zone (Creswell, 1973), but neutralization of potential acidity in tailings may require large quantities of limestone applied periodically as sulfide oxidation continues. Gold and Ag tailings from the Sheldon Mine in Arizona needed 75 t ha^{-1} of limestone for acid neutralization (Harwood, 1979), while Cu-molybdenum (Mo) tailings and spoils in Idaho required 44 to 72 t ha^{-1} for neutralization (Sorensen et al., 1980). Six sulfide-bearing metal mine tailings in Canada, ranging in pH from 1.9 to 2.9 yielded 10.8 to 37.0 cmol acid 1000 g^{-1} of sample when leached with 1 L water (MacLean and Dekker, 1976). The quantity of limestone required to neutralize these tailings in greenhouse studies ranged from 22.4 to 112 t ha^{-1}. Craze (1977a) used 63 t ha^{-1} of limestone to neutralize Cu, Zn, and Pb tailings in the Captains Flat Mining Area in New South Wales, Australia. The limestone neutralized acidity and the higher pH inhibited capillary rise of metal ions from tailings to overlying gravel and sand layers. Peters (1984) recommended 8.8 t ha^{-1} of limestone prior to seedling establishment on Cu and Ni tailings in Ontario, Canada, followed by additional liming in subsequent years. Ruppin (1987) described chemical changes over time in Cu tailings in New Guinea. The tailings were initially alkaline (pH 8.5) due to lime addition during processing, but became acidic (pH 4.5) within 4 to 6 years due to pyrite and chalcopyrite oxidation. As the tailings acidified, vegetation died and propagation of new vegetation became difficult. Foy and Oakes (1984) introduced a selection of limpograss [*Hermarthria altissima* (Poir) Stapf] for revegetation of acidic spoils. Although this grass is of tropical origin, it exhibited exceptional winter-hardiness, surviving temperatures of −20.5° C, and was recommended for the revegetation of strongly acid tailings and wastes at medium to high elevations in temperate climates.

Copper and Au tailings generally are low in available N and many also lack other plant essential nutrients. Gemmell (1977) showed that N, P, and K additions increased plant growth on Zn and Cu tailings in Great Britian, but sewage sludge addition further increased growth as compared with fertilization (Fig. 3). The author theorized that the added organic matter formed stable complexes with toxic metal ions and decreased their availability to plants. Verma et al. (1977) demonstrated that sewage sludge applied to Cu tailings in Arizona enhanced plant yield and survival compared with well water plus fertilizer N, P, and K added to equal these nutrients added in the effluent. MacLean and Dekker (1976) reported that a lime-digested sewage sludge mixture was an effective source of acid neutralization and nutrients on metal mine tailings in Canada. Day and Ludeke (1980a) stated that a deficiency of plant nutrients was a major limitation to revegetating Cu tailings in Arizona. Shetron (1983) showed that Cu tailings in Michigan were low in available N, P, K, Ca, and Mg. After 3 years of N (300 kg total N ha^{-1}) and P (200 kg total P ha^{-1}) additions, the tailings were still essentially devoid of available N and P. Crowder et al. (1982)

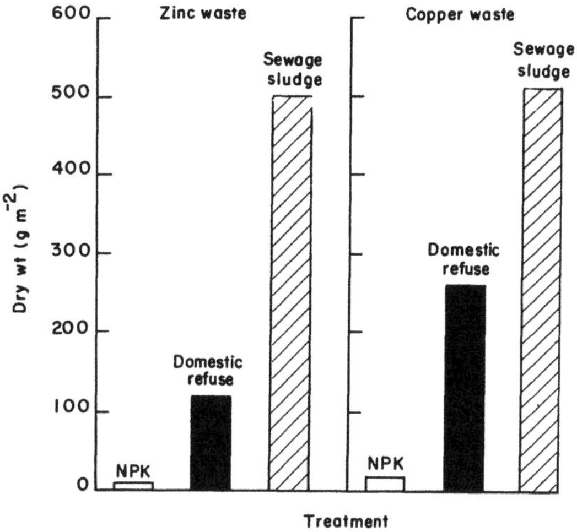

Figure 3. Effects of fertilizers and organic amendments on grass growth on copper smelter wastes (after Gemmell, 1977)

reported that regular fertilization of Cu mine tailings in Ontario, Canada, was necessary for vegetation establishment and long-term survival. McNeilly and Johnson (1978) recommended 119 kg N and 50 kg P ha^{-1} for the revegetation of Cu tailings in Great Britain. These authors further theorized that continued fertilization over time would likely be necessary for successful reclamation. Craze (1977a, b) found that 99, 53, and 84 kg N, P, and K ha^{-1} were optimal for establishing vegetation on Zn, Pb, and Cu contaminated tailings in Australia. Chambers et al. (1987) stated that heavy metal mining in alpine areas often results in wastes low in organic matter, with N often being the most limiting nutrient. Hume and August (1988) showed that white clover (*Triflolium repens*) responded to P addition up to 180 mg P kg^{-1} of Au and Ag tailings. Fuller and Lanspa (1975) suggested that tailings from a Pb-Zn-Cu mine in Arizona could serve as a Fe and Cu fertilizer source on calcareous soils.

Bradshaw (1952) reported on grass and herb species growing on tailings contaminated with high Cu, Pb, and Zn concentrations and suggested that these plants might be selected for revegetation of similar sites. McNeilly and Johnson (1978) researched the mineral nutrition of Cu tolerant browntop (*Agrostis tenuis* Sibth.) which had been developed commercially in Great Britain for the restoration of abandoned metalliferous spoils. Farrow et al. (1981) described the high Cu tolerance of *Agrostis canina* L. subsp. Montana Hartm. which was the only species to colonize Cu mine tailings in North Wales, even after 66 years. The addition of N, P, and lime increased seedling tillers and flowers, suggesting that correction of nutrient deficiencies could improve natural colonization of such sites. Shetron (1983) suggested additional research on potentially adapted species to determine the relationships between revegetation success and ions in tailings and to determine plant selectivity with respect to excesses of nutrients and metals. Desert tobacco (*Nicotinia glauca*) and Australian saltbush (*Atriplex semibaccata* R. Br.) were the tree and shrub species, respectively, resulting in the best growth and ground cover on Cu tailings in Arizona (Day and Ludeke, 1980a,b). Perennial ryegrass (*Lolium perenne*) and crested wheatgrass [*Agropyron desertorum* (Fisch. ex Link) Schult.] generally produced the least dry matter of all grasses grown on the above tailings, but resulted in the greatest ground cover (Day and Ludeke, 1982). Day and Ludeke (1973) showed in earlier work that vegetatively plugged giant bermudagrass (*Cynodon dactylon* var. ardius Harlan et de Wet) on Cu tailings produced greater vegetative cover 30 and 180 days after planting and greater tailings berm stabilization compared with broadcast seeding. Species that provided sufficient ground cover on Au mine tailings in South Africa included *Cynodon plectostachyus* (K. Schum.) Pilg., *Eragrostis curvula* (Schrad.) Nees, *Ehrharta calycina* Sm., and *Cynodon dactylon* (Creswell, 1973). Vegetative cover tended to decline over time, presumably because of decreased nutrient availability. Tree lucerne (*Cytisus proliferus*), Russell lupin (*Lupinus polphyllus*), and yellow tree lupin (*L. arboreus*) grew well on Au mine tailings in New Zealand, but dry matter yield of pasture legumes was increased by P fertilization. White clover dry matter increased from 5.4 to 8.9 t ha^{-1} and total N content increased from 180 to 310 kg N ha^{-1} with P addition. Tree growth was also improved by being associated with a vigorous legume species.

E. Iron

Iron tailings can be acidic, with problems similar to those previously described for metal tailings. Many Fe tailings, however, are neutral to alkaline, with infertility and low organic matter, cation exchange capacity (CEC), and water-holding capacity being major constraints to revegetation. Shetron and Spindler (1983) studied alfalfa establishment on alkaline ($pH \geq 8$) Fe tailings in the Iron Range of Michigan. Greatest concentrations of alfalfa roots and nodules were observed in fine textured strata within the tailings, presumably because of greater nutrient and water availability. Clay strata also exhibited lower bulk densities than sand or interstratified sand/silt tailings.

Shetron (1983) reported that 4-year-old alfalfa stands on Fe tailings were deficient in N, P, and Mg, while K and Ca were present in sufficient concentrations. Dry matter yields on sand-sized tailings were approximately 2 t ha^{-1}, while interstratified and clay-sized tailings produced 3 to 4 t ha^{-1}. Sand tailings retained less plant available water (5.7%) than clay (12.4%) or interstratified (23.4%) materials. The clay-sized tailings also had greater available Ca and a higher CEC compared with coarser tailings. A darker surface soil after 3 years of alfalfa indicated the formation of a weakly developed A horizon and demonstrated that pedogenic processes had commenced.

Jordan and Dewar (1985) reported on the reclamation of taconite (Fe) tailings in the Mesabi Range in Minnesota. Taconite is a lower grade of Fe ore which results in 2 tons of finely crushed tailings for each ton of taconite pellets produced. Approximately 8500 ha of tailings were at this location in 1980, with a projected area of 16 200 ha. These tailings were alkaline and not toxic, but were extremely infertile. Revegetation was also hampered by low cation exchange and water-holding capacities. Alfalfa and smooth brome (*Bromus inermis*) tended to dominate droughtier areas, while sweetclovers (*Melilotus* sp.), redtop (*Agrostis alba* L.), and Reed canarygrass (*Phalaris arundinacea* L.) were concentrated in wetter areas. Kochia (*Kochia scoparia*) rapidly colonized new sites, but quickly declined, possibly because of autotoxicity from decaying roots and leaves.

Vrabec (1985) described Fe ore mine spoils in Wisconsin which had been seeded with several legume species 12 years earlier. Approximately 87% of the spoil was schist with the remainder composed of sandstone, diabase, granite, hematite, magnetite, talc schist, and amphibolite. Plant available P and K were higher in the spoil than in native soils of the area. Exchangeable Ca and Mg and organic matter content were within acceptable limits for agricultural soils. Soil forming processes acting on the spoil during the 12-year period resulted in a fertile and productive mine soil.

F. Lead and Zinc

Mine wastes from Pb-Zn processing often have low pH, excessive concentrations of metals, and are devoid of vegetation. Rutherford et al. (1982) studied Pb-Zn tailings in Ontario, Canada, that had remained devoid of vegetation for >40 years. Tailings profiles were strongly acidic, low in organic matter, and exhibited profile characteristics of acid sulfate weathering (Van Breeman, 1982; Brinkman and Pons, 1973). Total S was about 20 times higher in the tailings compared with normal soils. Abandoned Pb-Zn mine sites in Oklahoma, characterized by pH values ranging from 7.0 to 7.4 and toxic levels of Pb (568 to 2575 mg kg^{-1}), Zn (2667 to 13 090 mg kg^{-1}) and cadmium (Cd) (11 to 27 mg kg^{-1}) were naturally colonized by a wide variety of species from Oklahoma forests and prairies (Gibson, 1982). These species were characterized as having efficient seed dispersal mechanisms and generally broad edaphic tolerance. Lead concentrations of sycamore (*Platanus occidentalis* L.) foliage ranged from 1.3 to 1120 mg kg^{-1} while that of twigs ranged from 1.8 to 320 mg kg^{-1} near Pb smelters and mines in eastern Missouri (Palmer and Kucera, 1980). Nitric acid extractable soil Pb ranged from 7 to 62 000 mg kg^{-1}. Particulate size, chemical form, level of atmospheric Pb, and the chemically active fraction of soil Pb were factors identified as playing important roles in the absorption of Pb into vegetation.

A common practice has been to attempt reclamation of abandoned Pb tailings with tolerant plants. Reclamation of Pb mine tailings at the Parc

Mine (North Wales) involved regulation of drainage in and around the spoil tip, mechanical redistribution of the spoil to improve the contours, spreading a shale/limestone amendment on the modified surface, fertilization, and planting a seed mixture containing a high proportion of tolerant grass cultivars (Firth et al., 1981). Studies of *Agrostis tenuis* populations from different habitats indicated that plants collected from an old Pb mine site in Wales grew on mine wastes, whereas plants from uncontaminated sites died (Bradshaw, 1952). Several native species required only minimal amendments when grown on Cu, Pb, and Zn tailings in Canada with *Arctagrostis latifolia* being most successful (Kuja and Hutchinson, 1979).

Bradshaw (1976) reported that the existence of tolerant races of plants on metal mines is the result of natural selection rather than innate physiological tolerance. A comparison of plant roots of mine and nonmine populations grown in a solution of metal salts suggested that a limited range of species existed on various metal mine wastes because they have evolved tolerant populations (Smith and Bradshaw, 1972).

Tolerant *Agrostis tenuis* plants from an acidic Pb mine were adapted to low levels of both Ca and P (Jowett, 1959). Thompson and Proctor (1983) found *Agrostis capillaris* and *Festuca rubra* to be the most common species on a mine spoil having potentially toxic concentrations of Cu, Pb, and Zn. Metal tolerant populations have the capacity to grow on toxic wastes without the addition of organic matter or lime to a point where the metal concentrations become too high (Gemmell, 1973). Clark and Clark (1981) reported that binding of Pb in organic forms, together with higher levels of the major plant nutrients, resulted in low levels of available Pb and a species-rich community on parts of a heavy metal mining complex on Grassington Moor at Yorkshire, England. Alvarez et al. (1974) concluded that the species colonizing Zn mine dumps in Utah should not be considered unique in their adaptation.

Recent studies have shown great interpopulation differences in the Zn tolerance of plants and have led to the selection of tolerant cultivars of *Festuca rubra* and *Agrostis tenuis* Sibth. (Smith and Bradshaw, 1979; Morrey et al., 1984). Morrey et al. (1984) suggested that carefully selected nontolerant cultivars can result in revegetation of calcareous metalliferous wastes even where the Zn concentration is as high as 1%. Karataglis (1980) showed that Cu and Zn toxicities may be additive for *Agrostis capillaris*. Populations of *Armeria maritima* from Zn and Cu mines are tolerant to those metals (Antonovics, 1972) and are found mixed with those few species known to have differentiated ecotypes highly tolerant to heavy metals, namely *Agrostis tenuis* Sibth., *Festuca ovina* L. and *Silene vulgaris* (Moench) Garcke (Antonovics et al., 1971). It appears that tolerant populations (or cultivars) of grasses accumulate Zn in their roots and therefore have relatively less in their shoots compared with intolerant ones (Reilly and Reilly, 1973; Brookes et al., 1981; Matthews and Thornton, 1982). *Armeria maritima* seeds from Plombieres, Belgium, were sown for revegetation trials on soils heavily polluted by Zn (15 000 ppm total Zn) in Palmerton (Pennsylvania, U.S.A.). Although no nutrients were used in the trials, growth was 3 to 4 times faster than at the original site apparently due to better soil conditions, higher temperatures, and higher rainfall (Lefebvre and Chandler-Mortimer, 1984).

Fertilizers are important inputs when metal tolerant plants are used. In the presence of fertilizer N, P, and K, Cu-tolerant *Agrostis tenuis* thrived in pure Cu waste. *Festuca rubra*, growing on Cu mine waste, which also had elevated levels of Pb and Zn, was found to be tolerant to all three metals (Wong, 1982). A simple screening technique for metal tolerant plants has been developed by Walley et al. (1974). Alfalfa, couchgrass (*Agropyron repens*), and red clover (*Trifolium pratense* L.) were the most successfully introduced species at high fertility rates to revegetate Pb-Zn tailings sites in Canada. A common cord moss (*Fumaria hygrometrica*) became established on the tailings wherever organic matter or fertilizer was applied but its persistence over time was not evaluated. In continental Europe, *Armeria maritima* (Mill.) Wild., a perennial tufted herb, frequently occurs as a phytosociologically characteristic species on old Zn-Pb mine sites scattered from East Belgium to South Poland (Lefebvre and Chandler-Mortimer, 1984) and has good potential for revegetation of heavy metal contaminated areas.

Decomposition of metal tolerant vegetation growing on mine waste containing high concentrations of Pb and Zn was retarded compared with that on an uncontaminated site as indicated by a greater accumulation of litter, less humus formation, reduced soil urease activity, and smaller microbial and microfaunal populations (Williams et al., 1977).

G. Phosphate

Mining operations for phosphate may produce a number of waste products: (1) normal overburden, (2) sand tailings, (3) phosphatic clay slimes, and (4) waste gypsum. Each waste has its own revegetation and reclamation problems. The chemical composition of phosphorites has been extensively studied and well documented. For reclamation purposes, two elements associated with phosphorites continue to be of environmental concern—uranium (U) and fluorine (F). These elements tend to be retained in phosphate products, byproducts, and wastes.

Radioactivity associated with phosphate tailings is a primary environmental concern. Radium activity, as ^{226}Ra, of phosphate slimes is reportedly 24 to 50 picocuries (pCi) g^{-1} of dry material (Roessler et al., 1979a, b; Mislevy et al., 1989). The radioactivity of the sand tailings averages about 5 pCi g^{-1} and is about five times higher than natural surface soil levels (Roessler et al., 1979a, b). Phosphate deposits throughout the world contain U and thorium (Th) series radionuclides in concentrations greater than natural background levels. All marine phosphorites contain U in the range of 0.005% to 0.05% and average about 0.01% (National Research Council, 1979). The ^{238}U series, including ^{226}Ra and its decay products, is the primary radiological concern. Generally, the upper 20 m of undisturbed overburden has relatively low radioactivity (Environmental Protection Agency, 1977). The radionuclide content below about 2 m increases, reaching a maximum in the leach zone which covers and contains more U and Ra than the phosphate matrix (Clark, 1975). Mixing of the deeper material with the surface material during overburden stripping results in the increased radionuclide concentrations in the disturbed overburden. These nuclides are redistributed during processing among the products and wastes and have potentially adverse effects on plants, animals, and man. Radium-226 in forage materials grown on dried phosphate clay ponds was nearly six times higher (0.23 pCi g^{-1}) than in plants from an unmined surface spodosol (0.04 pCi g^{-1}) (Mislevy et al., 1989). Recent developments in mixing of the sand and the clay to create a more stable land reclamation mix have resulted in a reduction of radioactive content of the waste. This procedure is being adopted, especially in new mines, since the ratio of sand to clay is favorable for this process. Older mining areas, where large volumes of clay slimes exist, could present a problem because of the presence of U and radioactive material near the surface. It is normal to reclaim land by covering with as much overburden as possible to promote a more stable, useful reclaimed land surface. This technique reduces the hazards of radon emissions since a cap of overburden retards the release of radon gas to the atmosphere. Capping with overburden and using construction techniques for ventilation of foundations to avoid radon gas build-up in structures are recommended procedures.

Mining of phosphate ore involves vast areas of land. In Florida, about 40% to 50% of mined land can be reclaimed immediately. The remainder is used for storing slimes. Slime wastes (clay plus water) from the beneficiation of phosphate ore comprise 30% to 40% by weight (dry) of the original input matrix. Slime disposal is a major problem in existing operations (Ryan and Cotter, 1980). Clay slimes contain no phytotoxic materials and are high in most plant nutrients (Mislevy et al., 1989). Dewatering takes years to accomplish and the maximum "dryness" achieved is only about 25% solids (Rogers and Nielson, 1980; Lamont et al., 1975). Mislevy et al. (1989) reported that phosphatic clay ponds can attain 40% to 50% solids after 10 to 15 years. Slime ponds can cover several hundred hectares and account for 50% to 60% of the land area mined for phosphate.

Dried phosphatic clay ponds have been shown to be quite productive for biomass production (Mislevy et al., 1989) with the only required input being N fertilizer. Dry biomass yields averaged over 4 years for erianthus (*Erianthus arundinaceum* (Retz) Jesw 'IK 76-63), leucaena (*Leucaena leucocephala* (Lam.) De Wet), and elephantgrass (*Pennisetum purpureum* L. "PI 300086") were 139.6, 58.5, and 56.5 Mg ha^{-1} yr^{-1}, respectively.

Quartz sand-tailings, which make up about 50% of the original matrix, amount to 90 million metric tons annually (Zellers and Williams, 1978). These sands contain no phytotoxic substances but are low in several nutrients, organic matter, and water retention capacity (Mislevy and Blue, 1981b). Relatively low tropical forage grass (Mislevy and Blue, 1981a), forage legume (Mislevy et al., 1981) and summer annual grass (Mislevy and Blue, 1981b) yields of good quality and adequate nutrient concentrations for animal production can be produced when these materials are properly fertilized and managed to increase clay and organic matter contents. Hortenstine and Rothwell (1972) demonstrated that addition of municipal compost along with N, P, and K to sand tailings significantly increased sorghum (*Sorghum bicolor* (L.) Moench) and oat (*Avena sativa* L.) yields. Phosphate sand tailings are quite low in U content.

A substantial amount of phosphogypsum byproduct from the phosphate industry is used as a soil amendment to improve saline and alkaline soils and as a source of nutrients in certain agricultural operations (Farina and Channon, 1988; Frenkel et al., 1989; Shainberg et al., 1988; Warrington et al., 1989). Lindeken and Coles (1977) indicated that the application of phosphogypsum at a rate of 6600 kg ha^{-1} tilled to a depth of 15 cm will result in about 0.45 pCi g^{-1} of ^{226}Ra being added to the soil. They reported that this concentration is less than the average background concentration of ^{226}Ra. Mays and Mortvedt (1986) reported that radiological assay of grain samples of corn (*Zea mays* L.), wheat (*Triticum aestivum* L.), and soybeans (*Glycine max* (L.) Merr.) showed no increase in ^{226}Ra in any of the crops due to phosphogypsum application. The phosphogypsum contained 25 nCi kg^{-1} of radioactivity and was applied at rates up to 112 Mg ha^{-1}.

H. Tin

Large acreages of tin tailings left after extraction of ore are found in Malaysia. Panton (1964), Wong (1970), and Kho (1970) estimated that there were 120000, 180000, and 200000 ha of tin tailings, respectively, in Peninsular Malaysia. The acreage has been estimated to be expanding at an average rate of more than 4000 ha yr^{-1} (Huan et al. 1981). The tailings are left largely unreclaimed. Tin mine tailings in southern Thailand are estimated to be 10000 to 30000 ha (Komes, 1973). Mining for tin on the Jos Plateau of central Nigeria has resulted in 321 km^2 of land being affected by open-cast mining that is largely unreclaimed (Alexander, 1989a).

Mining for tin is largely by dredging from alluvial deposits, but gravel pump mining and open-cast mining have also been used (Jin-Bee, 1955; Komes, 1973). Due to the different mining methods and deposition procedures employed, tailings show a wide range of physical and chemical characteristics. Early studies of Brickenshaw (1931) and Mitchell (1959) indicated that these tailings were completely devoid of humus, organic matter, and N. Shamshuddin et al. (1986) classified the tailings into three groups (sandy, clayey, and mixtures of sandy and clayey deposits). Basic cations, organic C, P, and N contents, and the CEC were very low. The pH was high with the dominant minerals in the clay fraction being kaolinite, mica, and chlorite. Furthermore, their physical and chemical properties were not normally conducive to plant growth without soil amendments. Dredging gives rise to two distinct types of tailings, sand and slime (Maene et al., 1973). Sand tailings are very difficult to reclaim because of the absence of organic matter and the predominance of coarse materials which have low water and nutrient retention capacities. The temperature of and evaporation rate of water from the surface are high. The slime areas have a high content of fine soil particles and consolidate upon drying into compact masses of low porosity. Infiltration is restricted and often water-logged conditions prevail. Acid pH conditions from the oxidation of sulfides are not uncommon. Higher clay and organic matter contents of the slimes compared with the sand tailings provide greater nutrient retention capacity (Huan et al., 1981). As a result, the slime areas are more fertile and natural vegetation establishes more quickly (Palaniappan, 1969). Mixtures of sand and slime tailings with the incorporation of various organic wastes appear to offer some promise for returning the sand tailings to agricultural cropping systems. Organics include chicken dung, peat, sewage sludge, skim latex, rice husks, sawdust, and palm oil mill effluent. Maene et al. (1977) and Komes (1973) reported that sand tailings can provide land for the production of vegetables and fruits, but large amounts of fertilizers, especially organic manures, must be added regularly in order to sustain production. Effective reclamation of sand tailings required the incorporation of 10% to 20% slime and 5% to 10% organic wastes on a volume basis (Mok and Lim, 1985). Water retention was only slightly improved by inputs, so frequent irrigation was necessary.

Cultivated plants grown for 6 years on tin spoils in Thailand were cashew nuts, coconut, eucalyptus (*Eucalyptus camaldulensis*), and singapore almond (Komes, 1973). Groundnut (*Arachis hypogea*) and mung bean (*Vigna radiata*) have been produced in Malaysia on sand and slime tailings when NPK fertilizers have been applied (Jahari and Yaacob, 1979). Eucalyptus plantings were established on tin-mine spoil in Nigeria starting in 1960 while *Acacia albida* has been a natural colonizer of spoil mounds (Alexander, 1989a, b). Although the eucalyptus trees have established and contribute a high litter input, large areas of the ground surface beneath the eucalyptus remains completely bare and ground vegetation is very limited in extent. Lack of ground cover is thought to be due to a combination of surface compaction caused during reclamation and the allelopathic properties of the eucalyptus leaves. Although total C and N have increased, a significant decrease in pH and base saturation indicate that the long-term effects of eucalyptus is one of progressive degradation of already poor soils (Alexander, 1989a). Organic C, total N, pH, percentage base saturation, and exchangeable Ca, Mg, Na, and K were all significantly higher in soils from beneath the canopy of *Acacia albida*.

I. Uranium

One of the largest uranium (U) ore reserves in the world is located in the Elliot Lake area of Ontario, Canada. It is estimated that U tailings from previous mining operations amount to about 80×10^6 t (Moffett and Tellier, 1977) covering more than 400 ha. Uranium tailings contain about 85% of the initial radioactivity of the source ore. It is estimated that 1 t of tailings has a radioactivity of a least 5 mCi (Moffett and Tellier, 1977). About 90% of this activity is due to the ^{238}U series (Bryant et al., 1977). The most troublesome daughter isotopes of ^{238}U are ^{230}Th, ^{226}Ra, and ^{210}Pb. The solid tailings at Elliot Lake are made up of quartz, pyrite, calcium sulfate (from the neutralized reaction caused by lime addition at discharge), metal hydroxides, precipitated Th, and rare earths (Vivyurka, 1975). The liquid portion of the tailings is a neutral to alkaline saturated calcium sulfate solution, containing Ra, NH_4 and NO_3. Fresh tailings areas have a pH of 6.5 to 9.0, whereas seepage and runoff from abandoned tailings have a pH of 2 to 3 due to oxidation of pyrite. Limiting percolation flow through tailings piles to reduce the bacterial oxidation responsible for the pH decrease and solubilization of radioactive constituents has been proposed. Revegetation with various crops (Murray and Moffett, 1977) and discharge with added thickeners to form cone-shaped piles as proposed by Rubinsky (1975) are possible methods for reducing water flow through tailings. Chemical fixation (solidification at high pH) of mine waste residues decreased the release of ^{226}Ra to effluent and leachate but the effluent and leachate were toxic to Daphnia (*Daphnia pulex*) and rainbow trout (*Salmo gairdneri*) (Bryant et al., 1977).

Wullstein (1980) suggested reclaiming U mine tailings with sandy textured soil in arid climates followed by stabilizing these coverings with rhizosheath-forming grasses such as Indian rice grass (*Oryzopsis hymenoides*). He reported N fixation (acetylene reduction) rates of about 20 μg N/plant/day on the U tailings. A long-term concern is the potential for intrusion of soil-covered U mill tailings by burrowing mammals as shown by Shuman and Whicker (1986). Radon flux from burrow entrances was significantly greater than that from undisturbed ground at a site in Colorado but not in Wyoming. They proposed that the difference in the two sites may have been due to time of mammal residence and differences in tailings characteristics such as texture and pH.

VI. Summary

Large areas of abandoned or recently generated mine tailings from a variety of mining and refining industries are common around the world. These tailings are unsightly, unproductive, and present a variety of environmental problems including air, land, and water pollution. Reclamation of mine tailings is a challenging task since they are largely devoid of organic matter

Table 6. Reported limiting parameters to reclamation of various types of mine tailings

	Limitation								
Tailing	Acidity	Salinity	Sodicity	Nutrients	Toxic metals	Texture	Structure	Organic matter	Radioactive
Asbestos	No	No	No	Yes	Yes	No	Yes	Yes	No
Bauxite	No	Yes	Yes	Yes	Yes	Yes	Yes	Yes	No
Clay	Yes	Yes	Yes	Yes	No	Yes	Yes	Yes	No
Copper-gold-nickel-Silver	Yes	Yes	No	Yes	Yes	Yes	Yes	Yes	No
Iron	Yes	No	No	Yes	Yes	Yes	Yes	Yes	No
Lead-zinc	Yes	No	No	Yes	Yes	Yes	Yes	Yes	No
Phosphate									
Quartz	No	No	No	Yes	No	Yes	Yes	Yes	Yes
Slimes	No	No	No	No	No	Yes	Yes	Yes	Yes
Tin									
Quartz	No	No	No	Yes	No	Yes	No	Yes	No
Slimes	Yes	No	No	Yes	Yes	Yes	Yes	Yes	No

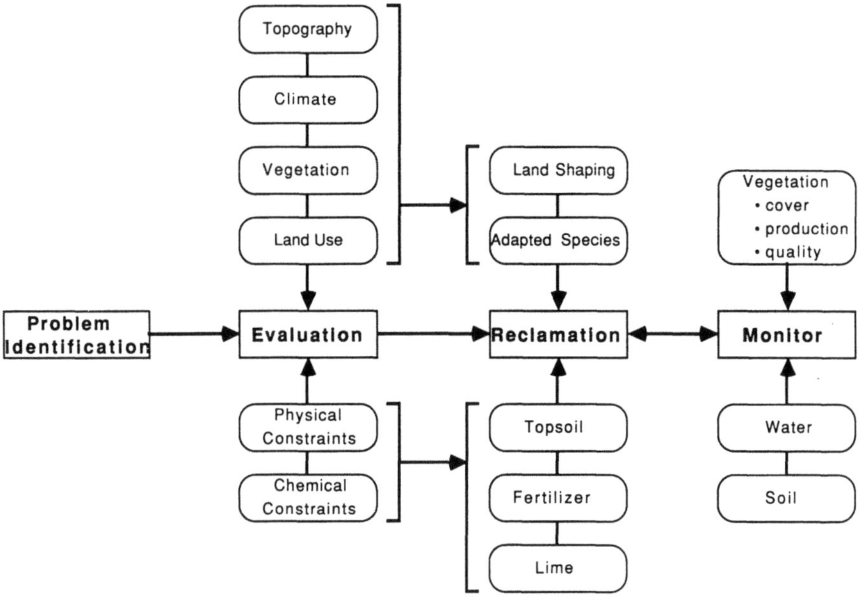

Figure 4. Systematic approach to evaluation and reclamation of mine tailings

and highly variable in terms of their chemical and physical properties. Each site must be evaluated and treated independently. Reclamation is largely controlled by measurable parameters such as acidity, salinity, sodicity, metal content, and nutrient deficiencies which are common chemical problems associated with mine tailings. Texture and structure of mine tailings can also limit successful reclamation. Radioactive nuclides are a concern in phosphate and uranium tailings reclamation.

One or more potential limitations to plant establishment and growth on mine tailings usually exist. Data presented in Table 6 summarize those parameters that are common limitations to reclamation for the various types of mine tailings discussed in this chapter. Low levels of organic matter and available nutrients and unfavorable texture and structure are limitations to revegetation of most tailings.

Early recognition and evaluation of potential problems associated with reclamation of mine tailings are essential to an orderly and successful reclamation effort. An idealized approach, showing the sequence of problem recognition, evaluation, reclamation, and monitoring, is presented in Fig. 4. Field evaluation is necessary and must be supported by laboratory analyses. Because tailings are so variable, those parameters that limit successful reclamation are expected to be site specific and must be evaluated on a case by case basis. Analytical methods for tailings evaluation and amendments to correct deficiencies of essential plant nutrients or soil salinity/sodicity are similar to those used in conventional agricultural situations. Potential acidity must be taken into account when determining lime rates of tailings containing metal sulfides.

Reclamation of tailings containing potentially toxic levels of metals is a particularly difficult problem. Metalliferous mine tailings commonly have several of the chemical and physical problems listed earlier and, in addition, have high concentrations of potentially toxic metals. For this reason they offer some of the more challenging reclamation problems. Revegetation with metal tolerant plants can stabilize tailings with moderate levels of metal contamination. This approach may provide vegetative cover and visual improvement but the tailings may still be subject to wind and water erosion and leaching of metals. Topsoiling of highly toxic tailings to provide a rooting medium for plants together with landforming to control or reduce leaching of metals and acid drainage appears to be a logical long-term solution. Waste products, such as sewage sludge and organic by-products, can provide nutrients and favorably influence adverse soil properties, but by themselves may not provide stable, long-term reclamation.

References

Alexander, M.J. 1989a. The long-term effect of Eucalyptus Plantations on tin-mine spoil and its implication for reclamation. *Landscape Urban Plann.* 17:347–60.

Alexander, M.J. 1989b. The effect of *Acacia albida* on tin-mine spoil and their possible use in reclamation. *Landscape Urban Plann.* 17:61–71.

Alvarez, H., J.A. Ludwig, and K.T. Harper. 1974. Factors influencing plant colonization of mine dumps at Park City, Utah. *American Midland Naturalist.* 92:1–11.

Antonovics, J. 1972. Population dynamics of the grass *Anthoxanthum odoratum* on a zinc mine. *J. Ecology.* 60:351–365.

Antonovics, J., A.D. Bradshaw, and R.G. Turner. 1971. Heavy metal tolerance in plants. *Adv. in Ecological Res.* 7:1–85.

Arora, H.S., J.B. Dixon, and L.R. Hossner. 1978. Pyrite morphology in lignitic coal and associated strata of east Texas. *Soil Sci.* 125:151–159.

Barrow, N.J. 1982. Possibility of using caustic residue from bauxite for improving the chemical and physical properties of sandy soils. *Australian J. Agric. Res.* 33:275–285.

Berg, W.A., E.M. Barrau, and L.A. Rhodes. 1975. Plant growth on acid molybdenum mill tailings as influenced by liming, leaching and fertility treatments. In: M.J. Chadwick and G.T. Goodman (eds.), pp. 207–222. *The Ecology of Resource Degradation and Renewal.* Blackwell Scientific Publications, Oxford.

Bergholm, J. and E. Steen. 1989. Vegetation establishment on a deposit of zinc mine wastes. *Env. Pollution.* 56:127–144.

Berry, C.R. and D.H. Marx. 1977. Growth of Loblolly pine seedings in strip-mined kaolin spoil as influenced by sewage sludge. *J. Environ. Qual.* 6:379–381.

Bradshaw, A.D. 1952. Populations of *Agrostis tenuis* resistant to lead and zinc poisoning. *Nature* 169:1098.

Bradshaw, A.D. 1976. Pollution and evolution. In: T.A. Mansfield (ed.). *Effects of Air Pollutants on Plants*, pp. 135–139. Cambridge Univ. Press, Cambridge.
Bradshaw, A.D. and M.J. Chadwick. 1980. *The Restoration of Land*. Blackwell Scientific Publications, Osney Mead, Oxford, England.
Bradshaw, A.D., M.S. Humphries, M.S. Johnson, and R.D. Roberts. 1978. The restoration of vegetation on derelict land producted by industrial activity. In: M.W. Holdgate and M.J. Woodman (eds.) *The Breakdown and Restoration of Ecosystems*, pp. 249–278. Plenum Press, New York.
Brickenshaw, F. 1931. Reclaiming old mining land for agriculture. *Malaysian Agr. J.* 21:470–476.
Brierly, J.K. 1956. Some preliminary observations on the ecology of pit heaps. *J. Ecol.* 44:383–390.
Brinkman, R. and L.J. Pons. 1973. Recognition and pridiction of acid sulfate soil conditions. In: H. Dost (ed.) *Acid Sulphate Soils*, vol. 1, pp. 169–203. ILRI Pub. 18. Institute for Land Reclamation and Improvement. Wageningen, The Netherlands.
Brookes, A., J.C. Collings, and D.A. Thruman. 1981. The mechanism of zinc tolerance in grasses. *J. Plant Nutr.* 3:695–705.
Bryant, D.N., D.B. Cohen, and R.W. Durham. 1977. Leachability of Radioactive Constituents from Uranium Mine Tailings: Status Report. Wastewater Technology Centre.Technology Development Report No. EPS 4-WP-79-4. Environmental Protection Service. Ottawa, Canada.
Caruccio, F.T., L.R. Hossner, and G. Geidel. 1988. Pyritic materials: acid drainage, soil acidity and liming. In: *Reclamation of Surface Mined Lands*, vol. 1, pp. 159–190. CRC Press, Boca Raton, FL.
Chambers, J.C., R.W. Brown, and R.S. Johnston. 1987. A comparison of soil and vegetation properties of seeded and naturally revegetated properties of seeded and naturally revegetated pyritic alpine mine spoil and reference sited. *Landscape Urban Plann.* 14:507–519.
Clark, R.K. and S.C. Clark. 1981. Floristic diversity in relation to soil characteristics in a lead mining complex in the Pennines, England. *New Phytol.* 87:799–815.
Clark, V. 1975. Natural radioactivity in the environment: the Florida phosphate study. *U.S. FDA 7th Annual Conference on Radiation Control*, pp. 166–169. Report No. DHEW/FDA-76-8026. Conf-7504107.
Coffey, P.S., W.D. Scott, and K.J. Summers. 1986. The effects of tailing dam profiles on relative wind erosion rates. *J. Environ. Qual.* 15:168–172.
Craze, B. 1977a. Investigations into the revegetation problems of Captains Flat mining area. *J. Soil Cons. New South Wales* 33:190–199.
Craze, B. 1977b. Restoration of Captains Flat mining area. *J. Soil Cons. New South Wales* 33:98–105.
Creswell, C.F. 1973. Changes in vegetational composition, nutritional status, soils, and microbial populations with the establishment of vegetation on gold-mine dumps on the Witwaterstrand. In: R.J. Hutnik and G. Davis (eds.) *Ecology and Reclamation of Devastated Land*, vol. 1, pp. 335–359. Gordon and Breach, New York.
Crews, J.T. 1984. Effect of minesoil compaction on growth and yield of KY-31 tall fescue and sericea lespedeza. United States Dept. of Agric., Northeastern Forest Exp. Sta. Res., Note NE-320.
Crowder, A.A., B.E. McLaughlin, G.K. Rutherford, and G.W. van Loon. 1982.

Site factors affecting semi-natural herbaceous vegetation on tailings at Copper Cliff, Ontario. *Reclamation Revegetation Res.* 1:177–193.

Day, A.D. and K.L. Ludeke. 1973. Stabilizing copper mine tailing disposal berms with giant bermudagrass. *J. Environ. Qual.* 2:314–315.

Day, A.D. and K.L. Ludeke. 1980a. Direct seeding of tree species on copper mine wastes in the southwestern United States. *J. Environ. Qual.* 9:304–306.

Day, A.D. and K.L. Ludeke. 1980b. Reclamation of copper mine wastes with shrubs in the southwest USA. *J. Arid Environments* 3:107–112.

Day, A.D. and K.L. Ludeke. 1981. The use of reclaimed copper mine wastes in the southwestern United States. *Minerals Environment* 3:21–31.

Day, A.D. and K.L. Ludeke. 1982. Stabilization of copper mine wastes in a semiarid environment with perennial grasses. *Arid Environment* 5:285–290.

Dean, S.A. and M.M. Fogel. 1982. Acid drainage from abandoned metal mines in the mountains of Southern Arizona. Symposium on Surface Mining Hydrology, Sedimentology and Reclamation, pp. 269–276. University of Kentucky, Lexington, KY.

Dollhopf, D.J. and B.J. Bauman. 1981. Bentonite mine land reclamation in the northern Great Plains. Res. Dept. 179, Montana Agric. Exp. Stn., Montana State University, Bozeman.

Dollhopf, D.J., E.J. DePuit, and M.G. Klages. 1980. Chemical amendment and irrigation effects on sodium migration and vegetation characteristics in sodic mine spoils in Montana. Bulletin 736. Montana Agricultural Experiment Station, Bozeman, MT.

Donovan, R.P., R.M. Felder, and H.H. Rogers 1976. Vegetative stabilization of mineral waste heaps. U.S. EPA Technol. Serv. EPA-600/a-76-087.

Drew, A. and C. Reilly. 1972. Observations on copper tolerance in the vegetation of a Zambian copper clearing. *J. Ecol.* 60:439–444.

Dwyer, P.J. 1986. Revegetation trials at Merrygoen New South Wales. *J. Soil Cons. New South Wales* 42:123–130.

Ellery, K.S. and B.H. Walker. 1986. Growth characteristics of selected plant species on asbestos tailings form Msauli Mine, Eastern Transvaal. *S. Afr. J. Bot.* 52:201–206.

Environmental Protection Agency (U.S.). 1977. Radiological surveys of Idaho phosphate ore processing. The thermal process plant. U.S. Office of Radiation Programs, Report No. ORP/LV-77-3.

Farina, M.P.W. and P. Channon. 1988. Acid-subsoil amelioration: II. Gypsum effects on growth and subsoil chemical properties. *Soil Sci. Soc. Am. J.* 52:175–180.

Farrow, S., T. McNeilly, and P.D. Putwain. 1981. The dynamics of natural selection for copper tolerance in *Agrostis canina* L. subsp. Montana Hartm. Heavy Metals in the Environment, pp. 289–295. International Conference, Amsterdam.

Firth, J.N.M., M.S. Johnson, and I.G. Richards. 1981. The reclamation of lead mine tailings at Parc Mine North Wales. In: D.D. Hemphill (ed.) *Trace subance in environmental health—XV*, pp. 333–339. University of Missouri, olumbia.

Foy, C.D. 1974. Effects of aluminum on plant growth. In: E.W. Carson (ed.) *The Plant Root and Its Environment*, pp. 565–600. Univ. Press of Virginia, Univ. Sta., Charlottesville, VA.

Foy, C.D. 1983. The physiology of plant adaptation to mineral stress. *Iowa J. Res.* 57:413–417.
Foy, C.D. and A.J. Oakes. 1984. A winter-hardy, aluminum tolerant, perennial pasture grass for reclamation of acid mine spoils. *J. Plant Nutrition* 7:929–951.
Frenkel, H., M.V. Fey, and A.D. Noble. 1989. Some factors limiting the dissolution rate of phosphogypsum in an acid soil. *Soil sci. Soc. Am. J.* 53:958–961.
Fuller, W.H. and K. Lanspa. 1975. Uptake of iron and copper by sorghum from mine tailings. *J. Environ. Qual.* 4:417–422.
Fuller, R.D. and C.J. Richardson. 1986. Aluminate toxicity as a factor controlling plant growth in bauxite residue. *Environ. Top. Chem.* 5:905–915.
Fuller, R.D., E.D.P. Nelson, and C.J. Richardson. 1982. Reclamation of red mud (bauxite residues) using alkaline-tolerant grasses with organic amendments. *J. Environ. Qual.* 11:533–539.
Gadgil, R.L. 1969. Tolerance of heavy metals and the reclamation of industrial waste. *J. Applied Ecol.* 6:247–259.
Gemmell, R.P. 1973. Revegetation of derelict land polluted by a chromate smelter. I. Chemical factors covering substrate toxicity in chromate smelter waste. *Environ. Pollution*. 5:181–197.
Gemmell, R.P. 1977. *Colonization of Industrial Wasteland.* Camelot Press, Southampton, England.
Gibson, D.J. 1982. The natural revegetation of lead/zinc mine spoil in northeastern Oklahoma. *Southwestern Naturalist* 27:425–436.
Goodman, G.T., C.E.R. Pitcairn, and R.P. Gemmell. 1973. Ecological factors affecting growth on sites contaminated with heavy metal. In: R.J. Hutnik and G. Davis (eds.) *Ecology and Reclamation of Devastated Land*, vol. 2, pp. 149–173. Gordon and Breach, London.
Griffiths, A.J., D.E. Hughes, and D. Thomas. 1974. Some aspects of microbial resistance to metal pollution. In: M.J. Jones (ed) *Minerals and the environment*, pp. 387–394. Inst. Mining and Metallurgy.
Harwood, G. 1979. A guide to reclaiming small tailings ponds and dumps. USDA Forest Service, General Technical Report INT-57, Ogden, UT.
Harwood, J.J., S.R. Koirtyohann, and C.J. Schmitt. 1987. Effects of cover materials on leaching of constituents from dolomitic lead mine tailings. *Water, Air, Soil Pollut.* 34:31–41.
Hogan, Gary. D., Gerald. M. Courtin, and W.E. Rauser. 1977. The effects of soil factors on the distribution of *Agrostis gigantea* on a mine waste site. *Can. J. Bot.* 55:1038–1042.
Hortenstine, C.C. and D.F. Rothwell. 1972. Use of municipal compost in reclamation of phosphate-mining sand tailings. *J. Environ. Qual.* 1:415–418.
Huan, L.K., L. Maene, G. Maesschalck, and W.S.W. Harun. 1981. Reclamation of tin tailings for agriculture in Malaysia. Soil Science Division Technical Bulletin, University Pertanian Malaysia, Kuala Lumpur, Malaysia.
Hume, L.J. and J.A. August. 1988. Effects of phosphorus application and mycorrhizal inoculation on white clover (*Trifolium repens*) growth in andesite tailings from Martha Mine, Waihi. *New Zealand J. Agric. Res.* 31:331–338.
Hunter, G. and P.C. Whiteman. 1974. Problems associated with the revegetation of metal-mining wastes. *J. Aust. Inst. Agric. Sci.* 40:270–278.
Hutchinson, T.C. and L.M. Whitby. 1974. Heavy metal pollution in the Sudbury mining and smelting region of Canada: I. Soil and vegetation contamination by

nickel, copper and other metals. *Environ. Conserv.* 1:123–132.
Jahari, R. and O. Yaacob. 1979. Performance of groundnut (*Arachis hypogea*) and mungbean (*Vigna radiata*) on tin tailings in Malaysia. *Pertanika* 2(1):66–68.
Jeffries, R.A., A.D. Bradshaw, and P.O. Putwain. 1981. Growth, nitrogen accumulation and nitrogen transfer by legume species established on mine spoils. *J. Applied Ecol.* 18:945–956.
Jennett, J.C. and S.M. Linnemann. 1977. Disposal of lead and zinc-containing wastes on soils. *J. Water Pollut. Control Fed.* 49:1842–1856.
Jin-Bee, O. 1955. Mining landscapes of Kinta. *Malayan J. Trop. Geog.* 4:1–40.
Johnson, M.S. and J.W. Eaton. 1980. Environmental contamination through residual trace metal dispersal from a derelict lead-zinc mine. *J. Environ. Qual.* 9:175–179.
Johnson, M.S., T. McNeilly, and P.D. Putwain. 1977. Revegetation of metalliferous mine spoil contaminated by lead and zinc. *Environ. Pollution* 12:261–277.
Jordan, J.E. and S.W. Dewar. 1985. Vegetation characterization of taconite tailing basin in Minnesota. Symposium on Surface Mining, Hydrology, Sedimentology and Reclamation. Dec. 9–13, University of Kentucky, Lexington, KY.
Jowett, D. 1959. Adaptation of a lead tolerant population of *Agrostis tenuis* to low soil fertility. *Nature.* 180:43.
Karataġlis, S.S. 1980. Zinc and copper effects on metal-tolerant and non-tolerant clones of *Agrostis tenuis* (Poaceae). *Plant Systematics and Evolution.* 134:173–182.
Keane, P. and B. Craze. 1978. Captains Flat mined area field revegetation trials. *J. Soil Cons. New South Wales* 34:37–42.
Kho, B.L. 1970. Tin tailings—a potential land resource. *Newsletter Malaysian Soil Sci. Soc.* March Issue, pp. 13–16.
Komes, A. 1973. The reclamation of some problem soils in Thailand, pp. 41–42. ASPAC Food and Fertilizer Technology Center. Technical Bulletin No. 14. Taipei City, Taiwan.
Kuja, A.L. and T.C. Hutchinson. 1979. The use of native species in mine tailings revegetation. Proceedings of the Canadian Land Reclamation Association. 4th conference. Saskatchewan, pp. 13–15.
Lamont, W.E., J.T. McLendon, L.W. Clements, and I.L. Feld. 1975. Characterization studies of Florida phosphate slimes. U.S. Bureau of Mines RI 8089. U.S. Dept. of Interior. Wahington, D.C.
Lampkin, A.J., III and M.R. Sommerfeld. 1981. Impact of reclamation on drainage from the Sheldon mine complex, Yavapai County, Arizona. *J. Arizona–Nevada Acad. Sci.* 16:65–68.
Lefebvre, C. and A. Chandler-Mortimer. 1984. Demographic characteristics of the perennial herb *Armeria maritima* on zinc-lead mine wastes. *J. Applied Ecol.* 21:255–264.
Lindeken, C.L. and D.G. Coles. 1977. Radium-226 content of agricultural gypsums. Univ. of California Lawrence Livermore Laboratory Rep. No. UCRL-78602.
Lindsay, W.L. 1979. *Chemical Equilibria in Soils.* Wiley, New York.
Lopez, J.M. and G.F. Lee. 1977. Environmental chemistry of copper in Torch Lake, Michigan. *Water, Air, and Soil Pollut.* 8:373–385.
Ludeke, K.L. 1977. Tailing reclamation. In: J.L. Thames (ed.) *Reclamation and Use of Disturbed Land in the Southwest*, pp. 262–271. University of Arizona Press, Tucson.

MacLean, A.J. and A.J. Dekker. 1976. Lime requirement and availability of nutrients and toxic metals to plants grown in acid mine taillings. *Can. J. Soil Sci.* 56:27–36.
Maene, L.M., C.K. Mok, and K.H. Lim. 1973. A study of the physical properties of tin tailing deposits in Peninsular Malaysia. Proceedings of Conference on Fertility and Chemistry of Tropical Soil, pp. 220–233. Malaysian Soc. Soil Sci., Kuala Lumpur.
Maene, L.M., C.L. Connor, and O. Yaacob. 1977. Food Production on tin tailings in the Seri Kembangan-Balakong area, Malaysia. In: H.F. Chin, I.C. Enoch, and W.O. Wan Mohamed (eds.), *Food and Agriculture Malaysia 2000*, Faculty of Agriculture, UPM, Serdang.
Matthews, H. and I. Thornton. 1982. Seasonal and species variation in the content of cadmium and associated metals in pasture plants at Shipham. *Plant and Soil.* 66:181–193.
Mays, D.A. and J.J. Mortvedt. 1986. Crop response to soil applications of phosphogypsum. *J. Environ. Qual.* 15:78–81.
McCawley, F.X. and L.H. Baumgardner. 1985. Aluminum. Mineral facts and problems. USDI, Bureau of Mines Publ., Bull. 675.
McNeilly, T. and M.S. Johnson. 1978. Mineral nutrition of copper-tolerant browntop on metal-comtaminated mine spoil. *J. Environ. Qual.* 7:483–486.
Meecham, J.R. and L.C. Bell. 1977a. Revegetation of aluminum refinery wastes. I. Properties and amelioration of the materials. *Australia J. Exp. Agric. Animal Husb.* 17:679–688.
Meecham, J.R. and L.C. Bell. 1977b. Revegetation of aluminum refinery wastes. II. Glasshouse experiments. *Australian J. Exp. Agric. Animal Husb.* 17:689–696.
Merrill, S.D., E.J. Doering, and J.F. Power. 1980. Changes of sodicity and salinity in soils reconstructed on strip-mined land. *North Dakota Agric. Exp. Sta. Farm Res*, 37:13–16.
Merrill, S.D., E.J. Doering, J.F. Power, and F.M. Sandoval. 1983. Sodium movement in soil-minespoil profiles: diffusion and convection. *Soil Sci.* 136:308–316.
Merrill, S.D., J.R. Deutsch, and M.W. Pole. 1987. Saturation Percentage. In: R.D. Williams, G.E. Schuman, (eds.) *Reclaiming Mine Soils and Overburden in the Western United States. Analytic Parameters and Procedures*, pp. 35–58. Soil Conservation of America. Ankeny, IA.
Meyer, D.R. 1980. Nutritional problems associated with the establishment of vegetation on tailings from an asbestos mine. *Environ. Poll.* Ser. A., pp. 287–298.
Mislevy, P. and W.G. Blue. 1981a. Reclamation of quartz sand-tailings from phosphate mining: I. Tropical forage grasses. *J. Environ. Qual.* 10:449–453.
Mislevy, P. and W.G. Blue. 1981b. Reclamation of quartz sand-tailings from phosphate mining: III. Summer annual grasses. *J. Environ. Qual.* 10:457–460.
Mislevy, P., W.G. Blue, and J.B. Brolmann. 1981. Reclamation of quartz sand-tailings from phosphate mining: II. Forage legumes. *J. Environ. Qual.* 10:453–456.
Mislevy, P., W.G. Blue, and C.E. Roessler. 1989. Productivity of clay tailings from phosphate mining: I. Biomass crops. *J. Environ. Qual.* 18:95–100.
Mitchell. B.A. 1959. The ecology of tin mine spoil heaps. *Malaysian For.* 22:111–132.
Moffett, D. and M. Tellier. 1977. Uptake of radioisotopes by vegetation growing

on uranium tailings. *Can. J. Soil Sci.* 57:417–424.

Mok, C.K. and K.H. Lim. 1985. Reclamation of tin tailings for crop production. In: A.M. Mokhtauruddin (ed.). *Advances in Soil Research in Malaysia: Proceedings of the Technical Workshop on Soil Science*, pp. 95–111.

Moore, T.R. and R.C. Zimmerman. 1977. Establishment of vegetation on serpentine asbestos mine waster, south-eastern Quebec. *Can. J. Appl. Ecol.* 14:589–599.

Morrey, D.R., M.S. Johnson, and J.A. Cooke. 1984. A comparison of metal tolerant and non-tolerant varieties of *Festuca rubra* for use in the direct hydraulic seeding of metalliferous fluorspar mine tailings. *J. Environ. Mgt.* 19:99–105.

Murray, D.R. 1977. Pit Slope Manual. Supplement 10-1. Reclamation by revegetation. Vol. 1. Mine Wastes Descriptions and Case Histories. CANMET (Can. Centre Min. Energy Technol.) Toronto, Canada. Report No. 77-31.

Murray, D. and D. Moffett. 1977. Vegetating the Elliot Lake uranium mine tailings. *J. Soil and Water Cons.* 32:171–174.

National Research Council. 1979. Redistribution of accessory elements in mining and processing. Part II. Uranium, phosphate, and alumina. National Academy of Sciences. Washington, D.C.

Nordstrom, Darrell Kirk. 1982. Aqueous pyrite oxidation and the consequent formation of secondary iron minerals. In: J.A. Kittrick, D.S. Fanning and L.R. Hossner (eds.). *Acid Sulfate Weathering*, pp. 37–56. Soil Science Society of America, Madison, WI.

Nunn, R.M., and J.R. Riches. 1978. Heavy metal mobilization from lead mine tailings, Northampton, W.A. In: J.E.D. Fox (ed.). *Rehabilitation of Mined Lands in Western Australia*, pp. 113–116. Western Australian Institute of Technology. Perth, Australia.

O'Shay, Tracey, L.R. Hossner and J.B. Dixon. 1990. A modified hydrogen peroxide oxidation method for potential acidity in pyritic overburden sediments. *J. Environ. Qual.* 19:778–782.

Palaniappan, V.M. 1969. "Studies on the several stages of plant succession on areas of tin mining industries around Kuala Lumpur." Master of Science Thesis. University of Malaya, Kuala Lumpur.

Palaniappan, V.M. 1974. Ecology of tin tailings areas: plant communities and succession. *J. Appl. Ecol.* 11:133–150.

Palmer, K.T. and C.L. Kucera. 1980. Lead contamination of sycamore and soil from lead mining and smelting operations in eastern Missouri. *J. Environ. Qual.* 9:106–111.

Panton, U.P. 1964. The soil map of Malaya. *J. Trop. Geog.* 18:118–124.

Peters, T.H. 1984. Rehabilitation of mine tailings: a case of complete ecosystem reconstruction and revegetation of industrially stressed lands in the Sudbury, Ontario, Canada. In: P.J. Sheehan, D.R. Miller, G.C. Butler, P. Bourdeau, and J.M. Ridgeway (eds.) *Effects of pollutants at the ecosystem level*. John Wiley and Sons, New York.

Peterson, H.B. and R.F. Nielson. 1973. Toxicities and deficiencies in mine tailings. In: R. Hutnik and G. Davis (eds.) *Ecology and Reclamation of Devastated Land*. vol. 1, pp. 15–25. Gordon and Breach Sci. Publ., New York.

Phillips, J.M. and A.E. Spooner. 1983. Dry matter production and percent ground cover during establishment of selected forage species on bauxite minesoils. Symposium on Surface Mining, Hydrology, Sedimentology, and Reclamation. Nov.

27–Dec. 2, University of Kentucky, Lexington, KY.
Pugh, C.E., L.R. Hossner, and J.B. Dixon. 1981. Pyrite and marcasite surface area as influenced by morphology and particle diameter. *Soil Sci. Soc. Amer. J.* 45:979–982.
Pugh, C.E., L.R. Hossner, and J.B. Dixon. 1984. Oxidation rate of pyrite as affected by surface area, morphology, oxygen concentration and autotrophic bacteria. *Soil Sci.* 137:309–314.
Proctor, J., S.R.J. Woodell. 1975. The ecology of serpentine soils. *Adv. Ecol. Res.* 9:256–366.
Reilly, A. and C. Reilly. 1973. Zinc, lead and copper tolerance in the grass *Stereochlaena cameronii* (Stapf.) Clayton. *New Phytologist.* 72:1041–1046.
Roessler, C.E., Z.A. Smith, W.E. Bolch, and R.J. Prince. 1979a. Uranium and radium- 226 in Florida phosphate materials. *Health Phys.* 37:269–277.
Roessler, C.E., Z.A. Smith, W.E. Bolch, and J.A. Wethington. 1979b. Management of low-level natural radioactive wastes of phosphate mining and processing. Proceedings of Health Physics Society 12th Midyear Topical Symposium, Report No. EPA 520/3-79-002, Williamsburg, VA.
Rogers, V.C. and K.K. Nielson. 1980. Tailings piles from uranium producers, vol. 1, pp. 163–172. DOE/Univ. Arizona Waste Management 80 Symposium, Tuscon AR.
Rogowski, A.S., H.B. Pionke, and J.G. Broyan. 1977. Modeling the impact of strip mining and reclamation processes on quality and quantity of water in mined areas: a review. *J. Environ. Qual.* 6:237–244.
Rubinsky, E.E. 1975. Thickened discharge a new approach to tailings disposal, pp. 47–53. Canadian Mining and Metallurgic Bulletin, December.
Ruppin, P.A. 1987. Tailings management at Bougainville copper Limited. In: T. Farrell (ed.) *Mining Rehabilitation '87*, pp. 71–87. Australian Mining Industry Council, Canberra, Australia.
Ruschena, L.J., G.S. Stacey, G.D. Hunter, and P.C. Whiteman. 1974. Research into the vegetation of concentrator tailing dams at Mount Isa. Reprinted from papers presented at regional meeting on: Recent technical and social advances in North Australia, Institute of Mining and Metallurgy.
Rutherford, G.K. and C.R. Bray. 1979. Extent and distribution of soil heavy metal contamination near a nickel smelter at Coniston, Ontario. *J. Environ. Qual.* 8:219–222.
Rutherford, G.K., D. Dimma, G.W. Van Loon, and W.G. Breck. 1982. The pedological properties of tailings derived from three mining operations on the sudbury Area, Ontario, Canada. *J. Environ. Qual.* 11:511–518.
Ryan, M.T. and S.J. Cotter. 1980. Integrated assessment of the phosphate industry. U.S. Dept. Energy Report No. ORNL–EV.
Salomons, W. and U. Forstner. 1988. *Chemistry and biology of solid wastes, dredge materials and mine tailings*. Springer-Verlag, New York.
Schafèr, W.M. 1979. Guides for estimating cover-soil quality and mine soil capability for use in coal strip-mine reclamation in the western United States. *Reclamation Rev.* 2:67–74.
Schuman, G.E. and T.A. Sedbrook. 1984. Sawmill wood residue for reclaiming bentonite spoils. *For. Prod. J.* 34:65–68.
Shainberg, I., M.E. Sumner, W.P. Miller, M.P.W. Farina, M.A. Pavan, M.V. Fey. 1988. Use of gypsum on soils: a review. *Adv. Soil Sci.* 9:1–111.

Shamshuddin, J., N. Mokhtar, and S. Paramananthan. 1986. Morphology, mineralogy and chemistry of an ex-mining land in Ipok, Perak. *Pertanika* 9:89–97.

Sheets, P.J., V.V. Volk, and E.H. Gardner. 1982. Plant and soil reactions to nickel ore processed tailings. *J. Environ. Qual.* 11:446–451.

Sheppard, M.I., S.C. Sheppard, and D.H. Thibault. 1984. Uptake by plants and migration of uranium and chromium in field lysimeters. *J. Environ. Qual.* 13: 357–361.

Shetron, S.G. 1983. Alfalfa, *Medicago sativa* L. establishment in mine mill tailings. I. Plant analysis of alfalfa grown on iron and copper tailings. *Plant and Soil* 73:227–237.

Shetron, S.G. and J.J. Sindler. 1983. Alfalfa, *Medicago sativa* L. establishment in mine mill tailings. II. Root patterns of alfalfa in iron tailings and natural soils *Plant and Soil* 73:239–246.

Shoemaker, H.E., E.O. McLean, and P.F. Pratt. 1961. Buffer methods for determining lime requirement of soils with appreciable amounts of extractable aluminum. *Soil Sci. Soc. Am. Proc.* 25:274–277.

Shuman, R. and F.W. Whicker. 1986. Intrusion of reclaimed uranium mill tilings by prairie dogs and ground squirrels. *J. Environ. Qual.* 15:21–24.

Sieg. C.H., D.W. Uresk, and R.M. Hansen. 1983. Plant-soil relationships on bentonite mine spoils and sagebrush-grassland in the northern high plains. *J. Range Mgt.* 36:289–294.

Simon, E. 1978. Heavy metals in soils, vegetation development and heavy metal tolerance in plant populations from metalliferous areas. *New Phytologist* 91:175–188.

Sims, H.P., C.B. Powter, and J.A. Campbell. 1984. Land surface reclamation: a review of the international literature. Alberta Land Conservation and Reclamation Council, Alberta Environment, Oxbridge Place, Edmonton, Alberta.

Smith, J.A., G.E. Schuman, E.J. DePuit, and T.A. Sedbrook. 1985. Wood residue and fertilizer amendment of bentonite mine spoils: I. Soil and general vegetation responses. *J. Environ. Qual.* 14:575–580.

Smith, R.A.H. and A.D. Bradshaw. 1972. Stabilization of toxic mine wastes by the use of tolerant plant populations. *Trans. Inst. Min. Metall.* 81:230–237.

Smith, R.A.H. and A.D. Brashaw. 1979. The use of metal tolerant plant populations for the reclamation of metalliferous wastes. *J. Appiled Ecol.* 16:595–612.

Sobek, A.A., W.A. Schuller, J.R. Freeman, and R.M. Smith. 1978. Field and laboratory methods applicable to overburdens and mine soils. U.S. EPA Rep. 600/2-78-054, Cincinnati, OH.

Soni, P. and H.B. Vasistha. 1986. Reclamation of mine spoils for environmental amelioration. *Indian Forester*. 112:621–632.

Sorensen, D.L., W.A. Kneib, D.B. Porcella, and B.Z. Richardson. 1980. Determining the lime requirement for the Blackbird Mine spoil. *J. Environ. Qual.* 3:162–166.

Street, H.E. and G.T. Goodman. 1966. Revegetation techniques in the lower Swansea Valley. In: K.J. Hilton (ed.) *The Lower Swansea Valley Project*, pp. 71–110. Longmans Green, London.

Thames, John L. 1977. *Reclamation and Use of Disturbed Land in the Southwest*. University of Arizona Press, Tuscon, AZ.

Thompson, J. and J. Proctor. 1983. Vegetation and soil factors on a heavy metal mine spoil heap. *New Phytologist* 94:297–308.

Tyler, G. 1975. Heavy metal pollution and mineralization of nitrogen in forest soils. *Nature* 255:701–702.

United States Department of Agriculture. 1979. User guide to soils. Mining and Reclamation in the West. General Techincal Report INT-68. Intermountain Forest and Range Experiment Station, Forest Service, Logan, UT.

Van Breeman, N. 1982. Genesis, morphology, and classification of acid sulfate soils in coastal plains. In: J.A. Kittrick, D.S. Fanning, and L.R. Hossner (eds.) *Acid Sulfate Weathering*, pp. 95–108. Soil Science Society of America. Madison, WI.

Verma, T.R. and E.N. Felix. 1977. Reclamation of orphaned mine sites and their effect on the water quality of the Lynx Creek watershed. *Hydrology and Water Resources in Arizona and the Southwest* 7:49–59.

Verma, T.R., K.L. Ludeke, and A.D. Day. 1977. Rehabilitation of copper mine tailing slopes using municipal sewage effluent. *Hydrology and Water Resources in Arizona and the Southwest* 7:61–68.

Vivyurka, A.J. 1975. Rehabilitation of tailings areas of uranium mines. 7th Annual Meeting of the Canadian Mineral Processors, January 21-23. Ottawa, Ontario, Canada.

Vrabec, S.H. 1985. Physiochemical characterization of vegetated iron ore mine spoils in central Wisconsin. Symposium on Surface Mining, Hydrology, Sedimentology and Reclamation, Dec. 9-13, Univ. of Kentucky, Lexington, KY.

Walley, K.A., M.S.I. Khan, and A.D. Bradshaw. 1974. The potential for evolution of heavy metal tolerance in plants. I. Copper and zinc tolerance in *Agrostis tenuis*. *Heredity*. 32:309–319.

Ward, S.C. 1983. Growth and fertilizer requirements of annual legumes on a sandy soil amended with fine residue from bauxite. *Reclamation Reveg. Res.* 2:177–190.

Ward, S.C. 1987. Reclaiming bauxite disposal areas in south-west Australia. In: T. Farrell (ed.) *Mining rehabilitation '87*, pp. 61–70. Australian Mining Industry Council, Canberra, Australia.

Warrington, I.S., M. Agassi, and J. Morin. 1989. Slope and phosphogypsum's effects on runoff and erosion. *Soi Sci. Soc. Am. J.* 53:1201–1205.

Warman, P.R. 1988. The Gays River mine tailing revegetation study. *Landscape Urban Planning* 16:283–288.

Whitby, L.M. and T.C. Hutchinson. 1974. Heavy metal pollution in the Sudbury mining and smelting region of Canada: II. Soil toxicity, tests. *Environ. Conserv.* 1:191–200.

Williams, S.T., T. McNeilly, and E.M.H. Wellington. 1977. The decomposition of vegetation growing on metal mine waste. *Soil Biol. Biochem.* 9:271–275.

Williamson, A. and M.S. Johnson. 1984. Reclamation of metalliferous mine wastes. In: N.W. Lepp (ed.). *Effect of heavy metal pollution on plants*, vol. 2, pp. 185–202. Applied Science, London.

Wong, I.F.T. 1970. The soil-crop suitability for Peninsular Malaysia. Soil Science Division Technical Bulletin. Department of Agriculture, Kuala Lumpur, Malaysia.

Wong, M.H. 1982. Metal co-tolerance (Cu, Pb and Zn) in *Festuca rubra*. *Environ. Res.* 29:42–47.

Wong, M.H. 1986. Reclamation of wastes contaminated by copper, lead, and zinc. *Environ. Mgt.* 10:707–713.

Wullstein, L.H. 1980. Nitrogen fixation (acetylene reduction) associated with rhizosheaths of Indian ricegrass used in stabilization of the Slick Rock, Colorado tailing, pile. *J. Range Mgt.* 33:204–206.

Zellers, M.E. and J.M. Williams. 1978. Evaluation of the phosphate deposits of Florida using the mineral availability system. Final Rep. Contract No. JO 377000. U.S. Dept. of Interior, Bureau of Mines, Pittsburg, PA.

Reclamation of Mine Land Using Municipal Sludge

W.E. Sopper

I. Introduction

A. Status of Land Disturbed by Mining

Disturbed land resulting from both surface and underground mining can result in major water quality problems as well as being unsightly and unproductive. The U.S. mining industry has disturbed over 1.48 million ha between 1930 and 1971, and only 40% of this has ever been reclaimed (Paone et al., 1978). Besides coal, sand, gravel, stone, clay, copper, iron ore, phosphate rock, and other minerals account for most of the mining. Table 1 shows the status of lands in the eastern U.S. disturbed by surface mining, including both land requiring reclamation by law and abandoned mine lands for which there is no legal requirement for reclamation. Pennsylvania, Ohio, Kentucky, and Illinois each have over 40 500 ha and West

Advances in Soil Science, Volume 17

Table 1. Status of land disturbed by surface mining in the eastern U.S. as of July 1, 1977

State	Reclamation not required by any law (ha)			Reclamation required by law (ha)
	Coal mines	Sand and gravel pits	Other mined areas	
Alabama	29 278	8 954	10 603	188 556
Connecticut	—	6 780	319	—
Delaware	—	1 179	25	—
Florida	—	5 883	103 932	9 836
Georgia	990	3 230	15 301	7 759
Illinois	64 642	11 709	7 593	21 885
Indiana	40 688	6 501	3 408	32 664
Kentucky	103 621	1 328	3 034	64 515
Maine	—	12 606	1 214	1 302
Maryland	4 906	6 954	1 180	6 957
Massachusetts	—	12 977	4 184	—
Michigan	57	22 310	11 135	7 992
Missouri	32 181	2 235	13 868	6 429
New Hampshire	—	5 154	169	—
New Jersey	—	9 967	2 256	—
New York	—	18 993	9 837	8 511
North Carolina	—	7 697	3 524	4 457
Ohio	100 872	15 908	11 077	4 137
Pennsylvania	121 500	10 530	18 427	40 500
Rhode Island	—	1 050	—	—
South Carolina	—	5 451	2 155	3 073
Tennessee	13 247	2 333	1 393	2 054
Vermont	—	1 723	866	177
Virginia	12 938	3 125	1 318	5 732
West Virginia	37 473	1 844	403	3 102
Wisconsin	—	21 664	4 220	5 973

From U.S. Department of Agriculture (1980)

Virginia, Alabama, and Missouri over 20 250 ha of unreclaimed coal-mined land. Florida has more than 81 000 ha of unreclaimed land after phosphate and other mining activities.

Surface mining, half of which is for coal, has disturbed over 1.6 million ha in the U.S. In the populated eastern half of the U.S., additional thousands of hectares will be disturbed each year (Table 2).

Although most legislative and technical experiences in land reclamation have involved land surface mined for coal, the information may be applied, with modifications, to other types of disturbed or marginal land. Forestry-related activities, roadway and other construction, and deposition of

Table 2. Projected regional land use for coal production from surface mining

Region	1975	1977	1980 (ha)	1985	1990
Northern Appalachia	8019	8302	9396	10813	14013
Southern Appalachia	5508	5710	6804	8505	10287
Midwest	7209	7411	8302	9922	11704
Gulf	810	1417	3888	5791	7209

From Paone et al. (1978)

dredge materials and fly-ash from coal-fired power plants create large tracts of wasteland which are often difficult to reclaim with conventional techniques and offer considerable potential for municipal sludge amendments. Large acreages of forest are harvested or devastated by forest fires, landslides, and other natural disasters each year which require reestablishment of trees for return to productivity.

About 400 million m^3 of sediment are dredged each year in the maintenance and establishment of waterways and harbors. Many sites where dredge spoils are deposited are highly acidic and of low productivity. Eroding sediment pollutes nearby waterways. Municipal sludge has been successfully used to stabilize and revegetate acidic dredge spoils along the Chesapeake and Delaware Canal (Palazzo, 1977).

Nearly 700 million metric tons of fly ash, cinders, and bottom ash from coal-fired power plants have been produced since the end of World War II (Palazzo, 1977), and with the increasing construction of new power plants, it is estimated that about 70 million metric tons of ash and 10 million metric tons of flue gas desulfurization sludge per year (U.S. EPA, 1979) will be produced. Fly ash contains some essential plant nutrients, and studies have evaluated the waste material in combination with municipal sludge for reclaiming derelict areas (Sutton, 1980; Covey, 1980).

Another area where sludge could be beneficially used is near construction sites and roadways. Over 10 million ha are occupied by public roads and highways in the U.S. Some work has been done using sludge to stabilize and revegetate these often eroded and unproductive soils (Gaskin et al., 1977; Palazzo et al., 1980), and sludge appears to have potential as an amendment along roads and right of ways.

B. Surface Mining Control and Reclamation Act of 1977

Revegetation of disturbed lands is currently an area of environmental concern and active research, as well as practical application, particularly since the 1977 federal Surface Mining Control and Reclamation Act established strict regulations for the revegetation of currently mined land. The act (PL 95-87, Sec. 515) requires that a diverse, effective, and permanent vegetative cover of the same seasonal variety native to the area of land to be affected must be established and must be capable of self-regeneration and plant succession at least equal in extent of cover to the natural vegetation of the area. In March 1982 amended regulations were published. These rules are currently set forth in 30 CFR 816 and 817 (Federal Register, 1982). Specifically, they state that:

1. The permanent vegetative cover of the area must be at least equal in extent of cover to the natural vegetation of the area and must achieve productivity levels compatible with the approved postmining land use. Both native and introduced vegetation species may be used.
2. The period of responsibility initiates after the last year of augmented seeding, fertilizing, irrigation, or other work which ensures revegetation success.
3. In areas of more than 66 cm of average annual precipitation, the period of extended responsibility will continue for not less than 5 years. In areas with 66 cm of precipitation or less, the period of responsibility will continue for not less than 10 years.
4. Normal husbandry practices essential for plant establishment would be permitted during the period of responsibility so long as they can reasonably be expected to continue after bond release.
5. In areas of more than 66 cm of precipitation, the vegetative cover shall be equal to the success standard only during the growing season of the last year of the responsibility period unless 2 years would be required by the regulatory authority. In areas with less than 66 cm, the vegetative cover must be equal to the success standard for the last 2 years of the responsibility period.
6. The ground cover, productivity, or tree stocking of the revegetated area shall be considered equal to the success standard approved by the regulatory authority when the parameters are fully equivalent with 90% statistical confidence.

It will be difficult to meet these requirements using current reclamation techniques. New methods will have to be developed and larger amounts of lime, fertilizer, and seed will undoubtedly be needed. Soil amendments, mulching, and even irrigation may be required on some sites.

In addition to land disturbed by coal mining, other areas continually in need of reclamation in the U.S. include borrow pits, dredge spoils, construction sites, quarries, gravel pits, clear-cut and burned forests, and shifting sand dunes.

The problem of disposing of ever-increasing amounts of municipal sewage sludge is one which must be faced by many U.S. municipalities, particularly with the increasing cost of incineration, the decreasing land available for landfills, and the current controversy and concern over ocean dumping. It is estimated that more than 7.7 million dry tons of municipal sludge are currently produced each year by the 15 300 public-owned treatment plants in the U.S. Approximately 25% of this is being land applied for its fertilizer and organic matter value (Federal Register, 1989).

Both problems, that of devastated lands and that of sludge disposal, may be alleviated by sludge utilization and recycling, i.e., using sewage sludge to aid revegetation.

During the past two decades, extensive research has been carried out in the U.S. on the feasibility of reclaiming disturbed land with sludge. The research as well as large-scale practical projects and commercial ventures have shown that stabilized municipal sludge is an excellent soil amendment and chemical fertilizer substitute. Consequently, there has been considerable use of sludge for the production of agricultural crops. One disadvantage, however, is that sludge may contain every element or compound found in wastes from domestic and industrial sources. Thus, some concern has been raised about the potential introduction of these elements, particularly heavy metals, into the human food chain. The U.S. Environmental Protection Agency as well as some states have developed guidelines and regulations governing sludge applications in agriculture (Bastian et al., 1982).

C. Federal and State Regulations Governing Use of Sludge on Mine Land

Most of these guidelines set limits on sludge application rates based on nitrogen and other nutrient requirements of the vegetation as well as trace metal loadings. For instance, the U.S. Environmental Protection Agency has developed guidelines concerning the maximum amounts of Pb, Zn, Cu, Ni, and Cd allowable on agricultural land used for growing food-chain crops (U.S. EPA, 1977). Food-chain crops are typically defined as those crops that can enter the human diet either with (wheat, corn) or without (leafy vegetables) processing. Researchers in the USDA and Agricultural Experiment Stations have proposed similar trace metal limits which would allow the growth of all crops after termination of sludge applications, provided the soil pH is maintained at 6.5 or above (Knezek and Miller, 1976, Baker et al., 1985). No federal guidelines have been issued specifically governing the use of sludge for reclamation purposes. Although the above guidelines were developed for agricultural applications, it is suggested that they also be considered for reclamation applications unless there are more specific state regulations.

The metal loadings suggested by the U.S. Environmental Protection Agency are given in Table 3. They are based upon the soil cation exchange capacity (CEC). The use of soil CEC was based on the assumption that metal solubility and thus plant availability tend to decrease with increasing CEC in most soils of the U.S.

These are guidelines and not regulations. The U.S. Environmental Protection Agency has only issued regulations for cadmium as part of the requirements under the Resource Conservation and Recovery Act of

Table 3. EPA recommended maximum amounts of trace metal loadings for agricultural cropland

	Soil cation exchange capacity ($cmol\ kg^{-1}$)		
	<5	5–15	>15
Metal		Amount of metal ($kg\ ha^{-1}$)	
Pb	560	1120	2240
Zn	280	560	1120
Cu	140	280	560
Ni	140	280	560
Cd	6	11	22

From U.S. EPA (1983)

1976 and the Clean Water Act of 1977 (Federal Register, 1979). These regulations are specifically for cadmium additions to cropland and can be summarized as follows:

1. The pH of the soil must be $\geqslant 6.5$ at the time of sludge application.
2. Annual cadmium additions are limited to 0.5 kg ha^{-1} $year^{-1}$ if leafy vegetables or tobacco are grown.
3. For other food-chain crops, the annual cadmium additions cannot exceed 0.5 kg ha^{-1} $year^{-1}$.
4. The cumulative cadmium applied must be <5 kg^{-1} ha^{-1} if the background soil pH is $\leqslant 6.5$.
5. The cumulative cadmium applied is as shown in Table 3 for soils with a background pH $\geqslant 6.5$ and for soils with a background pH $\leqslant 6.5$ provided the pH is 6.5 at the time food- chain crops are grown.

The U.S. Environmental Protection Agency, Food and Drug Administration, and the U.S. Department of Agriculture have also recommended that cumulative additions of lead to agricultural soils be limited to a maximum of 800 kg ha^{-1} rather than the values shown in Table 3 (EPA-FDA-USDA, 1981).

For soils used for growth of animal feed only, neither annual nor cumulative cadmium application limits have been established, but soil pH must be 6.5 and a plan is needed to show that the crop will not directly enter the human diet.

A general statement of federal policy and guidance in relation to land application of municipal sewage sludge for the production of fruits and vegetables has been published (EPA-FDA-USDA, 1981). The three federal agencies agree that the use of high quality sludges, coupled with proper management procedures, should safeguard the consumer from contaminated crops and minimize any potential adverse effect on the environment.

In addition, some states have even more stringent guidelines concerning sludge application on the land. For instance, in 1988 the Pennsylvania De-

Table 4. Pennsylvania recommended maximum trace element loading rates for land reclamation

	Maximum loading rate (kg ha^{-1})	
Constituent	Land reclamation	Land reclamation for farming
Cd	5.6	3.4
Cu	140	84
Cr	560	336
Pb	560	336
Hg	1.7	1.1
Ni	56	33
Zn	280	168

From Pennsylvania Department of Environmental Resources, 1988

partment of Environmental Resources (PDER) issued *Guidelines for Sewage Sludge Use for Land Reclamation* (Pennsylvania Department of Environmental Resources, 1988). These guidelines state that due to the high permeability of mine spoils and low retention of organic matter, sufficient nitrogen in excess of the crop requirement must be provided in order to establish growth. To provide sufficient nitrogen a maximum application rate of 134 metric tons ha^{-1} may be utilized for land reclamation. In addition, the application is further limited according to the trace metal content of the sludge and application rates may not exceed the limits given in Table 4.

The Pennsylvania state guidelines further require that the soil pH be adjusted to 6.0 during the first year of sludge application and maintained at 6.5 for 2 years following final sludge application. Liming is required to immobilize the trace metals in order to reduce their availability for plant uptake and to prevent their leaching into groundwater.

Other requirements include the following:

1. Sludge is to be incorporated within 24 h after application.
2. Sludge is not to be applied when the ground is saturated, snow covered, frozen, or during periods of rain.
3. Sludge is not to be applied within 30 m of streams, 90 m of water supplies, 8 m of bedrock outcrops, 15 m of property lines, or 90 m of occupied dwellings.
4. Sludge for revegetation of inactive mines or active coal refuse piles is not to be applied to slopes exceeding 15%.
5. Dairy cattle must not be allowed to graze land for at least 2 months after sludge application.

The potential for successful reclamation with municipal sludge is tremendous. Most of the highly beneficial properties of sludge as a soil amendment come from its high organic matter content. Although it has long been shown to increase productivity on good agricultural soils, sludge organic matter is extremely important where topsoil is inadequate in amount or quality, e.g., on sites that were previously in forest, and especially for areas mined and left barren for many years before the 1977 Act, where no topsoil exists. It is on land like this that the benefits of sludge are most evident. The establishment as well as the continued maintenance of cover for the minimum required 5 years is difficult on acid, eroded, infertile, compacted, and stony spoils that have no buffering capacity for temperature extremes and are incapable of retaining sufficient moisture for plants during periods of water stress. Although organic matter is the single most important component in the improvement of soil physical properties, sludges also contain neutralizing compounds and fertilizer elements that improve spoil pH and fertility.

While the benefits of using sludge to reclaim land seem obvious, there is still some reluctance on the part of landowners, local citizens, and local government officials to accept its use for reclamation. Due to the nature of devastated lands, larger amounts of sludge are used than for farmlands, but usually only a single application is made which allows the vegetation to become self-sustaining. The greatest obstacle appears to be the lack of knowledge on the part of the general public about the possible impacts of sludge on soils, plants, groundwater, surface water, and animal and human health. These impacts must be known in order to make rational decisions concerning the benefits and risks.

There is a large amount of research and demonstrated experience and information available on all types of land application projects, initiated to address public concerns. The increasing number of successful projects across the country clearly shows the value of sludge as a resource rather than disposing of it as a waste product.

Considerable detailed information is available on the planning, economics, social and legal problems, engineering, agricultural, ecological, and health-related aspects of reclaiming disturbed land with sewage sludge over a wide range of situations. Much of this information is summarized in the "Process Design Manual for Land Application of Municipal Sludge" (U.S. EPA, 1983) and in "A Guide for Revegetation of Mined Land in Eastern United States Using Municipal Sludge" (Sopper and Seaker, 1983).

II. Review of Land Reclamation Projects Using Municipal Sludge

A. Overview

During the past 20 years a considerable amount of research has been conducted on the feasibility of using municipal sludge for the revegetation of mine land. Some of the more significant projects are summarized in Table 5. An attempt has been made to review these publications and to summa-

Table 5. Recent land reclamation projects with municipal sludges

Type of Disturbed land	State	Sludge: Type[a]	Sludge: Application rates (mg ha^{-1})	Plant/animal studied	Parameters tested[b]	Reference
Acid strip-mine spoil	PA	Dig.-D + effluent	5–20 cm	Ryegrass Hybrid poplar	WA	McCormick and Borden (1973)
Deep mine anthracite refuse	PA	Dig.-D	0, 40–150	10 Tree species 5 Grass species 5 Legume species	GR PA SA WA	Kerr et al. (1979)
Acid strip-mine spoil and deep mine refuse	PA	Dig.-L + effluent	0–51 cm	8 Tree species 8 Grass species 8 Legume species	GR WA SA	Kardos et al. (1979)
Acid strip-mine spoil	PA VA	Dig.-L,D	198–730	Tall fescue Red top Ladino clover Birdsfoot trefoil Ryegrass	GR	Hill et al. (1979)
Acid strip-mine spoil and deep mine refuse	PA	Dig.-D,L,C	0, 7–202	Tall fescue Orchardgrass Birdsfoot trefoil Crownvetch	GR PA SA WA	Sopper and Kerr (1982); Sopper et al. (1981)
Acid strip-mine spoil and deep mine refuse	PA	Dig.-D,L,C	0, 7–202	Tall fescue Orchardgrass Birdsfoot trefoil Crownvetch	GR PA SA WA	Sopper and Seaker (1982)

Table 5 (Cont.)

Type of Disturbed land	State	Sludge		Plant/animal studied	Parameters tested[b]	Reference
		Type[a]	Application rates (mg ha^{-1})			
Acid strip-mine spoil and deep mine refuse	PA	Dig.-D,L,C	11–202	Birdsfoot trefoil Tall fescue Orchardgrass	PA	Seaker and Sopper (1982)
Deep mine anthracite refuse	PA	Dig.,D	0, 80, 108	Tall fescue Orchardgrass Birdsfoot trefoil Crownvetch	GR PA SA WA	Seaker and Sopper (1983)
Acid strip-mine spoil	PA	Dig.,D,L	0, 7, 11, 90, 184	Tall fescue Orchardgrass Birdsfoot trefoil Crownvetch	GR, PA, SA, WA	Seaker and Sopper (1984)
Acid strip-mine spoil	PA	Dig.,D,C/C	134	Rabbit	PA, SA, AH	Dressler et al. (1986)
Zinc smelter site	PA	Dig.,D	47	5 Grass species 5 Legume species 11 Tree species	GR PA SA	Sopper (1989a) Sopper (1989b) Sopper (1989c) Sopper (1989d) Sopper (1987) Sopper and McMahon (1987) Sopper and McMahon (1988a) Sopper and McMahon (1988b)

Acid strip-mine spoil	PA	Dig.,D,C/C	120–134	Micro-organisms	SO, SA	Sopper and Seaker (1987) Seaker and Sopper (1988a)
Acid strip-mine spoil	PA	Dig.,D,C/C	120–134	NA	SA	Seaker and Sopper (1988b)
Acid strip-mine spoil	PA	Dig.,D,C/C	128	Micro-organisms	SO, SA	Sopper and Seaker (1988)
Acid strip-mine spoil	PA	Dig.,D,C/C	134	Vole	PA, SA, AH	Alberici et al. (1989)
Coal mine spoil	CO	Dig.,D	0, 14, 28, 55	8 Grass species	SA, PA GR	Topper and Sabey (1986)
Coal spoil and topsoil	CO	Dig.,D	0, 40, 80, 120	Micro-organism activity	SA	Voos and Sabey (1987)
Copper mine Borrow pit Kaolin spoil Marginal land	TN GA	Dig.-D	0–275	Pine species Sweetgum	GR	Berry (1982)
Acid strip-mine spoil	IL	Dig.-L	0, 31–121	Tall fesue Weeping lovegrass	GR WA	Boesch (1974)
Acid strip-mine spoil	IL	Dig.-L	0, 78, 304	Tall fescue Weeping lovegrass	GR SA WA	Lejcher and Kunkle (1974)
Acid strip-mine spoil	IL	Dig.-L	0, 314, 627	Tall fescue Alfalfa	GR PA	Stucky and Newman (1977)
Strip-mine spoil	IL	Dig.-L	56	Corn	PA	Blessin and Garcia (1979)

Table 5 (Cont.)

Type of Disturbed land	State	Sludge		Plant/animal studied	Parameters tested[b]	Reference
		Type[a]	Application rates (mg ha^{-1})			
Strip-mine spoil	IL	Dig.-L	0–997	Blackbirds	AH	Gaffney and Ellertson (1979)
Strip-mine spoil	IL	Dig.-L	NI	NI	PO	Lue-Hing et al. (1979)
Strip-mine spoil	IL	Dig.-L	NI	NI	SO	Sundberg et al. (1979)
Strip-mine spoil	IL	Dig.-L	25–128/year	Corn Pheasant, swine	AH	Hinesly et al. (1979)
Strip-mine spoil	IL	Dig.-L	NI	Cattle	AH PO SA PA	Fitzgerald (1979)
Acid strip-mine spoil	IL	Dig.-L	448–997	8 Herbaceous species 18 Tree species	WA	Jones and Cunningham (1979)
Acid strip-mine spoil	IL	Dig.-L	448–997	7 Legumes 10 Grasses	GR PA SA	Stucky and Bauer (1979)
Acid strip-mine spoil	IL	Dig.-L	448–997	12 Tree species	GR	Roth et al. (1979)
Acid strip-mine	IL	Dig.-L	448–997	8 Tree species	PA	Svoboda et al., (1979)
Calcareous strip-mine spoil	IL	Dig.-L	0.8–85.8	Corn	GR PA	Peterson et at. (1979)

					SA WA	
Coal refuse	IL	D	225–900	Tall fescue Redtop Reed canarygrass	GR SA	Joost, et al. (1981)
Acid strip-mine spoil	KY	Dig.-D	0, 34–269	Corn Soybeans	GR PA SA	Feuerbacher et al. (1980)
Acid strip-mine spoil	KY	(NI)	28–96	European alder Blacklocust Cottonwood Loblolly pine Northern red oak	GR PA SA	Schneider et al. (1981)
Strip-mine spoil	IL	Dig.-D	0, 224–896	9 Grass species Corn Rye	GR PA SA	Hinesly et al. (1982)
Acid strip-mine spoil	IL	Dig.-L	448–997	5 Tree species	PA SA	Roth et al. (1982)
Acid strip-mine spoil	IL	Dig.-L	336–672	Annual rye Orchardgrass Tall fescue	WA	Urie et al. (1982)
Calcareous strip-mine spoil	IL	Dig.-L	0–453	Corn	GR PA WA	Peterson et al. (1982)
Strip-mine spoil	IL	Dig.-L	0, 11–45	Grasses Legumes Cattle	GR PA SA AH PO	Fitzgerald (1982)
Calcareous Strip-mine spoil	IL	Dig.,L.	174	Earthworms	SO	Pietz et al. (1984)

Table 5 (Cont.)

Type of Disturbed land	State	Sludge: Type[a]	Sludge: Application rates (mg ha^{-1})	Plant/animal studied	Parameters tested[b]	Reference
Acidic coal refuse	IL	(NI)	225, 450, 900	Tall fescue Reed canarygrass Red top	SA, PA, GR	Joost et al. (1987)
Acidic coal refuse	IL	Dig.,D,L	237, 305	Bromegrass Tall fescue Alfalfa	SA, PA WA, GA	Pietz et al. (1989a,b,c)
Acid strip-mine spoil	MD	Dig. com-posted	0, 56–224	Tall fescue Birdsfoot trefoil	GR PA SA	Griebel et al. (1979)
Gravel spoils	MD	C	0, 40–160	Corn Beans	GR PA WA	Hornick (1982)
Acid strip-mine spoil	OH	D	658	Forage	GR PA	Sutton and Vimmerstedt (1974)
Acid strip-mine spoil	OH	Dig.-D	11–716	Tall fescue	GR PA SA WA	Haghiri and Sutton (1982)
Zn smelter surroundings	OK	Dig.-L + effluent	2.5–34 cm	10 grass spp. 1 Legume	GR PA SA	Franks et al. (1982)
Lignite over-burden	TX	Dig.,D	56	NA	SA, WA	Hornby et al. (1986)
Acid strip-mine spoil	VA	C	0, 159–412	Virginia pine Tall fescue	GR SA	Scanlon et al. (1973)

				Perennial ryegrass Annual ryegrass		
Acid strip-mine spoil	VA	Dig.-D	Various (NI)	(NI)	SA	Younos and Smolen (1981)
Abandoned pyrite mine	VA	Dig.-D	82–260	Tall fescue Lespedeza Weeping lovegrass Wheat, rye, oats	GR PA SA WA	Hinkle (1982)
Sandstone and siltstone mine soil	VA	Dig.,D	22, 56, 112, 224	Tall fescue	GR SA, PA	Roberts et al. (1988)
Acid strip-mine spoil	WV	D	0–224	Tall fescue	GR PA SA	Mathias et al. (1979)
Acid strip-mine spoil	WV	Dig.-D,C	Various (NI)	Blueberries	GR PA	Tunison et al. (1982)
Overburden mine soil	WV	Dig.,D	0, 22.4, 44.8, 78.4	Red clover Tall fescue Orchardgrass Birdsfoot trefoil	SA, PA GR	Skousen (1988)
Iron ore tailings	WI	Dig.-D	42–85	5 Native prairie grasses 4 prairie forbes Foxtail	GR	Morrison and Hardell (1982)
Taconite tailings	WI	Dig.-D	28–115	4 Grass-legume Mixtures	GR	Cavey and Bowles (1982)

[a] Dig., digested; L, liquid; D, dewatered; C, composted; NI, no information
[b] GR, growth responses; PA, plant tissue analysis; SA, Soil analysis; WA, water analysis; SO, soil organisms; PO, pathogenic organisms; AH, animal health; NA, not applicable

rize the results in terms of evaluating the effects of sludge applications on vegetation growth responses, vegetation quality, physical, chemical, and biological properties of the mine soil, soil percolate and groundwater quality, and animal nutrition and health.

B. Effects on Vegetation

1. Growth Responses

a. Grass and Legume Species

The productivity and fertility of lands disturbed by mining activities have been substantially improved in most cases by sludge applications. Larger yield increases have been realized on sludge-amended mine land than on the same type of land amended with inorganic fertilizers.

In the Midwest, nine grass species were seeded on calcareous strip-mine spoil with pH between 6.0 and 7.5. Tall fescue,[1] perennial ryegrass, and western wheatgrass showed the most rapid establishment and most vigorous growth where relatively high sludge loading rates (224, 448, and 896 Mg ha^{-1}) were applied (Hinesly et al., 1982). In a greenhouse study, four different sludges were applied to acidic spoils. Tall fescue yield increased with increasing rate of two of the sludges, but high metal concentrations of the other two sludges inhibited fescue growth (Haghiri and Sutton, 1982). In another greenhouse study Stucky and Newman (1977) grew tall fescue and alfalfa for 2 years using 314 and 627 Mg ha^{-1} of sludge in acidic spoils. The high application significantly increased yields, especially for alfalfa. Where sludge was applied on Ohio mine spoil at 658 Mg ha^{-1} without lime, forage yield the first growing season was greater than 4.5 Mg ha^{-1} (Sutton and Vimmerstedt, 1974). On the Palzo strip-mine site in Illinois, 16 seed mixtures were evaluated on plots receiving 426 to 997 Mg ha^{-1} of sludge (Stucky and Bauer, 1979; Stucky et al., 1980). Perennial rye was a very successful initial cover crop, while Bermuda, orchard, reed canarygrass, and tall fescue grasses were recommended for permanent cover, along with the legumes red clover, lespedezas, and birdsfoot trefoil. Boesch (1974) gives a vivid description of the devastated Palzo site and the sequence of events that led to large-scale reclamation with Chicago sludge. In 1970, demonstration plots were seeded with tall fescue and weeping lovegrass, which did not germinate on control plots, but produced a complete groundcover in 2 months where 271 Mg ha^{-1} of sludge was applied. On coal refuse in Illinois, Joost and others (1981) observed a more than adequate cover with tall fescue, redtop, and reed canarygrass 2 months after seeding, even with high rates of high-metal sludge. Redtop was the most successful.

---

[1] A complete list of common names and scientific names of vegetation discussed in this chapter is given in Appendix Table A-1.

In the East, Kardos and others (1979) tested 10 grass and 10 legume species in boxes of bituminous mine spoil and anthracite refuse amended with liquid digested sludge. Sludge and effluent irrigation at 2.5 and 5.0 cm per week and totalling 59 to 147 cm per year detoxified the highly acidic materials and established a vegetation cover. The most successful species were weeping lovegrass, ladino cover, Iroquois alfalfa, sericea lespedeza, and red clover. Subsequently, Sopper and Kerr (1982) grew a lush cover of tall fescue, orchardgrass, birdsfoot trefoil, and crownvetch on several 4-ha demonstration sites in Pennsylvania. The sites were amended with several different types of municipal sludge applied at rates ranging from 7 to 202 Mg ha^{-1}. The rationale for the mixture is that the grasses provide quick cover while the legumes eventually take over to provide the permanent cover. Following the demonstrations, over 1500 ha have been successfully reclaimed using Philadelphia sludge, with phenomenal annual increases in dry matter production that surpass the Pennsylvania strip-mine revegetation requirements.

Yields of tall fescue on sludge-amended mine spoil (112 and 224 Mg ha^{-1}) in West Virginia were over 11 000 kg ha^{-1}, an 818% increase over controls (Mathias et al., 1979). Both Mathias and others (1979) and Hinesly and others (1982) observed some yield decreases over three growing seasons when sludge was applied only once and the crop was harvested annually with no additional amendments. Where the vegetation was not harvested, however, the recycling of nutrients and build-up of organic matter resulted in annual yield increases (Sopper and Seaker, 1982). The slow-release fertilizing action of sludge provided a "residual effect" for forage species. On anthracite refuse, dry matter production was greatest the first growing season, with 40–75 Mg ha^{-1} applications of sludge, but by the third growing season the 150 Mg ha^{-1} sludge treatment, with its added residual effect, gave the best herbaceous cover (Kerr et al., 1979). Composted Washington, D.C., sludge at 112 Mg ha^{-1} produced forage cover in accordance with Maryland strip-mine laws (i.e., 80% cover and 10% legume species after 2 years). Rates lower than 112 Mg ha^{-1} were not as effective, but rates greater than 112 Mg ha^{-1} resulted in superior performance (Griebel et al., 1979). In Virginia, the same sludge grew a very successful tall fescue and weeping lovegrass cover on two abandoned pyrite mines (Hinkle, 1982), but a less than adequate cover on another site (Hill et al., 1979) due to low pH and drought conditions. The same author, however, observed threefold yield increases over plots amended with a standard inorganic fertilizer when Pennsylvania coal spoils were amended with Williamsport sludge at 730 Mg ha^{-1}.

In Virginia, 159 and 412 Mg ha^{-1} of composted sludge without limestone amendments established a good cover of tall fescue and perennial and annual ryegrass on strip-mine spoil, along with volunteer native species. By the fourth growing season, results were excellent whereas those on plots amended with inorganic fertilizer were poor. In another study, tall fescue, annual ryegrass, and sericea lespedeza produced a significantly denser cover with sludge than with inorganic fertilizer. Again, volunteer species were more abundant on the sludged plots. At 58 Mg ha^{-1} the sludge was more effective than at 31 Mg ha^{-1} which points out the need to supply sufficient plant nutrients (Scanlon et al., 1973).

Two studies on iron ore tailings revegetation in Wisconsin found sludge to be far superior to inorganic fertilizers. Some of the successful species included sideoats grama grass, Canada wild rye, foxtail grass (Morrison and Hardell, 1982), smooth brome grass, alsike clover, and alfalfa (Cavey and Bowles, 1982). Soil contaminated in the vicinity of a zinc smelter in Oklahoma was successfully revegetated with sludge reinforced with urea-N, whereas inorganic fertilizer was ineffective. Switchgrass and kleingrass showed excellent response over three growing seasons (Franks et al., 1982).

Topper and Sabey (1986) reported that sewage sludge applications (0, 14, 28, 55, and 83 Mg ha^{-1}) on coal mine spoil in Colorado significantly increased the above-ground biomass and percent canopy cover of a pasture grass mixture over a control for two seasons of growth. Total production on the sludge treatments increased linearly up to 4300 kg ha^{-1} at the highest rate. The lowest sludge application rate (14 Mg ha^{-1}) produced 2400 kg ha^{-1} of above-ground biomass in comparison with 300 kg ha^{-1} on the control plots.

In Illinois, successful establishment of three forage grasses (reed canarygrass, tall fescue, and redtop) was achieved with applications of dried sludge at rates of 225, 450, and 900 Mg ha^{-1} on coal refuse. Most treatments maintained greater than 80% cover over the 4-year study period. Herbage yield was not significantly different among amended treatments for any of the three grasses. (Joost et al., 1987).

Roberts et al. (1988) studied the effects of sewage sludge applications on a 2:1 sandstone-siltstone spoil mixture to facilitate establishment of tall fescue in Virginia. Aerobically digested sludge was applied on plots at rates of 22, 56, 112, and 124 Mg ha^{-1}. Tall fescue production generally increased with increasing sludge application rates. Treatments amended with >56 Mg ha^{-1} sludge supported about twice as much fescue growth as other treatments for all 5 years of the study. First year standing biomass was 3.9, 7.7, 9.3, and 10.7 Mg ha^{-1} for the four sludge application rates, respectively in comparison to 5.8 Mg ha^{-1} on the control plot.

In Illinois, anaerobically digested sludge, lime, and gypsum, and various mixtures of these amendments were applied to plots on coal refuse material and seeded with alfalfa, bromegrass, and tall fescue (Pietz et al., 1989b). Plant yields were measured for 3 years (1978–80). Plant cover and dry matter yields increased each year on all sludge-treated plots. In 1980, the sludge (542 Mg ha^{-1}) and lime treatment had the highest percent cover (89%) and highest dry matter yield (6 Mg ha^{-1}).

Nurse crops including wheat, rye, and oats (Hinkle, 1982), and barley

(Cavey and Bowles, 1982) have been shown to promote vegetative cover significantly.

In general, studies show that good plant cover can be established on many types of disturbed land using municipal sludge, which is superior to inorganic fertilizer in such situations. Of course, plant performance varies considerably with species, and an appropriate seed mixture should be chosen carefully. Planting date is also crucial. On the Palzo site, winter rye seeded 1 month late resulted in a significant decrease in forage yield measured the following spring (Stucky and Bauer, 1979). Table 6 lists some successful plant species used in various sludge-reclamation projects. Differences in soil, climate, influence mixtures chosen for revegetation. For example, tall fescue is an excellent choice on highly acidic strip-mine spoil in Pennsylvania (Sopper and Kerr, 1982), but that species failed completely on sludge-amended soils contaminated by a nearby zinc smelter in Oklahoma (Franks et al., 1982). There, kleingrass and switchgrass were most successful.

b. Field Crops

Several studies have looked at yields of corn, beans, and even blueberries grown on sludge-amended disturbed land. On sand and gravel spoils with low pH and a heterogeneous profile, sweet corn, field corn, and bush bean biomass were significantly increased by applications of Washington D.C. sludge. In general, bean yields were increased by 40–80 Mg ha^{-1} applications, but corn ear yields were not (Hornick, 1982). In Illinois, the yields of corn, soybeans, small grains, and forages over 2 to 7 years were highly variable where Chicago sludge at rates up to 453 Mg ha^{-1} was applied to strip-mine soils. Yields ranged from poor to excellent, with moisture stress and nutrient availability being the controlling factors (Peterson et al., 1982). Yield fluctuations in corn were not related to heavy metal concentrations in the plants (Hinesly et al., 1979). In another study, an increase in corn production of 2666 kg ha^{-1} was observed when Chicago sludge was applied to strip-mine spoils at 56 Mg ha^{-1} (Blessin and Garcia, 1979). Spoil pH appeared to be the major factor governing yield of corn and soybeans on Kentucky mine sites amended with 15 to 120 Mg ha^{-1} of sludge. Some varieties of soybeans yielded twice that of others; corn varietal differences were not as dramatic (Feuerbacher et al., 1980).

In a greenhouse study, Tunison and others (1982) grew highbush blueberries using a low-metal sludge from Waynesburg, PA. Without composting, the plants were severely chlorotic and mortality was high, but composting resulted in healthier plants and increased berry production.

c. Trees

While ground cover is the crucial element in initial site stabilization, the potential of woody species for use in sludge management schemes is great,

Table 6. Some successful plant species and species mixtures used in various sludge-reclamation projects

State	Species	Seeding rate ($kg\ ha^{-1}$)	Reference
CO	Slender wheatgrass[a]	5.1	Topper and Sabey (1986)
	Intermediate wheatgrass[a]	4.8	
	Pubescent wheatgrass[a]	4.7	
	Crested wheatgrass[a]	3.8	
	Smooth brome[a]	4.6	
	Meadow brome[a]	2.6	
	Timothy[a]	1.5	
	Orchardgrass[a]	1.4	
IL	Tall fescue	22	Lejcher and Kunkle (1973)
	Weeping lovegrass	8	
IL	K-31 tall fescue	22	Boesch (1974)
	Weeping lovegrass	7.8	
IL	Common bermudagrass[a]	11	Stucky and Bauer (1979)
	Sericea lespedeza[a]	28	
	Kobe lespedeza[a]	11	
	Perennial rye grass[a]	22	
	Potomac orchardgrass[b]	17	
	Sericea lespedeza[b]	22	
	Kobe lespedeza[b]	11	
	Potomac orchardgrass[c]	22	
	Penngift crownvetch[c]	17	
IL	Tall fescue	25	Hinesly et al. (1982)
	Perennial ryegrass	25	
	Western wheatgrass	25	
IL	Reed canarygrass	34	Joost et al. (1987)
	Tall fescue	46	
	Redtop	17	
IL	Alfalfa[a]	22.9	Pietz et al. (1989b)
	Bromegrass[a]	9.5	
	Tall fescue[a]	9.1	
MD	Tall fescue	40	Griebel et al. (1979)
	Birdsfoot trefoil	10	
OH	Fall, balbo rye	9.6 ($bu\ ha^{-1}$)	Sutton and Vimmerstedt (1974)
	Spr., K-31 tall fescue[a]	11	
	Korean lespedeza[a]	3.4	
	Sweet clover[a]	3.5	
	Orchardgrass[a]	3.3	
OK	Switchgrass	154	Franks et al. (1982)
	Kleingrass	154	
PA	Reed canarygrass	224	Kerr et al. (1979)
	Tall fescue	224	
	Orchardgrass	224	

Table 6. (Cont.)

State	Species	Seeding rate (kg ha^{-1})	Reference
	Birdsfoot trefoil	224	
	Crownvetch	224	
PA	Ky-31 tall fescue[a]	39	Hill et al. (1979)
	Birdsfoot trefoil[a]	8	
	Rye grass[a]	6	
PA	Tall fescue[a]	22	Sopper and Kerr (1982)
	Orchardgrass[a]	22	
	Birdsfoot trefoil[a]	11	
	Crownvetch[a]	11	
VA	Tall fescue[a]	8.4	Scanlon et al. (1973)
	Perennial ryegrass[a]	8.4	
	Annual ryegrass[a]	8.4	
	Tall fescue[b]	22	
	Perennial rye[b]	22	
	Sericea lespedeza[b]	22	
	Black locust[b]	0.8	
VA	K-31 tall fescue[a]	67	Hill et al. (1979)
	Redtop[a]	5.6	
	Ladino clover[a]	5.6	
VA	Tall fescue	67.3	Hinkle (1982)
	Weeping lovegrass	22	
	Korean lespedeza	11.2	
VA	Ky-31 tall fescue	80	Roberts et al. (1988)
WI	Canada bluegrass[a]	11	Cavey and Bowles (1982)
	Red clover[a]	9.7	
	Smooth brome[b]	15.2	
	Alfalfa[b]	11	
	Western wheatgrass[c]	9.7	
	Alsike clover[c]	11	
	Barley[d]	16.5	
	Japanese millet (added to above mixtures)[d]	8.6	

[a] Species with the same superscript letter represent a seeding mixture

due to their relatively small input into the human food chain and their ability to differentially accumulate metals in specific plant organs (Schneider et al., 1981). In general, establishment of woody vegetation is enhanced by sludge. At the Palzo tract in Illinois (Roth et al., 1982), where 12 tree species were planted, the survival rate was 53% with sludge compared with 19% on the untreated area. On Illinois strip-mines and Pennsylvania anthracite refuse, hardwood species survival was greater when trees were planted along with herbaceous vegetation, but conifers did better without herbaceous cover (Kardos et al., 1979). Ground cover competition with conifers usually occurs when cover species are planted along with trees, but Berry (1982) found no weed competition problems with conifer establishment when sludge was applied at 34 Mg ha^{-1} on barren land in the southeast. Grass planting should be delayed until seedlings are established. Virginia pine was successfully established at a seeding rate of 1.4 kg ha^{-1} where there was no herbaceous cover (Scanlon et al., 1973), and the height of the trees was increased with increasing applications of composted sludge. The pines grown on spoils amended with ammonium nitrate fertilizer were markedly chlorotic, whereas those grown in composted sludge-amended spoils were green (Scanlon et al., 1973).

Of 10 tree species planted in sludge-amended spoils in Pennsylvania, hybrid poplar, black locust, and European alder gave the best results for survival and growth (Kardos et al., 1979). After 5 years of growth, the potential woody biomass produced was directly related to the amount of sludge applied, and was 10 times greater than trees not grown with sludge (Kerr et al., 1979). Hinkle (1982) noted a significant establishment of volunteer hybrid poplar on sludged pyrite mine refuse in Virginia, after an unsuccessful planting of loblolly pine during a period of drought.

Seedlings of four hardwood species and one pine, usually good performers on reclaimed sites, were grown in spoils amended with a high-metal and a low-metal sludge (Cd 710 vs 7 mg kg^{-1}; Zn 8010 vs 1075 mg kg^{-1}). Growth, vigor, and survival the first year were highest with the high-metal sludge, probably due to the higher amounts of organic matter and soil conditioning, since that sludge was put on at a higher rate. Species performance was in the following order: European alder > red oak > cottonwood > loblolly pine (Schneider et al., 1981).

2. Vegetation Quality

a. Macronutrients

Although increases in the major plant nutrients, N, P, K, Ca, and Mg, are usually observed for crops grown in sludge-amended spoils (Hinesly et al., 1982; Griebel et al., 1979; Seaker and Sopper, 1982), the effects are not always consistent. For example, Mathias and others (1979) found that foliar P always increased, K decreased, and N was not much different in plants grown on sludged spoils compared with nonsludged controls. Sludges are often deficient in K. Even 224 Mg ha^{-1} of sludge did not increase foliar K concentrations, but there was sufficient native K for plant growth in acidic Maryland spoils (Griebel et al., 1979). In Pennsylvania, where native feldspars and micas provide K, mine spoils have been successfully revegetated with municipal sludge without additional K, and no deficiency symptoms or yield reductions of forages were observed in five growing seasons (Seaker and Sopper, 1983; Seaker and Sopper, 1984).

Nitrogen levels in sweetcorn grain grown on sludge-amended sand and gravel spoils were significantly greater than fertilizer controls (Hornick, 1982), but sludge did not affect P, K, Ca, or Mg concentrations. Whole corn and soybean plants grown with sludge on acid spoils had higher levels of N, P, K, and Ca than control plants (Feuerbacher et al., 1980). If treatment differences are small (2.5 cm vs 5.0 cm of sludge), there may be no difference in forage growth response or in concentrations of N and P in the leaves (Franks et al., 1982). Schneider and others (1981) found lower foliar N concentrations in trees grown with sludge compared with those grown with inorganic fertilizers; they attributed the results to increased biomass and consequent N dilution in the leaves. Differences in sludge type and treatment affect the availability of nutrients to plants. For example, where composted sludge is used, total soil N remains high over a long period of time due to its organic nature and slow mineralization, 10% to 20% the first year (Griebel et al., 1979; Sopper and Kerr, 1981), while N applied in liquid sludges is more readily available (up to 40% the first year).

In Illinois, application of sewage sludge (542 Mg ha^{-1}) in combination with lime (89.6 Mg ha^{-1}) was an effective treatment in establishing vegetation growth on acidic coal refuse material (Pietz et al., 1989b). The effectiveness was related to the ability to supply N, P, and some K to the growing plants. Plant analyses for N, P, K, Ca, and Mg indicated that plant nutrients were in the adequate to high range for alfalfa and grasses, except for N and K. Tissue N concentration in a composite sample of grasses and alfalfa the third year after application was 15.0 g kg^{-1}, suggesting N deficiency (Martin and Matocha, 1973). Plant K concentration was 11.3 g kg^{-1}. A K concentration of $<$20.0 g kg^{-1} for cool season grasses is considered to be deficient (Martin and Matocha, 1973).

Topper and Sabey (1986) observed a linear increase in the tissue N and P of grasses with increasing levels of sludge applications ranging from 14 to 83 Mg ha^{-1} on coal mine spoil with a pH of 6.5. Roberts et al. (1988) also found that tall fescue tissue N increased as the sludge rate increased from 22 to 224 Mg ha^{-1}. Tissue N concentrations increased the second year on all sludge applications and then decreased slightly the third year after treatment. Authors concluded that the large cumulative N uptake by the grass over the first two growing seasons may have depleted the mineralizable N, thus limiting production the third year. The authors reported that P tissue concentrations were the highest on a plot treated with 56 Mg ha^{-1} sludge. As biomass increased with higher sludge applications (112 and 224 Mg ha^{-1}), P tissue concentrations were diluted. As with N, they found a decrease in the P tissue concentration the third year. The higher tissue P concentration the first 2 years after application may be related to the breakdown and release of organic and inorganic P in the sludge. The fixation of P into Fe forms in the mine spoil may also have contributed to the lower tissue concentrations. Plant tissue K content increased in all 3 years as the sludge application rate increased, but were not significantly different for the higher sludge rates (112 and 224 Mg ha^{-1}). Calcium and Mg are usually not limiting on mine spoils amended with sludge because lime is almost always added to raise the spoil pH. In the above study, tissue Ca and Mg concentrations increased as the rate of sludge addition increased.

Rock phosphate additions increased yields of fescue and birdsfoot trefoil on Maryland spoils in conjunction with one-time sludge applications (Griebel et al., 1979), but after repeated annual applications of sludge in Fulton Co., Illinois, a build-up of P in the soil of 15 times the optimum level reduced soybean yields (Hinesly et al., 1979a). Limestone incorporation along with sludge may increase concentrations of foliar Ca (Mathias et al., 1979). In most cases, the macronutrient concentrations of forages and other crops grown on sludge-amended spoils are within the range of concentrations in forages grown with inorganic fertilizers on agricultural soils.

b. Trace Metals

It is generally agreed that municipal sludge improves the capacity of spoil material to support vegetation, but questions often arise about the uptake of trace metals from sludges by forages, crops, trees, and other plants, which may be involved in food-chain dynamics.

1. Cover Species

Several studies indicate that a decrease in trace metal concentrations of sludge-grown vegetation over time may occur where a single application of sludge is used. Metal concentrations in tall fescue from Ohio mine spoils amended with up to 716 Mg ha^{-1} of sludge were considerably lower in the third growing season than they were in the first (Haghiri and Sutton, 1982). On anthracite refuse in Pennsylvania, Cu, Zn, and Cd increased in reed canarygrass tissues the first growing season after sludge was applied, but by the second and third years, with few exceptions, metal concentrations decreased to control levels or below (Kerr et al., 1979). As part of an extensive demonstration program in Pennsylvania, tall fescue, orchardgrass, birdsfoot trefoil, and crownvetch from two sludge-amended coal sites were analyzed for seven trace metals over a 5-year period (Seaker and Sopper, 1983; Seaker and Sopper, 1984). Results showed a definite decrease in heavy metal concentrations over time. With few exceptions, Cu, Zn, Cr, Pb, Co, Cd, and Ni remained well below suggested tolerance levels for agronomic crops as shown in Table 7 (Council for Agricultural Science and Technology, 1976; Melsted, 1973). Seaker and Sopper (1982) found the same trend occurring on five sites monitored for periods of 2 to 5 years. Other studies show no decline in metal concentrations, but these are often cases where sludge was applied annually for a number of years. In Illinois, metals in forage did not appreciably change, even after 4 successive years of Chicago sludge applications, although values were higher than on control plots (Fitzgerald, 1982). Hinesly et al. (1982) believe Cu and Zn may

Table 7. Suggested permissible tolerance levels of trace metals in agronomic crops

Element	Suggested tolerance level ($mg\ kg^{-1}$)
Fe	750
Mn	300
Al	200
Zn	300
Cu	150
B	100
Cr	2
Pb	10
Co	5
Ni	50
Cd	3

From Council for Agricultural Science and Technology (1976); Melsted, (1973), and University of Georgia Cooperative Extension Service (1979)

become less available with time because plant uptake decreased after sludge applications were terminated.

Several factors may account for decreasing metal concentrations over time. Iron and phosphorus added in sludge may complex with metals, forming sparingly soluble precipitates (Cunningham et al., 1975). Metals may bind with the humic fraction of sludge in the spoil-sludge mixture (Haghiri and Sutton, 1982). Although composted sludge additions increased Cu and Ni in forages after 2 years growth on an abandoned strip mine, the foliar concentrations decreased as compost application increased from 56 to 224 Mg ha^{-1} due to increased pH. When rock phosphate or limestone and sludge were added, uptake of Ni, Cu, Zn, and Cd decreased (Griebel et al., 1979). In another study, increasing the sludge rate from 314 to 627 Mg ha^{-1} resulted in a decrease of Mn, Zn, Ni, and Cd in tall fescue and alfalfa over 2 years. Cd and Zn reductions may be attributed to increased pH, and Mn and Ni reductions to the additional organic matter supplied. Another important factor involved in metal concentrations in plant tissues is the possible "dilution" effect that occurs as a result of increased biomass production where sludges are used (Kerr et al., 1979). The sampling date determines in part the concentrations of metals in plants, and seasonal differences may be greater than treatment differences. For example, in one study Pb concentrations of tall fescue were four times higher in January than they were the previous October (Mathias et al., 1979). Forages collected in the fall had Mn and Al concentrations high enough to be considered potentially toxic, but the following June the levels were safe for animal rations (Sutton and Vimmerstedt, 1974). This observation emphasizes the importance of sampling plant material at a predetermined growth stage or time of year. The part of the plant analyzed also influences metal concentration values. For example, significant differences in metal concentrations existed between grain and stover of winter rye for Cu, Pb, Zn, Mn, and Cd, but not for Ni and Cr (Stucky and Bauer, 1979). There are often differences in metal concentrations of different species grown under the same sludge regime, but this is not always the case. Stucky and Bauer (1979) observed little difference in Cu, Zn, Ni, Cd, Cr, Mn, or Pb among switchgrass, orchardgrass, and tall fescue, and Seaker and Sopper (1982) found no consistent differences in concentrations of the same elements in tall fescue, orchardgrass, and birdsfoot trefoil. McBride and others (1977), however, reported definite differences in Cd uptake between rye, Sudan grass, tall fescue, and reed canarygrass.

In general, sludge-grown vegetation usually contains higher concentrations of metals like Cu, Zn, Cr, Pb, Cd, and Ni than do plants grown without sludge, but the heavy metal increases are of varying degrees, and in most cases are not great enough to be a threat to human or animal health.

Within three growing seasons on sludge-amended spoils in Pennsylvania (Seaker and Sopper, 1982), concentrations of six metals in forages were continually below the suggested tolerance levels for agronomic crops (Council for Agricultural Science and Technology, 1976, 1980; Melsted, 1973). Annual mean concentrations from five sites over 2 to 5 years showed Cu and Zn concentrations in the forage below the required concentrations for total dairy rations. In other studies, Mn and Al accumulated to potentially harmful levels in forages the first year after sludge, but decreased to safe levels at the start of the second growing season (Franks et al., 1982). Cadmium concentrations in tall fescue grown with 224 Mg ha^{-1} sludge reached 2.3 mg kg^{-1} vs 0.5 mg kg^{-1} suggested as a safe food level (John et al., 1972; Mathias et al., 1979). On two sites near zinc smelter operations in Oklahoma, Cu concentrations of sludge-grown grasses were consistently below suggested tolerance levels, but Cd concentrations were usually in the yield-reducing range. The major source of metals, however, may not have been the sludge but rather the contaminated soil (Franks et al., 1982). In most cases, plant toxicity symptoms do not appear and metals are below suggested tolerance levels (Stucky and Newman, 1977; Seaker and Sopper, 1982; Hinkle, 1982; Scanlon et al., 1973).

Manganese is an element of concern in mine spoils since it may be accumulated by plants to phytotoxic levels. Manganese can become toxic to tall fescue growth in mine spoils when pH decreases to < 5 (Palazzo and Duell, 1974). Roberts et al. (1988) found that tall fescue tissue Mn levels were highest the first year after applying sludge at rates from 22 to 224 Mg ha^{-1}, but were not at phytotoxic levels. Tissue Mn levels then decreased after the first year and as sludge rates increased. The authors concluded that the sludge additions increased fescue production and appear to have diluted tissue Mn levels. They also reported that fescue tissue Zn, Cu, and

Cd concentrations increased as the sludge rate increased. Concentrations of Zn in fescue tissue decreased in treatments amended with ≥ 56 Mg ha^{-1} sludge over the 2 years following application. Cadmium tissue concentrations were the highest the first year, but well below phytotoxic levels, and then decreased with time.

Pietz et al. (1989b) reported the successful establishment of a vegetative cover on acidic coal refuse treated with sludge, lime, and gypsum, and various combinations of each. Sludge application rate was 542 Mg ha^{-1}. Concentrations of Cd, Zn, Cu, and Ni in composite plant samples ranged from 1.3 to 22.8, 62 to 758, 1.8 to 18.8 and 4.4 to 33.8 mg kg^{-1}, respectively.

The highest values for Cd and Zn exceeded the tolerance levels (3 mg Cd kg^{-1} and 300 mg Zn kg^{-1}) reported in the literature (Melsted, 1973; CAST, 1976). The high values were generally associated with the treatments that included gypsum. Authors also reported that concentrations of Al, Fe, and Pb in plant tissue from some treatments exceeded phytotoxic limits. Aluminum and Fe concentrations in most treatments exceeded the limits of 200 and 750 mg kg^{-1} proposed by Melsted (1973), respectively.

Joost et al. (1987) also treated coal refuse material with various sludge application rates (225 to 900 Mg ha^{-1}) and examined trace metal uptake by forage grass herbage. They reported that although Cd, Cr, Pb, and Ni were present in the coal refuse at levels considered toxic to plants, there appeared to be no detrimental effects on the grasses grown. In fact, tissue accumulation of all trace metals was within the range considered safe for animal consumption (Underwood, 1971). Highest concentrations of Cd, Cr, Pb, and Ni in forage grass herbage harvested over a 4-year period on a plot treated with 900 Mg ha^{-1} of sludge were 9, 11, 26, and 10 mg kg^{-1}, respectively. The maximum suggested safe in feed is 160, 50, 300, and 1100 mg kg^{-1}, respectively (Underwood, 1971; Church and Pond, 1974).

Little information is available on long-term effects of single applications of sludge on mine land and metal uptake by vegetation. However, a recent publication by Sopper and Seaker (1990) sheds some light on this subject. They resampled orchardgrass and crownvetch growing on an abandoned strip mine site amended with 184 Mg ha^{-1} of dewatered sludge 12 years after application. Foliar analyses for Zn, Cu, and Ni for the first 5 years after sludge application (1977–1981) and for 1989 are shown in Figs. 1 to 3. These are the elements commonly present in sludge that are most likely to cause phytotoxicity. Concentrations of Zn and Ni tended to be higher in crownvetch than in orchardgrass; whereas, Cu tended to be higher in orchardgrass.

In general, trace metal foliar concentrations tended to be highest the first year and then decreased over time. Except for Ni, foliar concentrations of trace metals in the sludge-grown orchardgrass plants were higher than in control plants. The 1989 values for Cu and Cd were quite similar to those of 1981. Foliar concentrations of Zn and Ni show a slight increase from 1981 to 1989. Although sludge application did increase some trace metal

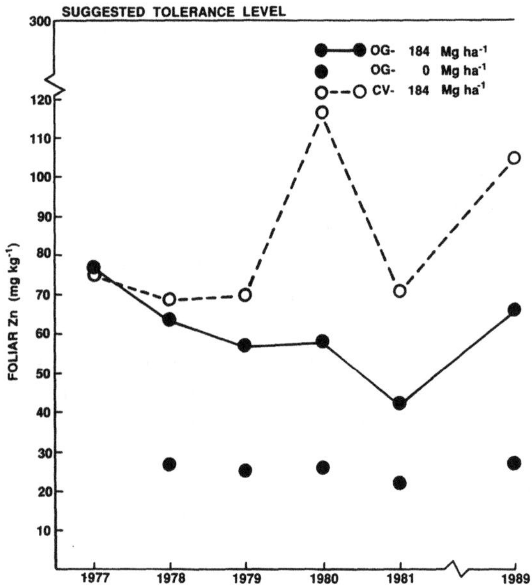

Figure 1. Mean foliar concentration of Zn in orchardgrass and crownvetch collected from the control and sludge-amended plots

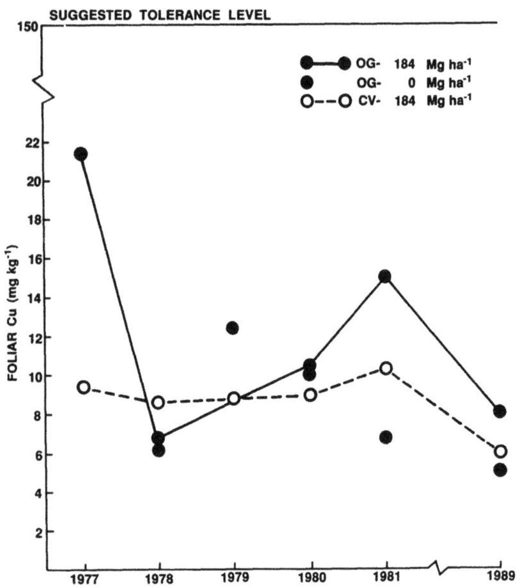

Figure 2. Mean foliar concentration of Cu in orchardgrass and crownvetch collected from the control and sludge-amended plots

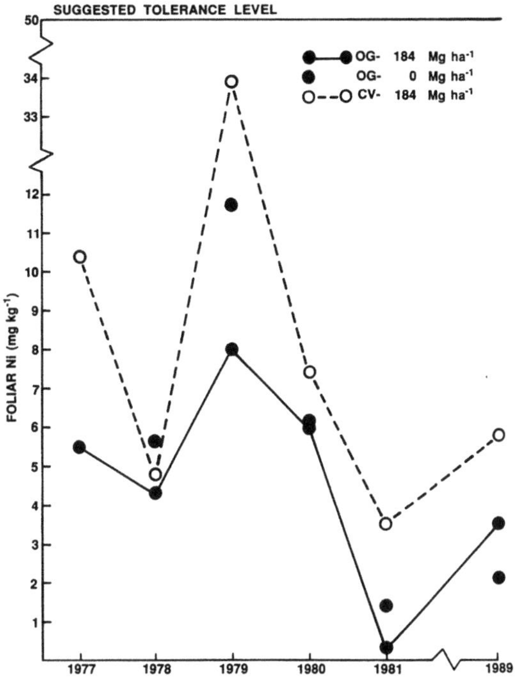

Figure 3. Mean foliar concentration of Ni in orchardgrass and crownvetch collected form the control and sludge-amended plots

concentrations in the foliage, these increases were minimal and well below the suggested tolerance levels, for agronomic crops (Melsted, 1973). No phytotoxicity symptoms were ever observed. The suggested tolerance levels are not phytotoxic levels but suggest foliar concentration levels at which decreases in growth may be expected.

2. Trees

Only a limited number of studies have been done with metal uptake by trees. Trees are beneficial for sludge-amended sites because they are not a significant food source, and because of their large biomass, they tend to retain heavy metals on the site through biotic storage in plant parts not grazed by herbivores, thereby limiting their entry into food chains (Roth et al., 1982; Morin, 1981). Two studies looked at accumulation of metals by component tree parts. Patterns of accumulation by roots, stems, and leaves varied with the metal and the tree species. Eastern white pine, silver maple, and green ash, for example, tended to accumulate Cu, Zn, and Cd in their roots as opposed to foliage and stems (Svoboda et al., 1979). Roots and leaves were higher accumulators than stems (Roth et al., 1982; Morin, 1981). In a study using a high-metal sludge, metals were sometimes lower in the leaves of sludge-grown trees than in the fertilizer controls, except Cd in hardwoods, which was consistently higher when sludge was used

(Schneider et al., 1981). Roth and others (1982) observed lower metal concentrations the third growing season after sludge application compared with the first, but some third-year concentrations exceeded those considered to be toxic. For example, the maximum Cd concentration of combined tree parts was 950 mg kg^{-1}. The ability of the tree to tolerate such high levels of metal may be due to a tie-up of metal ions in plant tissues in forms not readily movable throughout the plant.

3. Field Crops

Although field crops are not often grown on reclaimed sites, some studies have been made to assess metal uptake where sludge is used as an amendment. Hundreds of hectares on Fulton Co., Illinois, that were once strip mined for coal were returned to corn production after the soil was reconditioned with sludge from Chicago. Cadmium loadings as high as 135 kg ha^{-1} resulted in some fields, and Cd concentrations were increased in corn grain, with maximum values in 1979 of 0.46 mg kg^{-1} in controls and 0.81 mg kg^{-1} in sludge-treated corn. Such crops are used only for animal feed (Peterson et al., 1979; Peterson et al., 1982). In a similar study (Hinesly et al., 1982) with high sludge loading rates, Zn, Cd, and Ni were increased in corn grain and leaves, but Cu, Cr, and Pb were not increased. Plant-soil concentration ratios indicated that metals were less readily available from the sludge than from the original spoil material. Even though the grain accumulated more Cd and Zn than controls, there was no increase in concentration of these elements over seven growing seasons as a result of repeated annual applications (Hinesly et al., 1979a). On sludge-amended spoils in Kentucky, Ni, Cr, Cu, Cd, Mn, and Fe concentrations were not significantly increased in corn and soybean plants at 112 or 269 Mg ha^{-1} sludge rates (Feuerbacher et al., 1980), nor did Cu and Zn in corn and bush beans increase as sludge rates increased on sand and gravel spoils (Hornick, 1982). Copper and Zn values were similar to controls, but Cd increased with the sludge rate. All values for corn grain were well below suggested tolerance levels (Cd 0.27 vs 3.0 mg kg^{-1}; Zn 32 vs 300 mg kg^{-1}; Cu 4.5 vs 150 mg kg^{-1} (Hornick, 1982), and corn was considered a better metal excluder than soybeans (Feuerbacher et al., 1980). In another study corn grown on spoils amended with 56 Mg ha^{-1} of sludge showed no increased accumulation of Hg or Pb in the grain, and the protein content was increased over the control corn. Zinc concentrations in the grain were higher for the control than for the sludge-treated corn. Most of the metals were concentrated in the germ fraction rather than in the endosperm, and the corn products produced were found to add only minor quantities of Cd, Pb, and Hg to the human diet (Blessin and Garcia, 1979).

Few attempts have been made to produce horticultural crops on spoils, but Tunison and others (1982) grew highbush blueberries on spoils

Table 8. Elemental concentrations in plants in relation to suggested tolerance levels for agronomic crops.[a]

Reference	Years	Species	Tolerance level in agronomic crops		
			Below	Above	Variable
Haghiri and Sutton (1982)	3	Tall fescue	Cd, Cu, Ni	Zn	
Seaker and Sopper (1982)	3	Tall fescue Orchardgrass Birdsfoot trefoil	Al, Zn, Cu, Co, Pd, Ni, Cd		Cr
Kerr et al. (1979)	3	Reed canarygrass	Cu, Zn, Cd	Ni	
Fitzgerald (1982)	4	Forages	Cu, Ni, Zn	Cd, Cr	Pb
Hinesly et al. (1982)	—[b]	Corn leaf	Ni, Zn, Cu, Cr, Pb		Cd
Griebel et al. (1979)	—	Tall fescue Birdsfoot trefoil	Cu, Ni, Zn, Cd		
Stucky and Bauer (1979)	—	Tall fescue Switchgrass Orchardgrass	Cu, Ni, Pb	Zn, Cr	Cd
Mathias et al., (1979)	—	Tall fescue	Cu, Zn, Cd, Ni		Pb, Cr
Hinkle (1982)	3	Korean lespedeza Tall fescue Weeping lovegrass	Cu, Cr, Cd, Ni, Pb, Zn		
Franks et al. (1982)	2	Bermudagrass Switchgrass Kleingrass Bluestem	Cu	Zn, Cd	
Stucky and Newman (1977)	2	Alfalfa Tall fescue	Ni, Cd, Zn, Cu, Pb, Cr		
Hornick (1982)	—	Corn grain	Cd, Zn, Cu		
Tunison et al. (1982)	—	Blueberries	Zn, Cu, Cr, Ni, Pb, Cd		
Feuerbacher et al. (1980)	—	Corn plant Soybean plant	Ni, Cu, Cd, Mn	Cr	
Schneider et al. (1981)	—	Hardwood leaves Pine needle	Cu, Zn, Ni		
Seaker and Sopper (1984)	5	Tall fescue Birdsfoot trefoil	Zn, Pb, Cu, Cd, Ni		
Seaker and Sopper (1983)	5	Tall fescue Birdsfoot trefoil Panic grass	Cu, Zn, Cr Ni, Co, Cd		Pb
Sopper and Seaker (1990)	12	Orchardgrass Crownvetch	Zn, Cu, Pb, Ni, Cd		
Joost et al. (1987)	4	Reed canarygrass Tall fescue Redtop	Ni	Cd, Cr Pb	
Roberts et al. (1988)	3	Tall fescue	Zn, Cu, Cd		
Pietz et al. (1989b)	3	Alfalfa Bromegrass Orchardgrass Tall fescue	Cu, Ni, Pb, Zn	Cd	

[a]See Table 7 for tolerance levels
[b]Length of study uncertain

amended with sludge containing 1250 mg kg^{-1} Zn. High Zn levels and low Mg foliar levels resulted in severe chlorosis, but when the sludge was composted with bark, the symptoms were alleviated. The berries showed no significant accumulation of Cd, Cr, Cu, Ni, Pb, or Zn, and these elements were within limits considered safe for human consumption (Chaney, 1973).

Table 8 lists some effects of sludge applications on the levels of trace metals in plant tissue found in various studies. Most studies found trace metal concentrations in the vegetation to be below tolerance levels for yield suppression and well below phytotoxic levels.

C. Effects on Soil

1. Physical Properties

The high organic matter of sludge improves the physical condition of barren spoils tremendously. Parameters that benefit from sludge incorporation include water-holding capacity, bulk density, and surface temperatures. In Illinois, incorporation of digested sewage sludge at rates of 0, 224, 448, and 896 Mg ha^{-1} greatly improved the spoil physical properties (Hinesly et al., 1982). Ten months after sludge incorporation, untreated spoil contained 12.2% of water stable aggregates greater than 0.25 mm in diameter, as compared with 42.1% in spoil amended with 896 Mg ha^{-1} of sludge. The available water-holding capacity was increased from 14.8% in untreated spoil to 21.1% with 896 Mg ha^{-1} of sludge.

In Colorado, Topper and Sabey (1986) reported that application of digested sewage sludge on mine spoils at rates of 0, 14, 28, 55, and 83 Mg ha^{-1} significantly increased saturation water percentages. Percent saturation was increased from 27.9 on untreated spoil to 35.6, 41.9, and 43.2 on the three highest sludge application rates, respectively. This indicates an increase in water-holding capacity due to the added organic matter from the sludge.

Composted sludge additions of 160 Mg ha^{-1} doubled the percentage of moisture in sand and gravel spoils (Hornick, 1982), and increased water retention of coal refuse (Joost et al., 1981) and mine spoils (Schneider et al., 1981). Increased volumes of percolate water were reported on the Palzo mine site in Illinois after liquid sludge was disked 15 cm into the spoils (Urie et al, 1982). On Ohio strip-mine spoils (Haghiri and Sutton, 1982), however, the opposite effect occurred because of the rapid uptake and transpiration of soil water by increased plant biomass. In fact, the volume of percolate water was indirectly proportional to the sludge loading rate. Because sludge is about 50% organic matter (dry basis), it can improve water infiltration and retention (Griebel et al., 1979). Younos and Smolen (1981), reporting on simulation of the infiltration process in sludge-amended mine spoil, found that sludge promotes infiltration, rapid saturation, and increased hydraulic conductivity.

Sludge additions have resulted in decreased bulk density in Illinois mine spoils high in compacted clay (Peterson et al., 1979; Peterson et al., 1982) and coal refuse (Joost et al., 1981); in decreased temperatures on sand and gravel spoils (Hornick, 1982); and in reduced runoff, erosion, and sedimentation from coal mine spoils (Sutton and Vimmerstedt, 1974).

Joost et al. (1987) reported that applications of sewage sludge at 225, 450, and 900 Mg ha^{-1} reduced bulk density of acid coal refuse (gob). Sewage sludge application increased organic matter content of the gob by 2.0 to 2.5 times that of unamended gob. The sludge applications also increased the number of water-stable aggregates and the proportion of macropores present in the gob. Sludge-treated gob exhibited significantly higher water content at 0.03 MPa than untreated gob.

2. Chemical Properties

When sludges are applied to land, an increase in soluble salt content of the growing media may result. At a sludge application rate of 896 Mg ha^{-1}, as compared with a 224 Mg ha^{-1} rate, and an electrical conductivity (EC) of 6.6 mmho/cm, a 50% reduction in corn yield resulted. High soluble salts may also affect the establishment of some grass species (Hinesly, et al., 1982). Hinesly noted an increase in EC with increasing sludge levels (0, 224, 448, 896 Mg ha^{-1}) applied to a calcareous coal mine spoil in Illinois and speculated this caused a decline in vegetation growth. Electrical conductivities ranged from 2.2 mmho cm^{-1} in spoil samples from control plots to 6.6 mmho cm^{-1} in spoil amended with 896 Mg ha^{-1} of sludge. Topper and Sabey (1986) also reported significant increases in EC with applications of sludge at 28, 55, and 83 Mg ha^{-1} on coal mine spoil in Colorado. The greatest level of sludge application (83 Mg ha^{-1}) resulted in an observed mean EC of 5.5 dSm^{-1}. However, measurements of aboveground biomass seemed to indicate there were no adverse salinity effects on plant growth.

On another site, however, EC was significantly reduced after sludge incorporation in very stony spoil that was high in Fe, Al, and Mn (Schneider et al., 1981). Lejcher and Kunkle, (1974) also reported that after sludge incorporation into coal refuse EC at the 0 to 30 cm depth was decreased. Haynes and Klimstra (1975) reported that EC greater than 1.5 mmho cm^{-1} may be damaging to crops. In another study, EC of sludge-amended spoils ranged from 0.68 to 2.80 mmho cm^{-1}, but vegetative cover was not affected as much by EC as it was by pH (Stucky and Bauer, 1979). Potential soluble salt problems can usually be avoided by proper management techniques. Maximum effective loading rates should be based on the constituents in the sludge and the crop that is to be grown (Hinesly et al., 1982).

Limed sludges have a neutral to alkaline pH and this can raise the pH of acid spoils (Griebel et al., 1979). The pH of coal refuse gob was increased from 2.6 to 5.3 even without limestone when sludge was incorporated at rates of 450 to 900 Mg ha^{-1}. Sludge was more effective than 180 Mg ha^{-1} limestone in raising pH (Joost et al., 1981). Sludge applied at 67 to 269 Mg ha^{-1} with lime increased pH considerably, but 34 Mg ha^{-1} sludge with lime had no effect (Feuerbacher et al., 1980). Even after 195 Mg ha^{-1} of lime and commercial fertilizer were applied to spoils in Ohio with a 2.3 pH, rye cover was considerably poorer than on adjacent plots amended with sludge only (Sutton and Vimmerstedt, 1974). When anthracite refuse was irrigated with liquid digested sludge and sewage effluent, significant increases in pH to a 76 cm depth occurred (Kardos et al., 1979). After six consecutive lime and sludge applications in 4 years, pyrite mine spoils increased in pH from 2.4 up to 5.0, and after the seventh application pH ranged from 6.1 to 6.6 (Hinkle, 1982).

Limestone initially increases the pH of acid spoils, but the pH eventually declines as sulfur-bearing minerals are oxidized (Sutton and Vimmerstedt, 1974). When commercial fertilizer plus lime was applied on Virginia acid mine spoils, pH remained below 4.0 for 4 years, but where municipal compost (pH 8.5) was added once without any lime, pH ranged from 4.9 to 7.4 through the 4-year period (Scanlon et al., 1973). Little information is available on long-term effects of single high applications of sludge on mined land. However, Sopper and Seaker (1990) reported that a sludge application of 184 Mg ha^{-1} increased the pH of strip-mine spoil in Pennsylvania from 3.8 up to 6.2 in the surface 15 cm 4 months after application, and was still 5.4 after 12 years with excellent grass and legume cover (Table 9) (Sopper and Kerr, 1982, Sopper and Seaker, 1990). Sludge alone significantly raised the pH of spoils in West Virginia, and sludge plus lime was even more effective. Initial pH values of 3.0 to 4.0 were increased above 5.0 and usually above 6.0, with no decline after 3 years (Mathias et al., 1979). On Ohio strip-mine spoil where two sludges were applied at 179, 358, and 716 Mg ha^{-1} the pH did not decline over a 3-year period (Haghiri and Sutton, 1982). When Chicago sludge was applied annually to strip mines over a 6-year period, the pH decreased slightly the first 3 years, mainly because of the nitrification and organic-acid production in the soil. The decrease was minimal, however, and the pH appeared to stabilize during the next 3 years (Peterson et al., 1982). Hinesly and others (1982), however, observed a drop in pH from 7.5 to 6.0 in calcareous spoil amended with 896 Mg ha^{-1} of Chicago sludge. Topper and Sabey (1986) reported similar declines in pH on Colorado mine spoil having an initial pH of 7.1. Applications of sludge at rates of 14, 28, 55, and 83 Mg ha^{-1} resulted in spoil pH's for the 0 to 15 cm depth of 6.8, 6.5, 6.3, and 6.2, respectively. It has been recommended that sludge-amended spoil material be maintained at a pH of at least 6.5 so that potentially toxic metals will remain at relatively low solubility levels (Sopper and Seaker, 1983). Pietz et al. (1989a) also reported a pH decline with time after treating a coal refuse pile with 542 Mg ha^{-1} of sludge. The pH of the surface refuse (0 to

Table 9. Changes in spoil pH over a 12-year period on a Pennsylvania strip-mine spoil bank treated with 184 Mg ha^{-1} sludge

Depth (cm)	Soil pH					
	May 1977[a]	Sept. 1977	Nov. 1978	Oct. 1979	May 1981	Aug. 1989
0–15	3.8	6.2	6.7	7.3	5.8	5.4
15–30	3.8	4.2	4.6	5.1	3.4	5.6

From Sopper and Seaker (1990)
[a] Presludge samples

15 cm) was 2.8 on a control plot and was increased to 5.0 in 1976 following sludge applications. By 1981, the pH had dropped to 3.3.

In a zinc smelting area, lime and sludge raised soil pH, with an increase from 5.8 to 6.5 occurring within 2 years of the application (Franks et al., 1982). Soil pH is a critical factor for plant growth, and complete mixing of the lime and sludge is essential for uniform vegetative cover (Sutton and Vimmerstedt, 1974). Fibrous root systems, such as those produced by tall fescue, may have a stabilizing effect on pH, according to Stucky and Newman (1977).

Some trials have attempted to assess the effect of deep incorporation of sludge on plant establishment and growth, but results are vague. Incorporation of sludge to 60 cm as compared with a 30 cm incorporation did not result in better vegetative cover the first year (Joost et al., 1981). In another study incorporation from 15 cm to a 40 cm depth increased pH at the deeper level, and deeper root penetration as well as better root quality resulted (Feuerbacher et al., 1980). Root systems are often confined to nontoxic or treated layers when sludge is incorporated into spoils (Sutton and Vimmerstedt, 1974), but are usually much more prolific and deeper than when inorganic fertilizer is used (Scanlon et al., 1973).

Cation exchange capacity is normally improved by sludge addition to spoils because of the high CEC of organic matter (Stucky and Newman, 1977; Schneider et al., 1981; Jones and Cunningham, 1979). CEC is an important factor in the availability of cations, both nutrient (Ca, Mg) and trace metals (Cu, Zn, Pb, Cd), to plants and also affects their movement into groundwater.

The effect of sludge incorporation on the amount and availability of major plant nutrients varies. In general, sludges apply considerable N and P, but little K. The Ca, Mg, and S contribution varies with the composition of the sludge. Peterson et al. (1982) used soil N, P, and K as an index of soil rejuvenation by sludge, and found that over a 4-year period repeated application of Chicago sludge increased available N, P, and K each year on both agricultural soils and mine spoil. Significant increases in Kjeldahl-N and Bray-P occurred in spoils where sludge was applied at 75 Mg ha^{-1} in Pennsylvania. After 3 years, total N in control spoils for the 0 to 15 cm depth was 0.041% vs 0.219% in the sludge-treated spoils. Phosphorus concentration was 216 mg kg^{-1} in control spoils vs 472 mg kg^{-1} in sludge-treated spoils (Kerr et al., 1979). On abandoned pyrite mines, commercial fertilizer and sludge applied together increased spoil phosphate and K_2O, but K availability was much greater where good vegetative cover existed than where the site was bare, indicating the importance of plants in the uptake and recycling of nutrients (Hinkle, 1982). Sludge additions increased spoil P, Ca, and Mg, but not K in West Virginia (Mathias et al., 1979). Spoil K decreased after 3 years, probably because of removal of vegetation. At the Palzo mine site in Illinois, only K remained deficient after sludge application (Jones and Cunningham, 1979). Total and extractable N and P were increased by four different sludge application rates in Kentucky spoils, but there was no change in K content (Schneider et al., 1981). Sutton and Vimmerstedt (1974), however, observed a threefold increase in available K and an 18-fold increase in available P from applications of sludge at 658 Mg ha^{-1}. In another study extractable P was greater in sludge-amended spoils than in controls, but an increase in sludge loading form 112 to 224 Mg ha^{-1} did not further increase P. It is possible that the increased Ca at the higher rate neutralized some of the acid in the P extractant, since both Ca and Mg were markedly increased by sludge additions (Mathias et al., 1979).

Interpretation of reports on metal loadings to disturbed land after sludge application is difficult, mainly because the analytical methods used to determine metal concentrations are inconsistent. Values for extractable vs total metals are quite different, as are those for HCl-extractable vs DTPA-extractable metals. Roth and others (1982) found that Cd, Cu, Mn, and Ni concentrations were significantly different when four extractants were compared: DTPA (pH 7.3; pH 4.9) and 0.1 *N* HCl (pH 1.3; pH 4.9). No extraction method has yet been developed that adequately indicates the amount of metals available to plants. Plant uptake may be much less than indicated by DTPA extraction (Griebel et al., 1979). Such factors as pH, organic matter content, metal concentration of sludge, phosphorus and iron concentrations, and CEC all contribute to plant availability of trace metals, and further complicate the matter. For any one study, however, a comparison can be made of spoil metal concentrations before and after sludge additions, on the basis of a single known method of analysis.

Sludge addition usually results in increases in spoil heavy metal content, but which metals increase depends on the particular sludge, and in many cases, the increases are not significant. Sludge organic matter and pH effects often cause a decrease in availability of Fe, Al, and Mn, which are already found at extremely high concentrations in most acidic spoil materials. On anthracite refuse, liquid-sludge irrigation greatly increased phosphorus, which tended to tie up and detoxify Fe, Al, and Mn in the upper root zone (Kardos et al., 1979). In Pennsylvania bituminous spoil and anthracite refuse, sludge incorporation of 75 to 184 Mg ha^{-1} reduced extractable Fe, Al, and Mn in the plow depth but increased Cu, Zn, Cr, Pb, Cd, and Ni. The increases were minimal, however, and sludge did not affect concentrations below a 15 cm depth (Sopper and Kerr, 1982; Kerr et al., 1979; Seaker and Sopper, 1983; Seaker and Sopper, 1984). Cadmium and Cu, but not Ni and Zn, exceeded the normal soil ranges (Allaway, 1968) on sludge-amended spoils on the Palzo site (Roth et al., 1982).

In Fulton County, Illinois, Peterson and others (1982) reported that Chicago sludge, which is very high in metals, increased spoil Cd, Zn, Ni, and Cu. When applied annually, however, only Cd exceeded the values for normal ranges in agricultural soils reported by Allaway (1968). Another study (Fitzgerald, 1982) with the same sludge found large increases in Cd, Zn, Ni, Cu, Cr, Pb, and Hg in spoils as compared with controls. Copper, Zn, and Cd were at times above reported normal ranges. However, except for Zn and Pb, the concentrations did not appreciably increase over the 4 years of annual application. Unamended gob material may be high in Mn and Cr (Joost et al., 1981), and sludge additions to gob were found to cause large increases in Cr, Mn, Cu, Zn, Cd, and Pb based on maximum soil concentrations reported by Yopp et al. (1974) that might exert toxic effects on plants. However, grass establishment and growth apparently were not affected. Extractable Cu, Ni, Zn, and Cd increased as compost application rates increased on Maryland mine spoils (Griebel et al., 1979). Phosphate rock, which can contain significant trace metals, raised the values even higher, but limestone additions decreased them. Zinc, Cu, Pb, and Ni may increase in availability and potential toxicity as pH decreases (Council for Agricultural Science and Technology, 1976).

In some instances, the availability of metals in spoil was not increased by sludge additions, due to the metal composition of the sludge and increases in pH and organic matter. For example, in soils contaminated by zinc smelting, Zn, Cu, and Cd levels were extremely high and sludge addition improved plant cover without significantly adding to the metal load (Franks et al., 1982). On abandoned pyrite mines a dramatic drop in metal concentrations occurred as reclamation progressed, probably because of increases in spoil pH (Hinkle, 1982). Concentrations of Cu, Fe, Mn, and Zn were within normal soil ranges reported by Allaway (1968). Hinesly and others (1982) observed no increase in As, Mo, or Mn in sludge-amended spoils, due to low concentrations of these elements in the sludge.

Not much information is available relative to the long-term effects of single applications of sewage sludge on mine spoil or coal refuse. However, a recent publication by Sopper and Seaker (1990) provides some insight into long-term (12 years) effects on spoil chemical properties. In their study, dewatered sludge was applied at 184 Mg ha^{-1} on an abandoned strip mine spoil bank in 1977. Soil samples were collected over a 5-year period (1977–81) and then again in 1984 and 1989. Changes in concentrations of

Table 10. Changes in concentrations of Kjeldahl-nitrogen, Bray-phosphorus, and exchangeable cations in the spoil collected at the 0–15 cm depth on a sludge-amended mine site in Pennsylvania

Year	Kjeldahl nitrogen %	Bray-phosphorus	K	Ca	Mg
			($mg\ kg^{-1}$)		
May 1977[a]	0.04	2	12	541	452
Sept 1977	0.05	11	19	1222	32
1978	0.09	9	23	2600	40
1979	0.16	38	46	3873	53
1981	0.34	79	45	1298	99
1984	—	91	74	1440	108
1989	0.12	83	30	733	84

From Sopper and Seaker (1990)
[a]Presludge samples

Kjeldahl-nitrogen, Bray-phosphorus, potassium, calcium, and magnesium are given in Table 10. The nutrient status of the spoil shows a general increase in concentrations of Kjeldahl-N up to 1981 and up to 1984 for Bray-phosphorus, K and Ca. The application of lime and sludge initially resulted in a decrease in the concentration of Mg; however, since 1978 there has been a steady increase. The 1989 values are lower but still quite adequate to support plant growth.

Concentrations of extractable trace metals in the 0 to 15 cm spoil depth are given in Table 11. Concentrations of Cu, Zn, Cr, Pb, Cd, and Ni all show a steady increase for the first 5 years (1977–81). By this time, all the sludge organic matter was probably mineralized and all trace metals released to the surface spoil. Results of spoil analyses in 1984 and 1989 show a gradual decrease in concentrations of all trace metals.

Although the sludge application did increase the concentrations of extractable trace metals in the 0 to 15 cm spoil depth, these higher concentrations are still within the normal ranges for these elements in U.S. soils (Allaway, 1968). Analyses of spoil samples collected from the 15 to 30 cm depth showed a general increasing trend in trace metal concentrations from 1977 to 1989, indicating that some leaching of the trace metals through the spoil profile was occurring.

In another study, Seaker and Sopper (1983) applied 80 and 108 Mg ha^{-1} of dewatered sludge along with 11 Mg ha^{-1} agricultural lime to an anthracite coal refuse bank in northeastern Pennsylvania. Changes in refuse pH are given in Table 12. Addition of lime and sludge raised the initial pH of the mine refuse material (3.6 to 3.8), so that after 4 years the pH in the 0 to 15 cm depth was 5.8 to 5.9. However, it should be noted that 3 years after application the spoil pH ranged from 6.8 to 7.8 in the surface 5 cm. Con-

Table 11. Changes in concentrations of extractable trace metals from spoil collected at the 0–15 cm depth on a sludge-amended mine site in Pennsylvania

Sampling date	Cu	Zn	Cr[b] (mg kg^{-1})	Pb	Cd	Ni
May 1977[a]	2.5	2.9	0.2	0.5	0.02	1.1
Sept 1977	10.8	7.7	0.4	3.5	0.04	0.9
1978	8.8	7.7	0.2	2.3	0.02	1.2
1979	58.7	56.9	1.7	13.0	0.27	1.5
1981	87.3	74.6	3.5	22.7	0.95	2.8
1984	57.6	59.6	—	14.8	0.56	2.8
1989	51.9	37.8	—	13.5	0.42	2.0
Normal range for U.S. soils (Allaway, 1968)	2–100	10–300	5–3000	2–200	0.01–7.00	5–500

From Sopper and Seaker (1990)
[a] May 1977 values represent pretreatment conditions
[b] Values for Cr are total concentrations

Table 12. Mean pH of anthracite refuse material collected at the surface before and after application of sludge at the Lackawanna County demonstration project in Pennsylvania

		Sludge applied (Mg ha^{-1})	
Year	Depth (cm)	80	108
1978[a]	0–15	3.8	3.6
1979	0–15	3.8	3.6
1982	0–15	5.9	5.8
1981	0–5	6.8	7.8

From Seaker and Sopper (1983)
[a] Pretreatment values

centrations of trace metals in the 0 to 15 cm depth showed little effect of sludge 1 year after the application, but by the fourth year increases ranged from twofold for Cu and Co to 100-fold for Ni (Table 13). However, the actual concentrations of these metals were very low and, except for Cd, were lower than the mean for U.S. soils reported by Sommers et al. (1987). Although Cd concentration was higher than the mean for U.S. soils, the Cd concentration in the refuse material was, in fact, decreased after the application of lime and sludge. Similarly, concentrations of extractable Fe, Al, and Mn decreased with time.

Pietz et al. (1989a) reported similar results when sludge and lime were applied on coal refuse material in Illinois. The sludge and lime applications

Table 13. Changes in concentrations of extractable trace metals from anthracite refuse material collected at the 0–15 cm depth following sludge application at 108 Mg ha^{-1}

Year	Cu	Zn	Pb	Cd ($mg\ kg^{-1}$)	Ni	Co	Cr
1978[a]	7.6	1.5	1.1	0.500	0.01	0.38	<0.01
1979	7.8	3.8	0.9	0.003	0.97	0.35	0.05
1982	18.8	25.5	5.6	0.372	1.37	0.72	0.59
Mean for U.S. soils Sommers et al. (1987)	30	57	17	0.27	24	—	—

From Seaker and Sopper (1983)
[a] Pretreatment samples

were 542 Mg ha^{-1} and 89.6 Mg ha^{-1}, respectively. Results of coal refuse samples (0 to 15 cm) analyzed over a 5-year period are given in Table 14. The lower water-soluble Al and Fe concentrations appeared to reflect the ability of the applied sewage sludge to retain Al and Fe, and the ability of lime to decrease the solubility of these metals through an increase in pH and precipitation of Al and Fe compounds. Sommers et al. (1984) and Corey et al. (1987), reported that the primary mechanisms involved in the retention of metals by sludges are precipitation as carbonates, phosphates, sulfides, silicates, or hydrous oxides, and sorption by organic matter and hydros oxides.

3. Biological Properties

Although the immediate goal of reclamation is to establish a vegetative cover that will prevent soil erosion, the long-term goal is soil ecosystem development and stability. Mine spoils lack microbial activity and organic matter (Visser, 1985; Mills, 1985; Fresquez and Lindemann, 1982). Microbial processes such as humification, soil aggregation, and N cycling are essential in establishing productivity in mine spoils, and productivity should be evaluated not only on above-ground biomass, but also on the degree of development of functional microbial populations resembling those of an undisturbed soil. Microbial processes are so important to ecosystem recovery that the activity of micro-organisms may be used as an index of the progress of soil genesis in mine spoils (Schafer et al., 1980; Segal and Mancinelli, 1987).

If the premining organic layer (O horizon) has been destroyed, the only C source for microbial utilization is the plant biomass that is expected to accumulate over several growing seasons on the site. Until such accumulation occurs, microbial activity remains at a low level, with little improve-

Table 14. Concentrations of water-soluble Al and Fe in coal refuse samples collected from 1976 to 1981

Treatment	Al ($mg\ kg^{-1}$)	Fe ($mg\ kg^{-1}$)
1976		
Control	279	233
Sludge and lime	9	8
1978		
Control	255	190
Sludge and lime	3	1
1979		
Control	136	119
Sludge and lime	39	4
1980		
Control	103	68
Sludge and lime	2	1
1981		
Control	105	31
Sludge and lime	2	<1

From Pietz et al. (1989a)

ment of adverse soil physical and nutrient conditions. Vegetation growth and maintenance are also inhibited. On sites reclaimed with chemical fertilizers and lime, vegetation may initially be established, but poor physical conditions result in deterioration of the vegetative cover before it can begin to ameliorate the spoil (Stroo and Jencks, 1982). On both alkaline and acidic mine spoils, microbial activity, nutrient cycling, and spoil organic matter levels may take over 30 years to be reestablished (Segal and Mancinelli, 1987; Stroo and Jencks, 1982; Mills, 1985; Anderson, 1977; Schafer et al., 1980).

The use of sewage sludge as an organic amendment for mine spoil reclamation has been extremely successful (Varanka et al., 1976; Frequez and Lindemann, 1982; Visser, 1985; Seaker and Sopper, 1984) because of its immediate improvement of spoil chemical and physical conditions, acceleration of plant establishment and growth, and achievement of long-term productivity. The organic C and nutrient content of sludge is responsible for achieving a self-maintaining cover on mine spoils, but very few studies have quantitatively measured the effects of sludge application on microbial populations and activity, compared with sites reclaimed with lime and chemical fertilizer.

It has been proposed that heavy metals, many of which may be present in sludge, could potentially disturb the population dynamics and general ecology of soil microbes in natural habitats (Babich and Stotsky, 1977a). At high levels, inorganic salts of Zn, Cu, Cd, Cr, and Pb have been shown to interfere with microbial metabolism in laboratory cultures. Most studies involved metal concentrations far in excess of those found in land application systems using "typical sludges" with median metal concentrations at agricultural rates (Mathur et al., 1979; Bhuiya and Cornfield, 1974; Lighthart et al., 1983). Numerous studies have indicated that binding of metals to organic materials and clay minerals, precipitation, complexation, and ionic interactions significantly decrease their inhibitory effects on microbial activity (Gadd and Griffiths, 1978), so that inhibition by metals is substantially less in a soil system than in pure culture media (Babich and Stotsky, 1977b; Tomlinson, 1966).

A study on the Palzo tract in southern Illinois (Sundberg et al., 1979) found fungal populations in unreclaimed spoil to be only 1% to 2% of those in unmined agricultural soils. Application of sludge, and particularly incorporation, resulted in a tenfold increase in fungal activity, because of increase in pH and food supply, and better soil-moisture retention. Some fungi are introduced with the sludge, but with improved chemical and physical condition of the spoil and a vegetative cover that recycles organic matter and nutrients, natural successional changes and eventual stabilization of the fungal populations should occur.

A recent study reported by Seaker and Sopper (1988a) sheds some light on the value of sludge applications on mined land on the rejuvenation of microbial populations and activity. They conducted a field study of five strip-mine sites reclaimed with sewage sludge and one site reclaimed by conventional methods (chemical fertilizer and lime) to assess the effects of sludge amendments and time on populations of bacteria, fungi, and actinomycetes, and on microbial respiration and organic matter decomposition. The sludge-amended sites ranged in age from 1 to 5 years following sludge applications at rates of 120 to 134 Mg ha^{-1} (Table 15). The sludge amendment was from Philadelphia and consisted of a mixture of composted sludge (with wood chips) and digested dewatered sludge cake.

a. Aerobic Heterotrophic Bacteria

Seaker and Sopper (1988a) reported that bacterial populations on the sludge-amended sites ranged from 4 to 63×10^6 g^{-1} (Table 16). Bacterial counts were 5 to 15 times higher on site 1 than on the older sites, and were dramatically increased on all sludge-amended sites compared with the fertilizer-amended site. The first-year peak and subsequent stabilization of bacterial populations is a typical response following organic matter additions to mine spoils (Fresquez and Aldon, 1986). Considering the extremely low initial pH of the mine spoils in this study, commonly ranging from 3.0 to 5.0 prior to lime additions, the microbial populations achieved with lime and sludge amendments after only 1 year are remarkably high. They

Table 15. Site descriptions for Pennsylvania microbial community study

Site	Age (year)	Amendment	Application rate[a] (Mg ha^{-1})	Date of application	Lime application (Mg ha^{-1})	pH (1985)[b]
1	1	Sludge	120	Sept. 1984	18	6.9
2	2	Sludge	128	June 1983	18	7.0
3	3	Sludge	128	May 1982	12	6.8
4	4	Sludge	134	July 1981	18	6.7
5	5	Sludge	134	July 1980	11	7.3
Fertilizer-amended	5	Fertilizer (23-10-20, N-P-K)	0.5	Aug. 1980	11	6.3

From Seaker and Sopper (1988a)
[a] Dry weight basis
[b] At time of study

Table 16. Microbial populations on strip-mine sites 1 to 5 years following sludge application, and on the fertilizer-amended site (mean of six samples with standard error)

Site	Aerobic heterotrophic bacteria (10^6 g^{-1})	Fungi (10^5 g^{-1})	Actinomycetes (10^4 g^{-1})
1	63.67 ± 16.93[a,1]	18.14 ±5.45[a]	1.48 ± 1.04[b]
2	7.07 ± 1.32[b]	5.80 ± 2.35[b]	9.75 ± 5.48[b]
3	4.09 ± 0.77[b]	5.54 ± 1.32[b]	56.21 ± 26.71[a,b]
4	11.37 ± 3.64[b]	3.98 ± 0.46[b]	140.23 ± 59.57[a]
5	13.74 ± 3.58[b]	4.03 ± 1.05[b]	40.89 ± 22.68[b]
F value	***	**	*
Fertilizer-amended	3.06 ± 1.17	0.16 ± 0.04	6.94 ± 4.01

From Seaker and Sopper (1988a)
*, **, *** Significant effect at $P < 0.05$, 0.01, and 0.001, respectively.
[1] Means followed by different superscript letters are significantly different at the 0.05 level of probability, Waller-Duncan *k*-ratio *t*-test

compare favorably with estimates of 1 to 34 × 10^6 g^{-1} reported for undisturbed soils (Wilson, 1965; Segal and Mancinelli, 1987; Miller and Cameron, 1978; Alexander, 1977; Visser, 1985; Miller and May, 1979).

b. Fungi

Sludge application resulted in fungal populations in the range of 4 to 18 × 10^5 g^{-1} (Table 16) (Seaker and Sopper, 1988a). These compare favorably with fungal populations in undisturbed soils which have been reported to range from 0 05 to 9×10^5 g^{-1} (Miller and Cameron, 1978; Segal and Mancinelli, 1987; Alexander, 1977; Wilson, 1965; Miller and May, 1979). Fungal numbers were three to four times higher on site 1 than on the older sites, and were greatly increased on all sludge-amended sites compared with the fertilizer-amended site. Two other studies on reclamation with sludge failed to find increases in fungal numbers, but did report increased species diversity (Fresquez and Lindemann, 1982; Parkinson et al., 1980).

c. Actinomycetes

Sludge applications resulted in actinomycete populations ranging from 1.48 to 140.23×10^4 g^{-1} (Table 16) (Seaker and Sopper, 1988a), compared with actinomycete populations for undisturbed soils, reported in the range of 1 to 436×10^4 g^{-1} (Alexander, 1977; Miller and Cameron, 1978; Visser, 1985; Miller and May, 1979; Segal and Mancinelli, 1987; Wilson, 1965). Actinomycetes exhibited a different pattern of development than the bacteria and fungi. These microbes are less competitive than the other groups and their populations were significantly lower on sites 1 and 2 than on the older sites. The pattern follows that described by Alexander (1977), whereby the bacteria and fungi proliferate initially upon the addition of organic matter to the soil, and the actinomycete responses do not occur until later stages of decay, when competition has decreased. Actinomycete populations on sites 3, 4, and 5 were considerably higher than on the fertilizer-amended site.

d. Nitrifying Bacteria

Seaker and Sopper (1988a) found that *Nitrosomonas* populations were not significantly different on the five sludge-amended sites (Table 17), but were two to four orders of magnitude greater than on the fertilizer-amended site. *Nitrobacter* had a significantly larger population on site 1 than on the older sludge-amended sites, and were four to six orders of magnitude greater than on the control site. Populations of both genera ranged from 0.53×10^4 g^{-1} to 126.53×10^5 g^{-1}, compared with numbers of nitrifying bacteria in unamended soils, which have been reported to range from a few hundred to 10^5 g^{-1} (Stevenson, 1982). This indicates that nitrification was not inhibited on the sludge-amended sites. Because of the continuing release from the organic-N compounds in sludge, the supply of ammonium-N for nitrification remained at a high level (Seaker and Sopper, 1988b). On the fertilizer-amended site, however, *Nitrobacter* reached only 18 g^{-1}, and *Nitrosomonas* only 30 g^{-1}. The addition of N fertilizer to the site did not appear to be sufficient to provide a sustained supply of ammonium-N for a 5-year period. The low nitrifying populations suggest a severe lack of organic-N, and partly explain the sparse growth on the site even after 5 years.

The presence of vegetation on a developing mine spoil enhances nit-

Table 17. Populations of nitrifying bacteria on strip-mine sites 1 to 5 years following sludge application, and on the fertilizer-amended site (mean of six samples with standard error)

Site	*Nitrosomomas* (10^4 g^{-1})	*Nitrobacter* (10^5 g^{-1})
1	29.92 ± 11.57[a,1]	126.53 ± 45.44[a]
2	0.53 ± 0.29[a]	35.24 ± 23.15[b]
3	1.74 ± 0.43[a]	21.16 ± 8.21[b]
4	21.64 ± 13.55[a]	9.39 ± 3.56[b]
5	6.98 ± 2.63[a]	5.47 ± 0.88[b]
F value	NS	**
Fertilizer-amended	3.01×10^1 g^{-1}	1.79×10^1 g^{-1}

From Seaker and Sopper (1988a)
**Significant effect at $P<0.01$; NS, no significant effect
[1]Means followed by different superscript letters are significantly different at the 0.05 level of probability, Waller-Duncan *k*-ratio *t*-test

rification (Mills, 1985), as indicated by the high nitrifying bacteria populations on the densely vegetated sludge-amended sites. The pH, however, appears to have a stronger influence than plant cover (Wilson, 1965), with little nitrification occurring below pH 6.0, even on revegetated sites (Jurgensen, 1978). The lower pH value on the 5-year-old fertilizer-amended site may have contributed to the low level of nitrifying bacteria, compared with the sludge-amended sites that had pH values ranging from 6.7 to 7.3.

e. Soil Community Respiration

Soil community respiration has long been used as an indicator of biological activity in the soil profile, and may be a better estimator of relative microbial activities of mine spoils than population counts. Seaker and Sopper (1988a) reported that respiration was significantly higher on a 1-year-old site (site 1) than on the older sites due to the "flush" of microbial activity resulting from the readily available organic C addition (Table 18). On sites 2 to 5 years old, both the bacterial populations and the community respiration rate declined and stabilized. Production of CO_2 in the mine spoils was positively correlated with bacterial populations ($r = 0.64$). Respiration rates on all sludge-amended sites exceeded those on the fertilizer-amended site. Other workers found respiration to be consistently lower in barren mine spoils than in conventionally revegetated ones where chemical fertilizer was used, and highest in undisturbed soils (Wilson, 1965; Lawrey, 1977). On conventionally reclaimed spoils amended with chemical fertilizers, respiration increases with the age of the site (Visser, 1985; Stroo and Jencks, 1982), but extensive time periods are required for build-up of

Table 18. Soil community respiration and decomposition rates on strip-mine sites 1 to 5 years following sludge application, and on the fertilizer-amended site (means of six and three samples, respectively, with standard error)

Site	CO_2 evolution ($mg\ CO_2\ 100g^{-1}\ d^{-1}$)	Decomposition ($\%\ yr^{-1}$)
1	$138.08 \pm 28.12^{a,1}$	54 ± 8.1^{c}
2	43.92 ± 13.90^{b}	70 ± 5.4^{b}
3	27.22 ± 12.28^{b}	77 ± 1.2^{b}
4	70.66 ± 14.12^{b}	71 ± 2.6^{b}
5	78.69 ± 20.05^{b}	96 ± 0.7^{a}
F value	**	***
Fertilizer-amended	14.17 ± 2.97	—

From Seaker and Sopper (1988a)
, * Significant effect at $P < 0.01$ and 0.001 respectively
[1] Means followed by different superscript letters are significantly different at the 0.05 level of probability, Waller-Duncan *k*-ratio *t*-test

microbial populations, often as long as 20 years. On the sludge-amended sites, however, the immediate establishment of microbial communities through organic matter addition appeared to eliminate significant age effects. Establishment of a stable respiration rate appeared to occur within 1 year, with no significant decrease with site age.

Even the lowest mean respiration rate from the sludge-amended sites, which occurred on the 3-year-old site, was approximately double that of the 5-year-old fertilizer-amended site (Seaker and Sopper, 1988a). This indicates that without adequate organic matter input, microbial activity in conventionally reclaimed mine spoils remains extremely low.

f. Microbial Decomposition

Decomposition rate can be used as an indicator of the degree of soil ecosystem recovery, since it largely controls nutrient cycling (Miller and May, 1979). Seaker and Sopper (1988a) reported that on the sludge-amended sites, decomposition rate increased with site age ($r = 0.80$). After 1 year of exposure, little more than half the grass sample (54%) was decomposed on site 1, while almost all of the sample (97%) was decomposed on site 5 (Table 18). Decomposition was higher on the four older sites, even though microbial numbers were highest on site 1. This could be attributed to a younger microbial community having less diversity than on the older sites. Site 1 may have a greater proportion of sugar and starch decomposer species compared with the older sites that would probably have a greater proportion of species capable of decomposing cellulose and humus. In other studies, where mine spoils were reclaimed with chemical fertilizer, populations of cellulose-decomposing fungi and bacteria have been found to

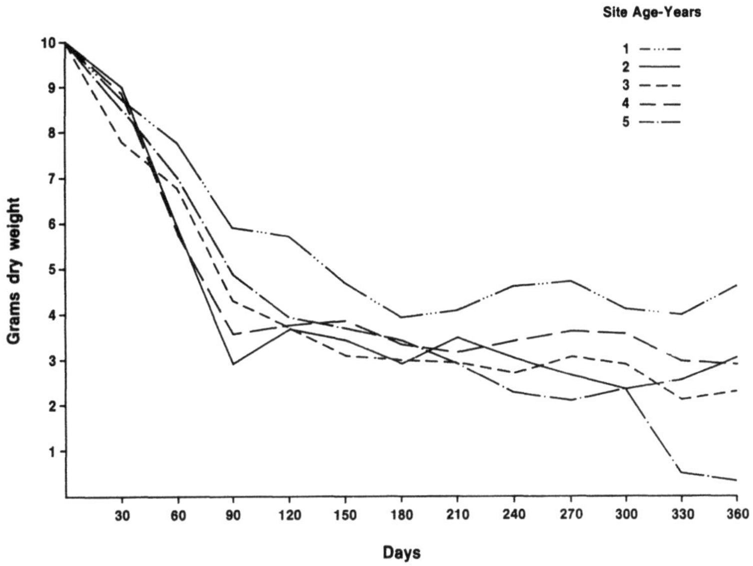

Figure 4. Amounts of orchardgrass remaining in nylon net bags at 30-day intervals over a 1-year exposure period for the five sludge-amended sites in Pennsylvania (from Seaker and Sopper, 1988a)

increase with time. Younger spoils had a predominance of sugar- and cellulose-decomposers, while older sites had more lignin-degrading species, which resulted in a higher decomposition rate (95%) (Miller and Cameron, 1978; Segal and Mancinelli, 1987).

Figure 4 shows the dry weight loss of the grass samples at 30-day intervals on the sludge-amended sites in Pennsylvania and supports the idea that more mature mine spoils possess a wider range of decomposers than do newly revegetated ones, Seaker and Sopper (1988a). Decomposition rates were similar for all sites during the first month, indicating a similar ability to decompose simple organic compounds. After the first 30-day period, site 1 began to lag behind the other sites. From day 300, the decomposition rate on site 5 increased significantly compared with sites 2, 3, and 4. Degradation of the more resistant components of the grass samples was most successfully achieved on the oldest site. Decomposition rate of cellulose has been found to be related to reclamation success, and was significantly increased by vegetative cover (Carrel et al., 1979) and by sewage sludge additions (Parkinson et al., 1980) in other studies. Microbial decomposition improves soil physical conditions by the formation of humus, and the leaching of decomposition products contributes to soil horizonation (Tate, 1985). It is probable that soil formation from mine spoil will occur at a faster rate when amended with sludge than when amended with chemical fertilizers.

Table 19. Extractable metal concentrations on Pennsylvania stripmine sites 1 to 5 years following sludge application, and on a fertilizer-amended site (means of three samples with standard error)

Site	Zn	Cu	Cd	Pb	Ni	Cr
			(mg kg^{-1})			
1	32 ± 8.1[d,1]	7 ± 3.2[c]	0.57 ± 0.09[d]	5 ± 2.1[c]	3 ± 0.32[c]	0.9 ± 0.33[c]
2	89 ± 8.3[c]	22 ± 2.7[b]	1.07 ± 0.09[c]	12 ± 1.3[a,b]	4 ± 0.24[c]	5.4 ± 0.79[a,b]
3	200 ± 7.7[a]	36 ± 4.5[a]	2.77 ± 0.07[a]	9 ± 1.5[b,c]	11 ± 0.30[a]	6.1 ± 0.91[a,b]
4	96 ± 4.2[c]	17 ± 1.8[b]	1.10 ± 0.06[c]	15 ± 1.9[a]	4 ± 1.02[c]	4.8 ± 0.55[b]
5	148 ± 10.4[b]	20 – 1.1[b]	1.47 ± 0.22[b]	13 ± 1.9[a,b]	7 ± 0.64[b]	7.0 ± 0.38[a]
F value	***	***	***	*	***	***
Fertilizer-amended	1.3 ± 0.2	1 ± 0	0.17 ± 0.03	3.2 ± 0.2	0.8 ± 0.24	0.3 ± 0.0

From Seaker and Sopper (1988a)
*, *** Significant effect at $P < 0.05$ and 0.001, respectively
[1]Means followed by different letters are significantly different at the 0.05 level of probability, Waller-Duncan *k*-ratio *t*-test

g. Trace Metals

Concentrations of HCl-extractable Zn, Cu, Cd, Pb, Ni, and Cr in the sludge-amended mine spoils and a fertilizer-amended spoil are given in Table 19 for the Pennsylvania sites. All metals were lowest on site 1, due to slightly lower metals concentrations in the sludge during that year, combined with a slightly lower application rate. Site 3 had significantly higher Zn, Cu, Cd, and Ni than the other sites, because the soil samples were taken from an area where sludge had been stockpiled prior to spreading. Although the absolute metal loadings on each site varied to some extent, they were all within the maximum allowed by Pennsylvania state regulations, which are very conservative compared with federal guidelines and to metal levels shown to inhibit microbial activities. The surface pH (0 to 15 cm) for sites 1 through 5 were 6.9, 7.0, 6.8, 6.7, and 7.3, respectively. The pH on the fertilizer-amended site was 6.3. A pH of 6.5 to 8.0 is optimum for the rapid decomposition of wastes in soils, and facilitates the growth of grass-legume forage as well as the immobilization of trace metals in sludge-amended soils.

At high levels, metals such as Zn, Cu, Cd, Pb, and Cr may interfere with microbial functions and have been shown to inhibit soil bacterial and fungal activity (Babich and Stotsky, 1977a). However, investigations in this area have generally utilized solution culture, pot studies, and plate culture techniques rather than land application systems, and purified metal salts rather than sewage sludges. Metal levels evaluated, which ranged from 100 to 10 000 mg kg^{-1} of Zn, Cu, Cd, Pb, Ni, or Cr, were far in excess of those normally encountered in land application systems employing median metal sludges (Babich and Stotsky, 1977b; Premi and Cornfield, 1969; Mathur et al., 1979; Bhuiya and Cornfield, 1974). Conclusions drawn from such studies are not directly applicable to field conditions because effects of metals depend on loading rates, sludge quality, and the form in which each metal occurs, and are strongly influenced by the complexity of the soil system as well as the complexity of the sludge. Binding of metals to humic or fulvic acids, proteins, or crystalline lattices of clay particles, as well as precipitation, complexation, and ionic interactions, influence metal inhibitory effects (Gadd and Driffiths, 1978). Studies have shown that microbial inhibition was reduced when metals were added to soil or organically complexed, and that a diverse soil population containing some tolerant forms would not be significantly affected by metal additions (Martin et al., 1966; Babich and Stotsky, 1977b; Doelman and Haanstra, 1979). Considering the high rates of metals involved in such studies, it is not surprising that with the low range of metal loading rates applied in the Pennsylvania study, microbial populations were within the ranges reported for undisturbed soils.

Very few studies have focused specifically on the effects of metals in land-applied sludge on microbial activity. Tomlinson (1966) suggested that complexing of sludge-borne metals with the components of the sludge/soil mixture may significantly reduce bacterial inhibition by metals. Soil amended with sludge to provide 100, 200, or 400 mg kg^{-1} Zn, Cu, Cd, Pb, and Cr, did not inhibit N transformations except at the highest level (Chang and Broadbent, 1982). Chicago sludge applied to mine spoils at rates up to 369 Mg ha^{-1} (dry wt basis) increased HC1-extractable Zn, Cu, Cd, Pb, Ni, and Cr above typical soil levels, but resulted in no significant reduction in microbial populations, percentage of denitrifiers, or specific enzyme activities (Varanka et al., 1976). Mean metal concentrations in the Chicago sludge were 333 mg kg^{-1} Zn, 70 mg kg^{-1} Cu, 16 mg kg^{-1} Cd, 70 mg kg^{-1} Pb, 11 mg kg^{-1} Ni, and 61 mg kg^{-1} Cr; pH was 6.6 to 6.7. These metal concentrations were considerably higher than on the Pennsylvania sites.

In fact, Zn, Cu, Ni, and Cr applied in sludge have been found in some instances to stimulate microbial activity and plant growth (Varanka et al., 1976; Premi and Cornfield, 1969) due to their role as cofactors for cellular enzyme systems. The fertilizer-amended site was deficient in Zn and Cu, which are essential for plant growth and microbial functioning.

A major problem with revegetating mine spoils is the extreme acidity that releases metals such as Zn, Cu, Cd, Pb, Al, and Mn from the spoil into the soil solution at concentrations inhibitory to soil micro-organisms (Mills, 1985). Reduced respiration and fungal populations in strip mine spoil have been attributed to high levels of metals coupled with low nutrient levels and acid pH (Lawrey, 1977). In sludge that has been processed by digestion and composting, sludge metals are bound to the organic components as sulfides, chlorides, carbonates, hydroxides, and other compounds not readily soluble. The rapidly established, dense vegetation achieved by sludge amendment would greatly increase water-holding capacity, reducing oxygen infiltration, acid formation, and release of metals from the spoil.

Although mine spoils can eventually recover "soil" characteristics through intensive reclamation and management techniques, annual fertilizer additions are usually required for several years. Such methods are rarely practical on vast acreages of nonagricultural land. Without annual maintenance, vegetative cover often deteriorates because microbial development is slow and nutrient cycling never becomes fully operative. The use of sludge as a spoil amendment eliminated the initial lag period that characterizes conventionally reclaimed sites, during which plant growth and microbial activity are at a low level, each one insufficient for maximum functioning of the other. Sludge amendments quickly increased the numbers and activity of micro-organisms, whose activities enhance the development of a soil environment conducive to continued plant growth. Development of an indigenous microbial community was achieved on all the Pennsylvania sites, which is a key factor in providing long-term site stability through biogeochemical cycling of energy and nutrients. Recovery of normal soil populations and processes in the surface 5 cm appeared to occur within 2 years, and did not show a tendency to deteriorate (Seaker and Sopper, 1988a).

h. Organic Matter

Reestablishing primary production, i.e., a vegetative cover that will persist for the 5-year-period required by federal surface mining regulations, is the principal aim of reclamation programs. This relatively short-term establishment of a ground cover, however, is not a sufficient guarantee that long-term ecosystem recovery will occur. Soil ecosystem stability results from continuous organic matter accumulation and cycling, and reclamation success may be measured by the degree of change in spoil characteristics toward characteristics indicative of a productive soil. These include detritus accumulation and decomposition; organic matter, organic C, and organic N contents; and root proliferation, all of which are determined by or influenced by soil microbial activity.

Recovery of native organic matter levels, soil structure, and A horizon development may require over 30 years in mine spoils from various environments, through natural succession or with conventional reclamation practices using only inorganic fertilizers (Anderson, 1977; Jenny, 1980; Leisman, 1957; Schafer et al., 1980). At best, reclamation with chemical fertilizers requires intensive management and annual fertilizer additions for several years. In practice, however, reclamation efforts are usually minimal, and after initial vegetation establishment, poor physical and biological conditions, not having been addressed, prevent the development of stable C and N cycles. As a result, the plant cover deteriorates before it has a chance to ameliorate the spoil (Stroo and Jencks, 1982; Tate, 1985). For example, even after 6 years, spent oil shales amended with fertilizer and revegetated did not recover their microbial community, and nutrient cycling and fertility remained low (Segal and Mancinelli, 1987). On fertilizer-amended spoil, organic matter and N levels tended to decrease with age, suggesting an eventual decrease in productivity and site stability after fertilizer amendments are discontinued (Stroo and Jencks, 1982). Nitrogen loss from the site due to poor retention and cycling indicated that site deterioration may not become apparent until after the 5-year bonding period required by federal regulations. It is for this reason that ecosystem recovery is seldom achieved with conventional reclamation techniques.

Organic amendments can be extremely important to successful reclamation (Sopper and Seaker, 1984; Down and Stocks, 1977; Visser, 1985). Numerous studies conducted over the past 60 years have shown that amending soils with organic materials such as sewage sludge increases soil organic matter content and improves soil structure and long-term fertility (Joost et al., 1987). Organic matter decomposition and cycling, processes difficult to initiate in disturbed soils, are quickly achieved by sludge addi-

Table 20. Organic matter content, C and N content, and C/N ratio of strip-mine spoil 1 to 5 years following sludge application, and of the fertilizer-amended site (means of six samples for organic matter, and three samples for C and N, with standard error)

Site	organic matter	Organic C (%)	Kjeldahl N	C/N
1	1.60 ± 0.09[c,1]	3.09 ± 0.44[b]	0.36 ± 0.06	8
2	2.93 ± 0.25[b]	6.13 ± 1.38[a,b]	0.41 ± 0.10	15
3	5.13 ± 0.30[a]	6.69 ± 0.35[a,b]	0.41 ± 0.08	17
4	3.72 ± 0.40[b]	3.80 ± 0.46[b]	0.30 ± 0.07	14
5	4.52 ± 0.34[a]	7.59 ± 1.72[a]	0.48 ± 0.12	16
F value	***	*	NS	NS
Fertilizer-amended	1.75 ± 0.38	1.57 ± 0.23	0.10 ± 0.01	15

From Seaker and Sopper (1988b)
*, *** Significant effect at $P < 0.05$ and 0.001, respectively; NS, no significant effect
[1] Means followed by different superscript letters are significantly different at the 0.05 level of probability, Waller-Duncan *k*-ratio *t*-test

tions. Depending upon the amount of topsoil material present, it is possible on some sites to achieve significant soil development in mine spoils in as few as 3 to 5 years through intensive management and the use of amendments that accelerate plant growth and soil-forming processes (Visser, 1985).

The reclamation success achieved with sludge is due to three factors related to its organic content: (1) the N content is in a slowly available organic form; (2) the high organic C content provides an immediate energy source for soil microbes; and (3) sludge organic matter improves the poor spoil physical conditions resulting from soil removal and compaction.

In the Pennsylvania study, previously cited, Seaker and Sopper (1988b) also reported on the effects of the sludge applications on spoil organic matter on the five sites ranging in age from 1 to 5 years after sludge application (Table 20).

The organic matter content of the spoil from sites 2 through 5 was greater than on the fertilizer-amended site; A more dense vegetative cover and higher decomposition rates (Seaker and Sopper, 1988b) on sites 2 through 5 increased the organic matter content up to threefold compared with the fertilizer-amended site. The older sludge sites had significantly higher organic matter content than site 1, and organic matter was correlated with site age ($r = 0.67$). The maximum was observed on site 3. Similar organic matter increases with site age have been reported for subalpine mine spoils (Visser et al., 1983), and 369 Mg ha^{-1} (dry wt basis) of sludge increased the organic matter content of silt loam mine spoil to 5.43% (Varanka et al., 1976).

Soil horizon development partially depends on such microbial processes as vegetation decomposition, incorporation of humus into the spoil, downward movement of organic material in the developing profile, and organic sorting (Ollier, 1969). Such processes are enhanced by the addition of sludge to mine spoil and by the subsequent increased biomass production compared with spoils amended with chemical fertilizers. The organic matter content of the fertilizer-amended site remained at a low level, suggesting that the initial stages of soil horizon development were inhibited.

There was no difference in N content between the sludge-amended sites, which indicates that N levels established by sludge application were maintained even on the 5-year-old site (Table 20). Nitrogenous organic additions, e.g., proteins contained in sludge, result in net mineralization of N that builds up the N pool. Because it is mainly in the organic form, the N in sludge becomes slowly available over time. Much of the N in this system is conserved and recycled. Losses of N due to nitrate leaching or ammonia volatilization were not measured but appear to be insignificant in comparison with the amount of N retained in the spoil.

An N level of 0.25% has been reported for undisturbed soils in the eastern USA (Wilson, 1965; Stevenson, 1982). The sludge-amended sites had N contents of 0.30% to 0.48%, while the fertilizer-amended site had an N content of 0.10%. This indicates that reclamation with sludge can restore and even improve soil fertility with respect to N for a period of time beyond that achieved with chemical fertilizers. Nitrogen accumulation is difficult to achieve with inorganic N fertilizers. Stroo and Jencks (1982) found mine spoils reclaimed with fertilizer and mulch initially productive, but little of the N remained in the soil, suggesting that self-sustaining ecosystems are not being achieved through current reclamation practices.

Seaker and Sopper (1988b) reported that organic C content was increased at least threefold on the sludge-amended sites compared with the fertilizer-amended site, and ranged from 3% to 7% (Table 20). In contrast, the C content of 20-year-old strip-mine sites, reclaimed without an organic amendment, were reported to average only 0.93% (Wilson, 1965). Organic C was lowest on site 1 and highest on site 5, but there was no significant correlation with site age. Chemical fertilizers can provide initial stimulation of plant growth, but unlike organic amendments, they do not directly stimulate microbial activity by providing a C source.

The higher values for organic C than for organic matter may be due to the difference in methodology. In the Walkley-Black method, organic matter is oxidized by chromic acid, while the Coulometric method is a dry combustion procedure. It is possible that the Coulometric method oxidized a higher percentage of the C present in soil-sized particles of coal, which have been reported to be a source of error in the determination of organic matter in mine spoil (Wilson, 1965; Schaefer et al., 1980). Despite the differences in organic C content achieved by the two methods, both sets of data follow the same general trend, with the lowest values for the fertilizer-amended spoil, and the highest values for sites 3 and 5 years old.

Seaker and Sopper (1988b) found that the C/N ratio approximately doubled after the first year due to accumulation of plant biomass (Table 20), and appears to have stabilized at about 16 by the second year. This ratio is between that of nutritionally balanced agricultural soils, which is about 25 according to Stevenson (1982), and that of most native soils, which is 10 to 12 (Reeder and Berg, 1977). The C/N ratio of the fertilizer-amended site was 15. However, the absolute amounts of C and N were very low compared with the sludge-amended sites, reflecting the low organic matter content and poor fertility that would largely account for the inferior plant growth. Both C and N were increased nearly fivefold on the sludge-amended sites compared with the control. There was no evidence of deterioration of either organic C or N in the sludge-amended spoils over time. Similarly, Hinesly et al. (1979) reported that residual C and N applied in sludge at a rate of 61 Mg ha^{-1} remained stable for 4 years after the sludge application stopped.

i. Nitrogen Mineralization

Lack of plant-available N may be a major problem in revegetation of some lands disturbed by surface mining. Sewage sludges contain 20 to 60 g N kg^{-1} or more, much of which is in the organic form (Sommers et al., 1976). The rate of mineralization of this organic N to NH_3-N and the subsequent nitrification to NO_3-N is important in supplying adequate N for revegetation establishment.

Disagreement exists among researchers concerning the effect of the rate of sewage sludge addition on the percent of added organic N mineralized. Terry et al. (1981) using sewage sludge application rates of 11.2, 22.4, and 44.8 Mg ha^{-1} found that the percent of added organic N mineralized was significantly greater at higher rates than the lower rates of sludge addition. Epstein et al. (1978) and Magdoff and Chromec (1977) observed no rate effect on the percent of added organic N mineralized at rates ranging from about 20 to 80 Mg ha^{-1}. Following this trend, Sabey et al. (1977) observed that the percent of added organic N mineralized decreased as the amount of N added increased.

Voos and Sabey (1987) conducted a 16-week laboratory incubation study to determine the rate of net N mineralization in sewage-sludge-amended coal mine spoil. Sewage sludge was added at rates of 0, 40, 80, and 120 Mg ha^{-1} which added 0, 1630, 3260, and 4890 kg N ha^{-1}, respectively. The total amount of inorganic N that accumulated during the experiments increased significantly as the rate of sewage sludge addition increased. Only small amounts of NO_3-N had accumulated in the mine spoil after 16 weeks. Ammonium-N increased with increasing rates of sewage sludge. The total net N mineralized in 16 weeks in the mine spoil treated with 0, 40, 80, and 120 Mg ha^{-1} of sludge were 274, 402, 505, and 617 mg kg^{-1}, respectively.

Hornby et al. (1986) also studied nitrogen mineralization potentials of lignite overburden in Texas. Nitrogen mineralization potentials were determined from laboratory data on a premined native soil and 4-year-old reclaimed mixed overburden that received 180 kg N ha^{-1} yr^{-1} as NH_4NO_3. In addition, the effect of anaerobically digested and dewatered sewage sludge on N^- mineralization potential of overburden was evaluated on samples (0 to 15 cm) collected from field plots. Samples were collected at 2, 26, and 52 weeks after amendment. Treatments were 0 kg N ha^{-1}, 212 kg N ha^{-1} as NH_4NO_3, 106 kg N ha^{-1} as NH_4NO_3 plus 106 kg N ha^{-1} as sludge, and 212 kg N ha^{-1} as sludge. Samples were incubated in the lab for 18 weeks. The N mineralization potentials for all sewage sludge plots remained higher than all other treatments at the end of 1 year. Release of NO_3-N was significantly higher in plots that received sewage sludge than in the control or fertilized plots. The total N in sludge-amended plots did not change significantly with time over the 52-week period. They concluded that a single application of sludge at 56 Mg ha^{-1} to overburden provided a greater supply of mineralizable N than that resulting from 4 years of mineral N fertilizer applications.

j. Earthworms

Sludge applications to agricultural soils and mine spoils can have a significant effect on earthworm populations. Earthworm activity is important in soil formation processes (Miller, 1974) and is a useful indicator of soil metal availability (Van Hook, 1974). On sludge-amended soils, Hartenstein et al. (1981) reported that earthworms accumulate high concentrations of metals. This may pose a potential hazard to the earthworms and their predators. Earthworm populations are usually very low or nonexistent on mine spoils, thus, few studies have been conducted.

A search of the literature resulted in finding only one study by Pietz et al. (1984) that was conducted at the Fulton County, IL land reclamation site. Anaerobically digested sewage sludge has been applied continually since 1972 to calcareous strip mine spoil (Peterson et al., 1982). The 3-year-study (1975–77) sampled mine soil and nonmined fields to determine the effect of land application of anaerobically digested sludge on the heavy metal accumulations in earthworms. The only earthworm species found on the mine soil fields was *Aporrectodea tuberculata*. On the nonmined fields the species *Lumbricus terrestris* was also found. Sewage sludge applications to fields on both land types (mine soil vs nonmined) resulted in significant accumulations of Cd, Cu, and Zn. Land type significantly affected earthworm Zn concentrations, with levels being higher in all nonmined fields sampled. Earthworm Cd and Cu accumulations in all fields sampled were significantly related to the current amounts of sludge-applied metals and the amount applied since the previous sampling. Concentrations of Ni, Cr, and Pb in earthworms were not significantly related to sewage sludge applications during the 1975–77 sampling period. Earthworm metal concentrations in all fields studied ranged from 19 to 506 mg Zn kg^{-1}, 0.6 to 98 mg Cd kg^{-1}, 0.1 to 8.8 mg Cr kg^{-1}, 1.1 to 25 mg Cu kg^{-1}, <0.1 to 10 mg Ni kg^{-1}, and <0.1 to 2.4 mg Pb kg^{-1}. The higher values all occurred on sludge-amended fields. Little is known about the toxicities of soil metals to earthworms. Hartenstein et al. (1981) determined the concentrations at which heavy metals added to activated sludge would induce a toxic effect on the growth of *Eisenia foetida*. All of the maximum metal concentrations cited above were from sludge-amended soils containing metal levels considerably lower than the threshold toxicity levels listed by Hartenstein et al. (1981). Since Cu, Cd, and Zn were accumulated in significant amounts by worms in sludge-amended soils at the Fulton County site, there may be a potential hazard to predators (primarily birds) with continued long-term sludge applications.

D. Effects on Water Quality

Some concern has been voiced over the effects of high rates of sludge application on the quality of groundwater and nearby streams, ponds, and lakes. The state of Pennsylvania, for example, prohibits the use of sludge for land reclamation directly on a watershed area that supplies drinking water to a community. But that is not to say that sludge application results in the deterioration of waters. In fact, the opposite is usually true. Stabilization of drastically disturbed lands with municipal sludge often improves the quality of the surrounding area in having an ameliorative effect on the ecosystem as a whole. Reports on the effects of sludge application on concentrations of NO_3-N, trace metals and on indicator organisms in soil percolate water, groundwater, nearby streams and lakes, and surface runoff indicate that a properly managed land application program will not cause deterioration of water quality on or near the site.

1. Soil Water and Groundwater

a. Nitrate–Nitrogen

There is a potential for nitrate build-up and eventual leaching into groundwater when sludge, particularly liquid sludge, is applied continuously (Kardos et al., 1979; Urie et al., 1982). Nitrate N of percolate water has increased in varying degrees after sludge application (Kardos et al., 1979; Urie et al., 1982; Sopper and Kerr, 1981). Irrigating with liquid sludge and sewage effluent at 2.5 and 5.0 cm per week, totaling 60 and 120 cm on bituminous spoil, and 75 and 150 cm on anthracite refuse, increased nitrate and ammonium in the leachate water. But nitrate N did not exceed the drinking water limit of 10 mg L^{-1} (Kardos et al., 1979). Nitrates below the drinking water limit were also reported for leachate 90 cm below the surface of a burned anthracite refuse bank amended with sludge at 40 to 150

$Mg\ ha^{-1}$ (Kerr et al., 1979). Some studies show an initial peak in nitrate-N concentration followed by a decrease to acceptable levels. Often, peak nitrate-N levels occur in late winter and spring when plants are not utilizing the nitrogen for growth (Haghiri and Sutton, 1982). Both C1 and nitrate levels in leachate from sludge-amended sand and gravel spoils were initially increased, but tended to decrease over an 8-week period (Hornick, 1982). The nitrate-N concentration of groundwater collected from wells at various depths on sludge-amended mine spoils in Pennsylvania was consistently within safe drinking water standards, when monitored monthly for up to 5 years (Sopper and Kerr, 1981).

Seaker and Sopper (1984) reported that an application of sludge of 184 $Mg\ ha^{-1}$ on an abandoned strip mine had little effect on groundwater quality, even though the water table was only 3 to 4 m below the surface. They reported that average monthly concentrations of NO_3-N were below 10 mg L^{-1} (maximum concentration for potable water) for all months during a 5-year period. The highest monthly value recorded was 2.4 mg L^{-1}. Groundwater samples were also analyzed for total and fecal coliforms. No fecal coliform colonies were observed for any monthly sample. Sopper and Seaker (1990) resampled groundwater in 1989, 12 years after sludge application, and found N03-N concentrations still at an extremely low level (0.06 Mg L^{-1}) (Table 21).

The application of lime and sludge and subsequent revegetation appears to have had a positive effect on groundwater pH. Groundwater pH increased from 4.6 (1977) to 6.0 by 1981. Results of the 1989 sampling indicate a pH of 6.6. There has also been a gradual increase in pH in the control well from pH 4.4 to pH 5.8. Since 1980, attempts were being made to reclaim the control area by conventional methods using lime and fertilizer. The amounts of lime and fertilizer applied and frequency of application are not known as the coal company is no longer in business. However, these applications and vegetation growth probably contributed to the increase in groundwater pH in the control well.

There appears to be no significant increase in any of the trace metal concentrations over the initial 5-year period (1977–81) in the groundwater samples from Well 2 compared with the control well. From 1977 to 1981 most of the monthly concentrations were within the U.S. Environmental Protection Agency drinking water standards. The only exception was Pb which exceeded the limit of 0.05 mg/L^{-1} for both the control well and Well 2, probably resulting from increased release of the element from the spoil material due to mining. The highest monthly Pb values were 0.28 mg/L^{-1} in the control well and 0.33 mg/L^{-1} in Well 2 in 1978, and the mean annual Pb concentrations were 0.19 and 0.20 mg/L^{-1} for control well and Well 2, respectively. By 1981, however, the mean annual Pb concentrations had decreased to 0.04 and 0.05 mg/L^{-1} for the two wells. Results of analyses of the groundwater samples collected in 1989 had extremely low concentra-

Table 21. Mean annual concentrations of nitrate-N and trace metals in groundwater collected on a strip mine site in pennsylvania amended with 184 mg ha^{-1} of sludge

Site	Year[a]	pH	NO_3-N	Cu	Zn	Cr	Pb	Cd	Ni
						(mg/l)			
Well 1 (control)	1977	4.4	1.4	0.22	4.13	0.02	0.14	0.006	3.67
	1978	4.3	<0.5	0.23	2.02	0.01	0.19	0.002	0.98
	1979	4.6	<0.5	0.17	1.48	0.03	0.13	0.001	0.50
	1989	5.5	0.6	0.05	0.89	0.05	0.09	0.003	0.50
	1981	5.7	0.7	0.06	0.83	0.03	0.04	0.001	0.31
	1989[b]	5.8	0.02	0.01	0.09	<0.001	0.01		0.06
Well 2 (sludge)	1977	4.6	1.1	0.10	3.39	0.03	0.09	0.001	2.67
	1978	4.5	<0.05	0.14	3.29	0.01	0.20	0.002	1.26
	1979	4.4	<0.05	0.18	1.49	0.03	0.13	0.001	0.97
	1980	5.7	0.6	0.05	1.05	0.04	0.11	0.001	0.76
	1981	6.0	0.6	0.05	0.57	0.02	0.05	0.001	0.31
	1989[b]	6.6	0.06	0.01	0.07	<0.001	0.01	0.001	0.04
EPA drinking water standard			10	1	5	0.05	0.05	0.010	—

From Sopper and Seaker (1990)
[a] Values are annual means of monthly samples
[b] Average of three samples collected in August 1989

tions of all trace metals in both wells in comparison with values for the initial 5 years (1977–81).

Similarly, on an anthracite coal refuse bank treated with 80 and 108 Mg ha^{-1} of sludge, Seaker and Sopper (1983) found little effect on groundwater quality. Over a 5-year monitoring period all monthly values of NO_3-N were less than 10 mg L^{-1}. Groundwater samples were also analyzed for total and fecal coliforms and no fecal coliforms were ever observed for any sample.

Pietz et al. (1989c) reported on the effects of applying sludge (542 Mg ha^{-1}), lime (89.6 Mg ha^{-1}), and gypsum (112 Mg ha^{-1}), and various combinations to coal refuse on percolate water quality at a depth of 1 m. Samples were collected monthly over a 5-year period. Yearly mean concentrations of NH_4-N, $[NO_3 + NO_2]$-N ranged from 0.8 to 225, and 0.0 to 278 mg L^{-1}, respectively. The high values were all associated with the sludge treatment. The NH_4-N in percolate from the sludge-amended treatments was initially high, but declined with time to near background levels after 5 years. The $[NO_3 + NO_2]$-N concentrations in the sludge-amended treatments increased the first 3 years reaching a maximum of 278 mg L^{-1}, and then declined rapidly in subsequent years to levels of 20 to 61 mg L^{-1} by the fifth year.

b. Trace Metals

On anthracite refuse and bituminous strip-mine spoils, leachate collected 107 cm below a grass, legume, and tree seedling cover was lower in Fe, Al, and Mn where liquid sludge and sewage effluent were applied biweekly than it was in control spoils (Kardos et al., 1979). Satisfactory renovation of the major constituents of sludge through acidic strip-mine spoil has been reported by McCormick and Borden (1973). The pH of the percolate water was related to the sludge rate and application method. Initially high sulfur levels resulting from the sludge decreased to below those in control areas. In Ohio, strip-mine spoils amended with sludge at rates up to 716 Mg ha^{-1}, leachate concentrations of Cu, Ni, and Mn did not increase, and even decreased with time. Zinc and Al initially increased, but then decreased, while Cd and Pb were below detection limits (Haghiri and Sutton, 1982). On a burned anthracite refuse bank, sludge applications of 75 to 150 Mg ha^{-1} did not degrade the quality of percolate water 90 cm below the surface (Kerr et al., 1979). In fact, Zn and Cd concentrations were lower in the sludge-treated plots than in the control plots. Where sludge applications were monitored for 2 to 5 years on three 4-ha sites in Pennsylvania, groundwater samples collected monthly showed no evidence of contamination (Sopper and Kerr, 1981); Sopper and Seaker, 1982; Seaker and Sopper, 1983, 1984). Copper, Zn, Cr, and Pb were, with very few exceptions, well within safe drinking water standards established by the U.S. EPA. Lead sometimes exceeded the limit by a minimal amount, even on unsludged areas, because of Pb-bearing minerals in the bedrock. In an ongoing program, over 1500 ha of strip-mined land in Pennsylvania have been reclaimed with Philadelphia sludge applied at 134 Mg ha^{-1}. Over a 5-year period, groundwater quality met drinking water standards for metals and fecal coliform bacteria (Sopper and Kerr, 1980a, b; Sopper 1982a–e; Sopper et al., 1981).

Pietz et al. (1989c), in the same study as previously cited, found that the concentrations of all metals in the percolate at 1 m, except Pb and Hg, were significantly affected by the sludge, lime, and gypsum treatments. Aluminum, Fe, Cu, Ni, Cd, and Zn levels were generally lowest in the sludge or control treatments. However, concentrations of these same metals increased with time, indicating a solubilization of these metals with time in both the sludge and sludge + lime treatments. The authors concluded that the applications of sludge (542 Mg ha^{-1}) and lime (89.6 Mg ha^{-1}) based on theoretical calculations to control acidity were too low for long-term reclamation. The authors recommend that for long-term reclamation (>5 years) of coal refuse, an application of lime or sewage sludge alone should be 189 and 1050 Mg ha^{-1}, respectively. If both materials are used, lime and sewage sludge application rates between 134 to 189 and 900 to 1350 Mg ha^{-1}, respectively, would be desirable for reclamation on a long-term basis.

2. Surface Water

Surface water runoff and a stream adjacent to the Palzo site were monitored following applications of liquid sludge (Urie et al., 1982; Jones and Cunningham, 1979). Reduction of surface runoff was related directly to the density of the vegetative cover. Ammonium-N, nitrate-N, and total-N concentrations were decreased in runoff analyzed 2 years after sludge application, compared with runoff from unsludged areas (Urie et al., 1982). Iron, SO_4, Al, and Cd concentrations in a nearby stream were drastically increased as a result of strip mining, but reductions in ion concentrations during sludge application did not degrade water quality during the period (Jones and Cunningham, 1979). The chemical and biological quality of Contrary Creek, adjacent to an abandoned pyrite mine in Virginia, was not affected within 1 year after the site was revegetated using municipal sludge. Because of runoff from barren areas and the toxic mine sediments already in the stream bed, it may take decades before site stabilization affects stream quality (Hinkle, 1982). On the Fulton County project in Illinois, the water-quality monitoring program calls for monthly sampling of 33 wells and nearby streams and reservoirs. Although sludge addition increased nitrate N minimally, mean annual concentrations of nitrate-N, Cd, Zn, Cu, Cr, and Pb were within U.S. EPA drinking water limits in the two watershed reservoirs tested during the 2 years after sludge was applied. Fecal coliform counts were not increased by sludge application; in fact, they decreased, probably because fewer livestock grazed on the site. The authors (Peterson et al., 1979) concluded that a properly managed, digested sludge application site will not adversely affect local surface waters.

Two lakes adjacent to an abandoned bituminous strip mine in Pennsylvania were monitored monthly for 5 years, after the site was reclaimed with liquid and dewatered sludges at rates up to 184 Mg ha^{-1}. For the entire 5-year period, nitrate N, Zn, Cu, Cr, and Cd were below the U.S. EPA maximum limits for drinking water. Nitrates were slightly increased the first several months after sludge application, but then decreased. Lead was often slightly above drinking water limits due to natural dissolution of Pb-bearing minerals in the underlying rock (Sopper and Kerr, 1981).

E. Effects on Animal Nutrition and Health

The quality of forage should be determined before livestock are grazed on lands reclaimed with municipal sludge, as it would be in the management of a normal farming operation. In a properly managed land application system, forages would be expected to be of good nutritional quality. On five (4-ha) sites in Pennsylvania where mine spoils were revegetated with several types of sludge at rates from (7 to 202 Mg ha^{-1}), tall fescue, orchardgrass, and birdsfoot trefoil were of excellent quality. Trace metal concentrations were low, and protein and fiber contents were comparable to those in forages grown on agricultural land amended with inorganic fertilizers (Seaker and Sopper, 1982). Sludge amendments on mine spoils in Illinois significantly improved corn grain quality as measured by protein content (Blessin and Garcia, 1979). Two studies showed some potential problems that may be encountered.

Nitrogen: sulfur ratios were low in tall fescue (5:1) grown in West Virginia strip-mine spoils amended with up to 224 Mg ha^{-1} of sludge, as compared with recommended ratios of 10:1 to 15:1 (Mathias et al., 1979). On soil contaminated by a nearby zinc smelter, forages grown with municipal sludge were not considered suitable for feed because of high nitrates, Zn, and Cd the first two growing seasons (Franks et al., 1982).

Plant uptake of metals and sludge deposits on leaves eaten by grazing animals can increase the potential for higher tissue concentrations. Proper use of sludge depends on the impact on soils, plants, and animals exposed to it. According to Fitzgerald (1982), animals exposed to excess heavy metals show toxic reactions rather quickly. In a study involving the Fulton County, Illinois, project the concentrations of seven trace metals in tissues from animals grazing in sludge-grown forage were not significantly different from those in the controls, except for increased Cd, Pb, Cu, and Zn in the liver and Cd in the kidney. Lead levels in the blood were increased fourfold. The diaphragm, heart, brain, bone, and milk showed no trace of sludge, nor was there any effect on the reproductive rate, or any evidence of disease. The growth of experimental cows was above average. In a similar study, pigs feeding in sludge-amended pens did not accumulate any more trace metals in the diaphragm, heart, or bone than did control animals, but Cd was increased in liver and kidney tissue.

Another study with Chicago sludge assessed metal accumulations in various organs of pheasants and swine which were fed for 100 and 56 days, respectively, corn grain harvested from reclaimed areas where liquid sludge had been applied annually for 5 to 6 years. Annual loading rates ranged from 25 to 128 Mg ha^{-1}. Because trace metal composition of muscle tissue was unaffected by a diet of sludge-grown corn, the authors concluded that the consumption of meat from these animals would present little, if any, potential health hazard to humans. Although Cd concentrations were increased significantly in the liver and kidney tissues of pheasants and swine fed the corn grain from the sludged fields as compared with concentrations in control animals, the maximum levels of Cd in the tissues were still comparable to those reported elsewhere for animals fed a normal diet (Hinesly et al., 1979b).

When red-winged blackbirds nesting on the sludge-reclaimed Palzo strip-mine were analyzed for tissue Cd, Zn, and Pb, it was found that Pb concentrations in their brain, liver, kidney, and muscle were no different than those in birds living in undisturbed areas or on strip-mine sites reclaimed with inorganic fertilizers (Gaffney and Ellertson, 1979). Higher kidney Cd was observed on the sludged sites, but in some tissues, Cd tissue

Table 22. Extractable trace metal concentrations (dry wt) in surface soil samples

Metal	Control ($mg\ kg^{-1}$) Mean SE	Treated ($mg\ kg^{-1}$) Mean SE	Normal range in soil[a]
Cu	2.92 ± 0.16	6.68 ± 2.47	2–100
Zn	1.50 ± 0.24	78.91 ± 23.41	10–300
Cr	0.24 ± 0.03	2.04 ± 0.91	5–3000
Pb	2.67 ± 0.15	3.53 ± 0.88	2–200
Co	1.37 ± 0.10	1.00 ± 0.15	1–40
Cd	0.06 ± 0.00	0.39 ± 0.09	0.01–7.0
Ni	2.21 ± 0.31	3.11 ± 0.77	10–1000

From Dressler et al. (1986)
[a] Allaway (1968)

concentrations were higher in birds from natural areas than in birds living on the Palzo site. Interpretation of such data is complicated due to Cd-Zn interactions that may occur in animal tissues and to the lack of a data base for normal metal levels in birds.

In Pennsylvania, two recent studies were reported which investigated the trace metal concentrations in the tissue of cottontail rabbits (*Sylvilagus floridanus*) and meadow voles (*Microtus pennsylvanicus*) trapped on an abandoned strip-mine site revegetated using Philadelphia sludge applied at 134 Mg ha^{-1}. The goal of both studies was to follow the transfer of trace metals from sludge to the soil and into the animals food chain (Dressler et al., 1986; Aberici et al., 1989). The amounts of trace metals applied in the sludge were 175 kg Zn ha^{-1}, 53 kg Pb ha^{-1}, 50 kg Cu ha^{-1}, 31 kg Cr ha^{-1}, 7 kg Ni ha^{-1}, and 2 kg Cd ha^{-1}. Extractable trace metal concentrations in surface soil samples (0 to 15 cm) were higher on the sludge-amended site but were not significantly ($P > 0.05$) different from a control site reclaimed by conventional methods using lime and chemical fertilizer (Table 22). Vegetation species on both sites were similar. Concentrations of Zn in all plant species and Cd and Cu in three of the four plant species were higher ($P < 0.05$) on the sludge-amended site compared with the control site (Table 23).

In the first study, ten adult rabbits were trapped on the treated site and eleven adult rabbits were trapped on the control site (Dressler et al., 1986). Trace metal concentrations found in rabbit tissues are given in Table 24. Levels of most metals in cottontail rabbits collected from both sites were not significantly different between males and females. Concentrations of Zn in cottontail rabbit femurs on the treated site were higher ($P < 0.05$) than those on the control site, corresponding to increased Zn concentrations in the vegetation on the treated site. Concentrations of Cu, Cr, Pb, Co, Cd, and Ni in all tissues (femur, kidney, liver, and muscle) were not significantly different between the two sites.

Table 23. Heavy metal concentrations (dry wt.) in vegetation foilar samples[a]

	Control (mg kg^{-1})			Treatment (mg kg^{-1})		
Element	$\bar{x}$	SD	Median	$\bar{x}$	SD	Median
			Bromus sp.			
Cu	5.8	0.7	5.5**	8.4	2.1	7.4**
Zn	21.9	2.1	21.5**	41.8	13.9	37.6**
Cr	0.0	0.0	0.0	0.6	1.5	0.0
Pb	2.7	1.4	2.9	2.0	0.9	1.9
Co	0.3	0.4	0.1	<0.1	0.1	0.0
Cd	0.02	0.02	0.01	0.06	0.07	0.03
Ni	0.5	0.4	0.5	1.6	2.9	0.6
			Orchardgrass			
Cu	9.4	0.8	9.4	10.5	1.4	10.4
Zn	23.7	2.0	23.5**	43.9	7.0	45.5**
Cr	0.0	0.0	0.0**	8.6	17.4	1.3**
Pb	4.2	0.7	4.0	3.2	1.1	3.3
Co	0.5	0.3	0.5	0.2	0.2	0.1
Cd	0.02	0.02	<0.01**	0.2	0.07	0.2**
Ni	1.0	0.6	1.0*	4.7	4.7	2.3*
			Tall fescue			
Cu	7.0	0.6	7.0*	8.0	0.5	7.9*
Zn	21.7	1.4	21.8**	41.4	8.8	37.6**
Cr	0.1	0.3	0.0	4.2	10.2	0.0
Pb	2.8	0.8	2.9	2.3	1.1	2.1
Co	0.1	0.2	0.0	0.3	0.4	0.0
Cd	0.02	0.02	0.02**	0.1	0.06	0.12**
Ni	1.4	0.9	1.4	4.4	2.4	1.1
			Trefoil			
Cu	10.7	0.4	10.8*	11.6	0.7	11.5*
Zn	43.0	1.5	43.0*	60.0	2.8	59.9*
Cr	0.3	0.6	0.0*	20.5	15.3	25.1*
Pb	4.3	0.4	4.5	3.6	0.3	3.7
Co	1.8	0.0	1.8*	0.5	0.5	0.4*
Cd	0.02	0.02	0.03*	0.2	0.04	0.16*
Ni	13.8	0.6	13.5	17.0	4.7	18.4

From Dressler et al. (1986)
*, ** Significant at $P \leq 0.05$ and $P \leq 0.01$, respectively
[a] All values based on six samples except control-trefoil ($n = 3$)

Table 24. Heavy metal concentrations (dry wt) of male and female cottontail tissues collected in March 1983[a]

	Control (mg kg^{-1})			Treatment (mg kg^{-1})		
Element	$\bar{x}$	SD	Median	$\bar{x}$	SD	Median
			Femur			
Cu	2.9	0.3	3.0	2.8	0.2	2.8
Zn	126	13	125*	148	22	147*
Cr	1.4	0.2	1.5	1.3	0.4	1.3
Pb	12.2	1.1	12.0	12.9	1.0	12.8
Co	5.3	0.5	5.5	5.4	0.6	5.4
Cd	0.01	0.01	0.01	0.01	0.01	0.01
Ni	6.8	0.6	6.8	6.5	0.4	6.3
			Kidney			
Cu	10.6	2.6	11.0	10.9	1.2	11.3
Zn	81.6	23.9	77.0	87.5	26.3	84.3
Cr	0.0	0.0	0.0	0.0	0.0	0.0
Pb	1.5	1.2	1.5	1.5	2.4	1.5
Co	0.0	0.0	0.0	0.0	0.2	0.0
Cd	9.6	7.1	5.3	17.8	14.1	12.3
Ni	0.5	0.8	0.0	0.8	1.0	0.0
			Liver			
Cu	11.0	1.9	10.8	11.8	2.2	11.4
Zn	110	20	120	115	26	114
Cr	0.0	0.0	0.0	0.0	0.0	0.0
Pb	0.2	0.5	0.0	0.8	0.7	0.9
Co	0.0	0.1	0.0	0.2	0.2	0.0
Cd	1.60	0.79	1.63	2.40	0.85	2.25
Ni	0.1	0.1	0.0	0.2	0.2	0.3
			Muscle			
Cu	4.3	0.5	4.0	4.0	0.5	4.3
Zn	41.1	3.7	41.8	44.1	5.5	43.6
Cr	0.1	0.4	0.0	0.5	1.7	0.0
Pb	0.3	0.4	0.0	0.1	0.3	0.0
Co	0.1	0.2	0.0	0.0	0.1	0.0
Cd	0.01	0.01	0.01	0.01	0.01	0.01
Ni	0.5	1.0	0.3	0.9	2.0	0.1

From Dressler et al. (1986)
*Significant at $P < 0.05$
[a]Control values based on 11 replicates; treatment values based on 10 replicates (replicate = one sample from one rabbit)

Table 25. Cadmium concentrations (dry wt) of tissues collected from female cottontails fed experimental diets for 126 days[1]

	Control (mg kg^{-1})			5 mg Cd kg^{-1}			25 mg Cd kg^{-1}		
Tissue	$\bar{x}$	SD	Median	$\bar{x}$	SD	Median	$\bar{x}$	SD	Median
Femur	1.0	<0.1	1.0[a,2]	1.3	0.4	1.2[b]	1.7	0.4	1.6[c]
Kidney	16	14	9[a]	98	15	95[b]	341	76	355[c]
Lever	1.3	0.9	0.0[a]	13.5	4.1	11.5[b]	59.6	19.2	57.5[c]
Muscle	0.00	0.00	0.00[a]	<0.01	0.01	0.00[a]	0.19	0.11	0.16[b]

From Dresster et al. (1986)
[1] Control and 5 mg Cd kg^{-1} values based on nine replicates, 25 mg Cd kg^{-1} values based on 12 replicates (replicate = one sample from one rabbit)
[2] All tests of treatment effects were significant ($P<0.01$); significant pairwise comparisons are indicated by different superscript letters ($P<0.01$)

In a companion study, 36 adult female rabbits were held in laboratory cages and fed a diet with additions of cadmium sulfate hydrate ($3CdSO_4 \cdot 8H_2O$), a highly soluble Cd compound. Treatments consisted of a control (Purina Lab Rabbit Chow HF5326), control with 5 mg Cd kg^{-1}, and control with 25 mg Cd kg^{-1}. The rabbits were necropsied after 126 days on the diet. Addition of the Cd salt to the diets resulted in an increase ($P<0.01$) in Cd in all cottontail rabbit tissues except muscle (Table 25). As expected, highest levels of Cd were found in the kidneys and liver and Cd levels increased with increased dietary intake of Cd.

Laboratory data using metal salts must be used cautiously and cannot be directly compared to field data where Cd is added in a sludge product. Logan and Chaney (1983) pointed out the many problems of comparing plant uptake of soluble metal salts with metals in digested sludge and comparing controlled laboratory studies with field experimental results. If soluble metal salts are taken up at a greater rate than sludge metals (Logan and Chaney, 1983), any field values above the laboratory levels might indicate a potential problem. Levels of Cd in rabbit tissue samples in this study from the mined sites (control and sludge-amended) and from nonmined sites, except muscle and liver tissues from the sludge-treated site, were comparable to or below ($P<0.05$) the laboratory control levels. Cadmium concentrations in muscle from the sludge-treated mine site were higher than the 5 mg Cd kg^{-1} laboratory treatment group. Although the muscle Cd concentrations from mined and nonmined sites were higher ($P<0.05$) than laboratory controls, there was no significant difference between the sites, and these levels were below mean levels reported in selected foods analyzed by the Food and Drug Administration, e.g., ground beef mean = 0.075 mg Cd kg^{-1} (Sharma, 1981). All Cd concentrations in muscle tissues from rabbits collected on the control and sludge-treated mined sites were below the corresponding values reported by Curnow et al.

(1977) in cottontails collected on southeastern Ohio watersheds. Cadmium levels in livers from the sludge-treated site (median = 2.25 mg kg^{-1}) were higher than those from the laboratory control group; however, no difference ($P < 0.05$) was noted when these were compared with cottontail rabbit livers collected from a nonmined area. Mean levels of Cd in rabbit livers from the sludge-treated site were above mean Cd amounts for selected foods evaluated by the FDA, e.g., raw beef liver mean = 0.183 mg kg^{-1}, but within the range of Cd values observed for several foods, e.g., ground beef, breakfast cereal, and sugar (Sharma, 1981). Authors concluded that the occasional consumption of cottontail rabbit muscle tissue from the sludge-treated site should pose no threat to human health.

In the second study conducted by Alberici et al. (1989), meadow voles were collected with snap traps and separated by sex and age. Liver, kidney, muscle, and bone tissue from 8 to 10 voles were pooled according to site, sex, and tissue for trace metal analyses. Concentrations of Cu, Zn, Co, Cd, and Ni in meadow vole tissues were not significantly different between the control and sludge-amended site (Table 26). However, Cr concentrations in kidney and bone, and Pb concentrations in liver and bone were significantly higher ($P \leq 0.05$) on the control site than on the sludge-amended site. The highest concentrations of Zn, Cd, Ni, Pb, and Co were found in bone tissue, whereas Cu tended to accumulate in the kidney. This is in contrast to a study by Johnson et al. (1978), who found that Cd tended to accumulate in the kidney. However, Anderson et al. (1982) demonstrated that Cd can accumulate differentially on the basis of age and sex. All the meadow voles in the Pennsylvania study were adults and Cd showed no differential accumulation between sexes.

Zinc, Cd, and Pb concentrations in vole tissues in the Pennsylvania study were lower than those found in meadow voles in a variety of polluted and nonpolluted areas studied by Johnson et al. (1978). Copper, Zn, Pb, and Cd levels in vole tissues in the Pennsylvania study were comparable to those found in meadow voles in a sludge-treated field studied by Anderson et al. (1982). Both Anderson et al. (1982) and Johnson et al. (1978) found no short-term toxic effects on the meadow vole as a result of the trace metals found in their studies. Long-term effects of trace accumulation in the meadow vole are not well documented. In the Pennsylvania study, there was no clear pattern evident of the movement of trace metals from the sludge to the soil and vegetation and finally to the meadow vole.

In addition to trace metal accumulation, another health-related subject of much concern is that of pathogenic organisms in sludge. There is considerable information available on the survival of pathogens in sludge and sludge-amended soils, but little on the occurrence of disease transmission to animals from organisms in sludge (Fitzgerald, 1979). With anaerobically digested and with composted sludge especially, pathogens have not been found to pose a serious health risk. Very few pathogenic organisms survive anaerobic digestion at 35° to 38° C, and even fewer survive composting at

Table 26. Extractable trace metal concentrations (dry wt) in meadow vole tissues

Tissue	Metal	Control ($mg\ kg^{-1}$)	Treated ($mg\ kg^{-1}$)
		Mean SE	Mean SE
Kidney	Cu	11.70 ± 0.15	12.29 ± 0.33
	Zn	55.07 ± 1.24	59.32 ± 0.31
	Cr*	0.68 ± 0.08	0.51 ± 0.07
	Pb	1.06 ± 0.26	2.27 ± 0.32
	Co	0.72 ± 0.07	0.77 ± 0.06
	Cd	0.48 ± 0.02	1.41 ± 0.22
	Ni	1.01 ± 0.10	0.59 ± 0.11
Liver	Cu	13.36 ± 0.20	13.60 ± 0.13
	Zn	81.26 ± 1.18	83.27 ± 0.88
	Cr	0.43 ± 0.30	0.37 ± 0.25
	Pb*	2.50 ± 0.20	0.41 ± 0.11
	Co	0.49 ± 0.04	0.40 ± 0.03
	Cd	0.23 ± 0.03	0.27 ± 0.04
	Ni	0.30 ± 0.14	0.21 ± 0.15
Muscle	Cu	7.50 ± 0.28	7.03 ± 0.10
	Zn	45.35 ± 0.77	49.16 ± 0.92
	Cr	0.43 ± 0.39	0.00 ± 0.00
	Pb	3.29 ± 0.13	3.00 ± 0.07
	Co	0.86 ± 0.06	0.91 ± 0.04
	Cd	0.38 ± 0.03	0.30 ± 0.01
	Ni	2.49 ± 0.35	2.38 ± 0.06
Bone	Cu	4.29 ± 0.53	3.09 ± 0.18
	Zn	179.28 ± 3.09	157.65 ± 2.72
	Cr*	4.53 ± 1.26	0.35 ± 0.16
	Pb**	12.90 ± 0.40	11.53 ± 0.42
	Co	5.77 ± 0.19	5.15 ± 0.23
	Cd	1.93 ± 0.08	1.47 ± 0.09
	Ni	8.80 ± 0.64	8.27 ± 0.35

From Alberici et al. (1989)
*, ** Significant at $P \leq 0.05$ and $P \leq 0.01$, respectively

temperatures above 55° C. However, with aerobically digested or raw sludge, there may be problems. A 4-year study related to the Fulton County, Illinois, project examined 100 cows that were grazed on anaerobically digested sludge-treated forage (Fitzgerald, 1979). No bacterial, viral, or fungal infections were observed in live animals or in blood or tissues at necropsy. No tissue parasites were found, and the incidence of intestinal parasites was the same in experimental and control animals. In a similar 4-month swine study, *Ascaris* species infected some of the pigs in the pens amended with 200 Mg ha^{-1} of digested sludge, but the number of worms was small. No other parasites or disease organisms were found, indicating that the transfer of pathogens from anaerobically digested sludge to grazing animals is "remote."

On the 6000 ha Fulton County project, land application of Chicago sludge resulted in no significant public health problems (Sedita et al. 1977). Actual cases of disease on such projects are extremely scarce, and problems can be eliminated with proper sludge processing and application. Four years of data on the project indicated no significant numbers of viruses or indicator organisms in groundwater or surface water, no differences in discharge of nematode eggs or coccidian oocysts from animals grazing on sludge-treated forage, and no significant difference in soil pathogen content from control and sludged areas. Health effects of digested sludge are quite different from those resulting from the application of raw sludge or of raw or treated waste-water effluents.

III. Summary

The use of municipal sewage sludge in reclamation and revegetation of drastically disturbed land has been extensively investigated. The results to date have been encouraging and show that stabilized municipal sludges, if applied properly according to present guidelines, can be used to revegetate mined lands in an environmentally safe manner with no major adverse effects on the vegetation, soil, or groundwater quality. Revegetation of coal strip-mine spoils, gravel spoils, coal refuse, clay strip-mine spoils, iron-ore tailings, abandoned pyrite mine spoils, and sites devastated by toxic fumes have been demonstrated by numerous studies using a variety of types and application rates of sludge. Results from many of the studies have substantiated that the present state and federal guidelines provide adequate protection of the environment.

Appendix

Table A-1. List of common names and scientific names of vegetation discussed in this chapter

Common Name	Scientific Name
Wheat	*Triticum aestivum*
Oats	*Avena sativa*
Canada bluegrass	*Poa compressa*
Red clover	*Trifolium pratense*
Smooth bromegrass	*Bromus inermis*
Alfalfa	*Medicago sativa*
Western wheatgrass	*Agropyron smithii*
Alsike clover	*Trifolium hybridum*
Barley	*Hordeum vulgare*

Table A-1. (Cont.)

Common Name	Scientific Name
Japanese millet	*Echinochola crusgalli var. frumentacea*
Tall fescue	*Festuca arundinacea*
Orchardgrass	*Dactylis glomerata*
Birdsfoot trefoil	*Lotus corniculatus*
Kleingrass	*Panicum coloratum*
Switchgrass	*Panicum virgatum*
Fescue	*Festuca megalura*
Perennial rye	*Lolium perenne*
Bermudagrass	*Cynodon dactylon*
Reed canarygrass	*Phalaris arundinacea*
Weeping lovegrass	*Eragrostis curvula*
Redtop	*Agrostis gigantea*
Ladino clover	*Trifolium repens*
Panic grass	*Panicum dichotomiflorum*
Sudan grass	*Sorghum vulgare sudanenese*
Serecia lespedeza	*Lespedeza cuneata*
Crownvetch	*Coronilla varia*
Annual ryegrass	*Lolium multiflorum*
Sideoats gramagrass	*Bouteloua curtipendula*
Canada wild rye	*Elymus canadensis*
Foxtail grass	*Setaria spp.*
Korean lespedeza	*Lespedeza stipulacea*
Sweetclover	*Melilotus spp.*
Kobe lespedeza	*Lespedeza striata*
Slender wheatgrass	*Agropyron trachycaulum*
Intermediate wheatgrass	*Agropyron intermedium*
Pubescent wheatgrass	*Agropyron trichophorum*
Crested wheatgrass	*Agropyron desertorum*
Meadow brome	*Bromus erectus*
Timothy	*Phleum pratense*
Corn	*Zea mays*
Soybean	*Glycine max*
Highbush blueberry	*Vaccinium corymbosum*
Bush bean	*Phaseolus vulgaris*
Virginia pine	*Pinus virginiana*
Hybrid poplar	*Populus spp.*
Black locust	*Robinia pseudoacacia*
European alder	*Alnus rugosa*
Red oak	*Quercus rubra*
Cottonwood	*Populus deltoides*
Eastern white pine	*Pinus strobus*
Silver maple	*Acer saccharinum*
Green ash	*Fraxinus pennsylvanica*
Loblolly pine	*Pinus taeda*

References

Alberici, T.M., W.E. Sopper, G.L. Storm, and R.H. Yahner. 1989. Trace metals in soil, vegetation, and voles from mine land treated with sewage sludge. *J. Environ. Qual.* 18:115–120.

Alexander, M. 1977. *Introduction to Soil Microbiology*. Wiley, New York.

Allaway, W.H. 1968. Agronomic controls over the environmental cycling of trace metals. *Adv. Agronomy* 20:235–271.

Anderson, D.W. 1977. Early stages of soil formation on glacial till mine spoils in a semi-arid climate. *Geoderma* 19:11–19.

Anderson, T.J., G.W. Barrett, C.S. Clark, V.J. Elia, and V.A. Majeti. 1982. Metal concentrations in tissues of meadow voles from sewage-sludge-treated fields. *J. Environ. Qual.* 11:272–277.

Babich, H. and G. Stotsky. 1977a. Sensitivity of various bacteria, including actinomycetes, and fungi to cadmium and the influence of pH on sensitivity. *Appl. Environ. Microbiol.* 33:681–688.

Babich, H. and G. Stotsky. 1977b. Effect of cadmium on fungi and on interactions between fungi and bacteria in soil: influence of clay minerals and pH. *Appl. Environ. Microbiol.* 33:1059–1066.

Baker, D.E., D.R. Bouldin, H.A. Elliott, and J.R. Miller (Eds.). 1985. Criteria and Recommendations for Land Application of Sludges in the Northeast. Bulletin 851, Pennsylvania Agricultural Experiment Station, Pennsylvania State University, University Park, PA.

Bastian, R.K., A. Montague, and T. Numbers. 1982. The potential for using municipal wastewater and sludge in land reclamation and biomass production as an I/A technology: an overview. In: W.E. Sopper, E.M. Seaker, and R.K. Bastian (ed.) *Land Reclamation and Biomass Production with Municipal Wastewater and Sludge*, pp. 13–54. Pennsylvania State University Press, University Park, PA.

Berry, C.R. 1982. Sewage sludge and reclamation of disturbed forest land in the Southeast. In: W.E. Sopper, E.M. Seaker, and R.K. Bastian (ed.) *Land Reclamation and Biomass Production with Municipal Wastewater and Sludge*, pp. 317–320. Pennsylvania State University Press, University Park, PA.

Bhuiya, M.R. and A.H. Cornfield. 1974. Incubation study of effect of pH on nitrogen mineralization and nitrification in soils treated with 1000 ppm lead and zinc oxides. *Environ. Pollut.* 7:161–164.

Blessin, C.W. and W.J. Garcia. 1979. Heavy metals in the food chain by translocation to crops grown on sludge-treated strip mine land. In: W.E. Sopper and S.N. Kerr (ed.) *Utilization of Municipal Sewage Effluent and Sludge on Forest and Disturbed Land,* pp. 471–482. Pennsylvania State University Press, University Park, PA.

Boesch, M.J. 1974. Reclaiming the strip mines at Palzo. *Compost Sci.* 15(1):24–25.

Carrel, J.E., K. Wieder, V. Leftwich, S. Weems, C.L. Kucera, L. Bouchard, and M. Game. 1979. Strip mine reclamation: production and decomposition of plant litter, pp. 670–676. In: M.K. Wali (ed.) *Ecology and Coal Resource Development*, vol. 2. Pergamon Press, New York.

Cavey, J.V. and J.A. Bowles. 1982. Use of sewage sludge to improve taconite tailings as a medium for plant growth. In: W.E. Sopper, E.M. Seaker, and R.K. Bastian (ed.) *Land Reclamation and Biomass Production with Municipal Waste-*

water and Sludge, pp. 400–409. Pennsylvania State University Press, University Park, PA.

Chaney, R. 1973. Crop and food chain effects of toxic elements in sludges and effluents. Recycling Municipal Sludges and Effluents on Land, pp. 129–141. National Assoc. of State Univ. and Land Grant Colleges, Washington, D.C.

Chang, F.H. and F.E. Broadbent. 1982. Influence of trace metals on some soil nitrogen transformations. *J. Environ. Qual.* 11:1–4.

Church, D.C. and W.G. Pond. 1974. *Basic Animal Nutrition and Feeding*. Schultz/Wack/Weir, Portland, OR.

Corey, R.B., L.D. King, C. Lue-Hing, D.S. Fanning, J.J. Street, and J.M. Walker. 1987. Effects of sludge properties on accumulation of trace elements by crops. In: A.L. Page et al. (eds.) *Land Application of Sludge*, pp. 25–51, Lewis Publ., Chelsea, MI.

Council for Agricultural Science and Technology. 1976. Application of sewage sludge to cropland: appraisal of potential hazards of the heavy metals to plants and animals. Report No. 64. Office of Water Programs, U.S. Environmental Protection Agency EPA-430/9-76-013.

Council for Agricultural Science and Technology. 1980. Effects of sewage sludge on the cadmium and zinc content of crops, CAST No. 83. Ames, Iowa.

Covey, J.N. 1980. Fly ash disposal with sludge. Proc. Municipal and Industrial Sludge Utilization and Disposal, Washington, D.C.

Cunningham, J.D., D.R. Keeney, and J.A. Ryan. 1975. Yield and metals composition of corn and rye grown on sewage-sludge-amended spoil. *J. Environ. Qual.* 4:448–454.

Curnow, R.D., W.A. Tolin, and D.W. Lynch. 1977. Ecological and land-use relationships of toxic metals in Ohio's terrestrial vertebrate fauna. In: H. Drucker and R.E. Wildung (ed.) *Biological Implications of Metals in the Environment*, pp. 578–594. Energy Research and Development Administration, Richland, WA.

Doelman, P. and L. Haanstra. 1979. Effect of lead on soil respiration and dehydrogenase activity, *Soil. Biol. Biochem.* 11:475–479.

Down, C.G. and J. Stocks. 1977. *Environmental Impact of Mining*. Applied Sciences Publ., London.

Dressler, R.L., G.L. Storm, W.M. Tzilkowski, and W.E. Sopper. 1986. Heavy metals in cottontail rabbits on mined land treated with sewage sludge. *J. Environ. Qual.* 15:278–281.

Epstein, E, D.B. Keane, J.J. Meisinger, and J.O. Legg. 1978. Mineralization of nitrogen from sewage sludge and sludge compost. *J. Environ. Qual.* 7:217–221.

Federal Register 44 No. 179 (13 September 1979): 53438-69. Criteria for Classification of Solid Waste Disposal Facilities and Practices: Final, Interim Final, and Proposed Regulations (40 CFR Part 257).

Federal Register 47 No. 56 (23 March 1982): 12596-604. Part VI, Surface Coal Mining and Reclamation Permanent Program Regulations; Revegetation.

Federal Register 54 No. 23 (6 February 1989): 5746-15902. Proposed Standards for the Disposal of Sewage Sludge; 40 CFR Parts 257 and 503.

Feuerbacher, T.A., R.I. Barnhisel, and M.D. Ellis. 1980. Utilization of sewage sludge as a spoil amendment in the reclamation of lands surface mined for coal. Proc. Symp. on Surface Mining Hydrol., Sedimentol., and Reclamation, pp. 187–192. University of Kentucky, Lexington.

Fitzgerald, P.R. 1979. Recovery and utilization of strip-mined land by application of anaerobically digested sludge and livestock grazing. In: W.E. Sopper and S.N. Kerr (ed.) *Utilization of Municipal Sewage Effluent and Sludge on Forest and Disturbed Land*, pp. 497–506. Pennsylvania State University Press, University Park, PA.

Fitzgerald, P.R. 1982. Effects of natural exposure of cattle and swine to anaerobically digested sludge. In: W.E. Sopper, E.M. Seaker, and R.K. Bastian (ed.) *Land Reclamation and Biomass Production with Municipal Wastewater and Sludge*, pp. 359–367. Pennsylvania State University Press, University Park, PA.

Frank, A. 1978. Mining waste, mine reclamation, and municipal sludge. *Sludge Magazine*, July–Aug. 1978, pp. 23–27.

Franks, W.A., M. Persinger, A. Iob, and P. Inyangetor. 1982. Utilization of sewage effluent and sludge to reclaim soil contaminated by toxic fumes from a zinc smelter. In: W.E. Sopper, E.M. Seaker, and R.K. Bastian (ed.) *Land Reclamation and Biomass Production with Municipal Wastewater and Sludge*, pp. 219–251. Pennsylvania State University Press, University Park, PA.

Fresquez, P.R. and E.F. Aldon. 1986. Microbial reestablishment and the diversity of fungal genera in reclaimed coal mine spoils and soils. *Reclam. Reveg. Res.* 4:245–258.

Fresquez, P.R. and W.C. Lindemann. 1982. Soil and rhizosphere microorganisms in amended coal mine spoils. *Soil Sci. Soc. Am. J.* 46:751–755.

Gadd, G.M. and A.J. Griffiths. 1978. Microorganisms and heavy metal toxicity. *Microb. Ecol.* 4:303–317.

Gaffney, G.R. and R. Ellerston. 1979. Ion uptake of redwinged blackbirds nesting on sludge-treated spoils. In: W.E. Sopper and S. N. Kerr (ed.) *Utilization of Municipal Sewage Effluent and Sludge on Forest and Disturbed Land*, pp. 407–516. Pennsylvania State University Press, University Park, PA.

Gaskin, D.A., W. Hannel, A.J. Palazzo, R.E. Bates, and L.E. Stanley. 1977. Utilization of sewage sludge for terrain stabilization in cold regions. Special Report pp. 77–37, U.S. Army Cold Regions Research and Engineering Laboratory, Hanover, NH.

Griebel, G.E., W.H. Armiger, J.F. Parr, D.W. Steck, and J.A. Adam. 1979. Use of composted sewage sludge in revegetation of surface-mined-areas. In: W.E. Sopper and S.N. Kerr (ed.) *Utilization of Municipal Sewage Effluent and Sludge on Forest and Disturbed Land*, pp. 293–306. Pennsylvania State University Press, University Park, PA.

Haghiri, F. and P. Sutton. 1982. Vegetation establishment on acidic mine spoils as influenced by sludge application. In: W.E. Sopper, E.M. Seaker, and R.K. Bastian (ed.) *Land Reclamation and Biomass Production with Municipal Wastewater and Sludge*, pp. 433–446. Pennsylvania State University Press, University Park, PA.

Hartenstein, R.E., E.F. Neuhausen, and A. Narahara. 1981. Effects of heavy metal and other elemental additives to activated sludge on growth of Eisenia foetida. *J. Environ. Qual.* 10:372–367.

Haynes, R.J. and W.D. Klimstra. 1975. Illinois lands surface mined for coal. Cooperative Wildlife Res. Lab., Southern Illinois University, Carbondale, IL.

Hill, R.D., K. Hinkle, and R.S. Klingensmith. 1979. Reclamation of orphaned mined lands with municipal sludges—case studies. In: W.E. Sopper and S.N. Kerr (ed) *Utilization of Municipal Sewage Effluent and Sludge on Forest and*

Disturbed Land, pp. 423–443. Pennsylvania State University Press, University Park, PA.

Hinesly, T.D., E.L. Ziegler, and G.L. Barrett. 1979a. Residual effects of irrigated corn with digested sewage sludge. *J. Environ. Qual.* 8:35–38.

Hinesly, T.D., L.G. Hansen, E.L. Ziegler, and G.L. Barrett. 1979b. Effects of feeding corn grain produced on sludge-amended soil to pheasants and swine. In: W.E. Sopper and S.N. Kerr (ed.) *Utilization of Municipal Sewage Effluent and Sludge on Forest and Disturbed Land*, pp. 483–496. Pennsylvania State University Press. University Park, PA.

Hinesly, T.D., D.E. Redborg, E.L. Ziegler, and I.H. Rose-Innes. 1982. Effects of chemical and physical changes in strip-mined spoil amended with sewage sludge on the uptake of metals by plants. In: W.E. Sopper, E.M. Seaker, and R.K. Bastian (ed.) *Land Reclamation and Biomass Production with Municipal Wastewater and Sludge*, pp. 339–352. The Pennsylvania State University Press, University Park, PA.

Hinkle, K.R. 1982. Use of Municipal sludge in the reclamation of abandoned pyrite mines in Virginia. In: W.E. Sopper, E.M. Seaker, and R.K. Bastian (eds.) *Land Reclamation and Biomass Production with Municipal Wastewater and Sludge*, pp. 421–432. Pennsylvania State University Press, University Park, PA.

Hornby, W.J., K.W. Brown, and J.C. Thomas. 1986. Nitrogen mineralization potentials of revegetated lignite overburden in the Texas gulf coast. *Soil Sci. Soc. Amer. Proc.* 50:1484–1489.

Hornick, S.B. 1982. Crop production on waste amended gravel spoils. In: W.E. Sopper, E.M. Seaker, and R.K. Bastian (ed.) *Land Reclamation and Biomass Production with Municipal Wastewater and Sludge*, pp. 207–218. Pennsylvania State University Press, University Park, PA.

John, M.K., C.J. Vanlaerhoven, and H.H. Chuah. 1972. Factors affecting plant uptake and phytotoxicity of cadmium adde d to soils. *J. Environ. Sci. Technol.* 6(12):1005–1009.

Jenny, H. 1980. *The Soil Resource*. Springer-Verlag, New York.

Johnson, M.S., R.D. Roberts, M. Hutton, and M.J. Inskip. 1978. Distribution of lead, zinc and environments. *Oikos* 30:153–159.

Jones, M. and R.S. Cunningham. 1979. Sludge used for strip mine restoration at Palzo: project development and compliance water quality monitoring. In: W.E. Sopper and S.N. Kerr (ed.) *Utilization of Municipal Sewage Effluent and Sludge on Forest and Disturbed Land*, pp. 369–378. Pennsylvania State University Press, University Park, PA.

Joost, R.E., J.H. Jones, and F.J. Olsen. 1981. Physical and chemical properties of coal refuse as affected by deep incorporation of sewage sludge and/or limestone. *Proc. Symp. on Surface Mining Hydrol., Sedimentol., and Reclamation*, pp. 307–312. University of Kentucky, Lexington.

Jurgensen, M.F. 1978. Microorganisms and the reclamation of mine wastes. In: C.T. Youngbert (ed.) *Forest Soils and Land Use*, pp. 251–286, Colorado State University, Fort Collins, CO.

Joost, R.E., F.J. Olsen, and J.H. Jones. 1987. Revegetation and minespoil development of coal refuse amended with sewage sludge and limestone. *J. Environ. Qual.* 16:65–68.

Kardos, L.T., W.E. Sopper, B.R. Edgerton, and L.E. DiLissio. 1979. Sewage effluent and liquid digested sludge as aids to revegetation of strip mine spoil and anthracite coal refuse banks. In: W.E. Sopper and S.N. Kerr (eds.) *Utilization of Municipal Sewage Effluent and Sludge on Forest and Disturbed Land*, pp. 315–332. Pennsylvania State University Press, University Park, PA.

Kerr, S.N., W.E. Sopper, and B.R. Edgerton. 1979. Reclaiming anthracite refuse banks with heat-dried sewage sludge. In: W.E. Sopper and S.N. Kerr (ed.) *Utilization of Municipal Sewage Effluent and Sludge on Forest and Disturbed Land*, pp. 333–352. Pennsylvania State University Press, University Park, PA.

Knezek, B.D. and R.H. Miller (eds.). 1976. Application of sludges and wastewaters on agricultural land: a planning and educational guide, North Central Regional Research Publication 235. Wooster, Ohio Agricultural Research and Development Center. (Reprinted by the U.S. Environmental Protection Agency as MCD-35).

Lawrey, J.C. 1977. Soil fungal population and soil respiration in habitats variously influenced by coal strip-mining. *Environ. Pollut.* 14:195–205.

Leisman, G.A. 1957. A vegetation and soil chronosequence on the Mesabi Iron Range spoil banks, Minnesota. *Ecol. Monogr.* 27:221–245.

Lejcher, T.R. and S.H. Kunkle. 1973. Restoration of acid spoil banks with treated sewage sludge. In: W.E. Sopper and L.T. Kardos (eds.) *Recycling Treated Municipal Wastewater and Sludge through Forest and Cropland*, pp. 184–199. Pennsylvania State University Press, University Park, PA.

Lighthart, B., J. Baham, and V.V. Volk. 1983. Microbial respiration and chemical speciation in metal-amended soils. *J. Environ. Qual.* 12:543–548.

Logan, T.J. and R.L. Chaney. 1983. Utilization of municipal wastewater and sludge on land-metals, In: A.L. Page et al. (eds.) *Utilization of Municipal Wastewater and Sludge on Land*, pp. 235–326, University of California, Riverside, CA.

Lue-Hing, C., S.J. Sedita, and K.C. Rao. 1979. Viral and bacterial levels resulting from the land application of digested sludge. In: W.E. Sopper and S.N. Kerr (eds.) *Utilization of Municipal Sewage Effluent and Sludge on Forest and Disturbed Land*, pp. 445–462. Pennsylvania State University Press, University Park, PA.

Magdoff, F.R. and F. W. Chromec. 1977. N mineralization from sewage sludge. *J. Environ. Sci. Health* A12:191–201.

Martin, J.P., J.O. Ervin, and R.A Shepherd. 1966. Decomposition of the iron, aluminum, zinc, and copper salts for complexes of some microbial and plant polysaccharides in soil. *Soil Sci. Soc. Am. Proc.* 30:196–200.

Martin, W.E. and J.E. Matocha. 1973. Plant analysis as an aid in the fertilization of forage crops. In: L.A. Walsh and J.D. Beaton (eds.) *Soil Testing and Plant Analysis*, pp. 393–426. Soil Sci. Soc. Am., Madison, WI.

Mathias, E.L., O.L. Bennett, and P.E. Lundberg. 1979. Use of sewage sludge to establish tall fescue on strip mine spoils in West Virginia. In: W.E. Sopper and S.N. Kerr (eds.) *Utilization of Municipal Sewage Effluent and Sludge on Forest and Disturbed Land*, pp. 307–314. Pennsylvania State University Press, University Park, PA.

Mathur, S.P., H.A. Hamilton, and M.P. Levesque. 1979. The mitigating effect of residual fertilizer copper on the decomposition of an organic soil in situ. *Soil Sci. Soc. Am. J.* 43:200–203.

McBride, F.D., C. Chavengsaksengkram, and D.H. Urie. 1977. Sludge-treated coal mine spoils increase heavy metals in cover crops. Research Notes NC-221, St. Paul, Minn., North Central Forest Experimental Station.

McCormick, L.H. and F.Y. Borden. 1973. Percolate from spoils treated with sewage effluent and sludge. In: R.J. Hutnick and G. Davis (ed.) *Ecology and Reclamation of Devastated Land*, vol. 1, pp. 239–250. Gordon and Breach, New York.

Melsted, S.W. 1973. Soil-plant relationships. Recycling Municipal Sludges and Effluents on Land, pp. 121–128. National Assoc. of State Univ. and Land-Grant Colleges, Washington, D.C.

Miller, R.H. 1974. Soil microbiology aspects of recycling sewage sludges and waste effluents on land. Recycling municipal sludges and effluents on land, pp. 79–90, National Assoc. of State Universities and Land-Grant Colleges, Washington, D.C.

Miller, R.M. and R.E. Cameron. 1978. Microbial ecology studies at two coal mine refuse sites in Illinois. Argonne National Laboratory Rep. ANL/LRP-3. Natl. Technical Information Serv., U.S. Dept. of Commerce, Springfield, VA.

Miller, R.M. and S.W. May. 1979. Staunton 1 reclamation demonstration project. Progress report II. Argonne National Laboratory Rep. ANL/LRP-4. Natl. Technical Information Serv., U.S. Dept. of Commerce, Springfield, VA.

Mills, A.L. 1985. Acid mine waste drainage: microbial impact on the recovery of soil and water ecosystems. In: R.L. Tate and D.A. Klein (eds.) *Soil reclamation processes*, pp. 35–81. Marcel Dekker, New York.

Morin, M. 1981. Heavy metal concentrations in three year old trees grown on sludge-amended surface mine spoil. Proc. Symp. on Surface Mining Hydrol., Sedimentol., and Reclamation, pp. 297–306. University of Kentucky, Lexington.

Morrison, D.G. and J. Hardell. 1982. The response of native herbaceous prairie species on iron-ore tailings under different rates of fertilizer and sludge application. In: W.E. Sopper, E.M. Seaker, and R.K. Bastian (eds.) *Land Reclamation and Biomass Production with Municipal Wastewater and Sludge*, pp. 410–420. Pennsylvania State University Press, University Park, PA.

Ollier, C. 1969. *Weathering*. Oliver and Boyd, Edinburgh, Scotland.

Palazzo, A.J. 1977. Reclamation of acidic dredge spoils with sewage sludge and lime at the Chesapeake and Delaware canal, Special Report 77-19, U.S. Army Cold Regions Research and Engineering Laboratory, Hanover, N.H.

Palazzo, A.J. and R.W. Duell. 1974. Responses of grasses and legumes to soil pH. *Agron. J.* 66:678–682.

Palazzo, A.J., S.D. Rindge, and D.A. Gaskin. 1980. Revegetation at two construction sites in New Hampshire and Alaska, CRREL Report 80-3, U.S. Army Cold Regions Research and Engineering Laboratory, Hanover, NH.

Paone, J., P. Struthers, and W. Johnson. 1978. Extent of disturbed lands and major reclamation problems in the United States. In: F.W. Schaller and P. Sutton (eds.) *Reclamation of Drastically Disturbed Lands*, pp. 11–22. Amer. Soc. of Agronomy, Madison, WI.

Parkinson, D., S. Visser, R.M. Danielson, and J. Zak. 1980. Restoration of fungal activity in tailing sand (oil sands). In: D.L. Dindal (ed.) *Soil Biology as Related to Land Use Practices*. Proc. 7th Int. Soil Zool. Colloq. Int. Soc. Soil Sci., Syracuse , NY. 29 July–30 Aug. 1979. p. 362–370. USEPA, Washington, DC.

Peterson, J.R., R.I. Pietz, and C. Lue-Hing. 1979. Water, soil, and crop quality of Illinois coal mine spoils amended with sewage sludge. In: W.E. Sopper and S.N. Kerr (ed.) *Utilization of Municipal Sewage Effluent and Sludge on Forest and Disturbed Land*, pp. 359–368. Pennsylvania State University Press, University Park, PA.

Peterson, J.R., C. Lue-Hing, J. Gschwind, R.I. Pietz, and D.R. Zenz. 1982. Metropolitan Chicago's Fulton County sludge utilization program. In: W.E. Sopper, E.M. Seaker, and R.K. Bastian (eds.) *Land Reclamation and Biomass Production with Municipal Wastewater and Sludge*, pp. 322–338. Pennsylvania State University Press, University Park, PA.

Pennsylvania Department of Environmental Resources. 1988. Guidelines for sewage sludge land reclamation. Rules and Regulations of the Department of Environmental Resources, Chapter 275. Commonwealth of Pennsylvania.

Pietz, R.I., C.R. Carlson, Jr., J.R. Peterson, D.R. Zenz, and C. Lue-Hing. 1989a. Application of sewage sludge and other amendments to coal refuse material: I. Effects on chemical composition. *J. Environ. Qual.* 18:164–169.

Pietz, R.I., C.R. Carlson, Jr., J.R. Peterson, D.R. Zenz, and C. Lue-Hing. 1989b. Application of sewage sludge and other amendments to coal refuse material: II. Effects on vegetation. *J. Environ. Qual.* 18:169–173.

Pietz, R.I., C.R. Carlson, Jr., J.R. Peterson, D.R. Zenz, and C. Lue-Hing. 1989c. Application of sewage sludge and other amendments to coal refuse material: III. Effects on percolate water composition. *J. Environ. Qual.* 18:174–179.

Pietz, R.I., J.R. Peterson, J.E. Prater, and D.R. Zenz. 1984. Metal concentrations in earthworms from sewage sludge-amended soils at a strip mine reclamation site. *J. Environ. Qual.* 13:651–654.

Premi, P.R. and A.H. Cornfield. 1969. Effects of additives of copper, manganese, zinc, and chromium compounds on ammonification and nitrification during incubation of soil. *Plant Soil* 31:345–352.

Reeder, J.D. and W.A. Berg. 1977. Nitrogen mineralization and nitrification in a cretaceous shale and coal mine spoil. *Soil Sci. Soc. Am. Proc.* 41:922–927.

Roberts, J.A., W.L. Daniels, J.C. Bell, and D.C. Martens. 1988. Tall fescue production and nutrient status on southwest Virginia mine soils. *J. Environ. Qual.* 17:55–62.

Roth, P.L., B.D. Jayko, and G.T. Weaver. 1979. Initial survival and performance of woody plant species on sludge-treated spoils of the Palzo Mine. In: W.E. Sopper and S.N. Kerr (eds.) *Utilization of Municipal Sewage Effluent and Sludge on Forest and Disturbed Land*, pp. 389–394. Pennsylvania State University Press, University Park, PA.

Roth, P.L., G.T. Weaver, and M. Morin. 1982. Restoration of woody ecosystem on a sludge-amended devastated mine-site. In: W.E. Sopper, E.M. Seaker, and R.K. Bastian (eds.) *Land Reclamation and Biomass Production with Municipal Wastewater and Sludge*, pp. 368–385. Pennsylvania State University Press, University Park, PA.

Sabey, B.R., N.N. Agbim, and D.C. Markstrom. 1977. Land application of sewage sludge: IV. Wheat growth, N content, N fertilizer value, and N use efficiency as influenced by sewage sludge and wood waste mixtures. *J. Environ. Qual.* 6:52–58.

Scanlon, D.H., C. Duggan, and S.D. Bean. 1973. Evaluation of municipal compost for strip mine reclamation. *Compost Sci.* 14:3.

Schafer, W.M., G.A. Neilson, and W.D. Nettleton. 1980. Minespoil genesis morphology in a spoil chronosequence in Montana. *Soil Sci. Soc. Am. J.* 44:802–807.
Schaller, F.W. and P. Sutton (eds.). 1978. Reclamation of Drastically Disturbed Lands. ASA-CSSA-SSA. Madison, Wis. 742 pp.
Schneider, K.R., R.J. Wittwer, and S.B. Carpenter. 1981. Trees respond to sewage sludge in reforestation of acid spoil. Proc. Symp. on Surface Mining Hydrol., Sedimentol., and Reclamation, pp. 291–296. University of Kentucky, Lexington.
Seaker, E.M. and W.E. Sopper. 1982. Production and quality of forage vegetation grown on municipal sludge-amended mine spoil. Proc. Symp. Surface Mining Hydrol., Sedimentol., and Reclamation, pp. 715–720. University of Kentucky, Lexington.
Seaker, E.M. and W.E. Sopper. 1983. Reclamation of deep mine refuse banks with municipal sewage sludge. *Waste Management Res.* (Copenhagen) 1:309–322.
Seaker, E.M. and W.E. Sopper. 1984. Reclamation of bituminous strip mine spoil banks with municipal sewage sludge. *Reclam. Reveg. Res.* 3:87–100.
Seaker, E.M. and W.E. Sopper. 1988a. Municipal sludge for minespoil reclamation: I. Effects on microbial populations and activity, *J. Environ. Qual.* 17:591–597.
Seaker, E.M. and W.E. Sopper. 1988b. Municipal sludge for minespoil reclamation: II. Effects on organic matter. *J. Environ. Qual.* 17:598–602.
Sedita, S.J., P. O'Brien, J.J. Bertucci, C. Lue-Hing, and D.R. Zenz. 1977. Public health aspects of digested sludge utilization. In: R.C. Loehr (ed.) *Land as a Waste Management Alternative*. Proc. 1976 Cornell, Agric. Waste Mgmt. Conf, pp. 391–410. Ann Arbor Science Publishers, Ann Arbor, MI.
Segal, W. and R.L. Mancinelli. 1987. Extent of regeneration of the microbial community in reclaimed spent oil shale land. *J. Environ. Qual.* 16:44–48.
Sharma, R.P. 1981. Soil-plant-animal distribution of cadmium in the environment. In: J.O. Nriagu (ed.) *Cadmium in the environment*, Part 1. Ecological Cycling, pp. 587–605. Wiley, New York.
Skousen, J.G. 1988. Effects of sewage sludge application on a revegetated minesoil in West Virginia. Mine drainage and surface mine reclamation, vol. 2, pp. 214–220. U.S.D.I. Bur. of Mines Information Circular 9184.
Sommers, L.E., D.W. Nelson, and K.J. Yost. 1976. Variable nature of chemical composition of sewage sludges. *J. Environ. Qual.* 5:303–306.
Sommers, L.E., D.W. Nelson, A.W. Kirleis, S.D. Strachan, J.C. Inman, S.A. Boyd, J.G. Graveel, and A.D. Behel. 1984. Characterization of sewage sludge and sewage sludge-soil systems. USEPA Rep. EPA—600/2-84-046. USEPA, Cincinnati, OH.
Sommers, L., V. Van Volk, P.M. Giordano, W.E. Sopper, and R. Bastian. 1987. Effects of soil properties on accumulation of trace elements in crops. In: A.L. Page, T.G. Logan, and J.A. Ryan (eds.) *Land Application of Sludge*, pp. 5–24. Lewis Publishers, Chelsea, MI.
Sopper, W.E. 1982a. Strip mine reclamation project—Kennell I site. Institute for Research on Land and Water Resources, Pennsylvania State University, University Park, PA.
Sopper, W.E. 1982b. Strip mine reclamation project—Mast site. Institute for Research on Land and Water Resources, Pennsylvania State University, University Park, PA.

Sopper, W.E. 1982c. Strip mine reclamation project—Leasure site. Institute for Research on Land and Water Resources, Pennsylvania State University, University Park, PA.

Sopper, W.E. 1982d. Strip mine reclamation project—Beachy I site. Institute for Research on Land and Water Resources, Pennsylvania State University, University Park, PA.

Sopper, W.E. 1982e. Strip Mine Reclamation Project—Black site. Institute for Research on Land and Water Resources, Pennsylvania State University, University Park, PA.

Sopper, W.E. 1987. Reclamation of the Palmerton superfund site. Proc. 1987 National Symposium on Mining, Hydrology, Sedimentology and Reclamation, pp. 171–178. University of Kentucky, Lexington, KY.

Sopper, W.E. 1989a. Revegetation of a contaminated zinc smelter site. *Landscape Urban Planning* 17:241–250.

Sopper, W.E. 1989b. Reforestation of the Palmerton zinc smelter superfund site. Healthy Forests, Healthy World, Proc. 1988 Soc. of Amer. Foresters National Convention, pp. 205–209. Bethesda, MD.

Sopper, W.E. 1989c. Revegetation of a contaminated zinc smelter site. *Landscape Urban Planning* 17:241–250.

Sopper, W.E. 1989d. Bio-remediation of the Palmerton superfund site. Proc. of 6th National Materials Conference on Hazardous Wastes and Hazardous Materials, pp. 322–326. Hazardous Materials Control Research Institute, Silver Springs, MD.

Sopper, W.E. and S.N. Kerr (eds.). 1979. Utilization of municipal sewage effluent and sludge on forest and disturbed land. Pennsylvania State University Press, University Park, PA.

Sopper, W.E. and S.N. Kerr. 1980a. Strip mine reclamation demonstration project—Decker site. Institute for Research on Land and Water Resources, Pennsylvania State University, University Park, PA.

Sopper, W.E. and S.N. Kerr. 1980b. Strip mine reclamation demonstration project—Soberdash site. Institute for Research on Land and Water Resources. Pennsylvania State University, University Park, PA.

Sopper, W.E. and S.N. Kerr. 1981. Revegetating strip-mined land with municipal sewage sludge. Project Summary EPA-600/52-81-182. U.S. Environmental Protection Agency.

Sopper, W.E. and S.N. Kerr. 1982. Mine land reclamation with municipal sludge—Pennsylvania's demonstration program. In: W.E. Sopper, E.M. Seaker, and R.K. Bastian (ed.) *Land Reclamation and Biomass Production with Municipal Wastewater and Sludge*, pp. 55–74. Pennsylvania State University Press, University Park, PA.

Sopper, W.E. and J.M. McMahon. 1987. Greening of Palmerton Blue Mountain. *Biocycle* 28(4):47–51.

Sopper, W.E. and J.M. McMahon. 1988a. Revegetation of a superfund site, Part I. *Biocycle* 29(7):57–60.

Sopper, W.E. and J.M. McMahon. 1988b. Revegetation of a superfund site, Part II. *Biocycle* 29(8):64–66.

Sopper, W.E. and E.M. Seaker. 1982. Strip mine reclamation with municipal sludge. Final Technical Report on Contract No. CR807408010. Municipal Environmental Research Lab, Environmental Protection Agency, Cincinnati, OH.

Sopper, W.E. and E.M. Seaker. 1983. A guide for revegetation of mined land in eastern United States using municipal sludge. Institute for Research on Land and Water Resources, Pennsylvania State University, University Park, PA.

Sopper, W.E. and E.M. Seaker. 1984. Use of municipal sludge to reclaim mined land. *CRC Crit. Rev. Environ. Control* 13:227–271.

Sopper, W.E. and E.M. Seaker. 1987. Sludge brings life to microbial communities. *Biocycle* 28(4):40–47.

Sopper, W.E. and E.M. Seaker. 1987. Development of microbial communities on sludge-amended mine land. In C.L. Carlson and J.W. Swisher (eds.) *Innovative Approaches to Mine Land Reclamation*, pp. 659–680. Southern Illinois University Press, Carbondale, IL.

Sopper, W.E. and E.M. Seaker. 1988. Rejuvenation of microbial communities on abandoned mine land amended with municipal sludge. Proc. Symp. on Mining, Hydrology, Sedimentology and Reclamation, pp. 199–206. University of Kentucky, Lexington, KY.

Sopper, W.E. and E.M. Seaker. 1990. Long-term effects of a single application of municipal sludge on abandoned mine land. In: *Proc. of the 1990 Mining and Reclamation Conf. and Exhibition*, (Skousen et al. (eds.) vol. 2, pp. 579–587. West Virginia University, Morgantown, WV.

Sopper, W.E., S.N. Kerr, E.M. Seaker, W.F. Pounds, and D.T. Murray. 1981. The Pennsylvania program for using municipal sludge for mine land reclamation. *Proc. Symp. Surface Mining Hydrol., Sedimentol., and Reclamation*, pp. 283–290. University of Kentucky, Lexington.

Sopper, W.E., E.M. Seaker, and R.K. Bastian (eds.). 1982. *Land Reclamation and Biomass Production with Municipal Wastewater and Sludge*. Pennsylvania State University Press, University Park, PA.

Stevenson, H.F. 1982. *Humus Chemistry*. Wiley, New York.

Stroo, H.F. and E.M. Jencks. 1982. Enzyme activity and respiration in mine spoils. *Soil Sci. Soc. Am. J.* 46:548–553.

Stucky, D.J. and J.H. Bauer. 1979. Establishment, yield, and ion accumulation of several forage species on sludge-treated spoils of the Palzo Mine. In: W.E. Sopper and S.N. Kerr (ed.) *Utilization of Municipal Sewage Effluent and Sludge on Forest and Disturbed Land*, pp. 379–388. Pennsylvania State University Press, University Park, PA.

Stucky, D.J. and T.S. Newman. 1977. Effect of dried anaerobically digested sewage sludge on yield and elemental accumulation in tall fescue and alfalfa. *J. Environ. Qual.* 6(3):271–274.

Stucky, D.J., J.H. Bauer, and T.C. Lindsey. 1980. Restoration of acidic mine spoils with sewage sludge. *Reclamation Rev.* 3:129–139.

Sundberg, W.J., D.L. Borders, and G.L. Albright. 1979. Changes in soil microofungal populations in the Palzo strip mine spoil following suldge application. In: W.E. Sopper and S.N. Kerr (eds.) *Utilization of Municipal Sewage Effluent and Sludge on Forest and Disturbed Land*, pp. 463–470. Pennsylvania State University Press, University Park, PA.

Sutton, P. 1980. Ohio Agricultural Research and Development Center, Wooster, OH, personal communication.

Sutton, P. and J.P. Vimmerstedt. 1974. Treat stripmine spoils with sewage sludge. *Compost Sci.* 15(1):22–23.

Svoboda, D., G. Smout, G.T. Weaver, and P.L. Roth. 1979. Accumulation of heavy metals in selected woody plant species on sludge-treated strip mine spoils at the Palzo Site, Shawnee National Forest. In: W.E. Sopper and S.N. Kerr (eds.) *Utilization of Municipal Sewage Effluent and Sludge on Forest and Disturbed Land*, pp. 395–406. Pennsylvania State University, University Park, PA.

Tate, R. L., III. 1985. Microorganisms, ecosystem disturbance and soil formation processes. In: R.L. Tate III and D.A. Klein (eds.) *Soil reclamation processes*. pp. 1–33. Marcel Dekker, New York.

Terry, R.E., D.W. Nelson, and L.E. Sommers. 1981. Nitrogen transformations in sewage sludge-amended soils as affected by soil environmental factors. *Soil Sci. Soc. Am. J.* 45:506–513.

Tomlinson, T.G. 1966. Inhibition of nitrification in the activated sludge process of sewage disposal. *J. Appl. Bacteriol.* 29:266.

Topper, K.F. and B.R. Sabey. 1986. Sewage sludge as a coal mine spoil amendment for revegetation in Colorado. *J. Environ. Qual.* 15:44–49.

Tunison, K.W., B.C. Bearce, and H.A. Menser, Jr. 1982. The utilization of sewage sludge: bark screenings compost for the culture of blueberries on acid minespoil. In: W.E. Sopper, E.M. Seaker, and R.K. Bastian (eds.) *Land Reclamation and Biomass Production with Municipal Wastewater and Sludge*, pp. 195–206. Pennsylvania State University Press, University Park, PA.

Underwood, E.J. 1971. *Trace elements in human and animal nutrition*. 3rd ed., Academic Press, New York.

University of Georgia Cooperative Extension Service. 1979. Plant Analysis Handbook for Georgia, 42. Bulletin No. 735. University of Georgia, Athens.

Urie, D.H., C.K. Losche, and F.D. McBride. 1982. Leachate quality in acid minespoil columns and field plots treated with municipal sewage sludge. In: W.E. Sopper, E.M. Seaker, and R.K. Bastian (eds.) *Land Reclamation and Biomass Production with Municipal Wastewater and Sludge*, pp. 386–398. Pennsylvania State University Press, University Park, PA.

U.S. Department of Agriculture. 1980. Soil and Water Resource Conservation Act: Appraisal 80. Review Draft.

U.S. Environmental Protection Agency. 1977. Municipal Sludge Management: Environmental Factors. Tech. Bul., EPA 430/9-76-004, MCD-28.

U.S. Environmental Protection Agency. 1979. Enviornmental impact statement: criteria for classification of solid waste disposal facilities and practices. SW-821.

U.S. Environmental Protection Agency, U.S. Food and Drug Administration, U.S. Department of Agriculture. 1981. Land Application of Municipal Sewage Sludge for the Production of Fruits and Vegetables—A Statement of Federal Policy and Guidance. Washington, D.C.

U.S. Environmental Protection Agency, 1983. Process Design Manual for Land Application of Municipal Sludge. EPA-625/1-83-016, Municipal Environmental Research Lab, Cincinnati, OH.

Van Hook, R.I. 1974. Cadmium, lead, and zinc distributions between earth worms and soils: potentials for biological accumulation. *Bull. Environ. Contaim. Toxicol.* 12:509–512.

Varanka, M.W., Z.M. Zablocki, and T.D. Hinesly. 1976. The effect of digestor sludge on soil biological activity. *J. Water. Pollut. Control Fed.* 48:1728–1740.

Visser, S. 1985. Management of microbial processes in surface mined land reclamation in western Canada. In: R.L. Tate III and D.A. Klein (eds.) *Soil reclamation processes*, pp. 203–341. Marcel Dekker, New York.

Visser, S., C.L. Griffiths, and D. Parkinson. 1983. Effects of mining on the microbiology of a prairie site in Alberta, Canada. *Can. J. Soil Sci.* 63:177–189.

Voos, G. and B.R. Sabey. 1987. Nitrogen mineralization in sewage sludge-amended coal mine spoil and topsoils. *J. Environ. Qual.* 16:231–237.

Wilson, H.A. 1965. The microbiology of strip mine spoil. West Virginia Univ. Agric. Exp. Stn. Bull. 506T.

Yopp, J.H., W.E. Schmid, and R.W. Holst. 1974. Determination of maximum permissible levels of selected chemicals that exert toxic effects on plants of economic importance in Illinois. Ill. Inst. for Env. Qual. IIEQ Doc. No. 74–33.

Younos, T.M. and M.D. Smolen. 1981. Simulation of infiltration in a sewage sludge amended mine soil. Proc. Symp. on Surface Mining Hydrol., Sedimentol., and Reclamation, pp. 319–324. University of Kentucky, Lexington.

Younos, T.M. (ed.). 1987. *Land application of wastewater sludge*. Am. Soc. Civil Eng., New York.

Research and Development Priorities for Soil Restoration

R. Lal and B.A. Stewart

I. Introduction

Some densely populated regions of the world have severe land shortages. These regions are also characterized by severe problems of soil and environmental degradation. Agriculturally marginal and unsuitable lands, used for food crop production because of land hunger, are being further degraded even to the point of no return from irreversible degradation. For these regions where land resources are insufficient to substantially meet needs, even with the judicious use of off-farm input and ameliorative soil amendments, it is necessary to enhance the soil resource base through restoration of degraded lands. For rcstorativc considcrations, degraded soils can be of three principal categories (Fig. 1). Some soils are unsuitable for agricultural use because of nonavailability of essential inputs, e.g., shortage of irrigation water in dry regions, nonavailability of lime for acidic soils, lack of essential plant nutrients in highly weathered soils. Technically, productivity of such soils can be restored by providing the input required, e.g., use of irrigation, supplemental fertilizer and amendments, choice of suitable crop species and cultivars. However, economic availability of these input and logistic problems may be major impediments. There are other soils with inherent crop-restrictive properties that limit their use for agricultural production. Some examples of these soils include exces-

Advances in Soil Science, Volume 17

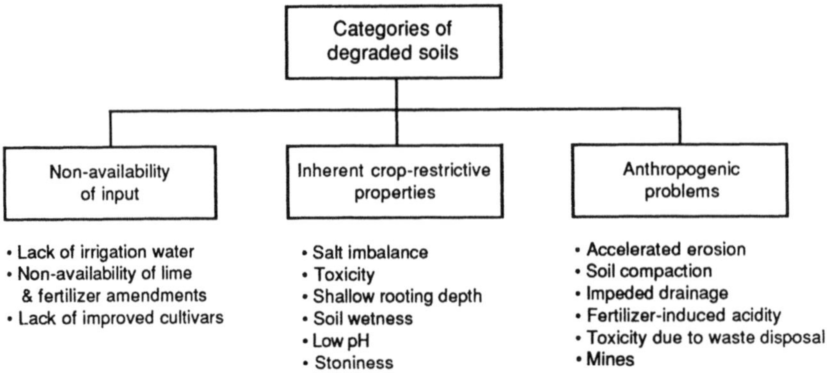

Figure 1. Categories of degraded soils

sively wet soils, low pH, and high P fixation capacity. Finally, there are soils that are degraded due to land misuse, e.g., eroded and compacted soils, and toxicity due to waste disposal, salt-affected soils, abandoned mines etc. Strategies for land restoration are different for different categories of degraded soils.

II. Approaches to Soil Restoration

Understanding processes, factors, and causes of soil degradation is a basic prerequisite toward successful restoration of the productivity of degraded soils. It is necessary to establish the cause-effect relationship, because elimination of the causative factors may reverse the degradative trend and set in motion the restorative processes. It is also important to establish and define critical limits of soil properties for adapting appropriate land use and cropping/farming systems once the soil is in the process of being restored. For example, knowledge of tolerable levels of salinity, acidity, or aluminum toxicity of different crops and cultivars may be necessary to obtain economic benefits by growing tolerable crops and to accelerate the process of restoration.

Knowing the category of soil degradation is an important stage because the approach to restoration depends on the category. Approaches to restoration for soils degraded by intensive farming are outlined in Fig. 2. There is a critical level of soil organic matter below which favorable structural attributes are difficult to maintain. Understanding the critical levels of soil organic matter, and adopting soil/crop management systems to achieve these levels is an important strategy for improving soil structure. Enhancing activity and species diversity of soil fauna is another useful strategy. The zonal tillage concept, limiting wheel compaction to the traffic zone, and adopting conservation tillage for the soil/water management zone, will reduce risks of soil compaction.

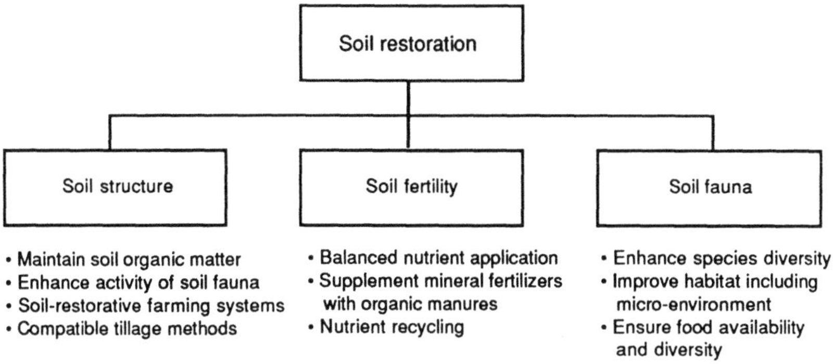

Figure 2. Approaches to restoration for soils degraded by intensive farming

Nutrient management in soils with fertility problems is crucial to sustained production. Reliable and quantitative information on the nutrient supplying ability of soils (both capacity and intensity factors) and response functions for different cultivars, crops, and cropping/farming systems is necessary for the judicious use of inherent and applied nutrients. Intensive land use and high yields on soils of low inherent fertility can only be achieved by raising the nutrient levels. Technological options for nutrient recycling should be explored. Excessive use of synthetic fertilizers can be avoided by decreasing losses (erosion, leaching, volatilization) and enhancing nutrient recycling. An adequate level of activity and species diversity of soil fauna is also essential for restoration of soil structure and enhancement of nutrient cycling. The effects of soil fauna on properties of soils are not well understood, especially with regard to different cropping/farming systems. There is a need for soil scientists and ecologists to work together to understand the interaction between fauna and soil properties.

Knowledge of basic processes and about the cause-effect relationship is also necessary for restoring soils with crop restrictive inherent characteristics (Fig. 3). Restoration of eroded lands, and prevention of degradation by new erosion are crucial to sustainable management of soil resources. Taking pressure off the marginal lands, by creating off-farm employment and developing income-generating opportunities are important policy considerations. Research is also needed on erosion-preventive measures and their effectiveness for diverse agro-eco regions. Productivity of eroded soils is constrained due to loss of soil organic matter, clay fraction and colloid complex, decrease in rooting depth, and reduction in plant-available water capacity. Exposure of unproductive subsoil is another major effect of accelerated erosion. Agronomic research is needed to develop package(s) of cultural practices to alleviate these constraints.

Salt-affected soils, totaling about 323 million hectares, are widely distributed throughout the arid and semi-arid regions. Predominance of sodium

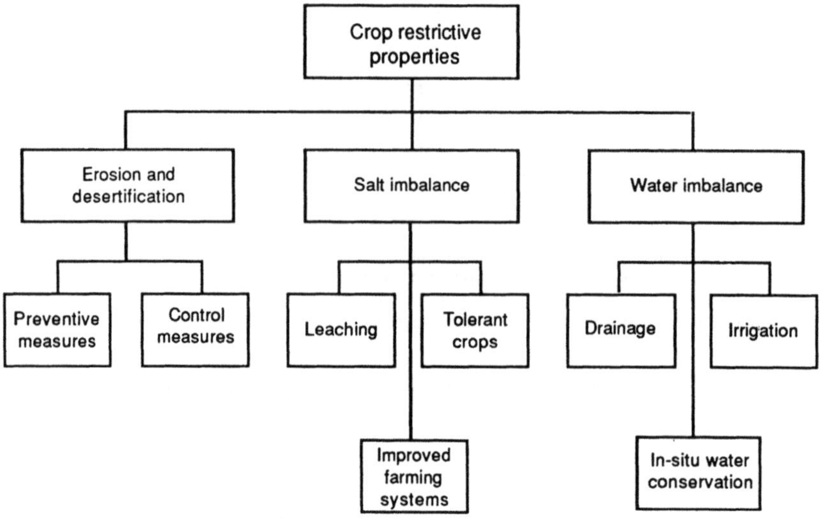

Figure 3. Approaches to land restoration for soils with severe crop-restrictive properties

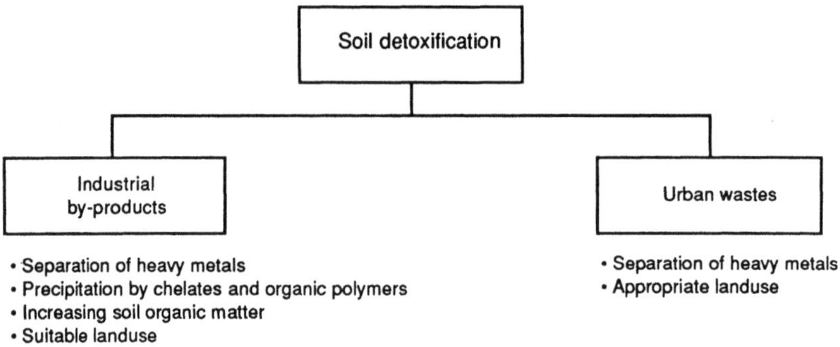

Figure 4. Restoration of soil degraded by nonagricultural uses

on the exchange complex can adversely affect soil structure and infiltration and percolation rate. In addition to physical measures of salt removal (leaching, scraping), research on cropping systems, with new crop species and improved cultivars, is crucial to alleviating production constraints. The problems of nutrient toxicity and deficiency can also be addressed through the judicious use of chemicals, adoption of ameliorative soil and crop management systems, and growth of adaptive crops and cultivars.

Soil is increasingly being used or abused for disposal of industrial and urban wastes. Indiscriminate disposal of toxic wastes is unwise and unethical. Legislative measures are needed to ensure safe disposal methods of industrial wastes; strategies to detoxify soil contaminated by industrial wastes are outlined in Fig. 4.

III. Strategies and Policies

Public participation is essential to any large-scale implementation of the restoration program. Public involvement is facilitated by institutional support and policy incentives. An external agency, e.g., a funding organization, has limited ability to influence meaningful and sustained development. Public participation can be fostered by policies that enhance feelings of national sovereignty, national pride, and self-interest. Progressive policies also provide guidelines and support for public officials in making and implementing intelligent decisions.

Discontinuities and inconsistencies in policies can lead to frustrations and lack of cooperation by people. Continuity in appropriate policies and development incentives, vital to public participation in sustainable development, is based on political stability. Political instability in sub-Saharan Africa has been a major deterrent to sustained agricultural development.

Institutional support requires making necessary inputs available on time, e.g., tree seedling for afforestation, seed for covercrops, fertilizers, pesticides, equipment, and implements. The lack of availability of such inputs at reasonable prices or the lack of credit at fair terms can be serious deterrents to the adoption of restorative technologies. Credit should be made available for inputs, e.g., for afforestation, development and installation of soil and water conservation measures, fertilizers, and amendments. The policy of providing credit at fair terms is a better choice than are subsidies. Subsidies are poor or no substitutes for making people realize that it is in their interest to restore degraded lands.

Land tenure regulations can be important towards restoration of degraded lands (FAO, 1987). Communal ownership of land can in some cases be used to an advantage. In other instances, communal tenurial rights are impediments to adoption of long-term improvement strategies, e.g., erosion control, fencing against uncontrolled grazing, planting of permanent crops. Insecurity of land ownership can be a serious disincentive to investment for land restoration. In many developing countries, prime agricultural land is held by large landowners. Marginal and fragile lands are cultivated by landless squatters—to the serious detriment of land and environment. These subsistence farmers, who carve out their living at a terrible price of land and environmental degradation, are passed over by the agricultural development programs and other incentives.

Although biophysical factors play a crucial role, the principal driving force in land and other natural resource degradation can be traced to socioeconomic, anthropogenic factors, and a variety of human traits, including greed, ignorance, short-sightedness, and poor planning. In developing countries, human despair and the tyranny of immediate necessity for basic human needs drives exploitation of the land beyond its carrying capacity. In developed economics, aspiration towards a high standard of living, changing social values, and the lure of leisure time often contribute to a marked shift from restorative and effective to land degradative systems.

Can legislation or appeals on a moral basis help? Can society intervene in the goals and objectives of individual owners/occupants of ecologically sensitive regions? Can the idea of stewardship of the land resources and land ethic be a sufficient force against economic and social pressures? The most highly evolved legislation of environmental protection has existed in the U.S. since as early as 1881 (Stern, 1982). And yet, the pollution of air and water and degradation of soil by erosion and other processes has steadily increased over the past century. Making a moral issue on the basis of stewardship of the land is hardly appealing to subsistence farmers worried about food for the next day rather than for generations to come. Nonetheless, developing countries have not promulgated adequate laws and regulations to control and plan the use of land and water resources and off-farm inputs. While the demand for agrichemicals in developing countries is rapidly increasing, development of an environmental ethic and environmental laws have not kept pace. Programs to educate farmers and their families about the dangers of ecologically unfriendly technologies are nonexistent.

Involving the farming community in understanding the problem and in developing legislation may improve the effectiveness of such legislation. Without adequate community involvement, enforcing regulation without mutual consent is bound to be of little use (TAC, 1988).

IV. Conclusions

Restoration of degraded lands is an ecological and socio-economic necessity, if the basic necessities of the earth's inhabitants, including its human and animal populations, are to be adequately met. Some of the processes leading to soil degradation cut across scientific disciplines but also political, ethnic, and cultural boundaries. The scientific community has to develop an interdisciplinary approach to address this serious challenge facing mankind. Similarly, the world community has to develop a coordinated effort to alleviate the serious constraints.

References

FAO. 1987. The effect of land tenure and fragmentation of farm holdings on agricultural development. Report of the Committee on Agriculture, FAO, Rome.

Stern, A.C. 1982. History of air pollution legislation in the United States. *J. Air Pollution Control* 32:44–61.

TAC. 1988. *Sustainable Agricultural Production: Implications for International Agriculture Research*. TAC Secretariat, FAO, Rome.

Index

Q

R

MIX
Papier aus verantwortungsvollen Quellen
Paper from responsible sources
FSC® C105338

If you have any concerns about our products, you can contact us on
ProductSafety@springernature.com

In case Publisher is established outside the EU, the EU authorized representative is:
Springer Nature Customer Service Center GmbH
Europaplatz 3, 69115 Heidelberg, Germany

Printed by Libri Plureos GmbH
in Hamburg, Germany